GLASSY MATERIALS AND DISORDERED SOLIDS

An Introduction to Their Statistical Mechanics

Revised Edition

GLASSY MATERIALS AND DISORDERED SOLIDS

An Introduction to Their Statistical Mechanics

Revised Edition

Kurt Binder
Johannes-Gutenberg Universität, Germany

Walter Kob
Université Montpellier 2, France

NEW JERSEY • LONDON • SINGAPORE • BEIJING • SHANGHAI • HONG KONG • TAIPEI • CHENNAI

Published by

World Scientific Publishing Co. Pte. Ltd.

5 Toh Tuck Link, Singapore 596224

USA office: 27 Warren Street, Suite 401-402, Hackensack, NJ 07601

UK office: 57 Shelton Street, Covent Garden, London WC2H 9HE

British Library Cataloguing-in-Publication Data
A catalogue record for this book is available from the British Library.

GLASSY MATERIALS AND DISORDERED SOLIDS (Revised Edition)
An Introduction to Their Statistical Mechanics

Desk Editor: Ryan Bong

ISBN-13 978-981-4350-17-4
ISBN-10 981-4350-17-6

Printed in Singapore.

We dedicate this book to our wives Marlies Binder and Martina Yague in appreciation of their patience and understanding for the many weekends and evenings we spent on this book project.

Preface

Understanding the physical properties of disordered materials and the nature of the transition from the supercooled fluid to the amorphous solid is one of the most difficult and at the same time fascinating "grand challenge problems" of our time. Consequently, there is an enormous scientific activity on the subject, both via experiments, analytical theories, as well as computer simulations. Despite these efforts a generally accepted and coherent view of this problem has not emerged yet. As a consequence, there is so far no "canonic" textbook on this subject, in contrast to most other topics in the physics of condensed matter (crystalline solids, simple liquids, and complex liquids such as polymers or liquid crystals). Most existing texts are rather restricted in scope, in that they either emphasize simple structural models of disorder or the phenomenological thermodynamic aspects, or focus on documenting the complex relaxation phenomena in these systems.

Hence, in the present monograph we fill a gap, by focusing on the statistical mechanics of the glass transition and the amorphous state of materials, and at the same time providing a pedagogical overview of important concepts for problems in the physics of disordered solids (percolation, fractals, disordered magnetic model systems such as spin glasses, etc.). We feel that the present selection of topics is quite natural, since there are many aspects of the statistical mechanics of disordered solids that are in common to glass-forming systems, and many of the concepts of the simpler problems have been invoked to explain the more complex phenomena occurring in the latter. E.g., there are numerous attempts to explain the structure of glasses and the glass transition in terms of percolation theory and similarly one tries to describe the physics of structural glasses by making use of the insight gained for the related problem of spin glasses.

The present book thus stresses these common aspects of such disordered condensed matter systems, emphasizing statistical thermodynamics as a general theoretical framework, and "classical" methods of computer simulation (such as Monte Carlo and molecular dynamics methods) as one of the central practical tools of study, in view of the difficulty of developing a viable analytical theory. Note that for problems such as critical phenomena associated with second order phase transitions one has well established methods, such as mean field theory and renormalization group methods, that describe the phenomena on a qualitative or even a quantitative basis. This is in contrast to the situation for glasses and the glass transition for which we might not even have a mean field theory that describes the phenomenon correctly.

Nevertheless, the intense research of the last decades has improved considerably our knowledge on glasses and other disordered solids and therefore the present book tries to put this progress in perspective and to give a coherent account of the basic facts that are now established. In turn, many of the ideas that are still on a very speculative level are not discussed in detail. Also interesting aspects such as electronic structure and electronic excitations, optical properties of amorphous solids etc. are not considered here. Thus, the selection of material that is presented is certainly strongly biased by the expertise and the interests of the authors and by no means we imply that topics that are not discussed here are irrelevant or unimportant. This restriction in the scope of the present book should have the advantage to make it easily understandable for graduate students and postdoctoral researchers who have an elementary knowledge of statistical thermodynamics since no further specialized knowledge is required to read it. In fact, one of the authors (K.B.) has given several times a course on this subject at the Johannes-Gutenberg Universität in Mainz and thus we feel that the book should be a very useful basis for similar courses. At the same time, the rich bibliography at the end of each chapter should make this monograph a useful source of references even for the advanced researcher. In fact, we do hope that the present book will to some extend influence the further research on the physics of glasses, since the knowledge of the facts and model descriptions summarized in this book should be very helpful for asking the right questions in future research.

Finally, it is a great pleasure to thank all the coworkers that have contributed significantly to our own research in this subject, to numerous colleagues from whose ideas and fruitful discussions we have profited, and last but not least we are grateful to Mrs. A. Chase for her valuable help in typing the manuscript.

Kurt Binder and Walter Kob

Preface to the second edition

The structure and dynamics of supercooled fluids, glasses and other solids with quenched disorder (e.g. spin glasses) have remained a very active field of research in the six years since the manuscript of the first edition has been completed. Although much progress in the theoretical understanding of these systems has been made, many important questions still remain open, and hence we foresee that research on this topic will continue at a very high intensity. At the same time, the physics of disordered systems, as it is comprehensively described in this book, takes up an increasingly prominent place in the repertoire of courses being offered to students of physics, physical chemistry, and materials science at universities.

Thus the present textbook will satisfy the strong need of researchers, university teachers, and advanced students to provide a coherent introduction and survey into this extremely rich subject. In the present edition, we have filled various small gaps by adding short sections on foams, on the relation of methods to deal with spin glasses to problems in computer science, etc. Moreover we have updated many sections by discussing (or at least mentioning) the most important recent developments. In particular, an additional (sixth) chapter has been written, entitled "Further Models for Glassy Dynamics; Dynamical Heterogeneities; Gels; Driven Systems", giving a survey of particularly important recent developments in these fields. Of course, minor inconsistencies were removed and typos corrected. Again we wish to thank many colleagues for their most valuable comments.

Kurt Binder and Walter Kob

Contents

Chapter 1

Introduction

1.1 Models of Disordered Matter: A Brief Overview

Although this text deals with the physics of condensed matter, its contents differs strongly from the one of most standard textbooks on this subject in that it focuses on materials such as window glass, plastics, rubber, amorphous metals, porous materials, magnets with frozen-in disorder, etc., i.e. on systems that are referred to as "amorphous materials" or "ill-condensed matter". Despite the fact that these materials have been used by man since ancient times, presumably the first glass was produced in ancient Egypt several thousand years ago, and are nowadays ubiquitous in daily life, most textbooks on the theory of condensed matter ignore this type of matter almost completely. The reason for this negligence is the lack of a coherent and elegant theoretical description for these systems, which in the case of crystals can be obtained by exploiting the periodicity of the crystal structure. Therefore, analytical theories for the properties of such strongly disordered matter are comparably scarce, and much of the theoretical knowledge that we have so far stems from computer simulations. Despite this lack of a well founded theoretical description, the physics and chemistry of glasses has become a prominent topic of research in the last decades: The structure of the glassy state has become a subject of many experimental investigations and numerical simulations, and in particular the "problem of the glass transition", i.e. the phenomenon that many liquids can be cooled below their melting temperature and solidify into an amorphous solid, is considered as one of the grand challenges in condensed matter physics, and hence studied intensively.

In this situation, the present book tries to fill a gap in the current literature by presenting an introduction to the subject under the point of view

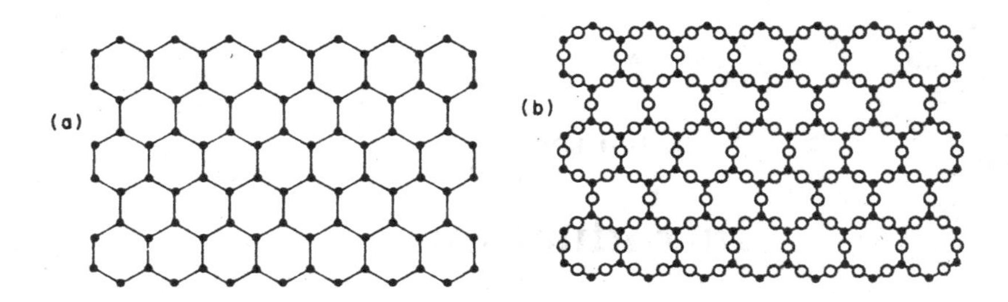

Fig. 1.1 (a) The honeycomb lattice and (b) the decorated honeycomb lattice. The topological structure of the honeycomb is the same as that of a layer of carbon atoms in graphite or of a layer of crystalline arsenic. The topology of the decorated honeycomb occurs in the structure of layers that make up crystalline As_2S_3 and As_2Se_3. The sulfur (or selenium) atoms are represented by open circles. After Zallen (1983).

of statistical thermodynamics. We feel that statistical thermodynamics is a particularly useful framework for the description of amorphous solids and the understanding of their properties as we will outline in this Introduction. More details on the various topics can be found in the subsequent chapters.

As we will see later, the nature of the structural disorder depends strongly on the type of disordered solids one considers, and obtaining a proper characterization of these structures is one of the main topics of this book. (More detailed discussions on structural disorder can be found in the texts of Cusack (1986), Elliott (1983), and Varshneya (1993).) A second main topic is to understand the implications of this "frozen-in disorder" (also called "quenched disorder") for the various physical properties. In addition to this frozen-in disorder, we have of course also the normal thermal disorder, which is one of the standard problems considered by statistical thermodynamics, but which wins new facets here in its interplay with the quenched disorder.

These general remarks can be illustrated with a specific example, the "Continuous Random Network" (CRN) model of covalent glasses, which dates back to Zachariasen (1932), cf. Figs. 1.1 and 1.2. In such an amorphous solid, one encounters disorder in the "equilibrium" positions of the atoms even if the *local* chemical order is the same as the one in the corresponding crystalline structure.

We put the word "equilibrium" in quotation marks, since we assume that the real equilibrium structure is the ordered crystal of Fig. 1.1, which has a lower free energy. However, often the rates for the nucleation and growth of crystal from the amorphous structure are so low, that for most

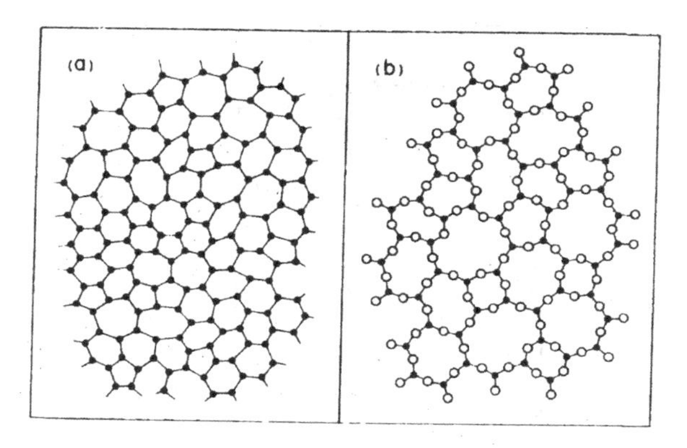

Fig. 1.2 Two-dimensional continuous random networks. A sketch of a (hypothetical) threefold-coordinated elemental glass is presented in (a), while Zachariasen's (1932) diagram for an A_2B_3 glass is shown in (b). After Zallen (1983).

practical purposes the latter is a metastable phase that is essentially indistinguishable from the thermal equilibrium.

The difference in the geometrical structure between the crystal (Fig. 1.1) and the glass (Fig. 1.2) has immediate consequences for the spectrum of elementary excitations of the respective solids, and the resulting thermodynamic properties at low temperatures. In the crystal as well as in the amorphous solid we find small-amplitude vibrations of the atoms around their equilibrium positions, which can be treated as harmonic oscillations at low temperatures, at least in a first approximation. Now the lattice periodicity of the crystal has an important physical consequence: the vibrations can be decomposed into non-interacting plane waves with a wave-vector $\vec{k}$ and a frequency $\omega(\vec{k})$, i.e. $\vec{k}$ is a "good quantum number". As one can read in more detail in any standard text book on theoretical solid state physics, one finds a homogeneous density of eigenstates $\vec{k}$ in the first Brillouin zone of the crystal, and this has an important consequence on the phonon density of states $\zeta(\omega)$ at small $k = |\vec{k}|$. Recalling the dispersion relation for acoustic phonons, $\omega = ck$, where c is the velocity of sound (for simplicity we ignore the distinction between longitudinal and transverse phonons here), one concludes immediately

$$\zeta(\omega)d\omega = \frac{\partial\omega}{\partial\vec{k}}\, d^d k \underset{k\to 0}{\longrightarrow} \text{const} \cdot k^{d-1}dk \propto \omega^{d-1}d\omega\,, \tag{1.1}$$

where d is the dimension of the crystal. This yields as a contribution to the

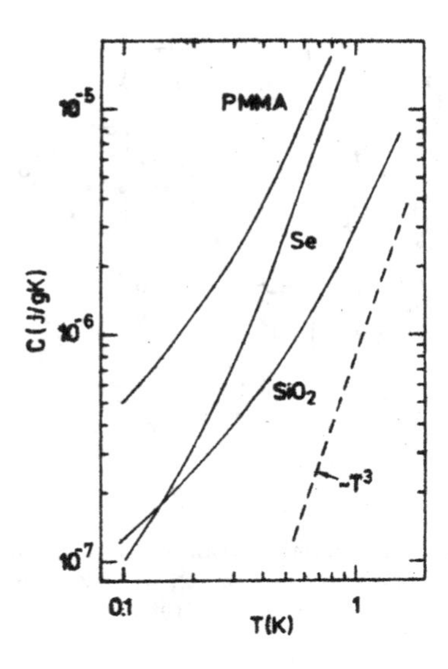

Fig. 1.3 Specific heat C_V of vitreous silica (SiO_2), amorphous selenium (Se) and polymethylmetacrylate (PMMA), also known as "plexiglass", as a function of temperature. In the case of amorphous silica, the Debye contribution due to sound waves Eqs. (1.1)-(1.3) is included as a broken straight line on this log-log plot. The variation of C_V at low temperature roughly follows a relation $C_V \propto T^{1+n}$ with $0 < n \leq 0.3$ in the various amorphous materials. After Stephens (1976).

internal energy due to phonons

$$\begin{aligned} U_{\text{phonons}} &= \int \hbar\omega\zeta(\omega)[\exp(\hbar\omega/k_BT) - 1]^{-1}\, d\omega \\ &\underset{T\to 0}{\longrightarrow} \text{const} \cdot T^{d+1} \int_0^\infty x^d[e^x - 1]^{-1} dx\,, \end{aligned} \tag{1.2}$$

where $\hbar$ and k_B are Planck's constant and Boltzmann's constant, and in the last step we have substituted $\hbar\omega/k_BT = x$ in the integral. From Eq. (1.2) we immediately recognize the famous Debye T^3 law for the specific heat at low T (in $d = 3$ dimensions), since

$$C_V = (\partial U/\partial T)_V \propto T^d\,. \tag{1.3}$$

(Here we have made the approximation that the internal energy U is given by U_{phonons}, an approximation which is reasonable if the contributions from conduction electrons or magnons can be neglected.)

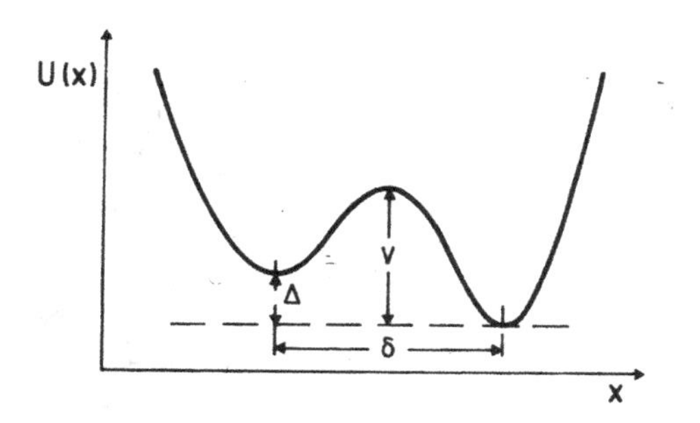

Fig. 1.4 Double-well potential with an asymmetry Δ, a potential barrier of height V, and width δ.

These considerations are no longer valid in amorphous materials: Since there is no crystal lattice, there is no Brillouin zone, and $\vec{k}$ is no longer a good quantum number. While acoustic waves with the dispersion relation $\omega = ck$ still exist for $k \to 0$, these waves are not dominant in the density of states for small ω. Instead the temperature dependence of the specific heat at low T, see Fig. 1.3, seems to hint for the presence of excitations that contribute to the density of states at low ω. In this context the model of the so-called "two-level systems" (TLS) has gained wide popularity to describe this excess. Within this model one postulates the existence of particular degrees of freedom (x) which experience a potential $U(x)$ with the shape of an asymmetric double well and which undergo a tunneling motion between the two wells, see Fig. 1.4. As we will see later, the TLS model (Anderson *et al.* 1972; Phillips 1972, 1981; von Löhneysen 1981; Kovalenko *et al.* 2001) is quite successful and widely used to account for many low temperature properties of various kinds of amorphous materials. E.g., if one assumes that the density of such TLS is independent of the gap Δ for $\Delta \to 0$, one can rationalize a linear variation of C_V with T. However, although the concept of two level systems in glasses is already more than 30 years old, it is still controversial what the physical meaning of the variable x in Fig. 1.4 really is. Figure 1.5 gives speculative examples of structures in which two-level systems could be present (Hunklinger and Arnold 1976). However, a microscopic theory that describes the precise physical nature of these two-level systems is still lacking. We will return to a more detailed discussion of these problems in Chap. 4 of this book.

While so far we have mostly considered solids with a disordered structure resembling the one of a liquid, as they are typically formed by supercooling a fluid sufficiently far below the melting/crystallization temperature

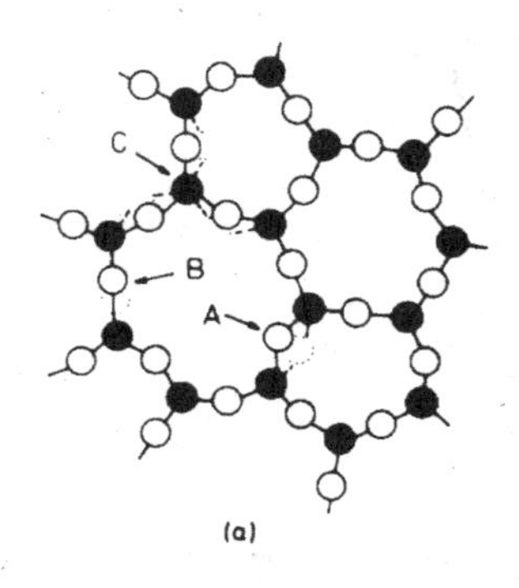

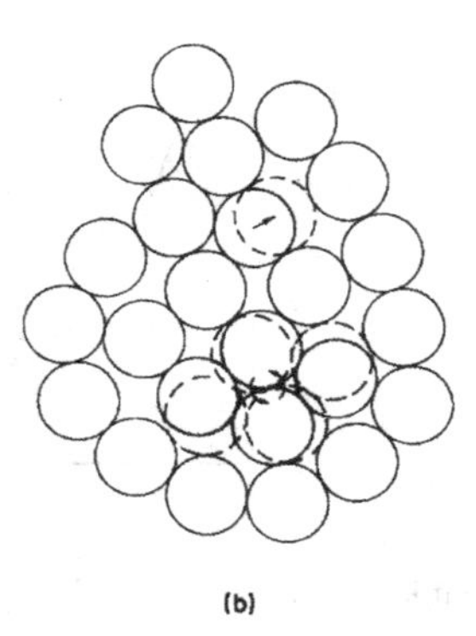

Fig. 1.5 (a) Two-dimensional projection of the structure of vitreous silica with possible double-well potential for an O atom (A and B) and a SiO_4 tetrahedron (C) (after Hunklinger and Arnold 1976). (b) Two-dimensional projection of a dense random packing of hard spheres with possible double-well potential for a single atom and a group of atoms. After von Löhneysen (1981).

T_m, very different disordered solid structures can be produced by aggregation of (nanoscopic) solid particles. As an example we show in Fig. 1.6 various aggregates of gold particles (Weitz and Huang 1984). The structure of these aggregates is rather loose, in that the colloidal particles do not fill the available space densely, and holes occur on all length scales. Thus such structures are reminiscent to fractals (Mandelbrot 1982) and as a result one can characterize them, at least approximately, by a (non-integer) dimension $d_f < d$, the so-called "fractal dimension". This can, e.g., be done by considering the relation how the total mass M of a cluster scales with its radius R

$$M \propto R^{d_f}, \quad R \to \infty. \tag{1.4}$$

Of course, this power law should be understood only as an asymptotic relation in that R should strongly exceed the radius of an individual gold particle (which is a compact, crystalline object; other colloidal particles, e.g. polystyrene spheres, may be amorphous). Figure 1.7 shows a typical plot to determine d_f for the type of aggregates shown in Fig. 1.6. Witten and Sander (1981) have proposed a simple theoretical model for the kinetics of the formation of such fractal aggregates, the so-called "diffusion-limited aggregation" (DLA), which in the past has been studied extensively (Meakin 1998) and which will be described in the Chap. 3. In Fig. 1.8 we show the computer-generated picture of such a DLA cluster, and the corresponding

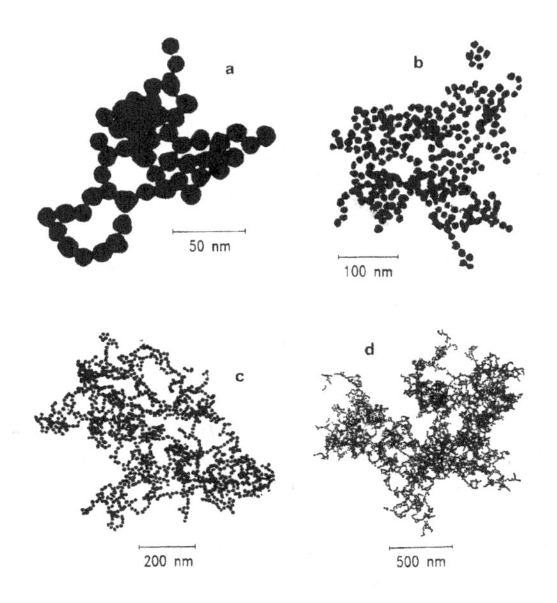

Fig. 1.6 Typical transmission electron micrographs of clusters of various sizes that have been generated by the aggregation of colloidal gold particles. After Weitz and Huang (1984).

plot to estimate a fractal dimension (Sander 1985).

One of the simplest fractal object is a percolation cluster right at the percolation threshold (Stauffer and Aharony 1992, Stauffer 1979). We will discuss the percolation concept extensively in Chap. 3. Further prominent examples for fractal objects are random walks and self-avoiding walks which will be considered in that chapter as well. For the moment, we only mention that the random structure of a fractal object like the aggregates shown in Figs. 1.6 and 1.8 has immediate consequences for its low-frequency excitations, e.g. their mechanical vibrations. E.g., if one counts for sound waves with wave-vector k, for which we thus have ($\omega = ck$), the number of corresponding eigenstates, one finds an anomalous density of states

$$\zeta_F(\omega) \propto \omega^{d_s - 1}, \tag{1.5}$$

where the so-called "spectral dimension" (Alexander and Orbach 1982) $d_s < d_f$ is a further characteristic of the fractal structure. As we will see in Chap. 4, fractal objects give rise to elementary excitations with rather anomalous properties, the so-called "fractons" (Alexander and Orbach 1982, Rammal and Toulouse 1983).

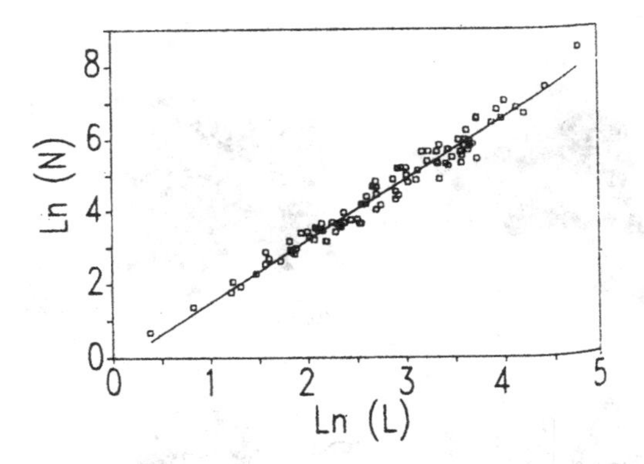

Fig. 1.7 Variation of mass M as a function of the radius R for gold colloidal aggregates. The mass is scaled by the mass of a single gold particle, and the radius is scaled by a single particle radius. After Weitz and Oliveria (1984).

So far we have dealt with disorder in the ("equilibrium") positions of the particles and which constitutes the most important case of frozen-in disorder in solids. Actually, a much simpler case of frozen-in disorder can be found in mixed crystals, in which the constituents still form a regular crystalline lattice, but the two species of atoms (in a binary system) occupy the available lattice sites at random. Let us consider the random statistical configurations of the two species (A and B) on the given lattice as frozen-in, i.e. we neglect all interdiffusion events (which would lead either to phase

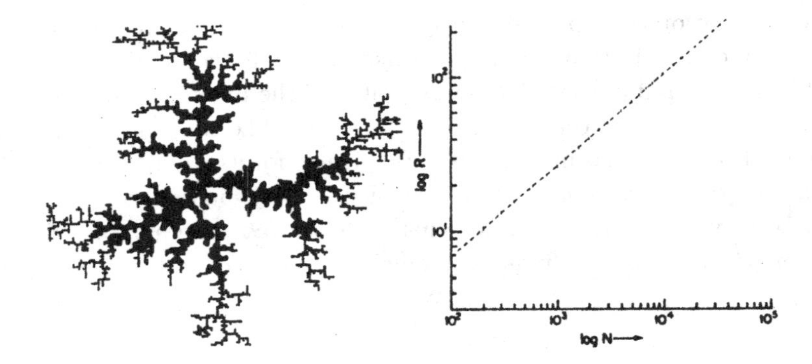

Fig. 1.8 Left: Snapshot picture of a DLA cluster grown via a Monte Carlo algorithm on the square lattice. After Sander (1985). Right: Log-log plot of the gyration radius R of DLA clusters on the square lattice versus the number N of lattice sites occupied by the cluster. After Sander (1985).

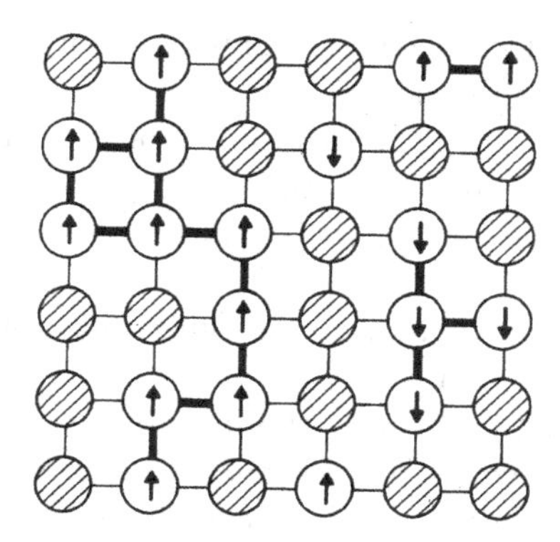

Fig. 1.9 Diluted Ising model above near the percolation threshold for the case of zero magnetic field and $T \to 0$. Nonmagnetic ions are shown by shaded circles, magnetic ions are shown by open circles, assuming temperature $T \to 0$ at zero magnetic field. Exchange interactions are indicated by thick lines. Arrows of the spins indicate a possible ground state configuration.

separation or to the formation of a periodically ordered superstructure, as discussed, e.g., by Binder and Fratzl 2001), an approximation that is indeed legitimate for all solids at sufficiently low temperatures. The disorder in the local composition of the binary system has very interesting consequences for various degrees of freedom such as, e.g., atomic displacements relative to their "equilibrium" positions in the mixed crystal - the treatment of the lattice dynamics of such substitutionally disordered crystals is a longstanding problem (Ziman 1979). Also the electronic states are strongly affected by this type of structural disorder, and it is a very important question to understand whether states which have energies near the Fermi energy are extended or localized. The theory of "Anderson localization" (Anderson 1958, Thouless 1974, Abrahams *et al.* 1979) has remained a longstanding challenge, and even the much simpler limit where electrical conductivity is described as a random hopping of carriers from lattice site to lattice site is difficult to treat near the percolation threshold. Although here we will not consider the scaling concepts on Anderson localization, percolation conductivity (Kirkpatrick 1973, Adler *et al.* 1990) will be discussed in Chap. 4.

A further type of randomly quenched disorder results if one considers binary mixed crystals in which one species is magnetic, while the other species is non-magnetic. The simplest case occurs when the exchange interaction between the magnetic ions is only nonzero for the nearest-neighbor distance on the lattice, and ferromagnetic. This problem leads once again to a variant of the percolation problem - if the concentration c of the ferromagnetic component exceeds the concentration c_p of the "site percolation problem", we have in the thermodynamic limit a connected cluster of spins of infinite size, and hence the system still maintains a spontaneous magnetization (Fig. 1.9). We will discuss the "percolation transition", which describes the vanishing of the spontaneous magnetization as c reaches c_p from above, in more detail in Chap. 3.

A particularly interesting problem is the magnetic relaxation of these diluted magnets near the percolation threshold. It is found that the relaxation times diverge faster than the usual Arrhenius law

$$\tau \propto \exp(E_{\mathrm{act}}/k_B T)\,, \tag{1.6}$$

in which the divergence of τ at $c = c_p$ occurs at zero temperature, but instead can be fitted by the so-called "Vogel-Fulcher-Tammann" law (VFT) (Vogel 1921, Fulcher 1925, Tammann and Hesse 1926)

$$\tau \propto \exp\{E_{\mathrm{act}}/[k_B(T - T_0)]\}\,, \tag{1.7}$$

and which implies a divergence of the relaxation time at a nonzero temperature T_0. Equation (1.7) is often taken as a phenomenological description for the temperature dependence of the viscosity η of supercooled fluids (note $\eta \propto \tau$, the structural relaxation time), although it lacks a real theoretical justification. In Chap. 5, we will return to Eqs. (1.6) and(1.7) in the context of the glass transition of supercooled fluids (more detailed treatments can be found in Zarzycki 1991, Gutzow and Schmelzer 1995, and Donth 2001). However, we emphasize already here that the description of the dynamics of glass-forming systems near the glass transition is not fully understood yet. Thus, the dynamics of diluted magnets near the percolation threshold is one of the rare examples of systems with cooperative activated dynamics for which we have a sound theoretical description (see Chap. 4).

Mixed crystals in which one species is magnetic and the other species is non-magnetic are not only examples for the percolation problem, but also provide examples for many other types of unconventional cooperative phenomena possible in disordered solids. E.g., a diluted antiferromagnet (i.e. a system in which the exchange interactions between nearest neighbors are negative) in an external magnetic field provides a realization of the so-called random field Ising model (Imry and Ma 1975, Fishman and

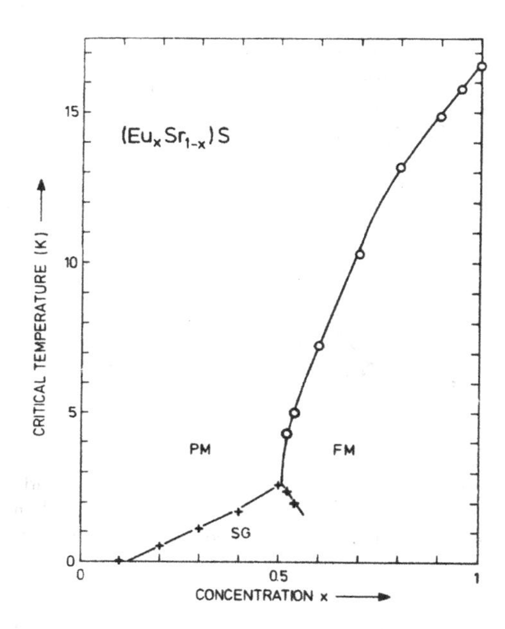

Fig. 1.10 Experimental phase diagram of $Eu_xSr_{1-x}S$ in the plane of variables, temperature T and Eu concentration x. Lines show critical temperatures of transitions between paramagnetic (PM), ferromagnetic (FM), and spin glass (SG) phases. After Maletta and Convert (1979).

Aharony 1979, Villain 1985, Nattermann and Villain 1988, Belanger 1988, 1998, Nattermann 1998). In this model, one introduces into a crystal a randomly quenched disorder via a local magnetic field that randomly points up or down at each lattice site, but which is zero on average. At low dimensionality such as $d = 2$, where a pure Ising magnet still exhibits long range magnetic order up to a critical temperature T_c (Stanley 1971, Cardy 2000), already arbitrarily weak random fields disrupt the long range order, causing the system to break up into an irregular arrangement of large magnetic domains. Thus there exists local order and even medium range order, but no long range order.

Even more complicated phenomena occur in diluted ferromagnets which have exchange interactions that are not uniformly ferromagnetic, but have competing ferromagnetic and antiferromagnetic exchange with different range. A simple example for such a system is the ferromagnet EuS, in which the magnetic Eu ions occupy the sites of a face-centered cubic lattice (and the non-magnetic sulfur atoms sit between two neighboring Eu

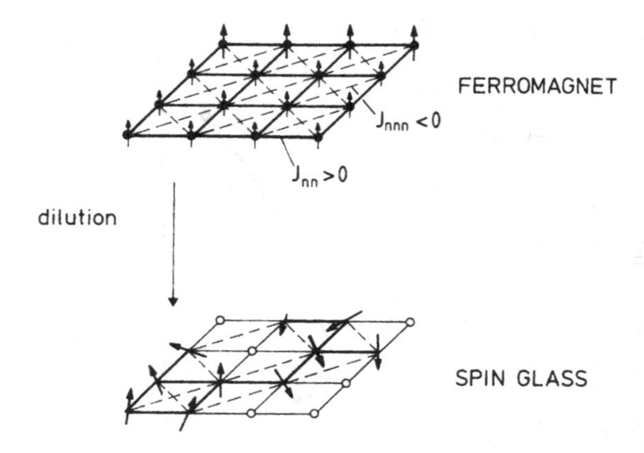

Fig. 1.11 Upper graph: Schematic picture of the order in one lattice plane for a ferromagnet with nearest neighbor ferromagnetic exchange J_{nn} (full lines) and next nearest neighbor antiferromagnetic bonds J_{nnn} (dashed lines). Lower graph: Schematic picture of the spin orientations in one lattice plane of the diluted system, in the spin glass phase. Nonmagnetic atoms are shown as open circles.

atoms, thus forming a rock salt structure). The nearest neighbors have a ferromagnetic coupling J_{nn}, while next nearest neighbors are antiferromagnetically coupled, with $J_{nnn} = -J_{nn}/2$ (Zinn 1976, Maletta and Zinn 1989). If this crystal is randomly diluted with nonmagnetic Sr, one finds in the phase diagram that the ferromagnetic order disappears already at an Eu concentration $x = 0.5$, i.e. long before the percolation threshold is reached (for nearest neighbors, the site percolation threshold on the fcc lattice is $x_c^{nn} = 0.198$ (Stauffer and Aharony 1992). For low temperatures, the experimental phase diagram (Fig. 1.10) for $Eu_xSr_{1\text{-}x}S$ (Maletta and Convert 1979) exhibits a phase, in which the spins are frozen-in in randomly oriented directions, a so-called spin glass phase (Cannella and Mydosh 1972; Edwards and Anderson 1975; Binder and Young 1986; Fischer and Hertz 1991; Young 1998). Figure 1.11 gives a qualitative idea why such a disordered phase is favorable: Due to the random fluctuations in the number of magnetic neighbors that a spin has in the diluted system, sometimes the ferromagnetic interactions "win" and sometimes the antiferromagnetic ones win, but overall no simple long range ordered state exists.

This type of behavior is by no means restricted to magnetic systems since it is also found in the so-called "dipolar glasses" (Höchli *et al.* 1990). These materials can be produced by random dilution of ferroelectric or antiferroelectric materials and the disordered phase corresponds to a state in

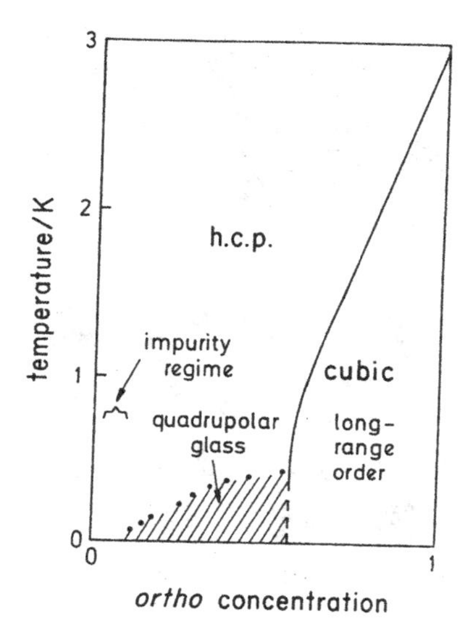

Fig. 1.12 Concentration-temperature phase diagram of ortho-para-hydrogen mixtures at low temperature. From Harris and Meyer (1985) as adapted by Höchli *et al.* (1990).

which the electric dipole moments are frozen in random directions. Similar phenomena occur in the so-called "quadrupolar glasses" or "orientational glasses" , i.e. diluted molecular crystals (e.g. para hydrogen diluted with ortho hydrogen, or N_2 diluted with Ar, etc.) where electrical quadrupole moments are frozen in random orientations, see Figs. 1.12 and 1.13 (Höchli *et al.* 1990, Binder and Reger 1992, Binder 1998). All these systems (spin glasses and orientational glasses , random field models) will be discussed in Chap. 4, since many concepts on structurally disordered non-crystalline solids and the glass transition from the fluid by which they are formed are "borrowed" from spin glass physics.

1.2 General Concepts on the Statistical Mechanics of Disordered Matter

1.2.1 *Lattice Models*

In the previous section we have already alluded to the fact that disordered systems of the "mixed crystal"- type are much simpler to handle than structural glasses (liquids, polymers, etc.). The reason for this is not only the presence of an underlying crystal lattice but also the fact that the degrees of

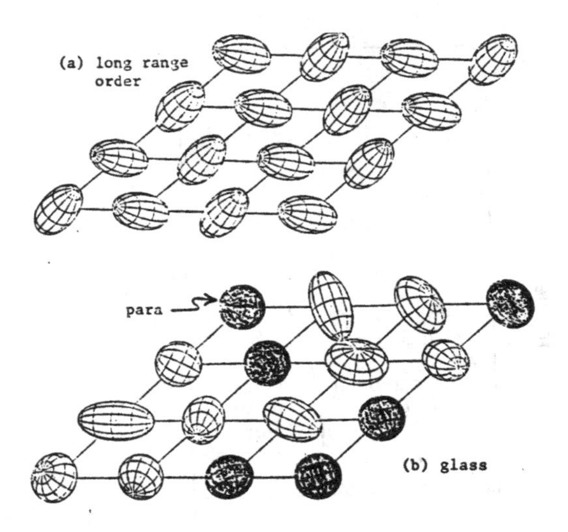

Fig. 1.13 (a) Schematic picture of the long range orientational order in one lattice plane of the pure ortho-hydrogen crystal. (b) Schematic picture of the molecular orientations in the quadrupolar glass phase. From Sullivan *et al.* (1987).

freedom that one wishes to treat within the framework of statistical thermodynamics (e.g. Ising spins $\{S_i = \pm 1\}$) are clearly separated from the variables characterizing the frozen-in disorder. This is in contrast to liquids in which the disorder is "self-generated", i.e. is intimately connected with the positions of the particles, i.e. the degrees of freedom one wants to describe theoretically. E.g., for the case of a disordered Ising model we can immediately write down a Hamiltonian

$$\mathcal{H} = -\sum_{i \neq j} J_{ij}\, S_i S_j - \sum_i H_i S_i, \quad S_i = \pm 1 \tag{1.8}$$

and characterize different models of disorder via the exchange couplings J_{ij} and/or the local fields H_i. The site percolation problem is, e.g., obtained by the choice

$$J_{ij} = \begin{cases} J\, x_i x_j, & \text{if } i, j \text{ are nearest neighbors} \\ 0, & \text{else,} \end{cases} \tag{1.9}$$

and $H_i = Hx_i$, where $J > 0$ and $\{x_\ell\}$ are random occupation variables of the lattice sites $\{\ell\}$, $x_\ell = 1$ if site ℓ is taken by a magnetic atom and $x_\ell = 0$ if not.

The random field Ising model is obtained by choosing a uniformly ferromagnetic nearest neighbor exchange $J_{ij} = J > 0$ and H_i is drawn either from a distribution function $P_\pm(H_i)$ that has two $\delta-$peaks or a gaussian distribution $P_G(H_i)$:

$$P_\pm(H_i) = \frac{1}{2}[\delta(H_i - H) + \delta(H_i + H)] \tag{1.10}$$

$$P_G(H_i) = (2\pi H^2)^{-1/2} \exp[-H_i^2/(2H^2)]. \tag{1.11}$$

Note that when we take an average $[\ldots]_{\rm av}$ over the quenched disorder

$$[H_i]_{\rm av} = 0, \quad [H_i^2]_{\rm av} = H^2. \tag{1.12}$$

Finally, the canonical model for an Ising spin glass, the "Edwards-Anderson model" (Edwards and Anderson 1975), is obtained by putting $H_i = H$ but drawing the exchange constants $\{J_{ij}\}$ from the distribution,

$$P_\pm(J_{ij}) = (1-x)\delta(J_{ij} - J) + x\delta(J_{ij} + J)\ , \quad \text{``}\pm J\text{model''}\,, \tag{1.13}$$

or

$$P_G(J_{ij}) = (2\pi)^{-1/2}(\Delta J)^{-1} \exp\{-(J_{ij} - J_0)^2/[2(\Delta J)^2]\}\,. \tag{1.14}$$

Note that in Eqs. (1.13) and (1.14) we have allowed for the possibility of a nonzero mean

$$[J_{ij}]_{\rm av} \equiv J_0\ \{= (1 - 2\,x)\,J \text{ in the case of the} \pm J\,\text{model}\} \tag{1.15}$$

in addition to the nonzero second moment

$$[J_{ij}^2]_{\rm av} - ([J_{ij}]_{\rm av})^2 \equiv (\Delta J)^2\{= 4J^2x(1-x)\,\text{in the case of the } \pm J\,\text{model}\}. \tag{1.16}$$

One of the key concepts about the statistical mechanics of glassy systems is known under the name "frustration" (Toulouse 1977), and is easily illustrated with the $\pm J$ model, Eq. (1.13). Consider, e.g., an isolated triangle of spins (Fig. 1.14) with an odd number of antiferromagnetic bonds: The ground state of this elementary triangle is 6-fold degenerate and the ground state energy is only $U_0 = -J$, while for a triangle with an even number of antiferromagnetic bonds the ground state energy is $U_0 = -3J$ and there is only a two-fold degeneracy. Such a triangle with an odd

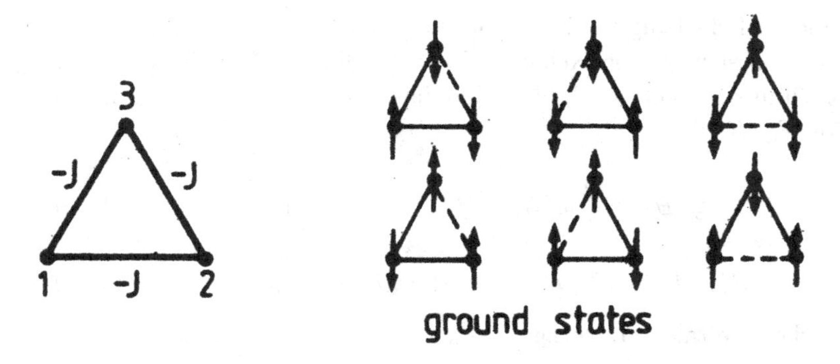

Fig. 1.14 Frustrated Ising triangle (left part) and its 6 ground states (signs of the Ising spins S_i are indicated by arrows, the "unsatisfied" bond is denoted by a broken line). From Binder (1980).

number of antiferromagnetic bonds is called "frustrated", since it is obviously impossible to find a spin configuration that would be satisfactory for all the bonds.

This effect carries over to closed loops with more than three spins. E.g. an elementary unit of the square lattice is also "frustrated" if one encounters an odd number of antiferromagnetic bonds going around such a square (the degeneracy of such a so-called "frustrated plaquette" is 8-fold and the energy $U_0 = -2J$ rather than $U_0 = -4J$, the energy of an "unfrustrated plaquette" with an even number of antiferromagnetic bonds, as the reader can easily work out). One can even imagine frustration without disorder. Consider, e.g., a regular square lattice that contains a periodic arrangement of frustrated plaquettes (Villain 1977). Although such models often show an enhanced ground state degeneracy, and in the thermodynamic limit there may even exist in the thermodynamic limit a nonzero ground state entropy, violating the third law of thermodynamics, they nevertheless may show a phase transition of conventional type at a nonzero temperature, or (depending on the details of the model) stay in the disordered phase at all temperatures. The Ising antiferromagnet on the triangular lattice, which has all elementary triangles frustrated, see Fig. 1.14, is an example of the latter case. Thus, such periodically frustrated models without disorder do *not* exhibit a glass transition, and the common view is that disorder as well as frustration are essential ingredients for glassy behavior.

1.2.2 *Averaging in Random Systems: Quenched versus Annealed Disorder*

Due to the presence of the frozen in disorder of the system, which for the cases treated in Eqs.(1.8)- (1.16) will be characterized by the random variables $\{x_i\}$, the averaging procedure of statistical mechanics implies that we have to do a double average: Over the spins $\{S_i\}$ as well as over the disorder variables $\{x_i\}$. This means that after the standard canonical average of an observable A, such as the magnetization per site, or the internal energy,

$$\langle A(\{S_i\},\{x_i\})\rangle_T = Z^{-1}\,Tr\big(\exp[-\beta\mathcal{H}(\{S_i\},\{x_i\})]A(\{S_i\},\{x_i\})\big), \quad (1.17)$$

where Z is the partition function, $Z = Tr\ \exp(-\beta\mathcal{H})$, and $\beta = 1/k_BT$, one has to take the average over the disorder:

$$[\langle A(\{S_i\},\{x_i\})\rangle_T]_{\rm av} = \int \Pi dx_i P(\{x_i\})\langle A(\{S_i\},\{x_i\})\rangle_T\,. \quad (1.18)$$

Equations (1.17) and (1.18) give a simple and well defined prescription on how thermodynamic quantities can be calculated for systems with frozen in disorder, as e.g., the spin systems discussed in the previous subsection. This prescription is certainly much simpler than the one needed for the disordered structure of a solid such as the one shown schematically in Fig. 1.2 since in that case it is not possible to write down the distribution function that describes the position of the atoms in the glassy structure at zero temperature and in fact it is rather difficult to come up with an appropriate description of such a structure. In Chap. 3 we will thus introduce and discuss various models and concepts to characterize the structure of disordered solids, while the difficult problem of the transition from the fluid to the glass is deferred to the Chaps. 5 and 6. The statistical mechanics of the models based on Eq. (1.8) will be discussed in Chap. 4.

We emphasize already at this point that for most observables A the averaging $[\ldots]_{\rm av}$ over the distribution of the disorder variables $\{x_i\}$ given by Eq. (1.18) is not really necessary since almost always a single realization $\{x_i\}$ yields a canonical average $\langle A(\{S_i\},\{x_i\})\rangle_T$ that is practically indistinguishable from the full average $[\langle A\rangle_T]_{\rm av}$, a property that is called "self-averaging". Note that in most cases experimentalists assume (implicitly) that their sample has this property since they usually study only one (or a small number) of samples (diluted crystal, fractal aggregate, spin glass,...). Similarly, for a given cooling history of a glass formed from a supercooled fluid one single sample is sufficient to get reproducible results. (But different cooling histories may give rise to systematically different structural disorder and hence also different physical properties of an amorphous solid.)

The theoretical justification for this self-averaging is the fact that a system with short range interactions can be divided into many subsystems, each of which is macroscopically large, but statistically independent of each other. Each such subsystem can be viewed as a realization of an ensemble generated according to the probability $P(\{x_i\})$. Therefore the average over the whole system $\langle A \rangle_T$ can be considered as an average over many subsystems with different $\{x_i\}$ and weight $P(\{x_i\})$, so that the average $[...]_{\rm av}$ is in fact realized. Of course, as usual in statistical thermodynamics, for any finite number N of the degrees of freedom in the system we expect fluctuations with relative magnitude $1/\sqrt{N}$. Let us consider these fluctuations for the case of the free energy density f: we assume a gaussian distribution for the probability $p(f)$, with average $[f]_{\rm av}$ and width $\Delta f/\sqrt{N}$:

$$p(f) = (2\pi)^{-1/2}(\Delta f)^{-1} \exp\{-N(f - [f]_{\rm av})^2/[2(\Delta f)^2]\}\,. \qquad (1.19)$$

Here the so-called "quenched average" $[f]_{\rm av}$ is given by the average over the logarithm of the partition function,

$$N[f]_{\rm av} \equiv [F]_{\rm av} = -k_B T[\ln Z\{x_i\}]_{\rm av}\,. \qquad (1.20)$$

At this point it is interesting to compare this quenched average of the free energy with the so-called "annealed average", which is defined by

$$F_{\rm ann} = -k_B T \ln[Z\{x_i\}]_{\rm av}\,, \qquad (1.21)$$

i.e., here one averages the partition function $Z\{x_i\}$ instead of the free energy $F = -k_B T \ln Z\{x_i\}$, with

$$Z\{x_i\} = Tr_{\{S_i\}} \exp[-\beta\mathcal{H}(\{x_i\}, \{S_i\})]\,. \qquad (1.22)$$

It is useful to realize that we can rewrite $[Z\{x_i\}]_{\rm av}$ as follows,

$$\begin{aligned}
[Z\{x_i\}]_{\rm av} &= \int \Pi dx_i P(\{x_i\})\, Tr_{\{S_i\}} \exp[-\beta\mathcal{H}(\{x_i\}, \{S_i\})] \\
&= \int \Pi dx_i Tr_{\{S_i\}} \exp[\ln P(\{x_i\})] \exp[-\beta\mathcal{H}(\{x_i\}, \{S_i\})] \\
&= Tr_{\{S_i, x_i\}} \exp[-\beta\mathcal{H}'(\{x_i\}, \{S_i\})] \equiv Z' \qquad (1.23)
\end{aligned}$$

where the Hamiltonian $\mathcal{H}'(\{x_i\}, \{S_i\})$ is defined as follows:

$$\mathcal{H}'(\{x_i\}, \{S_i\}) = \mathcal{H}(\{x_i\}, \{S_i\}) - k_B T \ln P(\{x_i\}) \,. \tag{1.24}$$

Thus $[Z\{x_i\}]_{\text{av}}$ can be considered as the partition function Z' of a system described by a Hamiltonian $\mathcal{H}'(\{x_i\}, \{S_i\})$, in which both the spins $\{S_i\}$ as well as the disorder variables $\{x_i\}$ are treated on the same footing in that both types of degrees of freedom are integrated over according to the Boltzmann weight $\exp(-\mathcal{H}'/k_B T)$. On the other hand, if one calculates a trace it does not matter in which order the various variables are integrated over. Therefore we can rewrite Eq. (1.23) as

$$Z' = Tr_{\{S_i\}} \exp[-\beta \mathcal{H}_{\text{eff}}(\{S_i\})] \,, \tag{1.25}$$

where

$$\exp[-\beta \mathcal{H}_{\text{eff}}(\{S_i\})] \equiv Tr_{\{x_i\}} \exp[-\beta \mathcal{H}'(\{x_i\}, \{S_i\})] \,. \tag{1.26}$$

Thus, if the disorder is thermally annealed, i.e. the disorder is due to some variables whose statistics is governed by a Boltzmann weight at the same temperature, the problem of averaging over the disorder can be done on the level of the Hamiltonian: the problem is equivalent to a "pure" system (i.e., not depending on disorder variables $\{x_i\}$ at all) with an effective Hamiltonian $\mathcal{H}_{\text{eff}}(\{S_i\})$. An example is a mixed disordered crystal in which A-and B-atoms are distributed over the available lattice sites without being fixed once and forever, but are allowed to change sites by interdiffusion.

What is the relation between the quenched and annealed disorder average? To understand this connection we use Eqs. (1.19) and (1.20) to calculate $[Z\{x_i\}]_{\text{av}}$. This yields

$$\begin{aligned}
[Z\{x_i\}]_{\text{av}} &= \int_{-\infty}^{+\infty} \exp[-Nf/k_B T] p(f) df \\
&\propto \int_{-\infty}^{+\infty} \exp\Big[-\frac{Nf^2}{2(\Delta f)^2} - \frac{Nf}{k_B T} + \frac{Nf[f]_{\text{av}}}{(\Delta f)^2} - \frac{N[f]_{\text{av}}^2}{2(\Delta f)^2} \Big] df \qquad (1.27) \\
&= \exp\left\{ \frac{N}{2} \left(\frac{\Delta f}{k_B T} \right)^2 - \frac{N[f]_{\text{av}}}{k_B T} \right\} \times
\end{aligned}$$

$$\int_{-\infty}^{+\infty} \exp\left[-\frac{N}{2(\Delta f)^2}\left\{f - [f]_{\mathrm{av}} + \frac{(\Delta f)^2}{k_B T}\right\}^2\right] df \tag{1.28}$$

$$= \sqrt{\frac{2\pi}{N(\Delta f)^2}} \exp\left\{\frac{N}{2}\left(\frac{\Delta f}{k_B T}\right)^2 - \frac{N[f]_{\mathrm{av}}}{k_B T}\right\}. \tag{1.29}$$

Hence we obtain with Eq. (1.20)

$$f_{\mathrm{ann}} = F_{\mathrm{ann}}/N = [f]_{\mathrm{av}} - (\Delta f)^2/(2k_B T). \tag{1.30}$$

Thus one can see that, as expected, the annealing of the disorder decreases the free energy,

$$f_{\mathrm{ann}} \leq [f]_{\mathrm{av}} , \tag{1.31}$$

and that the free energy of the annealed system is a lower bound for the quenched free energy. Since the effective Hamiltonian $\mathcal{H}_{\mathrm{eff}}(\{S_i\})$ possesses the full symmetry of the pure system, it is much easier to handle than the original Hamiltonian $\mathcal{H}(\{x_i\}, \{S_i\})$, for which the evaluation of the quenched free energy, Eq. (1.20), is often very difficult.

1.2.3 *"Symmetry Breaking" and "Ergodicity Breaking"*

In the model for a spin glass, as it has been defined in Eqs. (1.8), (1.13) or (1.14), the freezing transition at the temperature T_f from the disordered phase to the glass phase can be described by using the well known methods of statistical thermodynamics and thus for this case the conditions under which there occurs a thermodynamic phase transition can be clarified. It is thus, possible to determine, e.g., the order (in the sense of Ehrenfest) of the transition. While in the model mentioned above this transition is of second order, i.e. the glass order parameter, introduced and discussed in Chap. 4, increases *continuously* from zero for $T > T_f$ to non-zero for $T < T_f$, there exist also generalizations of this model in which in the "mean field"-limit of infinite-range interactions a *kind of first order* glass transition occurs in that at T_f a finite nonzero value of the glass order parameter appears *discontinuously*. In this case, this thermodynamic transition is preceded by a "dynamic transition" at which the system shows an ergodic-to-nonergodic transition at a temperature $T_D > T_f$. This scenario occurs, for instance, in the $p-$state Potts glass model with $p > 4$ (Gross *et al.* 1985) whose

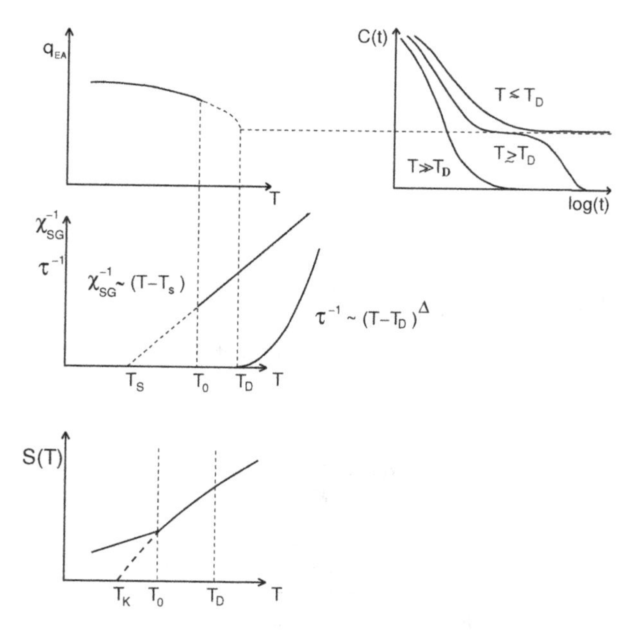

Fig. 1.15 Qualitative sketch of the mean field predictions for the $p-$state Potts glass model with $p > 4$. The glass order parameter $q_{\rm EA}$, which is given by the height of the plateau in the time correlation function $C(t)$ of the spins, is nonzero only for temperatures $T < T_0$ and jumps to zero discontinuously at $T = T_0$. A response function, i.e. the "spin glass susceptibility" $\chi_{\rm SG}$, see Chap. 4, follows a Curie-Weiss type relation with an apparent divergence at $T_s < T_0$. The relaxation time τ diverges at the dynamical transition temperature T_D. This divergence is due to the occurrence of a long-lived plateau of height $q_{\rm EA}$ in the time-dependent autocorrelation function $C(t)$ of the Potts spins. The entropy $S(T)$ has no jump at T_0, but shows only a kink, since there is no latent heat. The extrapolated (analytically continued) high temperature branch (dashed curve in the lower part of the figure) seems to vanish at the temperature T_K.

Hamiltonian is given by

$$\mathcal{H} = -\sum_{i<j} J_{ij}(p\delta_{\sigma_i\sigma_j} - 1)\,. \tag{1.32}$$

Here $\sigma_i \in \{1, ...p\}$ are the "Potts-spins", δ is the Kronecker-δ, and the coupling constants J_{ij} are again random variables. (It is easy to see that for $p = 2$ this model is identical to the Ising spin glass given by Eq. (1.8.) In Fig. 1.15 we sketch qualitatively the temperature dependence of various quantities of this model to illustrate the relative order of the different temperatures.

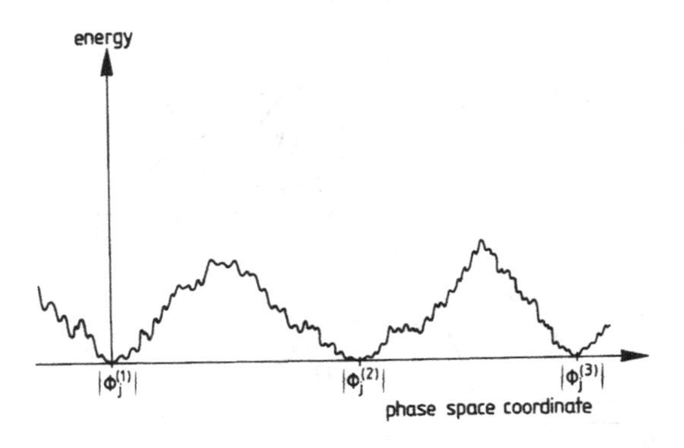

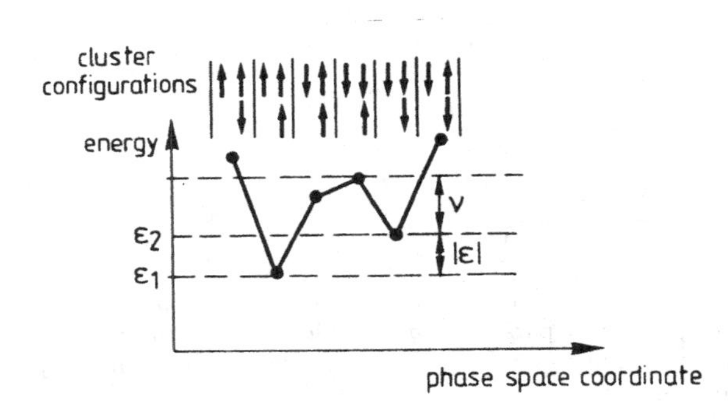

Fig. 1.16 Upper panel: Schematic plot of the coarse-grained free energy density of a glassy system at low temperatures plotted as a function of a phase-space coordinate. From Binder (1980). Lower panel: Schematic plot of the energy of a cluster of three spins in an Ising spin glass Eqs. (1.8) and (1.14) versus a phase space coordinate that represents the various states of this cluster. From Binder (1980), following Dasgupta *et al.* (1979).

In the context just mentioned "nonergodicity" means that the phase space of the model splits into distinct "valleys" separated by (infinitely high) free energy barriers (upper panel of Fig. 1.16). In an Ising spin glass this phase space means the set of all possible spin configurations $\{S_1 = \pm 1, \ldots, S_N = \pm 1\}$, as it is illustrated in the lower panel of Fig. 1.16 for a small cluster of three spins. It is clear that (free) energy barriers of infinite height can only arise in the thermodynamic limit $N \to \infty$ since for finite systems all barriers have finite height.

Note that this "breaking of ergodicity" is not at all specific for glassy systems since also any phase transition can be interpreted as an ergodic-to-nonergodic transition (Palmer 1982). In the disordered phase for $T > T_f$

the whole phase space is accessible, irrespective of the initial condition, and hence the system is ergodic. In contrast to this, in the low-temperature phase the system is confined to one of the valleys, which are also called "ergodic components", and it depends on the initial condition in which valley the system is. Initial conditions of the system may be specified, for instance, by a particular "cooling history" from the disordered phase.

A further concept that is important in the context of phase transitions is the one of "symmetry breaking". Consider, for instance, an Ising ferromagnet without any disorder in zero field, i.e., take Eq. (1.8) with $J_{ij} = J$, $H_i = 0$. The resulting Hamiltonian $\mathcal{H} = -J\sum_{\langle i,j\rangle} S_i S_j$ is invariant against a change of sign of all the spins,

$$\mathcal{H}(\{S_i \to -S_i\}) = \mathcal{H}(\{S_i\}) \,. \tag{1.33}$$

Due to this symmetry it is clear that the canonical average of the magnetization $M = 1/N\sum_i \langle S_i\rangle_T$ is *always* zero, i.e. the system does not seem to have a phase transition. This is, however, not the case since in fact this symmetry is broken spontaneously for temperatures below the Curie temperature T_c at which a spontaneous magnetization $M_{\rm sp}$ starts to set in. In order to define such a spontaneous magnetization, one first needs to "break the symmetry" by introducing a magnetic field H and subsequently one considers the limit $H \to 0^+$:

$$M_{sp} = \lim_{H\to 0^+} N^{-1}\sum_i \langle S_i\rangle_{T,H} \,. \tag{1.34}$$

One finds that M_{sp} is positive for $T < T_c$ whereas it is zero for $T > T_c$. If one takes the limit $H \to 0^-$, one obtains $-M_{\rm sp}$, since there exists of course a trivial degeneracy of the ordered phase as is obvious from the symmetry properties of the full Hamiltonian including the magnetic field H, the variable thermodynamically conjugate to the magnetization:

$$\mathcal{H}(\{S_i\}, H) = \mathcal{H}(\{-S_i\}, -H) \,. \tag{1.35}$$

These results concern the static properties of the systems. If one introduces a dynamics, e.g. by updating the spins via the rules of the "kinetic Ising model" (Glauber 1963), it is found that for $T < T_c$ the system has an infinitely high barrier between the phases with positive and negative spontaneous magnetization. This barrier appears at T_c, at which the phase space structure "bifurcates": Above T_c there is only one single valley and all the states are mutually accessible, and at T_c this single valley splits into the two valleys describing the two possible signs of the order parameter.

There exist also more complicated cases of symmetry in which, e.g., the phase space splits at T_c into infinitely many states. An example is the classical Heisenberg magnet,

$$\mathcal{H} = -J \sum_{\langle i,j \rangle} \vec{S}_i \cdot \vec{S}_j - H \sum_i S_i^z \,, \tag{1.36}$$

where $\vec{S}_i$ is a unit vector in the direction of the magnetic moment. For $T < T_c$ a symmetry breaking occurs in which, for $H = 0$, the spontaneous magnetization per spin $\vec{M} = (1/N) \sum \langle \vec{S}_i \rangle_T$ may point in any direction of space, and a particular direction of space is only singled out by the direction in which the field H is applied. Since in the absence of a magnetic field the Hamiltonian given by Eq. (1.36) is completely isotropic, no direction in space is singled out and the ordering is very degenerate. However, if we want to associate these ordered states again with "valleys" in phase space, we must realize that there exist certain directions in phase space where there is no barrier between one valley and its neighboring one, since one can reach another orientation of the magnetization via a continuous rotation of all the spins by the same angle and thus at zero cost in energy.

Thus, the typical situation for standard phase transitions is that at T_c a symmetry is broken and by the appearance of a nonzero order parameter. The resulting ordered phase has a degeneracy such that the various possible realizations of ordered states are related to each other precisely via that symmetry: In the case of the Heisenberg ferromagnet it is the full rotation symmetry, in the Ising ferromagnet it simply is the symmetry with respect to changing the sign of the magnetization. In stark contrast to this glassy systems have "ordered" states with a very different character: Although the mean field model of a spin glass has a freezing transition at which the phase space splits into many valleys (see Fig. 1.16), these valleys are not related by any known symmetry to each other, since they depend on the particular realization of the random quenched disorder. The description of this situation for the model given by Eq. (1.8) within the formulation based on Eqs. (1.17)-(1.20) is therefore much more intricate that for the one of standard ordered phases and in Chap. 4 we will discuss this in more detail.

1.2.4 *Configurational Entropy versus "Complexity", and the Kauzmann Paradox*

At the end of the preceding subsection we have mentioned the difficulty to understand the nature of the symmetry that is broken when a system with quenched disorder undergoes a glass transition. For the case of structural glasses the situation regarding this point is even worse and for the moment one can give only highly speculative answers. While for crystalline structures the symmetries due to their space groups are well understood, and the positional and orientational long range order of a crystal can be described quantitatively, the nature of the "order parameter" that appears at a glass transition of structural glasses (if there is one!) is not known. Despite this lack of understanding of the nature of the order parameter, it is certainly reasonable to assume that the free energy landscape has many local minima and maxima, see Fig. 1.16 (Goldstein 1969). If the valleys are separated by infinitely high barriers, it makes sense to decompose thermal averages according to contributions from the different valleys ("ergodic components" or "phases") which in the following we will label with an index ℓ while λ will denote a microstate of the system (so ℓ encompasses all the microstates λ that belong to a given "valley" in the upper panel of Fig. 1.16). We thus can write for any observable A

$$\begin{aligned}\langle A\rangle_T &= \frac{1}{Z}\sum_{\ell}\Big(\frac{1}{Z_\ell}\sum_{\lambda\in\ell} A(\lambda)\,\exp(-E(\lambda)/k_BT)\Big)Z_\ell \\ &= \sum_{\ell} P_\ell\langle A\rangle_T^{(\ell)}\,, \end{aligned}\tag{1.37}$$

where

$$P_\ell = Z_\ell/Z, \quad \langle A\rangle_T^{(\ell)} = \frac{1}{Z_\ell}\sum_{\lambda\in\ell} A(\lambda)\exp(-E(\lambda)/k_BT) \tag{1.38}$$

and

$$Z_\ell = \sum_{\lambda\in\ell}\exp(-E(\lambda)/k_BT) \equiv \exp(-Nf_\ell/k_BT)\,, \tag{1.39}$$

and we have

$$Z = \sum_{\ell} Z_\ell = \sum_{\ell}\exp(-Nf_\ell/k_BT)\,. \tag{1.40}$$

While this decomposition of the thermal averages into contributions from the different valleys seems at first sight rather trivial, it has an important consequence for the total free energy of the system since this energy, $F = -k_B T \ln Z$, cannot be written as an ergodic component average over all the valleys, as one might naively expect. This latter quantity is given by

$$\overline{F} = \sum_{\ell} P_\ell F_\ell = \sum_{\ell} P_\ell N f_\ell \,, \tag{1.41}$$

which, using $N f_\ell = -k_B T \ln Z_\ell$ and $P_\ell = Z_\ell / Z$ can be rewritten as

$$\overline{F} = F - k_B T \sum_{\ell} P_\ell \ln P_\ell = F + TI \,, \tag{1.42}$$

or

$$F = \overline{F} - TI \,. \tag{1.43}$$

Thus we see that $\overline{F}$ differs from F by an extra term TI, where the quantity I is usually called the "complexity":

$$I \equiv -k_B \sum_{\ell} P_\ell \ln P_\ell > 0 \,. \tag{1.44}$$

Due to its entropic nature, this complexity can be interpreted as the average additional information needed to specify a particular state, given the *a priori* probabilities P_ℓ. In analogy to the Boltzmann formula $S = k_B \ln \Omega$, where S is the entropy and Ω the number of states, one now can define an effective number K^* of states as

$$K^* = \exp(I/k_B) \tag{1.45}$$

which lies between unity ($K^* = 1$ if there is only one valley and hence no broken ergodicity) and the actual number of valleys $K \equiv \sum_{\ell} 1$ with $K^* = K$ if all valleys have the same probability, $P_\ell = 1/K$, because then $I = -k_B \ln(1/K)$. It is evident that in order for the complexity to be an extensive thermodynamic quantity a number of the order $K = \exp(\text{const} \times N)$ of valleys is needed. This type of reasoning can be exemplified nicely for mean field models of spin glasses. For instance, for the $p > 4$ Potts glass given by Eq. (1.32) it is known that we have only one valley for $T > T_D$, and the complexity I is strictly zero. For $T < T_D$, however, there are exponentially many valleys, resulting in a nonzero complexity (Kirkpatrick and Thirumalai 1988).

If one deals with structural glasses, it is very plausible to assume that a random structure such as the one shown in Fig. 1.2 is not unique, but that instead there exist many equivalent configurations that differ significantly in their (microscopic) structure. By "differ significantly" we mean that these configurations cannot be transformed into each other by either small displacements of all atoms or by rearrangement of a localized group containing a small number of atoms. Thus, it is tempting to associate with this diversity of random structures a "configurational entropy" S_{conf} and, if these configurations would indeed be separated by infinitely high barriers in phase space, S_{conf} would be precisely the complexity I in the above sense.

Unfortunately, this "complexity" or "configurational entropy" associated with the number of amorphous structures of a glass is not experimentally accessible. What is accessible to experiment, however, is the excess entropy $\Delta S = S - S_{\mathrm{crystal}}$ of the supercooled fluid or glass which can be obtained by integrating the specific heat, see Fig. 1.17 and Chap. 5. If one assumes that the vibrational contributions to the entropy of a glass or a fluid are identical, one can interpret ΔS in the temperature region $T < T_g$ as the configurational entropy of the glass. Thus ΔS as plotted in Fig. 1.17 for $T > T_g$ is often referred to as the "configurational entropy" of the fluid. However, in the fluid region the "vibrational entropy" is somewhat ill-defined since the atoms do not only vibrate but also relax, i.e. change their neighbors. For the same reason also the quantity ΔS is not really the number of valleys, since for $T \leq T_m$ there is only a single "valley" in the sense of Eqs. (1.37)- (1.45), and hence clearly the complexity $I = 0$ for $T > T_g$. Nevertheless it can be hoped that the nature of the vibrational motion in the fluid is similar to the one in the glass and also that the local (metastable) minima in the fluid phase have properties (number, local connectivity) that are not very different from the ones in the glass. If this is indeed the case it is reasonable to extrapolate the fluid data to lower temperatures in order to estimate the corresponding properties in the glass.

It is interesting that such an extrapolation of the liquid branch of the curve $\Delta S = \Delta S(T)$ to temperatures $T < T_g$ apparently seems to vanish at a temperature $T_2 \approx T_g - 40$ K, see Fig. 1.17. As emphasized by Kauzmann (1948), it is paradoxical if the total entropy of a supercooled fluid would, even at $T = 0$!, be less than the total entropy of a crystal since the crystal is the much more ordered state than the supercooled fluid. This constitutes the so-called Kauzmann paradox and to resolve it one usually assumes that the dashed extrapolation in Fig. 1.17 can only have a physical meaning for

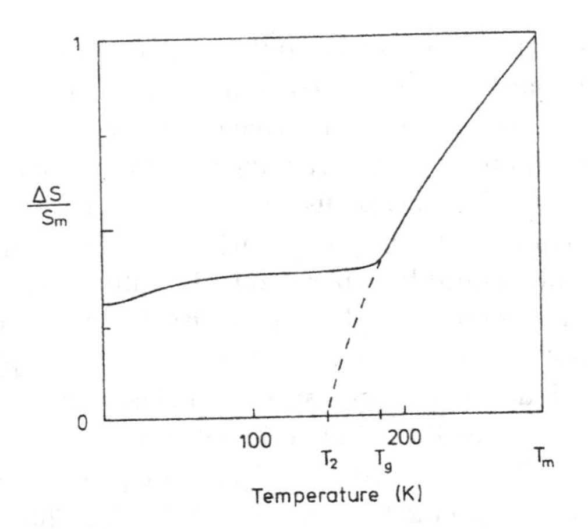

Fig. 1.17 Excess entropy $\Delta S = S - S_{\text{crystal}}$ in units of the entropy of melting $S_m = S(T_m) - S_{\text{crystal}}$ for glycerol as a function of temperature (solid line). The dashed line is an extrapolation of the data of the fluid branch and crosses the abscissa at the Kauzmann temperature T_2 ($T_m = 293$ K, $T_g = 186$ K, $T_2 = 150$ K). Data from Simon and Lange (1926).

$T > T_2$.[1] Thus it is often speculated that this "Kauzmann temperature" T_K (denoted as T_2 in Fig. 1.17) is a lower bound for the thermodynamic transition temperature from the (metastable!) supercooled fluid to a (metastable) "ideal glass". In this scenario, the actually observed glass transition at T_g is then just a precursor to this true transition, which in practice cannot be reached since the relaxation times of the system increase so strongly, see Eqs. (1.6) and (1.7), that it is not possible anymore to equilibrate at low T the fluid and which thus undergoes a glass transition at $T_g > T_K$. In Chaps. 5 and 6 we will discuss this point in more detail (see also Cavagna 2009).

A further remarkable feature of Fig. 1.17 is that the entropy of the glass branch stays nonzero down to $T = 0$, violating hence the third law of thermodynamics which states that $S(T = 0) = 0$. This nonzero ground state entropy is a clear indication that the actual glass that can be produced in the laboratory is a frozen-in metastable state far from equilibrium. The fact that the frozen-in states occurring for finite cooling rates are energetically unfavorable makes that below T_g the system shows the phenomenon

[1] Note that in a real experiment it is in principle possible to follow this dashed line by cooling the system *very* slowly since, as it will be discussed in Chap. 5, the glass transition temperature T_g depends weakly on the cooling rate.

of "aging", i.e. its properties (density, time correlation functions, etc.) depend on the time difference between the preparation of the sample, i.e. the crossing of T_g, and the start of the measurement (Struik 1978). Such a dependence of the properties of a material on the history of its preparation (and not just on the thermodynamic variables describing a state, as one is used to in the statistical mechanics of equilibrium phenomena) is also found in spin glasses and random field systems, see Chap. 4), and this fact underlines once more that all these strongly disordered systems have many features in common.

Despite this similarity we emphasize once more that for the moment it is not known whether or not for structural glasses an underlying quasi-equilibrium glass phase really exists, since the true equilibrium is certainly a crystal. In this respect the situation in spin glasses and orientational glasses is much more clean cut, at least from a theoretical point of view, since the variables defining the quenched disorder are clearly separated from the dynamical variables (magnetic dipole moments, electric multipole moments, etc.) and spin glass phases in thermal equilibrium are well-defined and can be studied by various methods (see Chap. 4).

Since a structural glass is formed by cooling a liquid, provided one can avoid crystallization, it is most natural to use the theory for the dynamics of liquids as a starting point for a theory of the glass transition. This approach is taken within the framework of the so-called "mode coupling theory of the glass transition" (MCT) (Götze 1990, 1999, 2009, Götze and Sjögren 1992) which so far is the only microscopic theory for the dynamics of glass-forming fluids, and it is found that, despite of certain limitations, this theory is surprisingly successful, see Chap. 5. In its idealized form, it predicts an ergodic-to-nonergodic transition at a temperature T_c, at which the structural relaxation time diverges with a power-law. Remarkably this transition is of the same type as was discovered later to occur in the mean field Potts glass for $p > 4$ (Fig. 1.15) and related models. The extended form of this theory, see Chap. 5, predicts that this critical singularity of the idealized MCT is rounded off, and that even for temperatures below the critical temperature T_c, which now has the meaning of a crossover temperature, the relaxation time is finite, although very large (it is predicted to increase according to an Arrhenius law). Figure 1.18 gives a qualitative sketch of this behavior as well as the predictions of the so-called "entropy theory of the glass transition" which relates the relaxation dynamics of a glass-forming system to the configurational entropy ΔS discussed above (Gibbs and Di Marzio 1958, Adam and Gibbs 1965). In Chap. 5 we will discuss these theoretical concepts in more detail and will also examine to

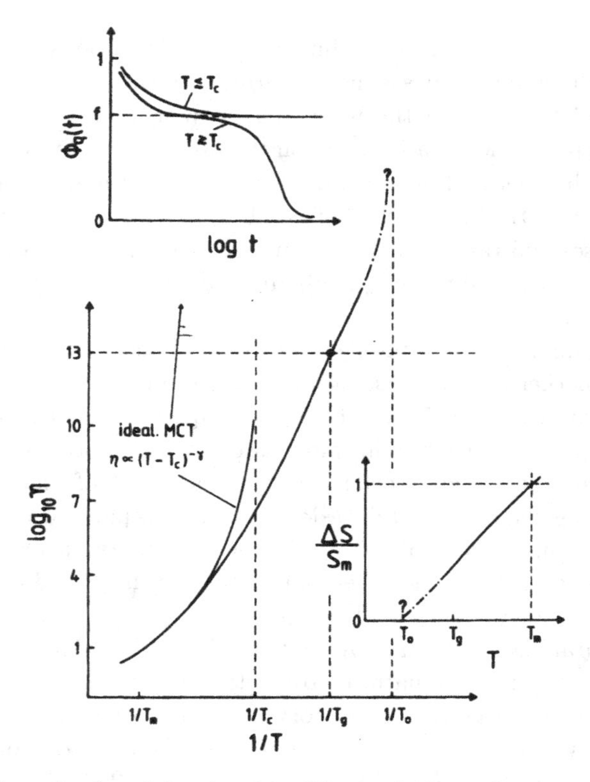

Fig. 1.18 Schematic plot of the viscosity $\eta(T)$ of a fluid as a function of inverse temperature $1/T$. The location of the melting temperature T_m, the critical temperature T_c of mode coupling theory, the glass transition temperature T_g, and the Vogel-Fulcher-Kauzmann temperature T_0 are shown on the abscissa. T_g is defined empirically via $\eta(T = T_g) = 10^{13}$ Poise. The upper inset shows the time dependence of the density correlation function $\Phi_q(t)$, where q is a wave vector near the first peak of the static structure factor $S(q)$. According to idealized MCT this function shows at temperatures slightly higher than T_c a plateau whose life time τ, with $\tau \sim \eta$, diverges like $\tau \sim (T - T_c)^{-\gamma}$ for $T \to T_c^+$. The lower right inset shows the entropy difference $\Delta S(T) = S(\text{fluid}) - S(\text{crystal})$, with $S_m = \Delta S(T_m)$. Its extrapolation for $T < T_g$ is used to define the Kauzmann temperature T_0 via $\Delta S(T = T_0) = 0$. This quantity is used in the entropy theory of the glass transition to predict the $T-$dependence of the relaxation time via $\tau \sim \exp(\text{const}/(T\Delta S))$.

what extent the concepts about mean field spin glasses, Fig. 1.15, really can be carried over to the structural glass problem, Fig. 1.18. For the latter it is clearly necessary to have a basic understanding of the static and dynamic correlations in dense liquids and therefore we will present in Chap. 2 an elementary discussion of this subject.

References

Abrahams, E., Anderson, P.W., Licciardello, D.C., and Ramakrishnan, T.V. (1979) *Phys. Rev. Lett.* **42**, 673.

Adam, G., and Gibbs, J.H. (1965) *J. Chem. Phys.* **43**, 139.

Adler, J., Meir, Y., Aharony, A., Harris, A.B., and Klein, L. (1990) *J. Stat. Phys.* **58**, 511.

Alexander, S., and Orbach, R. (1982) *J. Phys. (Paris) Lett.* **43**, L625.

Anderson, P.W. (1958) *Phys. Rev.* **109**, 1492.

Anderson, P.W., Halperin, B.I., and Varma, C.M. (1972) *Phil. Mag.* **25**, 1.

Belanger, D.P. (1988) *Phase Transitions* **11**, 53.

Belanger, D.P. (1998) in *Spin Glasses and Random Fields* (A.P. Young, ed.) p. 25 (World Scientific, Singapore).

Binder, K. (1980) in *Fundamental Problems in Statistical Mechanics V* (E.G.D. Cohen, ed.) p. 21 (North-Holland, Amsterdam).

Binder, K. (1998) in "*Spin Glasses and Random Fields*" (A.P. Young, ed.) p. 99 (World Scientific, Singapore).

Binder, K., and Fratzl, P. (2001) in *Phase Transformations in Materials* (G. Kostorz, ed.) p. 409 (Wiley VCH, Berlin).

Binder, K., and Reger, J.D. (1992) *Adv. Phys.* **41**, 547.

Binder, K., and Young, A.P. (1986) *Rev. Mod. Phys.* **58**, 801.

Cannella, V., and Mydosh, J.A. (1972) *Phys. Rev.* **B6**, 4220.

Cavagna, A. (2009) *Phys. Rep.* **476**, 51.

Cardy, J. (2000) *Scaling and Renormalization in Statistical Physics* (Cambridge University Press, Cambridge).

Cusack, N.E. (1986) *The Physics of Structurally Disordered Matter. An Introduction* (Adam Hilger, Bristol).

Dasgupta, C.S., Ma, S.-K., and Hu, C.K. (1979) *Phys. Rev.* **B20**, 3837.

Donth, E.-W. (2001) *The Glass Transition; Relaxation Dynamics in Liquids and Disordered Materials* (Springer, Berlin).

Edwards, S.F., and Anderson, P.W. (1975) *J. Phys.* **F5**, 965.

Elliott, S.R. (1983) *Physics of Amorphous Materials* (Longman, Essex).

Fischer, K.H., and Hertz, J.A. (1991) *Spin Glasses* (Cambridge University Press, Cambridge).

Fishman, S., and Aharony, A. (1979) *J. Phys.* **C12**, L729.

Fulcher, G.S. (1925) *J. Amer. Chem. Soc.* **77**, 3701.

Gibbs, J.H., and Di Marzio, E.A. (1958) *J. Chem. Phys.* **28**, 370.

Glauber, R.J. (1963) *J. Math. Phys.* **4**, 294.

Götze, W. (1990) in *Liquids, Freezing and the Glass Transition* (J. P. Hansen, D. Levesque and J. Zinn-Justin, eds). p. 287 (North-Holland, Amsterdam).

Götze, W. (1999) *J. Phys.: Condens. Matter* **11**, A1.

Götze, W. (2009) *Complex Dynamics of Glass-Forming Liquids* (Oxford University Press, Oxford).

Götze, W., and Sjögren L. (1992) *Rep. Prog. Phys.* **55**, 241.

Goldstein, M. (1969) *J. Chem. Phys.* **51**, 3728.

Gross, D.J., Kanter, I., and Sompolinsky, H. (1985) *Phys. Rev. Lett.* **55**, 304.

Gutzow, I., and Schmelzer, J. (1995) *The Vitreous State. Thermodynamics Structure, Rheology and Crystallization* (Springer, Berlin).

Harris, A.B., and Meyer, H. (1985) *Can. J. Phys.* **63**, 3.

Höchli, U.T., Knorr, K., and Loidl, A. (1990), *Adv. Phys.* **39**, 405.

Hunklinger, S., and Arnold, W. (1976) in *Physical Acoustics XII*, eds. W. P. Mason and R. N. Thurston (Academic Press, New York) p. 155.

Imry, Y., and Ma, S.-K. (1975) *Phys. Rev. Lett.* **35**, 1399.

Kauzmann, W. (1948) *Chem. Rev.* **43**, 219.

Kirkpatrick, S. (1973) *Rev. Mod. Phys.* **54**, 574.

Kirkpatrick, T.R., and Thirumalai D. (1988) *Phys. Rev.* **B37**, 5342.

Kovalenko, N.P., Krasny, Yu. P., and Krey, U. (2001) *Physics of Amorphous Metals* (Wiley-VCH, Berlin).

von Löhneysen, H. (1981) *Phys. Rep.* **73**, 161.

Maletta, H., and Convert, P. (1979) *Phys. Rev. Lett.* **42**, 108.

Maletta, H., and Zinn, W. (1989) in *Handbook on the Physics and Chemistry of Rare Earths, Vol. 12* (K.A. Gschneidner, Jr., and L. Eyring, eds.) p. 213 (Elsevier Science, Amsterdam).

Mandelbrot, B.B. (1982) *The Fractal Geometry of Nature* (Freeman, San Francisco).

Meakin, P. (1998) *Fractals, Scaling, and Growth Far From Equilibrium* (Cambridge University Press, Cambridge).

Nattermann, T. (1998) in *Spin Glasses and Random Fields* (A. P. Young, ed.) p. 277 (World Scientific, Singapore).

Nattermann, T., and Villain, J. (1988) *Phase Transitions* **11**, 5.

Palmer, R.G. (1982) *Adv. Phys.* **31**, 669.

Phillips, W.A. (1972) *J. Low. Temp. Phys.* **7**, 351.

Phillips, W.A. (1981) [ed.] *Amorphous Solids: Low Temperature Properties* (Springer, Berlin).

Rammal, R., and Toulouse, G. (1983) *J. Phys. (Paris) Lett.* **44**, L13.

Sander, L.M. (1985) in *Scaling Phenomena in Disordered Systems* (R. Pynn, A. Skjeltorp, eds.) p. 31 (Plenum, New York).

Simon, F., and Lange F. (1926) *Z. Phys.* **38**, 227.

Stanley, H.E. (1971) *An Introduction to Phase Transitions and Critical Phenomena* (Oxford University Press, Oxford).

Stauffer, D. (1979) *Phys. Rep.* **54**, 1.

Stauffer, D., and Aharony, A. (1992) *Introduction to Percolation Theory* (Taylor and Francis, London).

Stephens, R.B. (1976) *Phys. Rev.* **B13**, 852.

Struik, L.C.E. (1978) *Physical Aging in Amorphous Polymers and Other Materials* (Elsevier, Amsterdam).

Sullivan, N.S., Edwards, C.M., Lin, Y., and Zhou, D. (1987) *Can. J. Phys.* **65**, 1463.

Tammann, G., and Hesse, W. (1926) *Z. Anorgan. Allg. Chem.* **156**, 245.

Thouless, D.J. (1974) *Phys. Rep.* **13C**, 93.

Toulouse, G. (1977) *Commun. Phys.* **2**, 115.

Villain, J. (1977) *J. Phys.* **C10**, 1717.

Varshneya, A.K. (1993) *Fundamentals of Inorganic Glasses* (Elsevier, Amsterdam).

Villain, J. (1985) in *Scaling Phenomena in Disordered Systems* (R. Pynn and A. Skjeltorp, eds.) p. 423 (Plenum Press, New York).

Vogel, W. (1921) *Phys. Z.* **22**, 645.

Weitz, D.A., and Huang, J.S. (1984) in *Kinetics of Aggregation and Gelation* (F. Family and D.P. Landau, eds.) p. 19 (Elsevier, Amsterdam).

Weitz, D.A., and Oliveria, M. (1984) *Phys. Rev. Lett.* **52**, 1433.

Witten, T. A., and Sander, L.M. (1981) *Phys. Rev. Lett.* **47**, 1400.

Young, A.P. (1998) (ed.) *Spin glasses and random fields* (World Scientific, Singapore).

Zachariasen, W.H. (1932) *J. Am. Chem. Soc.* **54**, 3841.

Zallen, R. (1983) *The Physics of Amorphous Solids* (Wiley, New York).

Zarzycki, J. (Ed.) (1991) *Materials Science and Technology, Vol. 9*, (VCH Publ., Weinheim).

Ziman, J.M. (1979) *Models of disorder. The theoretical physics of homogeneously disordered systems* (Cambridge University Press, Cambridge).

Zinn, W. (1976) *J. Magn. Magn. Mater.* **3**, 23.

Chapter 2

Structure and Dynamics of Disordered Matter

In this chapter we mean by "disordered matter" both fluids and amorphous solids, as opposed to the crystalline solids in which the atoms occupy the sites of a perfect regular lattice. We will first proceed to discuss the geometrical structure of these systems as well as closely related static properties. Subsequently we will be concerned with the dynamical properties of matter, define suitable time correlation functions, and discuss transport coefficients. More detailed discussions of most of these topics can be found in the books by Barrat and Hansen (2003), Cusack (1986), Elliott (1983), and Hansen and McDonald (1986).

2.1 Pair Distribution Functions and the Static Structure Factor

Fluids and amorphous solids do not have the long range positional order found in crystals (Figs. 1.1 and 1.2). Nevertheless it would be quite wrong to assume that the positions of the atoms (or molecules, respectively) in a disordered material are simply random in space. Instead one finds a distinct short range order (SRO), which often resembles the SRO that one has in the crystalline material. Therefore the physical properties that depend mostly on the SRO and are not very sensitive to whether or not long range order is present, are often very similar to the corresponding properties of the crystal. One such property is the density of the material, which often changes only weakly if a fluid crystallizes and at the glass transition, in which the liquid changes from a fluid-like behavior to a solid-like one, the density ρ has no discontinuity at all. Instead one observes only a rounded kink in the ρ vs. T curve. The macroscopic properties of materials that are essentially controlled by their density and the strength of the forces between nearest neighbors, such as, e.g., the (average) sound velocity, are very similar in a crystal, fluid, or glass (of the same material).

By the word "average" we mean here an average over the direction of the wave-vector of the sound wave relative to the crystal axes, in the case when we consider a crystal, since the anisotropy of single crystals is a striking difference in macroscopic properties between crystals and disordered matter. The latter are perfectly isotropic, all directions in space are fully equivalent to each other, and for many applications this isotropy is a very desirable feature. E.g., the isotropic propagation of light in glasses is crucial for many of their applications in optics since birefringence of crystals makes them unsuitable for their use in lenses, mirrors, etc.

In order to understand the physical properties of fluids and amorphous solids in more detail, it is first of all necessary to characterize the structure and its SRO more precisely. Assuming a description of the system in terms of classical statistical mechanics, it is useful to define a local (particle) density $\rho(\vec{r})$ as follows

$$\rho(\vec{r}) = \sum_{i=1}^{N} \delta(\vec{r} - \vec{r}_i) . \tag{2.1}$$

In order to be specific we have assumed that we have N particles in a fixed volume at temperature T in a pure phase (i.e. at the moment we exclude the possibility of phase coexistence, e.g. part of the system being in the fluid phase, the other part crystallized). Thus, thermal equilibrium requires that the thermal average $\langle \ldots \rangle$ in the chosen statistical ensemble (which is the NVT ensemble under the specified conditions) yields a constant for the average local density, equal to the average density in the whole system:

$$\langle \rho(\vec{r}) \rangle_{\mathrm{NVT}} = \rho = N/V . \tag{2.2}$$

Note that in order to simplify the notation we will in the following write $\langle ... \rangle$ for $\langle ... \rangle_{\mathrm{NVT}}$.

We are now interested in the spacial correlation function between the fluctuations of the density at two points that are a distance $\vec{r} = \vec{r'} - \vec{r''}$ apart. In a homogeneous fluid or amorphous solid this correlation depends only on the displacement $\vec{r'} - \vec{r''}$ and thus we obtain:

$$G(\vec{r'} - \vec{r''}) = \langle [\rho(\vec{r}\,') - \rho][\rho(\vec{r}\,'') - \rho] \rangle \tag{2.3}$$

$$= \langle \rho(\vec{r})\,'\rho(\vec{r}\,'') \rangle - \rho^2 \tag{2.4}$$

$$= \sum_i \sum_j \langle \delta(\vec{r}\,' - \vec{r}_i)\delta(\vec{r}\,'' - \vec{r}_j) \rangle - \rho^2 \tag{2.5}$$

The double sum in the last equation can now be split into the part $i = j$ and $i \neq j$ and we obtain:

$$G(\vec{r}) = \frac{N}{V}\delta(\vec{r}) + \frac{1}{V}\sum_i \sum_{j(\neq i)} \langle \delta(\vec{r} + \vec{r}_i - \vec{r}_j)\rangle - \rho^2 \tag{2.6}$$

$$= \rho\delta(\vec{r}) + \rho^2 g(\vec{r}) - \rho^2 , \tag{2.7}$$

where we have introduced the "pair distribution function" $g(\vec{r})$:

$$\rho^2 g(\vec{r}) = \frac{1}{V}\sum_i \sum_{j(\neq i)} \langle \delta(\vec{r} + \vec{r}_i - \vec{r}_j)\rangle \tag{2.8}$$

$$= \rho \sum_{j(\neq i)} \langle \delta(\vec{r} + \vec{r}_i - \vec{r}_j)\rangle \text{ or} \tag{2.9}$$

$$\rho g(\vec{r}) = \frac{1}{N}\sum_i \sum_{j(\neq i)} \langle \delta(\vec{r} + \vec{r}_i - \vec{r}_j)\rangle . \tag{2.10}$$

This function can be interpreted as the (not normalized) conditional probability to find a particle a distance $\vec{r}$ away from the origin, given that there is a particle at the origin. Since a fluid or amorphous solid is isotropic, as emphasized above, $g(\vec{r})$ does not depend on the direction of $\vec{r}$, but only on the absolute value $r = |\vec{r}_i - \vec{r}_j|$ of the distance between the two considered particles. Therefore $g(\vec{r}) \equiv g(r)$ is often called the "radial distribution function".

It is also useful to consider the Fourier components $\rho_{\vec{k}}$ of the density $\rho(\vec{r})$, since this will allow us to recognize the relation between $g(r)$ and the (static) structure factor $S(\vec{k})$ that can be measured by elastic scattering of X-rays or neutrons.

Writing

$$\rho_{\vec{k}} = \int d\vec{r} \exp(-i\vec{k}\cdot\vec{r})\,\rho(\vec{r}) = \sum_{j=1}^{N} \exp(-i\vec{k}\cdot\vec{r}_j) , \tag{2.11}$$

we define $S(\vec{k})$ as follows:

$$S(\vec{k}) = \frac{1}{N}\langle \rho_{\vec{k}}\, \rho_{-\vec{k}}\rangle = \frac{1}{N}\sum_{j=1}^{N}\sum_{l=1}^{N}\langle \exp[-i\vec{k}\cdot(\vec{r}_j - \vec{r}_l)]\rangle . \tag{2.12}$$

Making use of the identity $1 = \int d\vec{r}\,\delta(\vec{r}+\vec{r}_j - \vec{r}_\ell)$ we can rewrite Eq. (2.12) as

$$S(\vec{k}) = 1 + \frac{1}{N}\sum_{j=1}^{N}\sum_{l(\neq j)}\int d\vec{r}\exp[-i\vec{k}\cdot(\vec{r}_j - \vec{r}_l)]\langle\delta(\vec{r}+\vec{r}_j - \vec{r}_\ell)\rangle \quad (2.13)$$

$$= 1 + \rho\int \exp[-i\vec{k}\cdot\vec{R}]g(\vec{R})d\vec{R}\,, \quad (2.14)$$

where in the last step we have made use of Eq. (2.10).
Exploiting now the isotropy of the system, i.e. using the fact that $g(\vec{R})$ is independent of the direction of $\vec{R}$, Eq. (2.14) yields a structure factor that is also independent of the direction of $\vec{k}$. Choosing spherical coordinates one readily obtains (in $d = 3$ dimensions)

$$S(k) = 1 + \rho\int_0^\infty g(R)\frac{\sin(kR)}{kR}\,4\pi R^2 dR\,. \quad (2.15)$$

Conversely, $g(\vec{R})$ can be expressed as the Fourier transform of $S(\vec{k}) - 1$,

$$\rho g(\vec{R}) = \frac{1}{(2\pi)^3}\int d\vec{k}\exp(i\vec{k}\cdot\vec{R})[S(\vec{k}) - 1]\,, \quad (2.16)$$

and by making again use of isotropy of $S(\vec{k})$ we obtain the relation

$$g(R) = 1 + \frac{1}{2\pi^2\rho}\int_0^\infty [S(k) - 1]\frac{\sin kR}{kR}k^2 dk\,. \quad (2.17)$$

Now we discuss the physical interpretation of the information that can be obtained from $g(R)$, see Fig. 2.1. It is obvious that for small enough R the distribution $g(R)$ is essentially zero since the electron shells of two atoms cannot overlap strongly. For simple fluids such as argon, for which $g(R)$ actually looks like the sketch Fig. 2.1, this repulsive interaction at very short distances can be described well by the Lennard-Jones potential

$$v_{LJ}(r_{ij}) = 4\,\varepsilon\left[(\sigma/r_{ij})^{12} - (\sigma/r_{ij})^6\right], \quad (2.18)$$

where the energy scale of the potential is given by ε, while σ describes its range {e.g. for Argon $\sigma \approx 3.4$ Å, $\varepsilon/k_B \approx 120$ K}. Thus $g(R \leq \sigma) \approx 0$ if the temperature or the density are not exceedingly high. On the other hand, the potential given by Eq. (2.18) favors energetically a structure in which the atoms are located close to $r_{\min} = 2^{1/6}\sigma$, the local minimum in

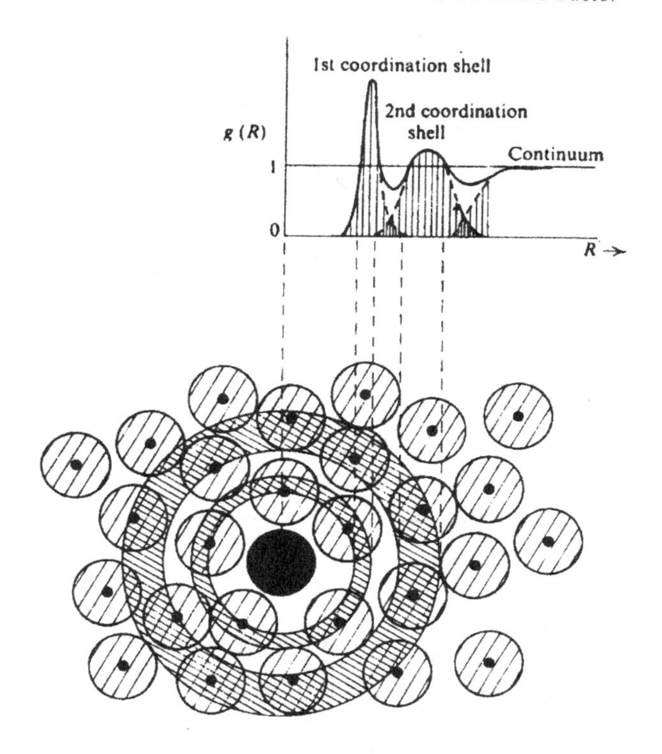

Fig. 2.1 Schematic explanation of the radial distribution function $g(r)$ of a monoatomic fluid. The atom at the origin is highlighted by a black sphere. The shaded regions between the concentric circles indicate which atoms contribute to the 1st and 2nd coordination shell, respectively. From Ziman (1979).

the potential. In the ground state structure of such a model, given by the face-centered cubic or hexagonal close-packed structure with a nearest neighbor distance exactly equal to $r_{\min}$, each atom would have 12 nearest neighbors, and each nearest-neighbor pair would contribute an energy $-\varepsilon$ to the internal energy. In contrast to this in the fluid phase not all the atoms around a considered central atom are at precisely the same distance as in the perfect crystal, but we still expect in $g(R)$ a rather pronounced peak due to this first coordination shell. Depending on the considered temperature and/or density, one can distinguish a second, third etc. coordination shell, but ultimately $g(R \to \infty) \to 1$. On the basis of Fig. 2.1, it is tempting to define the first coordination shell by the atoms between $r = 0$ and the first minimum between the peaks of the first and second maximum in $g(R)$. On

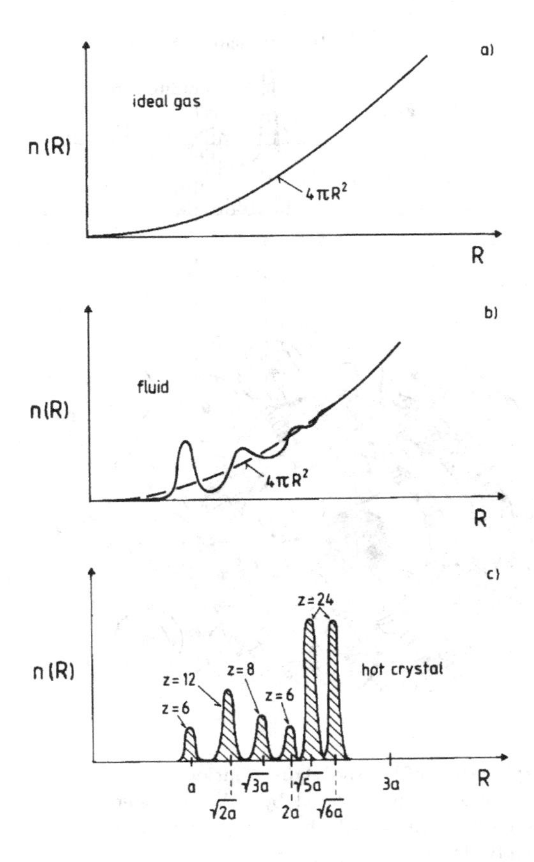

Fig. 2.2 Schematic sketch of the radial density function $n(R) = 4\pi R^2 g(R)$ for an ideal gas (a), a fluid (b), and a "hot crystal" (c).

this basis, an effective coordination number z can be defined by

$$z = \int_0^{R_1} g(R) 4\pi R^2 dR \,, \tag{2.19}$$

where R_1 is the position of the first minimum. Analogously one can define also coordination numbers for higher coordination shells.

At this point it is instructive to see how the function $n(R) \equiv 4\pi R^2 g(R)$, whose integral yields the coordination number, Eq. (2.19), behaves for an ideal gas, a fluid or glass, and a crystal at nonzero temperature. These cases are shown in Fig. 2.2. For the gas we have $g(R) \equiv 1$ and hence $n(R)$ is just a parabola. For the case of a liquid, this parabola is modulated by

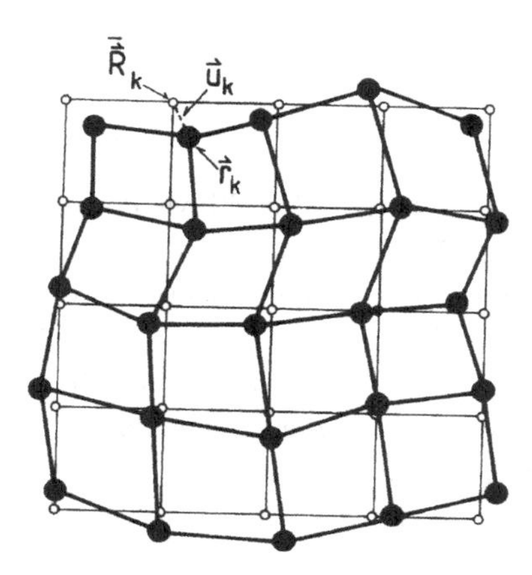

Fig. 2.3 Schematic sketch of atomic positions and nearest-neighbor distances of a solid crystallizing into a square lattice, at high temperatures but still in the solid phase. The ideal rigid lattice in the ground state structure $T = 0$ is also drawn as reference (ignoring thermal expansion of the crystal), and the displacements vector $\vec{u}_k = \vec{r}_k - \vec{\overline{R}}_k$ is indicated.

the $g(R)$ which is shown schematically in Fig. 2.1. Finally we have for the crystal a $g(R)$ that is a series of of $\delta-$functions which, due to the finite temperature, are smeared out. Note that in such a "hot crystal", shown schematically in Fig. 2.3, there is also considerable disorder in the positions of the atoms due to their thermal motion, but nevertheless there is long range order! Below we will return to a discussion on what type of disorder really is crucial to distinguish a fluid or glass from a crystalline solid. As is derived in many standard textbooks on theoretical solid state physics, for a solid at nonzero temperature the harmonic approximation yields for $G(\vec{R})$ a sum of gaussians, each one centered at the positions $\vec{\overline{R}}_k$ of the underlying ideal lattice, with $\overline{R}_{k\beta}$ its cartesian components ($\beta = x, y, z$):

$$g(\vec{R}) = \sum_{k=1}^{N} g_k(\vec{R})$$

$$g_k(\vec{R}) \propto \exp\left\{ -\frac{1}{4} \sum_{\beta\gamma} \{\underline{M}^{-1}\}_{\beta\gamma} (R_\beta - \overline{R}_{k\beta})(R_\gamma - \overline{R}_{k\gamma}) \right\}. \quad (2.20)$$

Here $\underline{M}^{-1}$ is the inverse of the matrix $\underline{M} = \{M_{\beta\gamma}\}$, with $M_{\beta\gamma}$ the correlation function of the atomic displacements $\vec{u}_k = \vec{r}_k - \vec{\overline{R}}_k$ from their average positions:

$$M_{\beta\gamma} = \langle u_\beta \, u_\gamma \rangle \,. \tag{2.21}$$

Note that in a crystal both the translational invariance and the isotropy of space are broken, and hence $g(\vec{R})$ depends explicitly on the direction of $\vec{R}$, via the anisotropy of the matrix $\underline{M}^{-1}$ which reflects the crystal symmetry. Hence the function $n(R)$ plotted in Fig. 2.2c, is obtained if one takes a spherical integral over the angular part of $\vec{R}$. As will be discussed later, it is important that we consider here three-dimensional crystals (contrary to what the schematic sketch of Fig. 2.3 suggests, where for simplicity a two-dimensional square lattice is shown), since only then $M_{\beta\gamma}$ as defined in Eq. (2.21) actually exists, whereas in $d = 1, 2$ one finds $M_{\beta\beta} = \infty$. Although for a crystal in real space $g_k(\vec{R})$ tends to a $\delta-$function $\delta(\vec{R} - \vec{\bar{R}}_k)$ only for $T \to 0$ and only when the quantum-mechanical zero-point motion is neglected, the crystalline positional long range order is clearly seen in reciprocal space in that the structure factor $S(\vec{k})$ shows δ-function singularities at all points $\vec{G}_h$ of the reciprocal lattice of the crystal lattice:

$$S(\vec{k}) = \frac{N}{(2\pi)^3} \sum_h \delta(\vec{k} - \vec{G}_h) \exp\Big(- \sum_{\beta\gamma} M_{\beta\gamma} k_\beta k_\gamma \Big) \,. \tag{2.22}$$

Note that in real space the thermal disorder in a crystal leads to a broadening of the δ-functions at $\vec{\bar{R}}_k$ into gaussians, see Eq. (2.20), but it does not lead to a broadening of the Bragg spots at $\vec{k} = \vec{G}_h$: only the intensity is reduced by the Debye-Waller factor $\exp[-2W(\vec{k})]$, with $W(\vec{k}) = \frac{1}{2}\langle(\vec{k}\cdot\vec{u})^2\rangle$.

In contrast to this, the disorder present in liquids or amorphous solids gives rise to a structure factor $S(k)$ that exhibits only peaks of finite height and nonzero width (apart from a δ-function singularity at $\vec{k} = 0$). Figure 2.4 shows, as a typical example, the static structure factor of polybutadiene melts and glasses. One sees that $S(k)$ for small k exhibits not much structure, and a distinct peak occurs at a wave-vector $k_m \approx 1.5$ Å^{-1}, which can be interpreted as $2\pi/r_{nn}$ where r_{nn} is the typical distance between nearest neighbor pairs of atoms in the material. The slight shift of k_m to larger values with decreasing temperature reflects the effect of thermal expansion (with decreasing temperature the density of the material increases). Apart from this effect, hardly any difference between the structure factor of the fluid polymer and of the glassy polymer can be seen, hence implying (because of Eq. (2.17)) that the radial distribution function $g(R)$ in the liquid

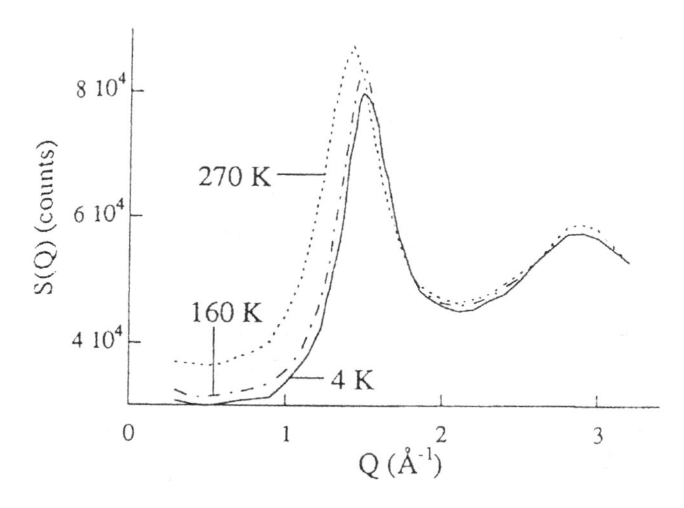

Fig. 2.4 Static structure factor of liquid (T = 270 K) and glassy (T = 160 K, T = 4 K) polybutadiene at ambient pressure plotted vs. wave-vector, as obtained from neutron scattering measurements. The scattering background is not subtracted here, thus the zero of the ordinate axis is not precisely known. The glass transition temperature T_g of this material is $T_g \approx 180$ K, and a fit of inelastic scattering data to the mode coupling theory (see Chap. 5) yielded a critical temperature of about $T_c \approx 220$ K. From Arbe *et al.* (1996).

and in the glass are very similar, justifying (at least to some extent) the view that the structure of a glass is just a "frozen liquid". From the figure we also see that the location and height of the second broad peak in $S(k)$, located at $k \approx 2.8$ Å^{-1}, seems to be almost independent of T. This observation can be understood from the fact that this peak reflects an intramolecular correlation along the backbone of the polymer chain and since these distances within the polymer result from very stiff covalent bonds, they show basically no temperature dependence.

As another example, Fig. 2.5 shows the static (neutron) structure factor of SiO_2 at room temperature, i.e. at T = 300 K [1]. Again the "first sharp diffraction peak", as the pronounced first maximum of $S(q)$ is called (sometimes it is also referred to as "amorphous halo"), appears at a wave-vector around 1.6 Å^{-1}, seemingly implying that in densely packed fluids and glasses the distances of closest approach between neighboring atoms are always rather similar, irrespective of the chemical structure and the nature of the interactions. However, one has to be cautious with such a conclusion:

[1] See Eq. (2.39) for the definition of the static neutron structure factor.

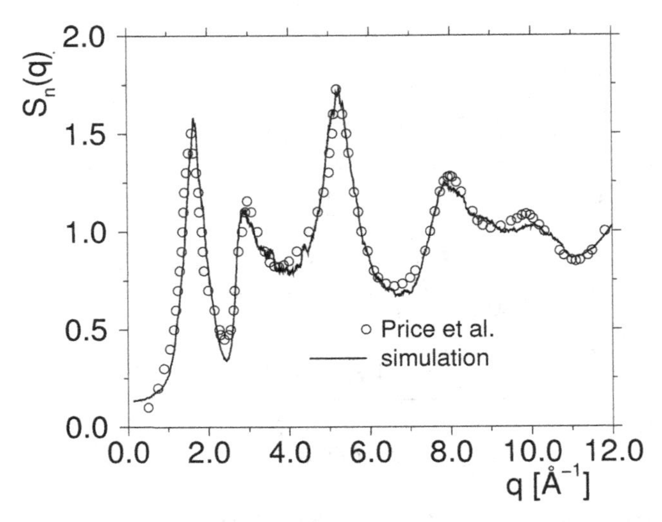

Fig. 2.5 Static structure factor of SiO_2 at $T = 300$ K, as obtained from neutron scattering experiments due to Price and Carpenter (1987) and from Molecular Dynamics simulations (Horbach and Kob 1999) plotted as a function of the wave-vector q. Since the experimental neutron scattering lengths for the Si and O atoms were used to predict the scattering intensity from the simulation, this comparison does not contain any adjustable parameter whatsoever.

As will be discussed in more detail in Sec. 3.6, silica is a prototype example of a network glass former ("continuous random network", CRN), since there are rather rigid covalent bonds linking each silicon atom (which sits in the center of an almost regular tetrahedron) to four oxygens (at the corners of the tetrahedron), and two tetrahedra share one such bridging oxygen, so the SiO_2 - stoichiometry is maintained. The strong local correlations in the mentioned network structure show up in the rather well-developed further peaks of $S(q)$ at larger q. Thus the first sharp diffraction peak is related to the distance between adjacent tetrahedra and *not* to the one between nearest neighbor atoms, i.e. the Si$-$O distance.

In contrast to this open network structure, polymer melts and glasses can be viewed as densely interpenetrating random walk like coils (see Sec. 3.1) and while one has strong covalent bonds along the "backbone" of the linear macromolecule, the binding between different chain molecules is only due to weak van der Waals interactions that can be described by Lennard-Jones potentials.

A further common feature of all fluids (disregarding fluids which are close to a critical point of the gas-liquid transition or a fluid-fluid unmixing) and glasses is the fact that for wave-vector $k \ll k_m$ the structure factor is flat and structureless. In fact, the limit $S(k \to 0)$ contains interesting physical information since it is related to the isothermal compressibility κ_T of the system,

$$S(k \to 0) = \rho k_B T \, \kappa_T \,, \tag{2.23}$$

where $\kappa_T \equiv -(1/V)\,(\partial V/\partial p)_T$. This "compressibility sum rule" can be derived as follows: First of all we use Eq. (2.14) to obtain

$$S(k \to 0) = 1 + \rho \int [g(\vec{R}) - 1] d\vec{R} \,. \tag{2.24}$$

(Note that here we have subtracted in the integrand the limiting value $g(\vec{R} \to \infty) = 1$ from $g(\vec{R})$, since this would give rise to a δ−function at $k = 0$ here, which is of no interest since we are interested in the limiting value of $S(k \to 0)$ and not in $S(k = 0)$.) We now recall a well known result from statistical mechanics which relates the fluctuations of particles to the compressibility (see, e.g., Pathria 1986):

$$\frac{\langle N^2 \rangle - \langle N \rangle^2}{\langle N \rangle} = \rho k_B T \kappa_T \,. \tag{2.25}$$

Note that although in the NVT ensemble we do not have any fluctuation $\langle N^2 \rangle - \langle N \rangle^2$ of the particle number N in a volume V, such fluctuations do occur in subvolumina $V_{\text{sub}} \ll V$ of the total volume. In the thermodynamic limit it is implied that first the limit $N \to \infty$ and hence also $V \to \infty$ is taken (keeping $\rho = N/V = \text{const}$), and only afterward the limit $k \to 0$. Thus, it is the limit of these subvolume fluctuations ($k \approx V_{\text{sub}}^{-1/3}$) which is picked up by Eq. (2.25) and hence we do not have any contradictions here.

Using Eq. (2.1) and the definition of $G(\vec{r})$ from Eq. (2.3) we can write the fluctuations as

$$\frac{\langle (N - \langle N \rangle)^2 \rangle}{\langle N \rangle} = \frac{1}{\langle N \rangle} \left\langle \int_V d\vec{r}\,'[\rho(\vec{r}\,') - \rho] \int_V d\vec{r}\,''[\rho(\vec{r}\,'') - \rho] \right\rangle \tag{2.26}$$

$$= \frac{1}{\langle N \rangle} \int_V d\vec{r}\,' \int_V d\vec{r}\,'' G(\vec{r}\,' - \vec{r}\,'') \tag{2.27}$$

$$= \frac{1}{\rho} \int_V d\vec{r} \, G(\vec{r}) \,. \tag{2.28}$$

Using Eq. (2.7), $G(\vec{r}) = \rho\delta(\vec{r}) + \rho^2 g(\vec{r}) - \rho^2$, we obtain thus immediately the right hand side of Eq. (2.24), and hence we have shown the validity of Eq. (2.23).

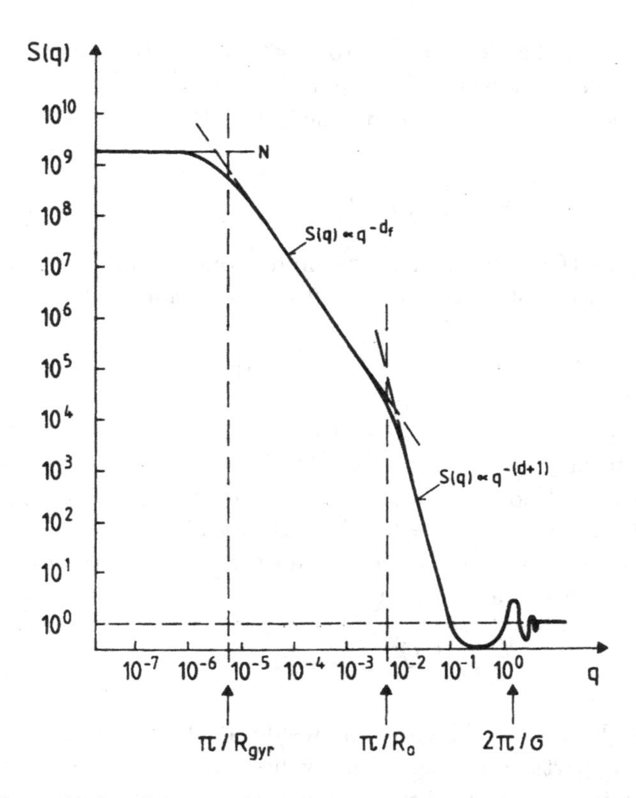

Fig. 2.6 Schematic plot of the static structure factor static of a fractal cluster of spheres of radius $R_0 \approx 10^3$ Å (cf. Fig.1.6), assuming that the cluster contains in total about $N \approx 2 \cdot 10^9$ scattering atoms. Arrows on the abscissa denote the inverse of the characteristic length scales in the system. Note the logarithmic scales of both axes.

We point out that one important physical ingredient of the considerations used to derive Eq. (2.23) is that we consider the structure of a fluid (or amorphous solid) that is macroscopically homogeneous. At this point it is instructive to see what happens if the condition that in all mesoscopic subvolumes $V_{\rm sub}$ should be the same is *not* fulfilled. E.g., let us discuss the scattering from a fractal aggregate of colloidal particles of radius R_0, such as shown in Fig. 1.6. For the sake of simplicity, we assume that the particles themselves are amorphous rather than crystalline, e.g. polystyrene balls. Figure 2.6 shows a schematic log-log plot of $S(q)$ for the scattering from an ensemble of such fractal objects. For wave-vectors q of the order of 1 Å, we probe the structure on an atomistic scale inside the spheres, and on this scale the picture is qualitatively the same as it is found for bulk fluids

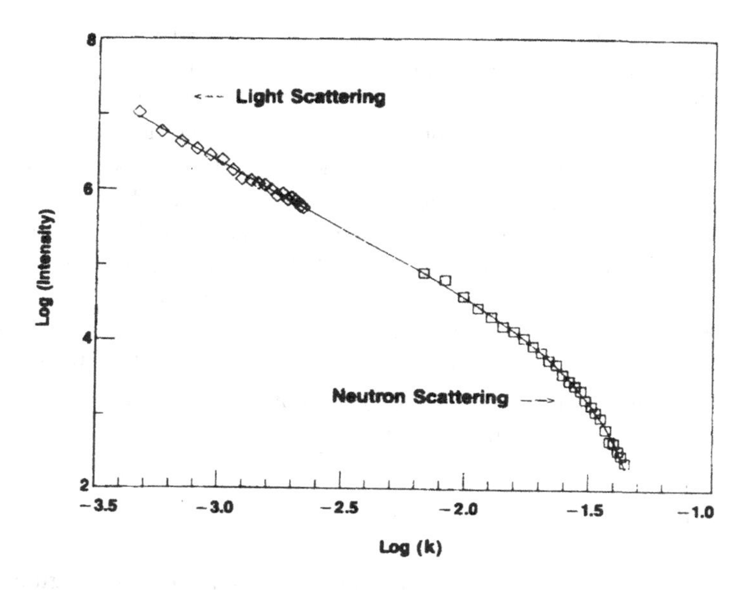

Fig. 2.7 Light scattering and small angle neutron scattering from a fractal clusters of gold colloids formed by diffusion limited aggregation (DLA). Scattering intensities are in arbitrary units. The units of k are Å^{-1}. The clusters typically contain about 5000 colloid particles with typical radii of 75 Å. Therefore the neutron scattering data probe the crossover from the fractal scattering at low k to the Porod regime. From Weitz *et al.* (1985).

(Fig. 2.4). In the range where $2\pi/R_0 \ll q \ll 1\ \text{Å}^{-1}$ we observe the scattering from the sharp interfaces between the spheres and the surrounding air (or vacuum, respectively). This scattering is of the form

$$S(q) \propto q^{-(d+1)}\ ,\ 2\pi/R_0 \ll q \ll 2\pi/\sigma\ ,\ \text{interfacial scattering}\,, \qquad (2.29)$$

where d is the dimension (=3 in our case), and σ is the typical size of the atoms that constitute the colloidal particles. This q−dependence is usually called "Porod's-law" (Porod 1951).

In the range $2\pi/R_{\rm gyr} \ll q \ll 2\pi/R_0$, where $R_{\rm gyr}$ is the radius of gyration of the fractal object, we find another power-law which reflects the fractal dimension of the object, see Eq. (1.4):

$$S(q) \propto q^{-d_f}\ ,\ 2\pi/R_{\rm gyr} \ll q \ll 2\pi/R_0\ ,\ \text{fractal scattering}\,, \qquad (2.30)$$

The reason for such a q−dependence is quite obvious, since the number of scattering particles increases like $R^{d_f} \sim q^{-d_f}$.

At the smallest wave-vectors, $q \ll 2\pi/R_{\text{gyr}}$, $S(q)$ crosses over to a constant, which is just given by the total number N of scattering particles.

Of course Fig. 2.6 is somewhat hypothetical, since in practice a wave-vector range spanning 7 decades is not available. But by combining light scattering with small angle scattering of X-rays or neutrons it is possible to span somewhat more than 2 decades and to study the part of $S(q)$ that contains information on d_f. An example for this is shown in Fig. 2.7 for a fractal cluster of gold particles of size $R_0 \approx 75$ Å in which one observes the crossover from Porod's-law at high wave-vectors to a fractal scattering with $d_f \approx 1.79$ at low k (Weitz *et al.* 1985).

From Eq. (1.4) it is easy to see that in real space the radial distribution function $g(r)$ that corresponds to Eq. (2.30) does not converge toward unity for $r \to \infty$, but instead also exhibits a power-law decay

$$g(r) \propto r^{d_f - d} \quad \underset{r\to\infty}{\longrightarrow} 0 \quad (\text{remember } d_f < d)\,. \tag{2.31}$$

Of course in reality Eq. (2.31) applies also only in an intermediate regime of interatomic distances, namely for $R_0 \ll r \ll R_{\text{gyr}}$, see Fig. 2.6.

Before we conclude this section we mention a slight generalization of our description, which, however, is very important in practice. In Eqs. (2.1)-(2.17) we have tacitly ignored the presence of different types of atoms and therefore the above description applies only to monoatomic fluids (liquid argon, silicon, or selenium, for instance: the latter two elements also occur as amorphous solids, while the former occurs, as all rare gas atoms, in the solid state only in the form of a crystal). Thus, for systems such as polymers (Fig. 2.4) or silica (Fig. 2.5) an extension of the formalism is required, which explicitly allows for the possibility that different types α of atoms are present in the system. Equation (2.1) is then replaced by (assuming ν distinct atomic species):

$$\rho_\alpha(\vec{r}) = \sum_{i=1}^{N_\alpha} \delta(\vec{r} - \vec{r}_i) \quad \alpha = 1, \ldots, \nu\,, \tag{2.32}$$

and we have the partial densities

$$\rho_\alpha = N_\alpha / V = \langle \rho_\alpha(\vec{r}) \rangle\,. \tag{2.33}$$

Similar to the one-component case, see Eq. (2.10), one can define the partial pair distribution functions $g_{\alpha\beta}$:

$$\rho g_{\alpha\alpha}(\vec{r}) = \frac{N}{N_\alpha^2} \sum_i^{N_\alpha} \sum_{j(\neq i)}^{N_\alpha} \langle \delta(\vec{r} + \vec{r}_i - \vec{r}_j) \rangle \tag{2.34}$$

$$\rho g_{\alpha\beta}(\vec{r}) = \frac{N}{N_\alpha N_\beta} \sum_i^{N_\alpha} \sum_j^{N_\beta} \langle \delta(\vec{r} + \vec{r}_i - \vec{r}_j) \rangle \quad \text{for} \quad \alpha \neq \beta \,, \tag{2.35}$$

where N is the total number of particles and ρ the total particle density. Obviously, the physical interpretation of $g_{\alpha\beta}(\vec{r})$ is the conditional probability to find a particle of type β at distance $\vec{r}$ from the origin given that at the origin there is a particle of type $\beta \neq \alpha$. Consequently, we arrive at a generalized definition of the coordination number, see Eq. (2.19):

$$z_{\alpha\beta}^{(n)} = \int_{R_{n-1}}^{R_n} g_{\alpha\beta}(r) 4\pi r^2 dr \,. \tag{2.36}$$

Here $z_{\alpha\beta}^{(n)}$ denotes the number of neighbors of type β in the n-th coordination shell of a particle of type α, and R_n is the n-th minimum of the corresponding radial distribution function $g_{\alpha\beta}(R)$. As an example we show in Fig. 2.8 the six partial radial distribution functions that can be defined for the sodium silicate melt (Na_2O) $(5\ SiO_2)$, and which have been obtained from a molecular dynamics simulation of a suitable model (Winkler 2002). Note that we have the obvious symmetry $g_{\alpha\beta}(r) = g_{\beta\alpha}(r)$. One can see that $g_{\alpha\beta}(r \to \infty) \to 1$, as it should be. Furthermore we recognize that only for $g_{\rm SiO}(r)$ we have $g_{\alpha\beta}(r) \approx 0$ around the first minimum ($R_1 \approx 2.5$ Å), showing that there is a well-defined covalent bond between Si- and O-atoms, and that there is a well-defined length of this covalent bond, at about 1.7 Å. All the distances between other types of pairs are much stronger fluctuating, at least at these high temperatures, and hence rather broad peaks in the corresponding $g_{\alpha\beta}(r)$'s occur, in particular if one goes beyond the first coordination shell. Thus for this system it is in fact difficult to obtain the various coordination numbers defined in Eq. (2.36) since the various shells are not well separated. We will return to a more detailed interpretation of data such as shown in Fig. 2.8 in the next chapter.

These distinctions referring to different types of atoms naturally carry over to the structure factor, since Eq. (2.11) also gets "decorated" with an

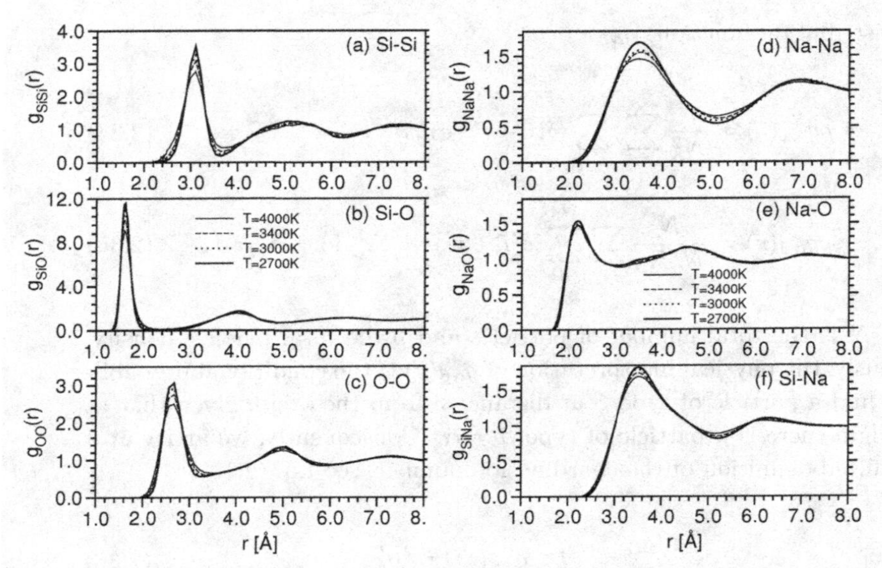

Fig. 2.8 Partial radial distribution functions for a sodium silicate melt of composition (Na_2O) $(5\ SiO_2)$ in the temperature region from $T = 2700$ K to $T = 4000$ K, as obtained from Molecular Dynamics simulations of a model system containing altogether 8064 atoms in a cubic box of linear dimension $L = 48$ Å. From Winkler (2002).

index α,

$$\rho_{\vec{k}}^{(\alpha)} = \int d\vec{r} \exp(-i\vec{k} \cdot \vec{r}) \rho_\alpha(\vec{r}) = \sum_{j=1}^{N_\alpha} \exp(-i\vec{k} \cdot \vec{r}_j)\,, \tag{2.37}$$

where it is understood that the sum is extended over the atoms of type α only.

Consequently Eq. (2.12) gets replaced by an expression for a partial structure factor

$$S_{\alpha\beta}(\vec{k}) = \frac{f_{\alpha\beta}}{N} \sum_{j=1}^{N_\alpha} \sum_{l=1}^{N_\beta} \langle \exp[-i\vec{k} \cdot (\vec{r}_j - \vec{r}_l)] \rangle \tag{2.38}$$

where $N = \sum_\alpha N_\alpha$ and $f_{\alpha\alpha} = 1$, while $f_{\alpha\beta} = 1/2$ for $\alpha \neq \beta$ since pairs $\alpha \neq \beta$ occur twice. In neutron scattering, for instance, the total scattering intensity then is a weighted average of these partial structure factors (Lovesey 1994, Hansen and McDonald 1986),

$$S^{\text{neu}}(\vec{k}) = \frac{N}{\sum\limits_\alpha N_\alpha b_\alpha^2} \sum_{\alpha\beta} b_\alpha b_\beta S_{\alpha\beta}(\vec{k})\,, \tag{2.39}$$

where the b_α are the experimentally determined neutron scattering lengths of the various atoms present in the sample. In fact, in Fig. 2.5 these quantities $\{b_{\text{Si}}, b_{\text{O}}\}$ were used in order to compare the simulation, which readily yields all the individual $S_{\alpha\beta}(q)$, with the experiment, which only yields the average $S^{\text{neu}}(\vec{k})$.

2.2 Topological Disorder and Bond Orientational Correlations

In this section we discuss various topics, all of which concern the characterization of disordered structures beyond the pair correlation function treated in the previous section. We start from the observation that the definition of coordination numbers in terms of the "counting of neighbors" that fall in between two adjacent minima of $g_{\alpha\beta}(R)$, see Eq. (2.36), is obviously somewhat arbitrary and ill-defined if these minima are rather shallow (in Fig. 2.8 this is, e.g., the case for the neighborhood of the Na-ions). Thus a less arbitrary definition of the local neighborhood is desirable in order to be able to deal with structures in which the disorder is very strong, and that hence have a broad distribution of coordination numbers.

One concept that helps to make the notion of a distribution $P(z)$ of local coordination numbers precise, is the so-called "Voronoi construction" (Voronoi 1908, Ziman 1979; Zallen 1983). (For the sake of simplicity we now restrict attention to the first coordination shell.) It is the analog of the well known concept of Wigner-Seitz cells in crystals (Fig. 2.9). The procedure to obtain the Wigner-Seitz cell is as follows: draw straight lines from an atom to all its neighbors in different directions (including the first few coordination shells only, of course). Construct the perpendicular plane (in $d = 2$ dimensions a line) at the midpoint of each of the connecting lines. The central polyhedron (polygon in $d = 2$) enclosed by these planes (lines) is the Wigner-Seitz cell. In a simple cubic (square) lattice the Wigner-Seitz cell is a cube (square) as well. However, this geometrical equality does not hold in general since, e.g., for the case of a triangular lattice the Wigner-Seitz cell is a hexagon (Fig. 2.9). If one applies the same geometrical construction to the case of a disordered structure a variety of polyhedra (or polygons in $d = 2$, Fig. 2.9) results. In this way, a well-defined local coordination number z_i for every atom (labeled by the index i) in a disordered amorphous structure (or fluid snapshot, respectively) is obtained, and averaging over all the atoms in the amorphous solid or fluid snapshots yields the desired

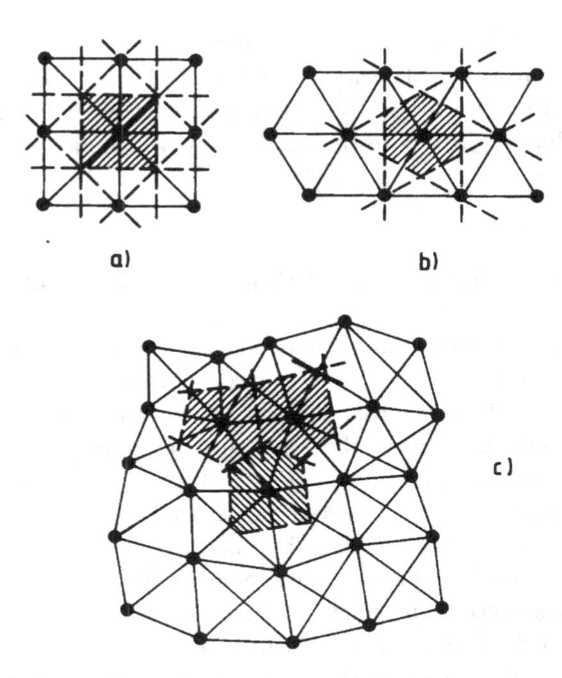

Fig. 2.9 Construction of the Wigner-Seitz cells for a square lattice (a) and a triangular lattice (b) and the corresponding Voronoi construction for the disordered square lattice shown in Fig. 2.3 (c). The positions of the atoms are shown by black dots. "Bonds" connecting nearest (and next nearest) neighbors are shown by straight lines, while the corresponding perpendicular lines at one half the distance to the respective neighbor are shown by broken lines. The resulting Wigner-Seitz cells are shaded in case a) and b). In case c) three Voronoi cells are shown. Note that in this case these cells are often pentagons, hexagons etc., although the undistorted crystal is a square lattice.

distribution $P(z)$. The properties of $P(z)$ for various random structures will be discussed in Chap. 3.

It must be stressed that the occurrence of a distribution of $P(z)$ is necessary but not sufficient to have a fluid or amorphous structure. A counterexample is the "hot crystal" of Figs. 2.3 and 2.9 which does have sharp Bragg spots that are described by a structure factor as given in Eq. (2.22). Although there is already a lot of disorder in a configuration such as the one shown in Fig. 2.3, another important kind of disorder in this structure is still missing: the so-called "topological disorder". It is this topological disorder that distinguishes crystals from fluids and amorphous materials. In fact, in the hot crystal we still can label all atoms by the lattice

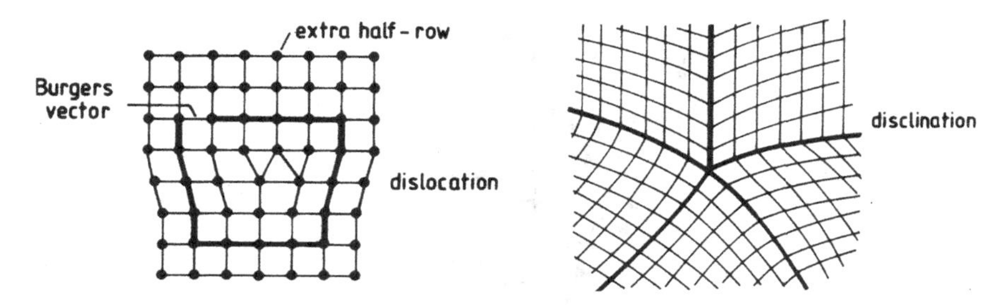

Fig. 2.10 Examples of topological defects disrupting long range order of two-dimensional systems. The left panel shows a dislocation in a square lattice, where an extra half row of atoms is inserted. The construction yielding the Burgers vectors is indicated. The right panel illustrates a disclination.

indices $\vec{\ell} \equiv (\ell_1, \ldots, \ell_d)$ in d dimensions, since, see Fig. 2.3, an atom at site $\vec{r}_\ell$ can be written as $\vec{r}_\ell = \vec{u}_\ell + \vec{R}_\ell$, with $\vec{R}_\ell = \ell_1 \vec{a}_1 + \cdots \ell_d \vec{a}_d$, $\vec{a}_1, \ldots, \vec{a}_a$ being the basis vectors of the ideal ($d-$dimensional) lattice. Thus by a continuous variation (all $\vec{u}_\ell \to 0$) one obtains from the disordered structure again the ideal crystal! In the presence of topological disorder this is no longer possible and hence we recognize the fundamental importance of this type of disorder.

In a computer simulation context, a useful concept for the discussion of disordered structures are the "inherent structures" (Stillinger and Weber 1982): one decreases the potential energy by a steepest descent procedure continuously from its value in the considered configuration of particles until one ends up in the (next) local minimum in configuration space. For a hot crystal without topological disorder the configuration thus found is the corresponding ideal crystal structure.

Actually, the concept of topological disorder already arises when one discusses "real crystals" in contrast to "ideal crystals" (which so far have been assumed exclusively). A real crystal has the property that although one can decompose $\vec{r}_\ell$ as $\vec{r}_\ell = \vec{u}_\ell + \vec{R}_\ell$ over restricted regions of the crystal volume, it is not possible to do so for the entire system since the presence of a network of dislocation loops and grain boundaries will destroy the order on very large scales. The basic elements of topological disorder are individual dislocations and disclinations, which are illustrated in Fig. 2.10 for the case of a square lattice. Let us consider the case of dislocations in more detail. In particular, let us construct the difference between coordinates $\Delta\vec{\ell} = \vec{\ell}^{\,(2)} - \vec{\ell}^{\,(1)}$ of two atoms (1), (2) in a perfect crystal, by taking an

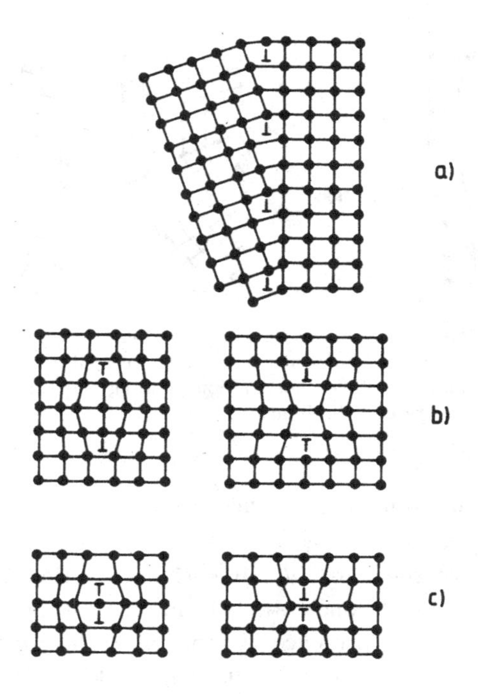

Fig. 2.11 Schematic description of various types of topological defects in a square lattice: (a) a grain boundary can be formed by a regular array of dislocations with parallel Burgers vectors. (b) Extended linear defects of finite length can be formed by dislocation pairs with opposite Burgers vectors: in the left part, the defect is a row containing three extra atoms, in the right part, three atoms in a row are missing. (c) Point defects (extra atoms or missing atoms, respectively) are equivalent to dislocations with opposite Burgers vectors with their cores on adjacent lattice sites.

arbitrary path from one atom (1) to one of its neighbors and so on until we reach atom (2). Obviously, the result cannot depend on which particular path is taken! In particular, if we take a closed loop, by combining a path from (1) to (2) and another path from (2) back to (1), the result must evidently be zero. Now in a crystal containing dislocations, this is no longer true if such a loop encloses a dislocation core (where the extra half row in the left panel of Fig. 2.10) ends): The resulting difference $\Delta\vec{\ell}$ is non-zero for a loop constructed in such a way defines the so-called "Burgers vector" (Nabarro 1967).

Of course, the theory of crystal dislocations is outside the scope of the present chapter. Therefore we mention only briefly that (in $d = 2$ dimensions) we can create a "grain boundary" by a linear array of Burgers vectors pointing all in the same direction (Fig. 2.11a), while pairs of opposite Burgers vectors create topological defects of finite extent, such as extra rows of atoms of finite length (Fig. 2.11b), or "missing rows" of finite extent. In the extreme case of a single interstitial or vacancy one has two dislocations with opposite Burgers vectors at adjacent lattice sites tightly bound together (Fig. 2.11c). Although even around such localized defects there is a lattice distortion of infinite extent, i.e., the elastic displacements $\vec{u}(\vec{r})$ are non-zero in the whole infinite crystal, these displacements decay sufficiently quickly with increasing distance r from the core of this defect to make the energy difference $\Delta \mathrm{E}$ between such a defective crystal and the corresponding ideal crystal finite. Therefore these bound dislocation pairs with opposite Burgers vectors are equivalent to ordinary point defects, vacancies or interstitials, and can hence be spontaneously created or destroyed at nonzero temperature simply by thermal fluctuations. As a consequence such defects play a crucial role to understand melting in two dimensions (Nelson 1983). In $d = 3$ dimensions, bound dislocation pairs correspond to whole lines of missing atoms (or extra atoms, respectively). Hence the excitation energy of these line defects is infinite, and therefore they cannot be created by thermal fluctuations. In fact, all the dislocations present in real three-dimensional crystals should not be considered as equilibrium structure but instead as frozen-in defects created because of imperfect crystallization: The velocity with which the crystal-melt interface (or crystal-gas interface, in crystallization from the vapor) moved was too fast to allow for the growth of perfect crystals.

We now briefly discuss another type of topological defect, the so-called "disclinations", see right panel of Fig. 2.10. Assume a perfect square lattice of size $L \times L$. We now cut out the lower left quarter (of size $L/2 \times L/2$) and glue together the remaining boundaries along this cut. From the atom in the center then lattice axes will run in *three* directions rather than in four as usual (namely the $\pm x$, $\pm y$ directions). We also may take this quarter piece of the square lattice that we have removed and glue it to an otherwise perfect lattice along a cut, to create a center from which now 5 rather than 4 lattice axes originate (right panel of Fig. 2.10). It turns out that one must put into a crystal both dislocations and disclinations in order to destroy completely the long range order of the crystal. In fact, one can show that in $d = 2$ dimensions an isolated dislocation can be considered as a tightly bound disclination-antidisclination pair. (In a square lattice, the local coordination of a disclination core is 5, while for the core of an antidisclination it is 3. Similarly, for a triangular lattice – where the regular

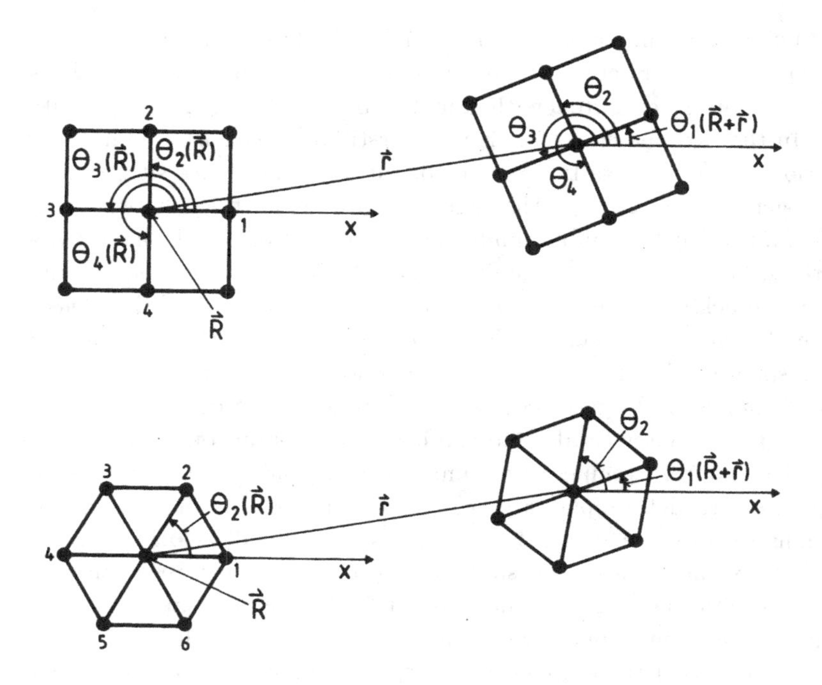

Fig. 2.12 Illustration of the concept of bond orientational order and disorder on the (distorted) square lattice (upper part) and triangular lattice (lower part). See text for further explanations.

coordination number is 6 – the local coordination of a disclination core is 7, and 5 for the antidisclination.) According to the theory of Kosterlitz, Thouless, Halperin, Nelson, and Young (Kosterlitz and Thouless 1973, Halperin and Nelson 1978, Young 1979) of defect-mediated melting, the fluid is characterized by freely moving topological defects, while in the solid only bound pairs of dislocations are possible. It is suggested that a phase intermediate in its properties between the solid and the fluid occurs, the so-called "hexatic phase", where free dislocations (but not yet any free disclinations) occur, i.e. there is long range order in the orientation but not yet in the position. The transition from the solid to the hexatic phase is then described as a "dislocation pair unbinding transition". Since a free dislocation can be viewed as a disclination – antidisclination pair, the "melting" of the hexatic phase into the fully disordered fluid is then interpreted as a "disclination pair unbinding transition".

For a deeper understanding of this concept, it is useful to discuss the notion of "bond orientational order" (Nelson 1983, Strandburg 1988, Kleinert 1989). Here we explain this concept only for the case of two-dimensional lattices, since this case is simpler, although the generalization to three dimensions exists and is useful. (Of course, in one dimension no topological disorder is possible, and also the concept of bond orientations is meaningless.)

The basic physical picture is that locally the short range order resembles always the one of an ideal crystal, but there can be disorder in the orientation of the local crystal axes.

In order to characterize this type of order, we first have to identify the nearest neighbors of a considered atom (e.g. by the Voronoi construction discussed above). We draw bonds from each atom to all its nearest neighbors. We now take one atom and one of its nearest neighbors and define this direction as the x-axis, see Fig. 2.12. We now measure the angles $\{\Theta_\nu(\vec{R})\}$ that all bonds make with this reference axis with $\nu = 1, \ldots, z_{\vec{R}}$, where $z_{\vec{R}}$ is the local coordination number of the atom at $\vec{R}$ and $\Theta_1(\vec{R})$ is the angle between the first neighbor, the central atom, and the last neighbor. If most coordination numbers are close to four, as would be the case for a square lattice, we consider the order parameter

$$\psi_4(\vec{R}) = \frac{1}{z_{\vec{R}}} \sum_{\nu=1}^{z_{\vec{R}}} \exp[i4\Theta_\nu(\vec{R})] \,, \qquad (2.40)$$

while in the case of a structure deriving from a triangular lattice the appropriate choice is

$$\psi_6(\vec{R}) = \frac{1}{z_{\vec{R}}} \sum_{\nu=1}^{z_{\vec{R}}} \exp[i6\Theta_\nu(\vec{R})] \,. \qquad (2.41)$$

Obviously, if the order of the structure is perfect all the angles $\Theta_\nu(\vec{R})$ would be multiples of $2\pi/4$ for the case of a square lattice and equal to $2\pi/6$ for the case of a the triangular lattice, and consequently we would have $\psi_4 = 1$ and $\psi_6 = 1$, respectively. In the case of a disordered structure, it thus makes sense to define again a pair correlation function of this local orientational order parameter (cf. Fig. 2.12):

$$G_4(\vec{r}) = \langle \psi_4^*(\vec{R})\psi_4(\vec{R}+\vec{r})\rangle \,, \; G_6(\vec{r}) = \langle \psi_6^*(\vec{R})\psi_6(\vec{R}+\vec{r})\rangle \,. \qquad (2.42)$$

Here the symbol * in ψ^* means complex conjugation. Of course, in terms of the actual positions of the atoms quantities such as $G_4(\vec{r})$ (and $G_6(\vec{r})$) are special 4-particle correlation functions. As is well known in the theory of fluids (Hansen and McDonald 1986), one can introduce a whole hierarchy of higher order correlation functions between point particles (three point, four point correlations). Although these functions can be determined in computer simulations, they are in most cases not directly accessible to experiments on real materials, and thus they are seldom considered explicitly. However, the quantities defined in Eqs. (2.40)- (2.42) are an exception, and as we will see now they are necessary to characterize simple fluids in two dimensions.

We illustrate this fact by showing, as a digression, some Monte Carlo simulation results for a two dimensional fluid of hard disks of diameter σ (Weber *et al.* 1995). In this model the pairwise potential between particles is thus given by

$$v(r_{ij}) = \infty \quad \text{if} \quad r_{ij} < \sigma\,, \quad v(r_{ij}) = 0 \quad \text{else}\,. \tag{2.43}$$

Hence, the control parameter of interest in this model is the reduced pressure $p/(\rho k_B T)$, where p is the pressure and ρ the particle density. Since there are no attractive forces between particles, no liquid-gas transition occurs. Nevertheless there is agreement in the literature that at low densities the model is in a disordered fluid phase and at high densities in a crystalline phase with long range orientational order of triangular type, as described by Eq. (2.41). As we will discuss below, in low-dimensional systems the long wavelength phonon excitations destroy positional long range order, although long range orientational order can still persist in two dimensions (Mermin and Wagner 1966, Mermin 1968). Thus the ordered phase has long range orientational order but no long range positional order.

Despite the consensus on these results, there is controversy in the literature whether the transition from fluid to solid upon an increase of the density is a first order transition (as it happens for the crystallization/melting transition in $d = 3$ dimensions), or whether one has a sequence of two continuous transitions of the Kosterlitz-Thouless (KT) (1973) type, as suggested by Halperin and Nelson (HN) (1978) and Young (Y) (1979), see Fig. 2.13. Between the liquid (with only short range orientational order) and the solid (with true long range orientational order) intrudes the "hexatic phase", in which $G_6(r)$ exhibits a power-law decay at large distances:

$$G_6(r \to \infty) \propto r^{-\eta(\rho)}\,, \quad \rho_\ell \le \rho < \rho_s\,. \tag{2.44}$$

This equation thus implies that for the KTHNY scenario the order parameter $\langle \Psi \rangle_\infty$, defined as the absolute value of the average of $\psi_6(\vec{R})$ in

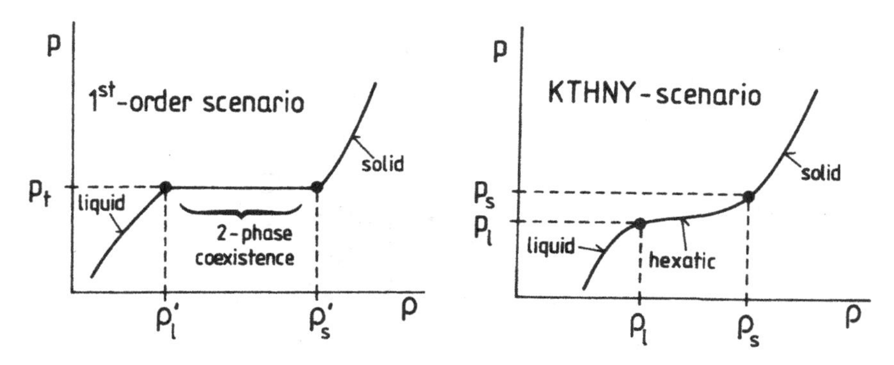

Fig. 2.13 Qualitative behavior of the isotherms of a two-dimensional fluid, plotting pressure p versus density ρ. The left panel shows the scenario appropriate for a first order transition in which there exists a region where the isotherm is strictly horizontal, for $\rho'_\ell < \rho < \rho'_s$, corresponding to the two-phase coexistence between the liquid and the solid phase. The right panel shows the scenario proposed in the KTHNY theory in which there occurs first a continuous transition (at density ρ_ℓ and pressure p_ℓ) from the liquid phase to the hexatic phase, and then at a higher density ρ_s (and pressure p_s) a second continuous transition from the hexatic phase to the solid.

Eq. (2.41) over the total system, should be zero in the regime $\rho_\ell \le \rho < \rho_s$. However the $r-$dependence of G_6 also predicts that the bond orientational "susceptibility" χ_6,

$$\chi_6 = \int d\vec{r}\, G_6(\vec{r}) \,, \tag{2.45}$$

diverges for $\rho \to \rho_\ell$ from the liquid side, and stay infinite throughout the hexatic phase. In contrast to this, the susceptibility stays finite in a first order scenario. In this case the order parameter $\langle \Psi \rangle_\infty$ increases linearly in the interval $\rho'_\ell \le \rho \le \rho'_s$, see Fig. 2.14, a dependence that is due to the linear increase of the solid-like volume fraction in this density range.

If one tests this behavior by means of computer simulations, one finds a very strong dependence of the function $\langle \Psi \rangle_L(\rho)$ on L, where $L = S/M_b$ is the size of a subbox in a system of size $S \times S$ that has been divided into $M_b \times M_b$ subboxes, see Fig. 2.14. For densities $\rho \le \rho_\ell \approx 0.899$ an analysis of χ_6 as well as higher moments of the distribution function of the bond orientational order parameter implies that one stays in the liquid phase (Weber *et al.* 1995, Jaster 1999a,b, Jaster 2004, Mak 2006). Although the simulations (Jaster 1999a,b) are consistent with the predicted behavior $\ln \chi_6 \propto (\rho_\ell - \rho)^{-1/2}$, they have so far failed to give *direct* evidence for the presence of a hexatic phase. (An indirect method approaching the

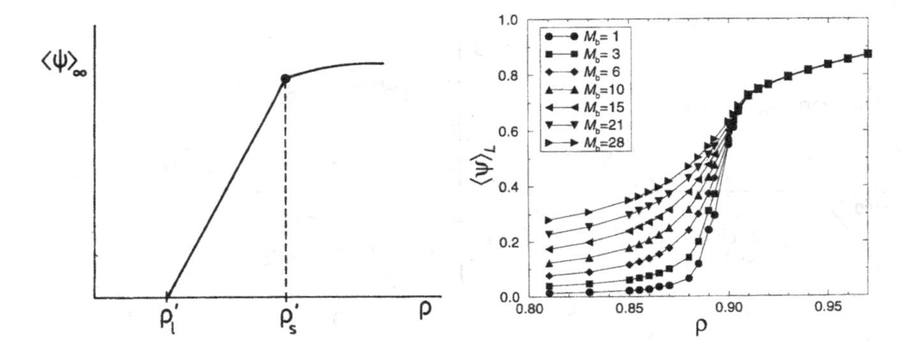

Fig. 2.14 Left: Schematic variation of the average bond orientational order parameter $\langle\Psi\rangle_\infty$ in the thermodynamic limit (infinite linear dimension of the system), if the first order scenario applies. Right: Monte Carlo results for the bond-orientational order parameter of the hard disk fluid as a function of density for selected subbox sizes $L = S/M_b$ (values of M_B are given in the legend), for an $S \times S$ system with periodic boundary conditions containing $N = 16384$ particles. Note that all lengths are measured in units of the hard disk diameter $\sigma \equiv 1$, and that the shape of the simulation box is a parallelepiped compatible with the triangular lattice structure. The parameter M_b is thus simply the number of subboxes along the edge of the total system. From Weber *et al.* (1995).

transition density ρ_s from the solid phase has yielded a preliminary estimate $\rho_s \approx 0.914$ (Sengupta *et al.* 2000)). However, experiments on colloidal particles with a repulsive r^{-3} interaction have provided evidence for the KTHNY scenario of melting in $d = 2$ dimensions (Zahn *et al.* 1999, Zahn and Maret 2000).

All these results make it clear that the hard disk system is a good example for a system for which (for $\rho \to \rho_\ell$) the standard pair distribution function $g(r)$ stays short ranged, while a higher order correlation function, namely $G_6(r)$, shows a divergent correlation length ξ_6

$$G_6(r \to \infty) \propto \exp(-r/\xi_6) \ , \ \ln \xi_6(\rho) \propto (\rho_\ell - \rho)^{-1/2} \ . \qquad (2.46)$$

Later on in this book we will consider other examples of systems, for which the pair correlation function does not contain the relevant information about the growing correlations in the system and therefore higher order correlations have to be considered. One prominent example for such systems are spin glasses which will be discussed in Chap. 4.

As a last point of this section we consider the effect of spatial dimensionality on the stability of crystalline order in solids, in order to give a motivation why for the case of a two-dimensional solids it has been necessary to consider bond orientational order, rather than restricting attention to the standard description of crystalline order in terms of the usual Bragg scattering, Eq. (2.22). For this purpose we consider a crystal with N identical atoms of mass M within the harmonic approximation. Following Peierls (1935) we can write the relative mean squared displacement of two atoms whose distance from their ideal crystal positions is $\vec{R}$ (recall that $\vec{r}_k = \vec{R}_k + \vec{u}_k$, see Fig. 2.3) as:

$$\langle|\vec{u}_{\vec{k}} - \vec{u}_{\vec{k}+\vec{R}}|^2\rangle = \frac{2\hbar}{MN} \sum_{\vec{q}\in 1.\mathrm{B.Z.}} \omega^{-1}(\vec{q})[1-\cos(\vec{q}\cdot\vec{R})]/\{\exp[\hbar\omega(\vec{q})/k_BT]-1\}\,. \tag{2.47}$$

Here we have used the decomposition of the displacements $\vec{u}_{\vec{k}}$ in terms of the amplitudes of phonon modes with energy $\hbar\omega(\vec{q})$ and (quasi-) momentum $\hbar\vec{q}$, admitting all wave-vectors $\vec{q}$ that lie in the first Brillouin zone (1.B.Z) of the considered crystal lattice. (For a derivation of Eq. (2.47) any standard textbook on theoretical solid state physics may be consulted.) We are now interested in values of $R = |\vec{R}|$ that are much larger than the lattice spacing a. In the following we show that in $d = 1$ and $d = 2$ dimensions the sum in Eq. (2.47) diverges for $R \to \infty$ and that hence there is no long range positional order. This divergence is due to the contribution of low-lying acoustic phonons, which can be described as sound waves, $\omega(\vec{q}) = cq$, c being the sound velocity (for simplicity, the distinction between longitudinal and transverse phonons here is ignored). Then the Bose occupation factor (inverse of the last term in the curly brackets in Eq. (2.47)) can be approximated as $k_BT/\hbar\omega(\vec{q})$, and converting the sum over $\vec{q}$ into an integral over the first Brillouin zone we get

$$\langle|\vec{u}_{\vec{k}} - \vec{u}_{\vec{k}+\vec{R}}|^2\rangle \approx \frac{2k_BT}{Mc^2}\left(\frac{a}{2\pi}\right)^d \int d\vec{q}\,[1-\cos(\vec{q}\cdot\vec{R})]/q^2\,. \tag{2.48}$$

In $d = 1$ the first Brillouin zone simply means $-\pi/a < q < +\pi/a$, and since

$$\int_0^{\pi/a} dq(1-\cos qR)/q^2 \underset{R\to\infty}{\longrightarrow} R\int_0^{\infty} dx(1-\cos x)/x^2 = R\pi/2\,, \tag{2.49}$$

we find (Emery and Axe 1978)

$$\langle|\vec{u}_k - \vec{u}_{\vec{k}+\vec{R}}|^2\rangle \underset{R\to\infty}{\longrightarrow} \frac{k_BTaR}{Mc^2} \equiv n\delta^2 \quad \text{with} \quad n = \frac{R}{a}\,. \tag{2.50}$$

This result shows that for large R the mean squared displacements of an atom from its position in the classical ground state (at $T = 0$) diverge, and hence there is no positional long range order. If one works out the static structure factor, one finds instead of Eq. (2.22) that the Bragg peaks for a harmonic chain of atoms are broadened into gaussians centered at $G_h = 2\pi h/a$, where $h = 0, \pm 1, \pm 2, \ldots$, see e.g. Axe (1980):

$$S(q) = \sum_n e^{iqna} \langle e^{iqu_n} e^{-iqu_0} \rangle \approx \sum_n e^{iqna} \exp[-\frac{1}{2} q^2 \langle (u_n - u_0)^2 \rangle]$$

$$= \sum_n e^{iqna} \exp[-\frac{1}{2} q^2 n \delta^2] = \frac{\sinh(\frac{1}{2}\delta^2 q^2)}{\cosh(\frac{1}{2}\delta^2 q^2) - \cos(qa)} . \quad (2.51)$$

This is a typical liquid-like scattering function, with $S(q \to \infty) \to 1$. Actually, a behavior compatible with such a one-dimensional liquid structure described via the theory of one-dimensional harmonic solids, i.e. Eq. (2.51), is experimentally observed in certain quasi-one-dimensional crystals, such as the mercury chain compound $Hg_{3-x}AsF_6$ (Axe 1980). Thus we can conclude that a (classical) one-dimensional solid has its melting transition at $T = 0$, whereas for $T > 0$ one has a liquid whose structure shows a gradual onset of positional long range order as $T \to 0$ with a sharpening of the peaks of $S(q)$ at $q = G_h$.

In $d = 2$ the linear divergence with R of the integral in Eq. (2.49) is replaced by a much weaker logarithmic one. This can be seen by using in Eq. (2.48) polar coordinates (q, φ):

$$\int dq/q \int_0^{2\pi} d\varphi [1 - \cos(qR\cos\varphi)] = 2\pi \int_0^{\infty} dq [1 - J_0(qR)]/q$$

$$\approx 2\pi \int_{1/R}^{\infty} dq/q \sim 2\pi \ln R , \quad (2.52)$$

where J_0 is a Bessel function of the first kind. Thus also in this case there is no long range positional order.

It turns out that in $d = 2$ the δ–function singularities at the Bragg positions $\vec{G}_h$ of Eq. (2.22) are not completely wiped out, but replaced by power-law singularities (Halperin and Nelson 1978, Nelson 1983):

$$S(\vec{q}) \propto |\vec{q} - \vec{G}_h|^{-(2-\eta(\vec{G}_h))} . \quad (2.53)$$

This result can be derived by taking the Fourier components $\rho_h(\vec{R}) \equiv \exp[i\vec{G}_h \cdot \vec{u}(\vec{R})]$ as order parameter components of the crystal, and invoking continuum elasticity theory to write the free energy functional of the solids as follows:

$$\mathcal{F} = \int d\vec{r}\Big(2\mu \sum_{ij} u_{ij}^2 + \lambda \sum_{k} u_{kk}^2\Big)\,. \tag{2.54}$$

Here μ, λ are elastic constants and u_{ij} is the $ij-$component of the strain tensor. In linearized approximation, u_{ij} is obtained from the displacement field $u_i(\vec{r})$ via

$$u_{ij} = \frac{\partial u_i(\vec{r})}{\partial x_j} - \frac{\partial u_j(\vec{r})}{\partial x_i}\,, \quad \vec{r} = (x_1, \ldots, x_d)\,. \tag{2.55}$$

The statistical mechanics of the harmonic solid in this long wave length limit can then be treated by considering Eq. (2.54) as an effective Hamiltonian. The free energy F therefore becomes

$$F = -k_B T \, \ln\{\int \mathcal{D}\vec{u}(\vec{r}) \exp[-\mathcal{F}\{\vec{u}(\vec{r})\}/k_B T]\}\,. \tag{2.56}$$

Since $\mathcal{F}$ is harmonic, see Eq. (2.54), the functional integration in Eq. (2.56) of the resulting gaussian integrals can be carried out analytically, and one can also calculate the correlation function

$$C_h(\vec{R}) \equiv \langle \rho_h(\vec{R}) \rho_h^*(0) \rangle \tag{2.57}$$

and show that $C_h(|\vec{R}| \to \infty) \propto R^{-\eta(\vec{G}_h)}$. The Fourier transform of $C_h(\vec{R})$ then yields Eq. (2.53). The resulting exponent $\eta(\vec{G}_h)$ are found to be

$$\eta(\vec{G}_h) = \frac{k_B T (3\mu + \lambda)}{4\pi\mu(2\mu + \lambda)} \, |\vec{G}_h|^2\,. \tag{2.58}$$

As a corollary, we thus can conclude that on the basis of phonons alone a crystal in $d = 2$ would never melt into a real fluid, in which $C_h(R \to \infty)$ decays exponentially fast for all reciprocal lattice vectors $\vec{G}_h$. Relative to the crystal structure, smooth displacements $\vec{u}_k$ do not give rise to sufficient disorder and one needs to consider the topological disorder in order that the correlation functions and the structure factor $S(q)$ are that of a real fluid. In contrast to this, in $d = 1$ melting of the crystal into the fluid occurs already at $T = 0$, and at finite temperature no phase transition is possible (for short range interaction between the particles). In later chapters we shall discuss the concept of a "lower critical dimension" d_ℓ, below which thermal fluctuations destabilize the long range order, for other types of orderings, and hence the reader should keep in mind that $d_\ell = 2$ for solid-like crystalline long range order. While crystals in $d = 3$ dimensions normally exhibit both positional and orientational long range order, we draw attention to the existence of "quasicrystals" (Janssen *et al.* 2007) which have only orientational long range order (the positional "order" can be characterized by an irrational projection from a lattice in $d = 6$ dimensions to a $d = 3$ subspace, leading to an aperiodic arrangement of the atoms).

2.3 General Aspects of Dynamic Correlation Functions and Transport Properties

In order to be specific, let us consider a classical fluid (or solid) composed of N identical structureless particles. The phase space point $\vec{X}(t) = \{\vec{r}_1(t), \dots, \vec{r}_N(t), \vec{p}_1(t), \dots, \vec{p}_N(t)\}$, where $\vec{r}_j(t)$ is the spatial position and $\vec{p}_j(t)$ the momentum of the j−th particle at time t, then moves through phase space Γ according to the Newtonian equations of motion. According to the ergodicity hypothesis of statistical mechanics, time averages along this trajectory $\vec{X}(t)$ are then equivalent to the ensemble averages considered so far in this chapter.

It is useful to introduce the probability density $f(\vec{X}, t)$ that at time t the physical system is in the $6N$-dimensional volume element $d\vec{r}^{\,(N)} d\vec{p}^{\,(N)}$, centered around $\vec{X}(t)$, of phase space. As is well known, the Newtonian equations of motion for the N particles are equivalent to the Liouville equation for $f(\vec{X}, t)$,

$$\frac{\partial f(\vec{X},t)}{\partial t} + i\mathcal{L} f(\vec{X},t) = 0\,, \tag{2.59}$$

where the Liouville operator $i\mathcal{L}$ is defined by

$$i\mathcal{L} = \sum_{j=1}^{N} \Big(\frac{\partial \mathcal{H}}{\partial \vec{p}_j} \frac{\partial}{\partial \vec{r}_j} - \frac{\partial \mathcal{H}}{\partial \vec{r}_j} \frac{\partial}{\partial \vec{p}_j} \Big)\,, \tag{2.60}$$

and $\mathcal{H}(\vec{X})$ is the Hamiltonian of the N-particle system.

The time dependence of an arbitrary dynamical variable A that depends on $\vec{X}$, e.g. a Fourier component $\rho_{\vec{k}}(t) = \sum_{j=1}^{N} \exp(-i\vec{k}\cdot\vec{r}_j(t))$ of the particles, cf. Eq. (2.11), can be written in a compact form fully analogous to Eq. (2.59) namely,

$$\frac{dA}{dt} = i\mathcal{L} A\,, \tag{2.61}$$

from which one obtains immediately the formal solution

$$A(t) = \exp(i\mathcal{L})A(0)\,. \tag{2.62}$$

In this section we are now interested in time correlation functions of two dynamical variables A and B in thermal equilibrium,

$$C_{AB}(t) = \langle A(t'+t)B(t')\rangle = \langle A(t)B(0)\rangle\,, \tag{2.63}$$

where we have already implied in our notation that the system is in equilibrium and hence time translation invariant. (In equilibrium the origin of time, $t = 0$, is irrelevant, and hence the correlations written in Eq. (2.63) do not depend on the two times t' and $t'+t$ separately, but only on their difference.) Using the fact that the derivative of the correlator with respect to t' in Eq. (2.63) must therefore vanish,

$$\frac{d}{dt'}\langle A(t'+t)B(t')\rangle = \langle \dot{A}(t'+t)B(t')\rangle + \langle A(t'+t)\dot{B}(t')\rangle = 0\,, \tag{2.64}$$

and using once more the time translation invariance we obtain immediately

$$\langle \dot{A}(t)B(0)\rangle = -\langle A(t)\dot{B}(0)\rangle \quad \text{and} \quad \langle \dot{A}(0)A(0)\rangle = 0\,. \tag{2.65}$$

It is also clear that equal-time correlations $\langle A(t')B(t')\rangle$ are nothing else than the static correlations $\langle AB\rangle$ considered before, while for $t \to \infty$ correlations will vanish and hence the correlators factorize,

$$\lim_{t\to\infty}\langle A(t'+t)B(t')\rangle = \langle A\rangle\langle B\rangle\,. \tag{2.66}$$

For this reason, one often redefines the correlation functions such that they contain only the "fluctuating part",

$$\widetilde{C}_{AB}(t) = \langle [A(t) - \langle A\rangle][B(0) - \langle B\rangle]\rangle\,, \tag{2.67}$$

since then $\widetilde{C}_{AB}(t \to \infty) \to 0$. It is also useful to consider the Fourier transform with respect to time, $C_{AB}(\omega)$, which often is called the "spectral function",

$$C_{AB}(\omega) = \frac{1}{2\pi}\int_{-\infty}^{+\infty} \exp(i\omega t)\widetilde{C}_{AB}(t)dt\,, \tag{2.68}$$

or its Laplace transform, z being a complex frequency

$$\widetilde{C}_{AB}(z) = i\int_{0}^{\infty} \exp(izt)\widetilde{C}_{AB}(t)dt\,. \tag{2.69}$$

Since $C_{AB}(t)$ is bounded, $\widetilde{C}_{AB}(z)$ is analytic in the upper half of the complex z plane ($\mathrm{Im}(z) > 0$). Using then

$$\widetilde{C}_{AB}(z) = i \int_{-\infty}^{+\infty} \frac{C_{AB}(\omega)}{\omega - z}\, d\omega \tag{2.70}$$

together with the standard relation $\lim_{\varepsilon\to 0}(x - i\varepsilon)^{-1} = \mathcal{P}(1/x) + i\pi\delta(x)$, where $\mathcal{P}$ denotes the principal part, and noting that the spectral function of an autocorrelation is real and an even function of ω, we obtain a relation between the spectrum of an autocorrelation function and its Laplace transform:

$$C_{AA}(\omega) = \lim_{\varepsilon\to 0} \frac{1}{\pi} \mathrm{Re}\widetilde{C}_{AA}(\omega + i\varepsilon)\,. \tag{2.71}$$

Furthermore one can show that $C_{AA}(\omega)$ is nonnegative.

The importance of these formal definitions is that the spectra measured by various spectroscopic techniques (dielectric spectroscopy, dynamical magnetic susceptibility, inelastic scattering of neutrons, X-rays, etc.) are directly related to the spectral functions of the corresponding dynamical variables. In addition, the linear transport coefficients of hydrodynamics are related to time integrals of certain autocorrelation functions. We recall here only some salient features, since excellent treatments can be found in the literature (e.g. Hansen and McDonald 1986).

As is well known, experimental information on the dynamics of the particles in a fluid can be obtained by incoherent neutron scattering (Lovesey 1994). If $\vec{k}_i$ and $\vec{k}_f$ are the wave-vector of the incoming and scattered neutron beam, respectively, that have energies $\hbar\vec{k}_i$ and $\hbar\vec{k}_f$, the scattering intensity for energy transfer $\hbar\omega = \hbar(\omega_f - \omega_i)$ and momentum transfer $\hbar\vec{k} = \hbar(\vec{k}_f - \vec{k}_i)$ is given by the double differential cross section

$$\frac{d^2\sigma_{\mathrm{inc}}}{d\Omega d\omega} = \frac{k_f}{k_i}(\langle b_j^2\rangle - \langle b_j\rangle^2) N S_s(\vec{k},\omega)\,. \tag{2.72}$$

Here $d\Omega$ is the element associated with the (solid) scattering angle, and b_j are the scattering lengths of the individual nuclei. In deriving Eq. (2.72) one makes use of the fact that there is no correlation between the positions of the N scattering particles and the value of their scattering length b_j. The function $S_s(\vec{k},\omega)$ is called "self-dynamic structure factor" and it can be shown that it is given by the space-time Fourier transform of $G_s(\vec{r},t)$, the so-called "self-part of the van Hove function", which can be expressed in terms of the positions of the particles as follows:

$$G_s(\vec{r},t) = \frac{1}{N}\sum_j \langle \delta(\vec{r} - (\vec{r}_j(t) - \vec{r}_j(0)))\rangle \,. \tag{2.73}$$

Thus $G_s(\vec{r},t)$ is just the probability that within a time interval t a particle has made a displacement $\vec{r}$. From this definition it therefore follows that

$$S_s(\vec{k},\omega) = \frac{1}{2\pi}\int_{-\infty}^{+\infty} dt \exp(i\omega t)\int G_s(\vec{r},t)\exp(-i\vec{k}\cdot\vec{r})d\vec{r}\,. \tag{2.74}$$

The space Fourier transform of $G_s(\vec{r},t)$ is called "incoherent intermediate scattering function" and is thus given by

$$F_s(\vec{k},t) = \int G_s(\vec{r},t)\exp(-i\vec{k}\cdot\vec{r})d\vec{r} \tag{2.75}$$

$$= \frac{1}{N}\sum_j \langle \exp\{i\vec{k}\cdot[\vec{r}_j(0) - \vec{r}_j(t)]\}\rangle\,, \tag{2.76}$$

from which we obtain the relation

$$S_s(\vec{k},\omega) = \frac{1}{2\pi}\int_{-\infty}^{+\infty} dt \exp(i\omega t) F_s(\vec{k},t)\,. \tag{2.77}$$

Note that if one assumes that the displacements $\delta\vec{r}_j(t) = \vec{r}_j(t) - \vec{r}_j(0)$ of the particles are gaussian distributed, one can express $F_s(\vec{q},t)$ in terms of the mean-squared displacement:

$$F_s(\vec{k},t) = \exp\{-\frac{1}{2d}k^2\langle[\delta\vec{r}_j(t)]^2\rangle\}\,. \tag{2.78}$$

Studying the q-dependence of $F_s(\vec{q},t)$ hence allows to test for the presence or absence of a non-gaussian behavior. A gaussian behavior of displacements occurs both in the case of phonons, see Eq. (2.20), and for the case of self diffusion, see Eq. (2.97) below. However, as we will see later, a supercooled fluid near its glass transition temperature shows a nontrivial non-gaussian behavior over time-scales of the order of the structural relaxation time.

The total intensity of scattered neutron contains in general both coherent and incoherent parts (Lovesey 1994):

$$\frac{d^2\sigma}{d\Omega d\omega} = \frac{d^2\sigma_{\rm coh}}{d\Omega d\omega} + \frac{d^2\sigma_{\rm inc}}{d\Omega d\omega} \tag{2.79}$$

with

$$\frac{d^2\sigma_{\rm coh}}{d\Omega d\omega} = \frac{k_f}{k_i}\langle b_j\rangle^2 N S(\vec{k},\omega)\,. \tag{2.80}$$

Here the function $S(\vec{k},\omega)$ is the so-called "dynamic structure factor" and is defined as the space-time Fourier transform of the van Hove function $G(\vec{r},t)$ which is given by

$$G(\vec{r},t) = \frac{1}{N}\sum_j\sum_l \langle\delta(\vec{r}-(\vec{r}_j(t)-\vec{r}_l(0)))\rangle\,, \tag{2.81}$$

i.e.

$$S(\vec{k},\omega) = \frac{1}{2\pi}\int_{-\infty}^{+\infty} dt\exp(i\omega t)\int G(\vec{r},t)\exp(-i\vec{k}\cdot\vec{r})d\vec{r}\,. \tag{2.82}$$

Thus if one introduces the so-called "coherent intermediate scattering function" $F(\vec{k},t)$ as the space Fourier transform of $G(\vec{r},t)$,

$$F(\vec{k},t) = \int G(\vec{r},t)\exp(-i\vec{k}\cdot\vec{r})d\vec{r} \tag{2.83}$$

$$= \frac{1}{N}\sum_j\sum_\ell \langle\exp\{i\vec{k}\cdot[\vec{r}_j(0)-\vec{r}_\ell(t)]\}\rangle \tag{2.84}$$

$$= \frac{1}{N}\langle\rho_{\vec{k}}(t)\rho_{-\vec{k}}(0)\rangle\,, \tag{2.85}$$

one obtains the relation

$$S(\vec{k},\omega) = \frac{1}{2\pi}\int_{-\infty}^{+\infty} dt\exp(i\omega t)F(\vec{k},t)\,. \tag{2.86}$$

Thus the comparison of Eq. (2.85) with Eq. (2.12) shows that $F(\vec{k},t)$ is just the time dependent generalization of the structure factor $S(\vec{k})$.

Note that Eq. (2.72) implies that if all b_j are the same, the scattering is purely coherent. Different values of b_j arise from the orientation of the nuclear spin (this happens e.g. for hydrogen) or when an element contains a mixture of isotopes.

From Eq. (2.81) we see that the correlation function $G(\vec{r},t)$ can, within a description of scattering in the limiting case of classical physics, be interpreted as a (conditional) probability that a particle is found at a position $\vec{r}$ and time t, given that there was a particle (either the same one, or a different one) at the origin $\vec{r} = 0$ at time $t = 0$. However, it is necessary to add at this point a word of caution. Within the framework of molecular dynamics simulations (Allen and Tildesley 1987, Binder and Ciccotti 1996), the world is indeed purely classical, and this has been assumed in Eqs. (2.59)- (2.62) as well. However, when one discusses neutron scattering experiments, the quantum nature of physical phenomena needs to be taken into account. It turns out that Eqs. (2.72)- (2.86) are still valid, but that strictly speaking the interpretation of the functions $G_s(\vec{r}, t)$ and $G(\vec{r}, t)$ in terms of probabilities gets lost, because these functions acquire imaginary parts. This happens, because the operators associated with $\vec{r}_j(0)$ and $\vec{r}_j(t)$, or $\vec{r}_l(t)$, do not commute with each other. While $S(\vec{k}, \omega)$ and $S_s(\vec{k}, \omega)$ must be real functions, since they describe scattering intensities, this is not guaranteed for $G(\vec{r}, t)$ or $G_s(\vec{r}, t)$, and is actually not the case. E.g., $G(\vec{r}, t)$ would be real if $S(\vec{k}, \omega)$ were an even function in ω, but in fact one can show that this property holds only in the limit $\hbar \to 0$, since

$$S(\vec{k}, \omega) = \exp(\hbar\omega/k_B T) S(\vec{k}, -\omega) . \tag{2.87}$$

This relation can be derived from the principle of detailed balance which requires that the ratio of cross sections for the scattering processes $|\vec{k}_1, 1\rangle \to |\vec{k}_2, 2\rangle$ and $|\vec{k}_2, 2\rangle \to |\vec{k}_1, 1\rangle$ must be equal to the ratio of the statistical weights $\exp(-E_1/k_B T)$ and $\exp(-E_2/k_B T\rangle$ of the states $|1\rangle$ and $|2\rangle$ of the scattering system (remember $\hbar\omega = E_2 - E_1$).

For systems which have inversion symmetry, which is trivially true in fluids, we have

$$S(-\vec{k}, \omega) = S(\vec{k}, \omega), \quad S_s(-\vec{k}, \omega) = S_s(\vec{k}, \omega) , \tag{2.88}$$

and analogous properties hold for the incoherent intermediate scattering function. For $G(\vec{r}, t)$ and $G_s(\vec{r}, t)$ one can show the relations

$$G(-\vec{r}, -t) = G^*(\vec{r}, t) \quad , \quad G_s(-\vec{r}, -t) = G^*(\vec{r}, t) . \tag{2.89}$$

Only in the classical limit the imaginary part of $G(\vec{r}, t)$ and $G_s(\vec{r}, t)$ vanishes. Note that to reach this limit, it is not enough to be able to treat the scattering system within the framework of classical statistical mechanics. E.g. for the scattering from a classical ideal gas the function $G(\vec{r}, t)$, which is identical to $G_s(\vec{r}, t)$, is still complex and the imaginary part describes the perturbation of the system due to the observation via scattering events, since the neutrons transfer recoil energy to the gas atoms. A real $G(\vec{r}, t)$ is obtained only in the limit $\hbar/k_B T \to 0$.

After this digression on the observation of correlation functions such as $G_s(\vec{r},t)$ and $G(\vec{r},t)$ by means of scattering techniques we illustrate the behavior of these functions on a simple example: The self diffusion of a tagged particle. For this we define $\rho^{(s)}(\vec{r},t)$ and $\vec{j}^{(s)}(\vec{r},t)$ as the macroscopic density and current of the tagged particles. Fick's law states that these two functions are related to each other via:

$$\vec{j}^{(s)}(\vec{r},t) = -D\nabla\rho^{(s)}(\vec{r},t)\,, \tag{2.90}$$

where D is the diffusion constant. Since the total number of particles is conserved, we have also a continuity equation

$$\partial\rho^{(s)}(\vec{r},t)/\partial t + \nabla\vec{j}^{(s)}(\vec{r},t) = 0\,, \tag{2.91}$$

and together with Fick's law we thus obtain the diffusion equation:

$$\frac{\partial}{\partial t}\rho^{(s)}(\vec{r},t) = D\nabla^2\rho^{(s)}(\vec{r},t)\,. \tag{2.92}$$

We first consider the solution of Eq. (2.92) in reciprocal space by introducing the Fourier components $\rho^{(s)}_{\vec{k}}(t) = \int \rho^{(s)}(\vec{r},t)\exp(-i\vec{k}\cdot\vec{r})d\vec{r}$.

The Fourier transform of Eq. (2.92) is thus given by

$$d\rho^{(s)}_{\vec{k}}(t)/dt = -Dk^2\rho^{(s)}_{\vec{k}}(t)\,, \tag{2.93}$$

from which we obtain immediately the solution

$$\rho^{(s)}_{\vec{k}}(t) = \rho^{(s)}_{\vec{k}}(0)\exp(-Dk^2t)\,, \tag{2.94}$$

where $\rho^{(s)}_{\vec{k}}(0)$ is the Fourier component of the tagged particle density at $t = 0$.

From this result we can conclude that in a diffusive system the density fluctuations decay with a relaxation time $\tau_{\vec{k}} = (Dk^2)^{-1}$. The fact that for long wave-length we have $\tau_{\vec{k}} \to \infty$ illustrates an important principle, namely the so-called "hydrodynamic slowing down" (Kadanoff and Martin 1963). On sufficiently large length scales this behavior is found for all liquids, since it is a simple consequence from the fact that in order to relax a density fluctuation for k small, one has to move *many* particles by *large* distances.

The expectation value for the density-autocorrelation function is obtained by multiplying Eq. (2.94) by $\rho^{(s)}_{-\vec{k}}(0)$ and by taking the thermal average. We thus obtain the incoherent intermediate scattering function

$$F_s(\vec{k},t) = \langle \rho_{\vec{k}}^{(s)}(t)\rho_{-\vec{k}}^{(s)}(0)\rangle = \langle \rho_{\vec{k}}^{(s)}(0)\rho_{-\vec{k}}^{(s)}(0)\rangle \exp(-Dk^2 t) = \exp(-Dk^2 t)\,. \tag{2.95}$$

Note that from Eqs. (2.93) and 2.94) one obtains that

$$\langle \frac{d\rho_{\vec{k}}^{(s)}(0)}{dt} \rho_{\vec{k}}^{(s)}(0)\rangle = -Dk^2\,, \tag{2.96}$$

a result that obviously violates the general result that for any observable A we have $\langle \dot{A}(0)A(0)\rangle = 0$, see Eq. (2.65). The reason for this contradiction is the approximative character of Fick's law which does not hold on microscopically small times since there the particles move ballistically and thus the particle current is not given by the gradient of the particle density, as implied by Eq. (2.90).

From Eq. (2.95) one can calculate immediately the self part of the van Hove function, see Eq. (2.75) and one obtains

$$G_s(\vec{r},t) = (4\pi Dt)^{-d/2} \exp[-|\vec{r}\,|^2/4Dt]\,, \tag{2.97}$$

or by calculating the time Fourier transform the self-dynamic structure factor from Eq. (2.77):

$$S_s(\vec{k},\omega) = \frac{1}{\pi}\frac{Dk^2}{\omega^2 + (Dk^2)^2}\,, \tag{2.98}$$

i.e. we find that $S_s(\vec{k},\omega)$ is just a Lorentzian.

From Eq. (2.97) one can also calculate easily the mean squared displacement of the tagged particles and one obtains

$$\langle [\delta\vec{r}(t)]^2\rangle = \langle [\vec{r}(t) - \vec{r}(0)]^2\rangle = 2dDt, \quad t \to \infty\,. \tag{2.99}$$

Eq. (2.99) is nothing else but the well known Einstein relation. Although we have here emphasized that it is expected to hold only asymptotically for large enough times, Eq. (2.97) actually gives this relation for all times $t > 0$. However, for systems that do not obey the diffusion equation for all times, the relation (2.99) can still be expected to hold if the time t is larger than the typical structural relaxation time τ of the fluid. While for simple fluids like liquid argon under normal conditions we have $\tau \approx 10^{-12}$ s, for glass-forming fluids near the glass transition τ becomes macroscopically large, and the Einstein relation holds only for macroscopic time scales.

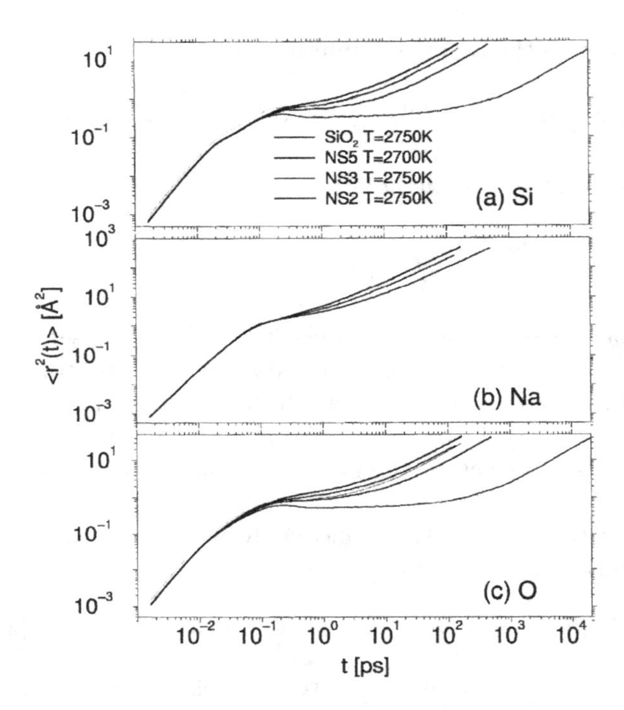

Fig. 2.15 Log-log plot of the mean square displacement (denoted as $\langle r^2(t)\rangle$ in the figure) of Si atoms (a), Na atoms (b) and O atoms (c) versus time, as obtained from Molecular Dynamics simulation of the same model as used in Fig. 2.8 for the description of SiO_2, (Na_2O) $(5\ SiO_2)$ [NS5], (Na_2O) $(3SiO_2)$ [NS3] and (Na_2O) $(2SiO_2)$ [NS2], from bottom to top, at the temperatures $T = 2750$ K or $T = 2700$ K, as indicated in the figure. For further explanations see text. From Winkler (2002).

As an example, Fig. 2.15 shows the time dependence of the mean squared displacements of Si, Na, and O atoms in SiO_2 melts (bottom curves in (a) and (c)) and in sodium silicate melts. We recognize that for t of the order of 10^{-2}ps, the mean squared displacements increase proportional to t^2 (ballistic motion), then the curves flatten out (due to the cage effect that will be discussed in Chap. 5), and only after 10 to 1000 ps we observe in these systems the onset of diffusion described by Eq. (2.99). We also note that Si and O atoms in the silicon dioxide melts move slower than in the sodium silicate melts at the same temperature, a result that is related to the structure of these systems. The behavior seen in Fig. 2.15 will be discussed in more detail in Chaps. 5 and 6.

Before we end this chapter we discuss briefly the so-called Green-Kubo relations (Green 1954 and Kubo 1957). These relations make a connection between transport coefficients and time integrals over appropriate correlation functions. As a prominent example we start with the velocity autocorrelation function of a tagged particle $Z(t''-t') \equiv d^{-1}\langle \vec{v}(t'') \cdot \vec{v}(t') \rangle$. For this purpose we note that

$$\vec{r}(t) - \vec{r}(0) = \int_0^t \vec{v}(t')dt' \,, \tag{2.100}$$

and hence

$$\langle [\vec{r}(t) - \vec{r}(0)]^2 \rangle = \int_0^t dt' \int_0^t dt'' \langle \vec{v}(t'') \cdot \vec{v}(t') \rangle = d \int_0^t dt' \int_0^t dt'' Z(t'-t'') \,. \tag{2.101}$$

Note that to obtain the last equality one has to make use of the time translation invariance of the system. Using now $s = t'' - t'$ instead of t'' as an integration variable yields

$$\langle [\vec{r}(t) - \vec{r}(0)]^2 \rangle = 2d \int_0^t ds\, Z(s)(t-s) = 2d\, t \int_0^t (1 - s/t) Z(s) ds \,. \tag{2.102}$$

From the Einstein relation we know that for $t \to \infty$ the left hand side must be given by $2dDt$ and hence we can conclude that

$$D = \int_0^\infty Z(s)ds = \frac{1}{d} \int_0^\infty \langle \vec{v}(s) \cdot \vec{v}(0) \rangle ds \,. \tag{2.103}$$

Thus we find that the self-diffusion constant is the time integral of the velocity autocorrelation function which is, as mentioned above, an example of a Green-Kubo relation. In fact, such relations can be derived for all transport coefficients (Hansen and McDonald 1986, Boon and Yip 1980). We quote some of them here without derivation: The shear viscosity can be related to the time correlations of the off-diagonal components of the stress tensor σ^{xy},

$$\eta = \frac{1}{k_B T V} \int_0^\infty dt \langle \sigma^{xy}(t) \sigma^{xy}(0) \rangle \,, \tag{2.104}$$

where, for the case that the forces in the system are due to pair potentials, σ^{xy} can be written as

$$\sigma^{xy} = \sum_{i=1}^N \left[m v_i^x v_i^y + \frac{1}{2} \sum_{j \neq i} x_{ij} F_y(r_{ij}) \right] . \tag{2.105}$$

Here m is the mass of a particle and $F_y(r_{ij})$ is the y-component of the force between particles i and j, $F_y = -\partial u(r_{ij})/\partial y_i$, where $u(r_{ij})$ is the pair potential.

Similarly, for the thermal conductivity λ_T we need to consider correlations of the energy current:

$$\lambda_T = \frac{1}{k_B T^2 V} \int_0^\infty dt \langle j_z^e(t) j_z^e(0) \rangle \,, \tag{2.106}$$

where the energy current in direction z is given by

$$j_z^e = \sum_{i=1}^N v_i^z \left[\frac{1}{2} m |\vec{v}_i|^2 + \frac{1}{2} \sum_{j \neq i}^N u(r_{ij}) \right] - \frac{1}{2} \sum_{i=1}^N \sum_{j \neq i}^N \vec{v}_i \cdot \vec{r}_{ij} \frac{\partial u(r_{ij})}{\partial z_{ij}} \,. \tag{2.107}$$

Finally, the electrical conductivity of a system where the particles carry charges q_i can be obtained from correlations of the electrical current j^{el},

$$\sigma_{\text{el}} = \frac{1}{k_B T V} \int_0^\infty dt \langle j_x^{\text{el}}(t) j_x^{\text{el}}(0) \rangle, \quad \text{with} \quad j_x^{\text{el}} = \sum_{i=1}^N q_i v_i^x \,. \tag{2.108}$$

In addition one can derive many more useful properties of these various correlation functions that appear in all the above expressions (Hansen and McDonald 1986). E.g., the velocity autocorrelation function $Z(t)$ has a short time expansion

$$Z(t) = \frac{k_B T}{m} (1 - \Omega_0^2 t^2/2 + - \cdots) \tag{2.109}$$

where the frequency Ω_0 can be related to the interparticle forces (assuming continuous interparticle potentials here throughout) as follows,

$$\Omega_0^2 = \frac{1}{3} \left(\frac{m}{k_B T} \right) \langle \dot{\vec{v}}(0) \cdot \dot{\vec{v}}(0) \rangle = \frac{1}{3 m k_B T} \langle |\vec{F}_i|^2 \rangle \,, \tag{2.110}$$

$\vec{F}_i$ being the total force acting on particle i from all other particles in the fluid. For pairwise potentials $u(r_{ij})$, Eq. (2.110) can be rearranged to give

$$\Omega_0^2 = \frac{\rho}{3m} \int g(r) \nabla^2 u(r) d\vec{r} \,, \tag{2.111}$$

where ρ is again the density of the fluid and $g(r)$ the pair distribution function, see Eq. (2.10). Sometimes Ω_0 is called the "Einstein frequency", since it represents the frequency at which the tagged particle would vibrate if it were to undergo small oscillations in the potential well produced by the particles of the fluid when kept at their mean equilibrium positions around the tagged particles. A typical order of magnitude for Ω_0 is 10^{13} s^{-1}.

Note that Eqs. (2.109)- (2.111) hold only for systems in which the interactions are smooth functions of the distance between the particles. Hence for the important case of hard spheres they are no longer valid since the short time expansion of the trajectory is not quadratic in t. Instead one find (Lebowitz *et al.* 1969)

$$Z(t) = \langle v^2 \rangle (1 - \Omega_0 t + \cdots) \quad \text{with} \quad \Omega_0 = \frac{8\rho\sigma^3}{3}\left(\frac{\pi k_B T}{m}\right)^{1/2}, \tag{2.112}$$

where σ is the hard sphere diameter of the particles.

Another important concept in the dynamics of liquids is the one of Brownian motion: A large and massive particle moves randomly in a bath of much smaller and lighter particles. Thus the collisions of the small particles with the large one subject the latter to a random force and a friction. If we assume that this friction is proportional to the velocity of the large particle one obtains a so-called Langevin equation for this motion:

$$m\dot{\vec{v}}(t) = -\zeta m\vec{v}(t) + \vec{f}_r(t), \tag{2.113}$$

where ζ is a friction coefficient. The average of the random force $\vec{f}_r(t)$ is assumed to vanish, to be uncorrelated with the velocity $\vec{v}$, and δ-correlated in time:

$$\langle \vec{f}_r(t) \rangle = 0, \ \langle \vec{v}(t) \cdot \vec{f}_r(t) \rangle = 0, \quad \langle \vec{f}_r(t+t_0) \cdot \vec{f}_r(t_0) \rangle = 2\pi C_{ff}\delta(t). \tag{2.114}$$

Here the notation C_{ff} was chosen for the sake of consistency with Eq. (2.67). Note that a time correlation function that is given by a $\delta-$function corresponds to a constant in the frequency domain. Hence this type of random force is also often referred to as "white noise".

It is important to realize that the two terms on the right-hand side of Eq. (2.113) are not independent. To see their connection, we start from the formal solution of Eq. (2.113)

$$m\vec{v}(t) = m\vec{v}(0)\exp(-\zeta t) + \exp(-\zeta t)\int_0^\infty \exp(\xi s)\vec{f}_r(s)ds. \tag{2.115}$$

By squaring and taking the thermal average, one finds with the help of Eq. (2.114) that

$$m^2 \langle |\vec{v}(t)|^2 \rangle = m^2 \langle |\vec{v}(0)|^2 \rangle \exp(-2\zeta t) + \frac{\pi C_{ff}}{\zeta}[1 - \exp(-2\zeta t)]. \tag{2.116}$$

In the limit $t \to \infty$ the Brownian particle should be in thermal equilibrium with the bath, regardless of the initial conditions, and should have a mean squared velocity in accord with the Maxwell-Boltzmann distribution,

$$\langle |\vec{v}(t \to \infty)|^2 \rangle = 3k_B T/m \,, \tag{2.117}$$

which implies that

$$\zeta = \frac{\pi C_{ff}}{3k_B T m} = \frac{1}{3k_B T m} \int_0^\infty \langle \vec{f}_r(t) \cdot \vec{f}_r(0) \rangle dt \,. \tag{2.118}$$

Eq. (2.118) is a well known special case of the so-called "fluctuation-dissipation theorem" since it relates the strength of the friction with the amplitude of the fluctuating random force. We also mention that, with a little algebra, it is easy to derive a further relation for the self-diffusion coefficient, e.g. by using Eq. (2.115) in Eq. (2.103), to find (using also Eq. (2.114) and Eq. (2.118))

$$D = k_B T/\zeta m \,. \tag{2.119}$$

We point out that it is possible to estimate ζ from a hydrodynamic calculation of the frictional force on a sphere of diameter σ moving in a fluid that has a shear viscosity η. For the case of a "stick boundary condition", i.e. at the surface of the sphere the fluid has the same velocity as the surface, the result is $\zeta = 3\pi\eta\sigma/m$ (Landau and Lifshitz 1987). Inserting this result in Eq. (2.119) leads to the famous Stokes-Einstein relation for the self-diffusion coefficient:

$$D^{\mathrm{S.E}} = k_B T/(3\pi\eta\sigma) \,. \tag{2.120}$$

Although this result derives from considerations on a quasi-macroscopic scale, it is found that it applies not only to Brownian particles but is also a quite reasonable approximation for particles with a microscopic size. In this case the diameter σ becomes the effective diameter of the diffusive fluid particle.

If the dimensions and the mass of the diffusing particle are comparable or even the same as those of the neighbors with which it collides, it is obviously a poor approximation to assume that the friction force depends only on the velocity at precisely the same time. The implication of such an assumption is that the motion adjusts instantaneously, and all memory on the previous history is lost, i.e. we have a Markov process. A more reasonable assumption is then clearly to allow some "memory" associated with the motion of the particle. One possibility to achieve this is to introduce a friction coefficient $\zeta(t-t')$ that is nonlocal in time, and thus Eq. (2.113) is replaced by

$$m\dot{\vec{v}}(t) = -m \int_0^t \zeta(t-t')\vec{v}(t')dt' + \vec{f}_r(t) \,. \tag{2.121}$$

The random forces $\vec{f}_r(t)$ are however still supposed to have the properties given in Eq. (2.114). If we multiply Eq. (2.121) by $\vec{v}(0)$ we obtain a simple equation for the velocity autocorrelation function $Z(t)$:

$$\dot{Z}(t) = -\int_0^t \zeta(t-t')Z(t')dt' \,. \tag{2.122}$$

The quantity $\zeta(t)$ can hence be considered as a "memory function" of the velocity autocorrelation function $Z(t)$. By taking the Laplace transform (LT) of Eq. (2.122), we obtain a simple relation between $\widetilde{Z}(z)$ and $\widetilde{\zeta}(z)$ [2],

$$\widetilde{Z}(z) = \frac{k_B T/m}{-z + i\widetilde{\zeta}(z)} \,. \tag{2.123}$$

Note that Eqs. (2.121)– (2.123) can be considered as being exact since they can be used as definitions of the unknown functions $\zeta(t)$ (or $\widetilde{\zeta}(z)$). In practice one tries to approximate $\zeta(t)$ (or $\widetilde{\zeta}(z)$) by a simple function in order to have a good description of $Z(t)$. In Chap. 5 we will return to this approach in more detail when we discuss the Zwanzig-Mori projection operator formalism.

[2]Here one has to make use of the convolution theorem which states that $\mathrm{LT}[\int_0^t X(\tau)Y(t-\tau)d\tau](z) = -i\mathrm{LT}[X(t)](z)\ \mathrm{LT}[Y(t)](z)$.

References

Allen, M.P., and Tildesley, D.J. (1987) *Computer Simulation of Liquids* (Clarendon Press, Oxford).

Arbe, A., Richter, D., Colmenero, J., and Farago, B. (1996) *Phys. Rev.* E **54**, 3853.

Axe, J.D. (1980), in *Ordering in Strongly Fluctuating Condensed Matter Systems* (T. Riste, ed.) p. 399 (Plenum, New York).

Barrat, J.-L., and Hansen, J.-P. (2003) *Basic Concepts for Simple and Complex Liquids* (Cambridge University Press, Cambridge).

Binder, K., and Ciccotti, G. (1996) *Monte Carlo and Molecular Dynamics of Condensed Matter Systems* (Società Italiana di Fisica, Bologna).

Boon, J.P., and Yip, S. (1980) *Molecular Hydrodynamics* (McGraw Hill, New York).

Cusack, N.E. (1986) *The Physics of Structurally Disordered Matter. An Introduction* (Adam Hilger, Bristol).

Emery, V.J., and Axe, J.D. (1978) *Phys. Rev. Lett.* **40**, 1507.

Elliott, S.R. (1983) *Physics of Amorphous Materials* (Longman, Essex).

Green, M.S. (1954) *J. Chem. Phys.* **22**, 398.

Halperin, B.I., and Nelson, D.R. (1978) *Phys. Rev. Lett.* **41**, 121.

Hansen, J.-P., and McDonald, I.R. (1986) *Theory of Simple Liquids* (Academic Press, San Diego).

Horbach, J., and Kob, W. (1999) *Phys. Rev.* **B60**, 3169.

Janssen, T., Chapuls, G., and de Boissieu, M. (2007) *Aperiodic Crystals: From Modulated Phases to Quasicrystals* (Oxford University Press, Oxford).

Jaster, A. (1999a) *Europhys. Lett.* **42**, 277.

Jaster, A. (1999b) *Phys. Rev.* **E59**, 2594.

Jaster, A. (2004) *Phys. Lett. A* **330**, 120.

Kadanoff, L.P., and Martin, P.C. (1963) *Ann. Phys. (NY)* **24**, 419.

Kleinert, H. (1989) *Gauge Fields in Condensed Matter* (World Scientific, Singapore).

Kosterlitz, J.M., and Thouless, D.J. (1973) *J. Phys. C: Solid State Phys.* **6**, 1181.

Kubo, R. (1957) *J. Phys. Soc. Jpn.* **12**, 570.

Landau, L.D, and Lifshitz E.M. (1987) *Fluid Mechanics: Volume 6*, (Butterworth-Heinemann Publisher, Boston).

Lebowitz, J.L., Percus, J.K., and Sykes, J. (1969) *Phys. Rev.* **188**, 487.

Lovesey, S.W. (1994) *Theory of Neutron Scattering from Condensed Matter* (Oxford University Press, Oxford).

Mak, C.H. (2006) *Phys. Rev.* **E73**, 065104.

Mermin, N.D. (1968) *Phys. Rev.* **176**, 250.

Mermin, N.D., and Wagner H. (1966) *Phys. Rev. Lett.* **17**, 1133.

Nabarro, F.R.N. (1967) *Theory of Crystal Dislocations* (Oxford University Press, Oxford).

Nelson, D.R. (1983) in *Phase Transitions and Critical Phenomena,* Vol. 7 (C. Domb and J. L. Lebowitz, eds.) (Academic Press, London).

Pathria, R.K. (1986) *Statistical Mechanics* (Pergamon Press, Oxford).

Peierls, R.E. (1935) *Ann. Inst. Henri Poincaré* **5**, 177.

Porod, G. (1951) *Kolloid. Zeit.* **124**, 83.

Price, D.L., and Carpenter, J.M. (1989) *J. Non-Cryst. Solids* **92**, 153.

Sengupta, S., Nielaba, P., and Binder, K. (2000) *Phys. Rev.* **E61**, 6294.

Stillinger, F.H., and Weber, T.A. (1982) *Phys. Rev. A* **25**, 978.

Strandburg, K.J. (1988) *Rev. Mod. Phys.* **60**, 161.

Voronoi, G. (1908) *J. reine angew. Math.* **134**, 198.

Weber, H., Marx, D., and Binder, K. (1995) *Phys. Rev.* **B51**, 14636.

Weitz, D.A., Lin, M.Y., Huang, J.S., Witten, T.A., Sinha, S.K., Gethner, J.S., and Ball, R.C. (1985) in *Scaling Phenonema in Disordered Systems* (R. Pynn and A. Skjeltorp. eds.) p. 171 (Plenum Press, New York).

Winkler, A. (2002) *Ph.D thesis* (Johannes Gutenberg Universität Mainz).

Young, A.P. (1979) *Phys. Rev.* **B19**, 1855.

Zahn, K., and Maret, G. (2000) *Phys. Rev. Lett.* **85**, 3656.

Zahn, K., Lenke, R., and Maret, G. (1999) *Phys. Rev. Lett.* **82**, 2721.

Zallen, R. (1983) *The Physics of Amorphous Solids* (Wiley, New York).

Ziman, J.M. (1979) *Models of disorder. The theoretical physics of homogeneously disordered systems* (Cambridge University Press, Cambridge).

Chapter 3

Models of Disordered Structures

3.1 Random Walks: A Simple Model for the Configurations of Flexible Polymers

Solid polymers are ubiquitous in daily life: plastic bags, containers, toys, compact discs, the keyboards and cases of computers, phones, etc. Nevertheless, the understanding of the structure of these materials on the atomic scale is a complicated problem, and here we will present only a very simplified view, restricting attention to flexible neutral homopolymers with linear (as opposed to branched) architecture. "Homopolymer" means that the building blocks of the polymer which are repeated along the chains, i.e. the "monomers", are all identical (apart, of course, from the endgroups; e.g., in polyethylene, C_nH_{2n}, the $(n-2)$ inner monomers are CH_2 groups, while at the chain ends one has CH_3 groups). Although special polymerization techniques allow also to form copolymers with two (or more) chemically different monomers, such systems will not be considered, neither the so-called "ring polymers", which, by forming closed loops, have no chain ends, nor the so-called "star-polymers" which are single stranded polymers that are tethered together at a special chemical group. Thus the typical examples of the polymers that we will discuss here are polyethylene or polystyrene (see Fig. 3.1 for a sketch of the respective monomers) or polycarbonate. The degree of polymerization, n, is typically of the order $10^3 \leq n \leq 10^5$. In real life, polymer melts and glasses do not contain chains of completely uniform "chain length" (i.e., uniform degree of polymerization), but rather one finds a distribution $P(n)$. This spread of chain lengths is called "polydispersity", while a melt of chains with exactly uniform chain length is called "mono-disperse" (Flory 1953, des Cloizeaux and Jannink 1990, Strobl 1996, Rubinstein and Colby 2003, Graessley 2008).

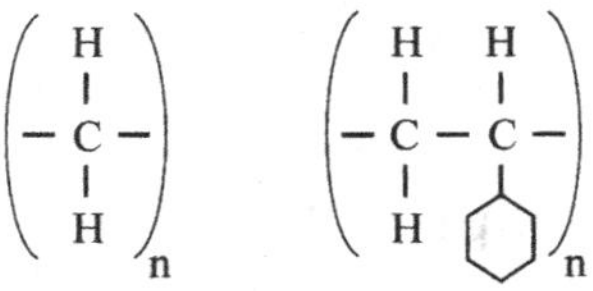

$$\left(-\overset{\displaystyle H}{\underset{\displaystyle H}{C}}- \right)_n \qquad \left(-\overset{\displaystyle H}{\underset{\displaystyle H}{C}}-\overset{\displaystyle H}{C}- \right)_n$$

Fig. 3.1 Repeat unit of polyethylene (left) and polystyrene (right). The hexagon stands for a benzene ring.

While on large length scales the structure of a polymer chain in a dense melt or amorphous solid is a statistical coil with random walk-like statistics, see Fig. 3.2, one has on small length scales a very pronounced short range order, in particular along the backbone of the chain, where the lengths of the covalent bonds are typically controlled by quite stiff potentials, as well as the bond angles between successive covalent bonds along the chain. Therefore it is not surprising that bond lengths as well as bond angles in a polyethylene melt have essentially the same values as in a polyethylene crystal.[1]

The configurational disorder present in the configuration of a polymer chain in the melt is due to the degeneracy between the various minima in the torsional potential that controls rotations of the bond $i + 2$ out of the plane formed by bonds i and $i + 1$ (labeling the successive bonds along the chain backbone from $i = 1$ to $i = n$). Figure 3.3 shows a schematic sketch of such a torsional potential: While the torsional angle $\phi = 0$ is energetically preferred (called "trans state configuration") and hence the ground state of the polyethylene chain is the linear zigzag structure, at nonzero temperature there is also a nonzero probability to find the angles near the "gauche+" ($g+$), $\phi = 2\pi/3$ and "gauche–" ($g-$), $\phi = -2\pi/3$ minima (Fig. 3.3).

In the literature one finds various simplified models differing in the level of chemical detail that is accounted (Flory 1953, 1969; de Gennes 1979; des Cloizeaux and Jannink 1990; Grosberg and Khokhlov 1994). The simplest (and most abstract) model is the "freely jointed chain", in which one considers n bonds of a fixed length ℓ, and treats the bond angles Θ_i as well as the torsional angles ϕ_i as completely random. A slightly more realistic model chooses the bond angles Θ_i as being fixed at the location of

[1] Polymers with large degree of polymerization are of course very difficult to crystallize, since there are huge free energy barriers to cross in a transition of a polymer chain from a random coil-like configuration to a stretched out linear zigzag-type configuration, as would be required for a single crystal. In practice one therefore usually rather finds semi-crystalline polymers, i.e. thin crystalline lamellae with amorphous regions in between. See Strobl (1996) for more details.

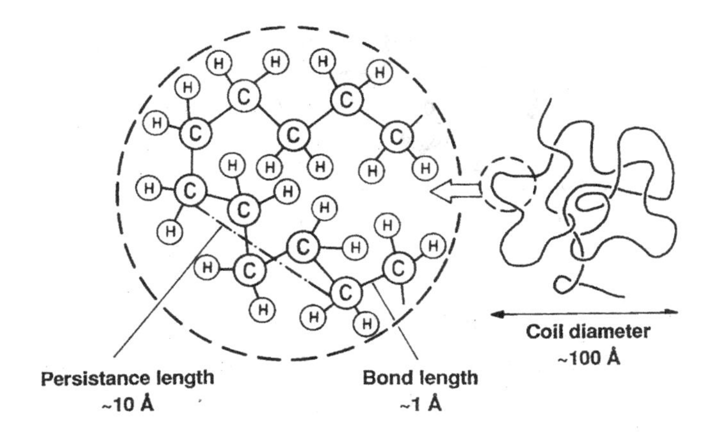

Fig. 3.2 Hierarchy of length scales for a polymer in a dense melt, using polyethylene as an example. The "persistence length" describes the range of bond orientational order along the backbone of a chain.

the minimum of the bond angle potential, the torsional angles still being arbitrary. We also can take the torsional potential very simply into account, if we assume that successive torsional angles are statistically completely independent from each other (of course, this cannot be strictly true, because then monomers could "sit" on top of each other, which is energetically very unfavorable due to repulsive forces, the so-called "excluded volume"-interaction). If we nevertheless make this approximation, we can write the conformational part of the Hamiltonian and the partition function of the chain as

$$\mathcal{H}_{\rm conf} = \sum_{i=1}^{n} U_{\rm rot}(\phi_i), \quad Z_n = (Z_{\rm mon})^n \,, \tag{3.1}$$

where the partition function of a monomer is

$$Z_{\rm mon} = \frac{1}{2\pi} \int_{-\pi}^{+\pi} d\phi \, \exp[-U_{\rm rot}(\phi)/k_B T] \approx 1 + 2\exp(-\Delta U_g/k_B T) \,. \tag{3.2}$$

In Eq. (3.1), chain end effects are ignored, and in Eq. (3.2), we have assumed a potential of the form shown in Fig. 3.3, with 2 excited gauche minima, higher in energy by an amount ΔU_g (labeled ΔU in the figure) than the central t-minimum with $\phi = 0$. Equation (3.1) reduces the partition function of the chain to the partition functions of the monomers, in

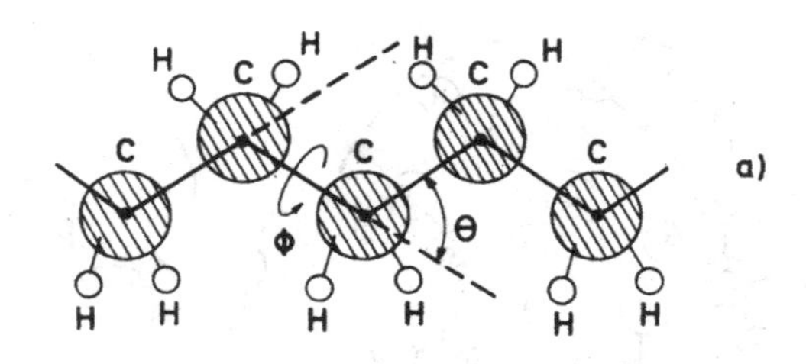

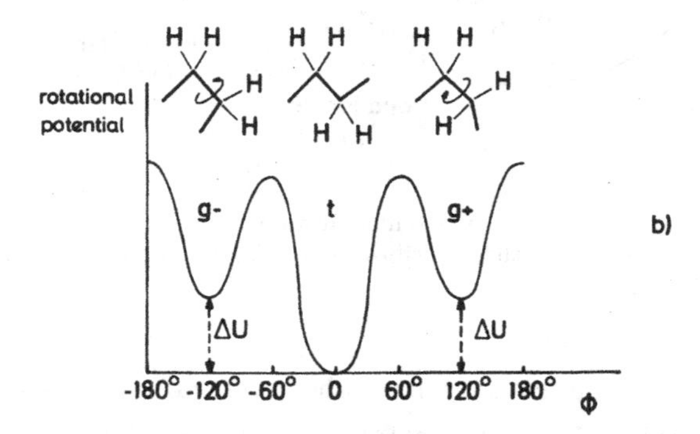

Fig. 3.3 a) Schematic model of a piece of polyethylene. Carbon atoms are shown as shaded circles, Hydrogen atoms as small white circles, and covalent bonds as straight lines. The definition of the bond angle Θ and of the torsional angle ϕ is indicated. Note that in the "all trans" configuration, where the third successive C-C bond also lies in the plane defined by the two previous C-C bonds, one has $\phi = 0$. b) Qualitative sketch of the torsional rotational potential for alkane chains, indicating the three energetically preferred states, gauche minus ($g-$), trans (t), and gauche plus ($g+$). The minimum of the trans configuration is deeper by an energy ΔU.

analogy to other problems where there are no interactions (ideal gases, ideal paramagnets, etc.). In addition to Z_n and the corresponding free energy $F_n = -k_B T \ln Z_n$, it is also interesting to consider the typical linear dimensions of the polymer coil. As a first example, consider the mean squared end-to-end distance $\langle R^2 \rangle$ of a freely jointed chain formed by n bonds of length ℓ, where $\langle \cdots \rangle$ denotes the thermal average. We obtain thus

$$\vec{R} = \sum_{i=1}^{n} \vec{\ell}_i, \quad \vec{\ell}_i = \vec{r}_i - \vec{r}_{i-1}, \quad |\vec{\ell}_i| = \ell \text{ for all } i\,, \tag{3.3}$$

$$\langle R^2 \rangle = \sum_{i=1}^{n} \langle \vec{\ell}_i^{\,2} \rangle + \sum_{i \neq j} \langle \vec{\ell}_i \cdot \vec{\ell}_j \rangle = n\ell^2\,, \tag{3.4}$$

since $\langle \vec{\ell}_i \cdot \vec{\ell}_j \rangle = 0$ for $i \neq j$ for a freely jointed chain.

This result can easily be generalized to models with rigid bond angles $\Theta_i = \Theta$. For this it is convenient to define a transformation tensor $\underline{T_j}$ which rotates the vector $\vec{\ell}_j$ into the direction of $\vec{\ell}_{j+1}$, i.e., $\vec{\ell}_{j+1} = \underline{T_j}\vec{\ell}_j$ (Flory 1969). From this it follows that for $j > i$ one has

$$\vec{\ell}_i \cdot \vec{\ell}_j = \vec{\ell}_i \cdot \underline{T_{j-1}}\,\vec{\ell}_{j-1} = \cdots = \vec{\ell}_i \underline{T_{j-1}}\;\underline{T_{j-2}} \cdots \underline{T_i}\vec{\ell}_i\,. \tag{3.5}$$

Since in this model successive transformations are not correlated, we can factorize the averages needed to calculate $\langle R^2 \rangle$ as follows:

$$\begin{aligned}\langle \vec{\ell}_i \cdot \vec{\ell}_j \rangle &= \langle \vec{\ell}_i \cdot \underline{T_{j-1}T_{j-2}} \cdots \underline{T_i}\vec{\ell}_i \rangle = \vec{\ell}_i \cdot \langle \underline{T_{j-1}} \rangle \langle \underline{T_{j-2}} \rangle \cdots \langle \underline{T_i} \rangle \cdot \vec{\ell}_i \\ &= \vec{\ell}_i \cdot (\langle \underline{T} \rangle^{|j-i|}) \cdot \vec{\ell}_i\,. \end{aligned} \tag{3.6}$$

Disregarding chain end effects, the resulting geometric series is summed from $k = 1$ to infinity and thus we obtain from Eq. (3.4), using $\sum_{i \neq j} = 2 \sum\limits_{j>i}$,

$$\langle \vec{R}^2 \rangle = \sum_{i=1}^{n} \left[\vec{\ell}_i^{\,2} + 2\vec{\ell}_i \cdot \sum_{k=1}^{\infty} \left\{ \langle \underline{T} \rangle^k \right\} \cdot \vec{\ell}_i \right] = n\vec{\ell}_1 \frac{\underline{1} + \langle \underline{T} \rangle}{\underline{1} - \langle \underline{T} \rangle} \vec{\ell}_1\,. \tag{3.7}$$

Comparing Eq. (3.4) and Eq. (3.7), we see that the result for the freely jointed chain gets enhanced by a numerical factor C_∞,

$$\langle \vec{R}^2 \rangle = C_\infty n \ell^2 \quad \text{with} \quad C_\infty = \frac{\underline{1} + \langle \underline{T} \rangle}{\underline{1} - \langle \underline{T} \rangle}\,. \tag{3.8}$$

In order to calculate C_∞ one thus has to average the matrix $\underline{T_i}$

$$\underline{T_i} = \begin{pmatrix} \cos\Theta_i & \sin\Theta_i & 0 \\ \sin\Theta_i \cos\phi_i & -\cos\Theta_i \cos\phi_i & \sin\phi_i \\ \sin\Theta_i \sin\phi_i & -\cos\Theta_i \sin\phi_i & -\cos\phi_i \end{pmatrix} \tag{3.9}$$

over the torsional angles ϕ_i. Using $U_{\text{rot}}(\phi_i) = U_{\text{rot}}(-\phi_i)$, which implies that $\langle \sin\phi_i \rangle = 0$, one obtains

$$\begin{aligned}\langle \underline{T} \rangle &= \int_{-\pi}^{+\pi} \underline{T}_i(\phi) \exp[-U_{\text{rot}}(\phi)/k_B T] d\phi \Big/ \int_{-\pi}^{+\pi} \exp[-U_{\text{rot}}(\phi)/k_B T] d\phi \\ &= \begin{pmatrix} \cos\Theta & \sin\Theta & 0 \\ \sin\Theta \langle \cos\phi \rangle & -\cos\Theta \langle \cos\phi \rangle & 0 \\ 0 & 0 & -\langle \cos\phi \rangle \end{pmatrix}. \end{aligned} \tag{3.10}$$

From this one now derives easily that

$$C_\infty = \frac{1+\cos\Theta}{1-\cos\Theta} \; \frac{1+\langle\cos\phi\rangle}{1-\langle\cos\phi\rangle} \,. \tag{3.11}$$

It is clear that this model with separable energy is extremely simplified and therefore has some unphysical features. E.g. the model predicts that the energy of a configuration with two consecutive gauche states $(g+g+)$ is the same as the one for which these two states are replaced by a $g+g-$sequence, and therefore these two configurations have the same statistical weight. In reality one would, however, expect that the former configuration occurs less frequent than the latter, since the forth nearest neighbors should not approach each other too much due to the excluded volume interaction. One possibility to avoid this problem is to replace Eq. (3.1) by a model that allows for a nearest-neighbor interaction between successive bonds along a chain:

$$\mathcal{H}_{\text{conf}} = \sum_{i=1}^{n-1} U(\phi_i, \phi_{i+1}) \,. \tag{3.12}$$

Despite the added complexity the partition function as well as the mean squared end-to-end distance $\langle R^2\rangle$ of such a model can still be computed exactly using transfer matrix methods, as described in detail by Flory (1969). Particularly convenient is the use of the so-called "rotational isomeric state" (RIS) model, in which one restricts attention to discrete angles $\phi_i(=0, \pm 2\pi/3)$ instead of a continuous distribution of ϕ_i.

So far we have discussed only interactions that are short ranged. However, if we consider an isolated polymer chain (e.g. in a dilute solution, neglecting the explicit effects of the solvent molecules), it is clear that also excluded volume interactions between monomers which are rather remote from each other, as measured by the monomer index i along the chain backbone, will be very important if the chain folds back on itself so that the geometric distance between the considered monomers is small. Thus if we describe such polymer chains as random walks (RW) on a lattice by requiring that, as a consequence of the excluded volume interaction between monomers, each lattice site is occupied at most by one monomer, we obtain the so-called "self-avoiding walk" (SAW) model of a polymer chain. Although the excluded volume interaction is short range in real space, it is long range in a coordinate system that follows the backbone of the polymer chain. Therefore one finds that, for dimensions $d \leq 4$, $\langle R^2\rangle$ increases faster than linear in n for the SAW, as we will discuss in more detail below. In contrast to this it turns out, however, that in dense melts and solid amorphous polymers the excluded volume interaction between effective monomers is essentially screened out, and hence the statistical properties of polymers are qualitatively the same as those of a simple random walk (RW) . For the latter case the central limit theorem of probability theory can immediately be applied to infer that the distribution function of the end-to-end vector $\vec{R}$ is a gaussian,

$$P(\vec{R})d\vec{R} = A\exp[-b^2R^2]d\vec{R}\,, \tag{3.13}$$

where A is a normalizing constant, and the constant b can be related directly to the value of $\langle R^2 \rangle$. In particular one finds for $d = 3$

$$\langle R^2 \rangle = \int R^2 P(\vec{R})d\vec{R} = 4\pi A \int_0^\infty R^4 \exp(-b^2R^2)dR\,, \tag{3.14}$$

and since by definition of A we have $4\pi A \int_0^\infty R^2 \exp(-b^2R^2)dR = 1$, one obtains:

$$b^2 = 3/(2\langle R^2 \rangle) \qquad (= d/(2\langle R^2 \rangle) \text{ in } d \text{ dimensions})\,. \tag{3.15}$$

From this result it follows that the position at which $\rho(R) = 4\pi R^2 P(R)$ has its maximum and the width of this distribution are of the same order of magnitude, since both of them are proportional to $\langle R^2 \rangle^{1/2}$. Thus the (relative) fluctuations of $\vec{R}^2$ around its mean value $\langle R^2 \rangle$ do not become small in comparison with $\langle R^2 \rangle$ itself, even if $n \to \infty$. It is also of interest to note that these considerations not only apply to the end-to-end distance of the chain, but also to interior distances $\vec{r}$ between monomers that are $m \gg 1$ steps along the chain apart. Thus

$$\langle r^2 \rangle = C_\infty m\ell^2\,, \tag{3.16}$$

where the constant C_∞ is the same as in Eq. (3.8) [while $C_\infty = 1$ for a freely jointed chain]. It turns out that for properties that concern large distances between monomers one can always absorb the factor C_∞, which contains the microscopic information on the bond angle Θ, torsional potential $U_{\text{tot}}(\phi)$, cf. Eqs. (3.8) - (3.10), by redefining the chain of n bonds of length ℓ as effective freely-jointed chain of n' bonds of length ℓ',

$$n'\ell' = n\ell, \quad \langle R^2 \rangle = C_\infty \ell^2 n = \ell'^2 n' = \ell'(n\ell)\,, \tag{3.17}$$

i.e.

$$\ell' = C_\infty \ell\,. \tag{3.18}$$

Of course, one can also allow for fluctuations in the lengths ℓ' of the effective chain segments. Such random walks are then clearly very similar to the random walks that are familiar from the observation of the motions of Brownian particles: The index i then does not label a position along the backbone of a chain molecule, but rather the tick of a clock. While no physical principle forbids that the trajectory of a Brownian particle intersects itself, the situation is clearly different for a polymer chain in solutions. Thus, for a self-avoiding walk model one finds that $P(\vec{R})$ for $d \leq 4$ deviates significantly from a gaussian distribution, and one has a nontrivial exponent ν that characterizes the mean-square end-to-end distance, see de Gennes (1979) and des Cloizeaux and Jannink (1990):

$$\langle R^2 \rangle \propto n^{2\nu} \text{ with } \nu = 3/4 \text{ for } d = 2 \text{ and } \nu \approx 0.588 \text{ for } d = 3 \,. \qquad (3.19)$$

For $d = 1$ it is obvious that the SAW degenerates to a rigid rod and we have $\langle R^2 \rangle = \ell^2 n^2$, i.e. $\nu = 1$, while for $d = 4$ one finds $\langle R^2 \rangle \propto n$, but with logarithmic corrections.

A second nontrivial exponent appears in the partition function: Note that for the model of Eqs. (3.1) and (3.2) with $\Delta U_g = 0$, i.e. all three choices of the angle are equally probable, we simply have $Z_{\text{mon}} = 3$, i.e. $Z_n = 3^n$. Analogously, if we consider a simple random walk on a lattice with coordination number $q \geq 3$, we have

$$Z_n^{\text{RW}} = q^n \,. \qquad (3.20)$$

For the self-avoiding walk, on the other hand, q is replaced by a nontrivial constant $q_{\text{eff}} \leq q - 1$, and in addition there is a power-law prefactor,

$$Z_n^{\text{SAW}} \propto n^{\gamma-1} q_{\text{eff}}^n \quad \text{for } n \to \infty \,. \qquad (3.21)$$

While the constant q_{eff} (just as the prefactors that are not even written in Eqs. (3.19) and (3.21)), are "nonuniversal" (i.e. they do depend on the chosen lattice, and other details that the model chain possibly could display, e.g. if step lengths different from a nearest neighbor spacing are admitted as well, etc.), exponents such as ν, Eq. (3.19), and γ, Eq. (3.21), are "universal", i.e. they depend only on the lattice dimensionality.

There have been various methods that have yielded information on these universal exponents. Historically, the first ones were based on extrapolation of exact enumerations of short SAW's and on simple sampling Monte Carlo methods (Sokal 1995). Let us briefly mention the salient features of these methods. For the exact enumeration methods, we first note that the ratio of partition functions can be written as

$$Z_n^{\text{SAW}}/Z_{n-1}^{\text{SAW}} \underset{n\to\infty}{=} [n/(n-1)]^{\gamma-1} q_{\text{eff}} \approx (1 + \frac{\gamma-1}{n}) q_{\text{eff}} \,. \tag{3.22}$$

This result suggests that a plot of $Z_n^{\text{SAW}}/Z_{n-1}^{\text{SAW}}$ versus $1/n$ yields both estimates for q_{eff} and for $(\gamma-1)q_{\text{eff}}$, from the intercept and the slope of the straight line that one should be able to "fit" to the numerical results for these ratios for large enough n.

Within the framework of simple sampling Monte Carlo method one gives up the idea of systematically enumerating SAW's up to a given order $n_{\max}$ (obviously the computational effort to do this increases exponentially, since it is proportional to $\exp[n \ln(q_{\text{eff}})]$) but instead one generates only a statistical sample of $\mathcal{N} \gg 1$ such walks, using random numbers. One constructs each SAW configuration step by step: forbidding immediate folding back of the path, one chooses one of the remaining $q-1$ directions on the lattice at random. If this choice would lead to a configuration that violates the excluded volume constraint, the whole construction attempt is abandoned, and one starts it all over again. Although this "simple sampling" method deteriorates in efficiency also exponentially with n, namely proportional to $[q_{\text{eff}}/(q-1)]^n$, the method has been used with some success. Today, much more useful "importance sampling" Monte Carlo algorithms exist, as reviewed by Sokal (1995).

An analytical understanding of the SAW exponents is best gotten from the renormalization group approach (des Cloizeaux and Jannink 1990, Schäfer 1999). However, in the context of the structure and dynamics of amorphous materials this really would be a side track of the topic of this book, and moreover many excellent introductions to the renormalization group can be found in the literature (Cardy 2000, Schäfer 1999). Therefore, we will here not describe this approach, but mention only the simple argument by Flory (1953), which has subsequently been generalized by Fisher (1969) to general dimensionality d.

One starts by considering the restricted partition function $Z_n(R)$ which yields the number of configurations of the chain at fixed $|\vec{R}| = R$,

$$Z_n(R) = \rho(R)\langle\exp(-U/k_BT)\rangle_R = S_d R^{d-1} P(\vec{R})\langle\exp(-U/k_BT)\rangle_R \,, \tag{3.23}$$

where S_d is the surface area of a d-dimensional unit sphere, and U represents the excluded volume interaction. Assuming that even in the case of excluded volume interaction $P(\vec{R})$, i.e. the distribution of the end-to-end vectors is given by a gaussian, $P(\vec{R}) \propto \exp(-dR^2/2n\ell^2)$, see Eqs. (3.13) and (3.15), we find

$$Z_n(R) \propto R^{d-1} \exp(-dR^2/2n\ell^2) \langle \exp(-U/k_BT) \rangle_R \,. \tag{3.24}$$

Thus the task is to estimate the effect of the excluded volume potential. We try to do this with a mean field type of approach only, similar to the treatment of short range interactions between particles in a fluid when one derives the van der Waals equation. Thus, one assumes that the monomers of the polymer are essentially distributed at a constant density $c = n/R^d$ in the sphere of volume R^d (all prefactors of order unity being suppressed here), and the average effect of the excluded volume interaction is of order $U \propto R^d c^2 \propto n^2/R^d$. This result is particularly plausible, if one replaces the excluded volume potential by a softer potential, describing a finite (rather than infinite) energy penalty if two monomers occupy the same volume region, which on average should have a similar effect. Thus we obtain from Eq. (3.24) that

$$Z_n(R) \propto R^{d-1} \exp(-dR^2/2n\ell^2) \exp(-n^2\varepsilon/R^d k_BT) \,, \tag{3.25}$$

where ε is a constant that is setting the energy scale. Hence the constrained free energy $F_n(R)$ becomes

$$\begin{aligned} F_n(R) &= -k_BT \ln Z_n(R) = \\ &-k_BT(d-1) \ln R + k_BT dR^2/(2n\ell^2) + n^2\varepsilon/R^d + \text{const}\,. \end{aligned} \tag{3.26}$$

Now the thermal equilibrium state of the coil is reached when $F_n(R)$ is at a minimum, $\partial F_n(R)/\partial R = 0$, and hence one obtains

$$\frac{R^2}{n\ell^2} = \frac{\varepsilon}{k_BT} \frac{n^2}{R^d} + \frac{d-1}{d} \,. \tag{3.27}$$

We now make the Ansatz that $R \propto n^\nu$ with $\nu > 1/2$. Then $R^2/n\ell^2 \propto n^{2\nu-1} \to \infty$ for $n \to \infty$, and hence in this limit the term $(d-1)/d$ in Eq. (3.27) is negligible. Then Eq. (3.27) yields

$$R = [\ell^2(\varepsilon/k_BT)]^{1/(d+2)} n^{3/(d+2)}, \quad \text{i.e. } \nu = 3/(d+2) \,. \tag{3.28}$$

It turns out that this result for ν is correct in $d = 1, 2$, and 4 ($\nu = 1$, $\frac{3}{4}$, and $\frac{1}{2}$, respectively) and in $d = 3$ ($\nu = 3/5$) it is a rather good approximation of the true value of 0.588. Of course, Eq. (3.28) also shows that for $d \geq 4$

the assumption $R \propto n^\nu$ with $\nu > 1/2$ is no longer true, and thus for $d \geq 4$ the term $(d-1)/d$ can no longer be neglected. In fact, for $d > 4$ the term involving the excluded volume interaction (ε/k_BT) is negligible for $n \to \infty$, while at the "marginal dimensionality" $d^* = 4$ both terms on the r.h.s. of Eq. (3.27) are of the same order of magnitude.

Thus, we see that (in $d = 2$ and $d = 3$ dimensions) an (isolated) polymer chain (or, alternatively, a polymer chain in a dilute solution so that interactions with other chains can be neglected) takes a random coil configuration that is "swollen" in comparison with a standard polymer. This swelling of the coil is no longer present in a dense polymer melt, however: Only an isolated chain can reduce its free energy $f_n(R) = U_n(R) - TS_n(R)$ by swelling the coil since $U_n(R)$ decreases more strongly than the loss of entropy (as long as $R < R_{\text{min}}$ where the minimum of $f_n(R)$ occurs). In a dense melt, however, most of the interaction energy (for $d > 2$) is not due to the interaction with monomers of the same chain, but due to interactions with monomers from other chains. E.g., in $d = 3$ the density of monomers belonging to a particular chain in the volume R^3 that it occupies is of the order of $\rho_{\text{self}} \approx n/R^3 \approx n^{-1/2}\ell^{-3} \to 0$ as $n \to \infty$. (Recall that for a random walk we have $R \propto n^{1/2}\ell$.) This density is negligibly small in comparison with the density of other chains. In fact, it turns out that the total energy is strictly independent of R, due to the so-called correlation hole effect: Near the center of gravity of a chain, the density of monomers of other chains is slightly reduced so that the same density is maintained irrespective of the fluctuations in the magnitude of R that still may occur. Even in $d = 2$, where coils cannot interpenetrate each other, a similar correlation hole effect has the consequence that $R \propto \ell n^{1/2}$ holds in dense systems. However, all these arguments do not imply that the prefactor in the relation between $\langle R^2 \rangle$ and n is simply given by $C_\infty \ell^2$, Eq. (3.8), the value obtained from the consideration of a single chain in the absence of excluded volume interactions, and numerical studies of various models have indeed shown this prefactor is actually different and is not understood in general. Thus the quantitative description of the structure of polymer coils in dense melts is a nontrivial problem and still a matter of research (Baschnagel *et al.* 2000a, Praprotnik *et al.* 2008, Baeurle 2009).

Of course, for a polymer in solution the description in terms of a SAW is qualitatively reasonable only for good solvent conditions. In fact, there exist also the so-called "Theta solvents" for which the excluded volume interactions are essentially canceled by the monomer-solvent interactions, and then the chain configuration is essentially the same as that of a gaussian random coil, with $\langle r^2 \rangle \propto n$ (apart from logarithmic corrections, which we disregard here). Also poor solvent conditions can occur, and then isolated chains in dilute solution take a collapsed globular configuration, $\langle r^2 \rangle^{1/2} \propto n^{1/d}$ ($= n^{1/3}$ in $d = 3$ dimensions.) In this case there is then no longer a distinction between d_f and d.

As a final topic of this section, we will now consider the "fractal dimensionality" (Mandelbrot 1982) d_f of random walks and self-avoiding walks. This is a geometric concept in which one relates the linear dimension of an object (in the present case the radius of a polymer coil) to the number of its basic building blocks (the number of monomers n in the present case). For this one makes the power-law Ansatz

$$n \propto R^{d_f} , \tag{3.29}$$

where the exponent d_f is the fractal dimension. For a simple random walk we always have $d_f = 2$, while for the SAW we have $d_f = 1/\nu \approx 5/3$ ($d = 3$) or $d_f = 4/3$ ($d = 2$). On the other hand, the "topological dimensionality" of random and/or self-avoiding walks is $d_t = 1$, of course, as appropriate for the description of "linear polymers". The fractal dimension $d_f > d_t$ describes the structure in which the considered object is embedded in the d-dimensional Euclidean space.

The concept of the "fractal dimension" (Mandelbrot 1982) is also useful for understanding why $d^* = 4$ is the marginal dimension for the excluded volume interaction between monomers in self-avoiding walks. Instead of focusing on the probability of self-intersection of a chain, one can consider the probability that two objects with fractal dimensions d_1, d_2 intersect if they are located in the same volume region of the embedding space (which has dimensionality d). The simple rule that emerges says that self-intersections can be neglected if $d_1 + d_2 < d$. E.g. a straight line has $d_f = 1$, and two straight lines in general intersect if $d = 2$ but do not if $d = 3$. Similarly, a flat plane has $d_f = 2$, and in $d = 3$ two planes do generally intersect along a line. Thus one can argue that random walks, which have $d_f = 2$, will intersect, i.e. if they represent polymers they will have a strong interaction, if $d = 3$, but the interaction will be negligible in $d = 5$, while $d = 4$ is the marginal case. The rule of thumb which results from this consideration is

$$2d_f^{\mathrm{MF}} = d^* . \tag{3.30}$$

Here we have emphasized that it is the "mean field" (MF) value of the fractal dimensionality d_f which has to be used (this value d_f^{MF} always applies for $d > d^*$, i.e. $d_f^{\mathrm{MF}} = 2$ for SAW's). Equation (3.30) also applies for other geometrical objects such as randomly branched polymers and the related "lattice animals", percolation clusters, etc., and we will make use of such arguments in later sections.

The "fractal dimensions" d_f is also useful for the interpretation of scattering experiments. It turns out that d_f can be directly measured in a suitable region of momentum transfer in elastic scattering (see also Sec. 2.1). In order to verify this statement, we consider the single-chain scattering function $S_s(q)$ of a polymer that follows the gaussian statistics applicable to simple random walks (Strobl 1996):

$$\begin{aligned}\frac{1}{n}S_s(\vec{q}) &= \frac{1}{n^2}\sum_{i,j}\langle\exp[-i\vec{q}\cdot(\vec{r}_i-\vec{r}_j)]\rangle = \frac{1}{n^2}\sum_{i,j}\exp\left\{-\frac{1}{2}\langle[\vec{q}\cdot(\vec{r}_i-\vec{r}_j)]^2\rangle\right\}\\ &= \frac{1}{n^2}\sum_{i,j}\exp\left[-\frac{1}{6}q^2\langle(\vec{r}_i-\vec{r}_j)^2\rangle\right]\\ &= \frac{1}{n}+\frac{2}{n^2}\sum_{i<j}\exp\left[-\frac{1}{6}q^2\langle(\vec{r}_i-\vec{r}_j)^2\rangle\right]. \end{aligned} \tag{3.31}$$

Denoting now $j-i=k$, and using then $\langle(\vec{r}_i-\vec{r}_j)^2\rangle = k\ell^2$, we can transform the double sum in the last equation into a single sum by noting that each value of k occurs $n-k$ times:

$$\begin{aligned}\frac{1}{n}S_s(q) &= \frac{1}{n}+\frac{2}{n}\sum_{k=1}^{n}\left(1-\frac{k}{n}\right)\exp\left(-\frac{1}{6}q^2\ell^2k\right)\\ &\approx \frac{1}{n}+2\int_{1/n}^{1}\left(1-x\right)\exp\left(-\frac{1}{6}q^2\ell^2nx\right)dx\,,\end{aligned} \tag{3.32}$$

Here we have assumed that the sum can be transformed into an integral, using $x \equiv k/n$. With straightforward algebra one now obtains as a final result the so-called "Debye function"

$$\frac{1}{n}S_s(q) = \frac{2}{z^2}(e^{-z}-1+z), \quad \text{with } z \equiv n\ell^2q^2/6 = q^2\langle R^2\rangle/6 = q^2\langle R_g^2\rangle\,. \tag{3.33}$$

Here $\langle R_g^2\rangle$ is the mean squared gyration radius of the chain which is defined as

$$\langle R_g^2\rangle = \frac{1}{2n^2}\sum_{i,j}\langle(\vec{r}_i-\vec{r}_j)^2\rangle\,, \tag{3.34}$$

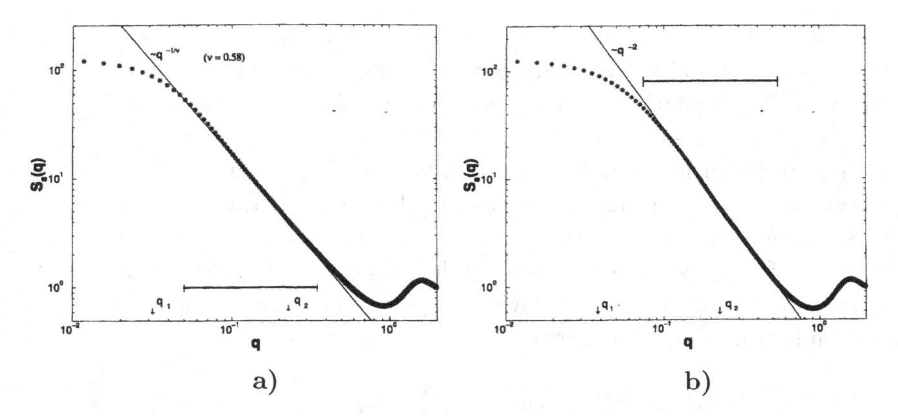

Fig. 3.4 a) Log-log plot of the normalized single chain structure factor $S_s(q)$ vs. q as determined from a computer simulation of a bead-spring model for a polymer. q is measured in units of $1/L$ where $L = 64$ is the linear dimension of the simulation box using the maximum bond length of a spring as length unit. The chain length is $n = 128$ and two densities of monomers, $\rho = 0.125$ (a) and $\rho = 2.0$ (b) are shown. Arrows show $q_1 = (2\pi \langle R_g^2 \rangle^{1/2})^{-1}$ and $q_2 = (2\pi\ell_0)^{-1}$, where $\ell_0 = 0.7$ is the average length of a spring between two consecutive effective monomers. The horizontal bar indicates the range over which a power-law $S_s(q) \propto q^{-1/\nu}$ was fitted to the data. Resulting estimates for ν are $\nu = 0.58$ (a) and 0.5 (b), respectively. From Yamakov *et al.* (1997).

from which it follows that for a gaussian chain one has $\langle R_g^2 \rangle = \langle R^2 \rangle / 6$. From Eq. (3.33) we can can see immediately that for $z \ll 1$ we have

$$S_s(q) \approx n(1 - z/3) = n(1 - \langle R_g^2 \rangle q^2/3), \quad z \ll 1, \tag{3.35}$$

while in the opposite limit $z \gg 1$ one has

$$S_s(q) \approx 2n/z = 12/(\ell^2 q^2), \quad z \gg 1. \tag{3.36}$$

The result for large q can also be written as

$$S_s(q) \propto q^{-d_f}, \quad \langle R_g^2 \rangle^{-1/2} \ll q \ll 1/\ell, \tag{3.37}$$

and in fact the power-law of Eq. (3.37) is not only found for gaussian random walks, but also for general fractal objects, and thus also the SAW (with $d_f = 1/\nu \approx 5/3$ in $d = 3$). Figure 3.4 shows typical results for the structure factor $S_s(q)$ of single chains in a dilute solution and a melt, as obtained from Monte Carlo simulations of a coarse-grained off-lattice model of flexible polymer chains (Yamakov *et al.* 1997). This model is a bead-spring model, where successive beads are coupled by a finitely extensible spring (with equilibrium extension $\ell_0 = 0.7$ and a maximum extension

$\ell_{\max} = 1$, the latter being the unit of length), while non-bonded beads interact with a short range potential of Morse type, to represent the excluded volume interaction between the beads representing the effective monomers. The figure shows that indeed for polymers in a dense melt the power-law $S_s(q) \propto q^{-2}$ for ordinary random walks, Eq. (3.36), is observed, while in dilute solution one finds the expected SAW behavior. Similar findings have also been made in corresponding experiments. Note that although in dilute solution the contrast between solvent molecules and monomers allows a rather direct measurement of $S_s(q)$, this is not the case in melts. In order to obtain information on the coherent scattering from all the monomers of a single chain, rather than from all monomers of all chains in the melt, one uses neutron scattering in which a small fraction of chains is deuterated while the majority of the chains is left protonated. Since hydrogen atoms for neutrons are a strong incoherent scatterer whereas they are only a weak coherent scatterer, the only coherent scattering in the system that is observed is due to the small fraction of deuterated chains (since deuterons are a relatively strong coherent scatterer). Thus as in the dilute solution, interference effects between different chains that are non-interacting vanish, and hence in this way one can measure the single-chain structure factor $S_s(q)$ of a polymer in a melt (Kirste *et al.* 1975). These experiments proved compellingly that polymers in a melt behave indeed like simple random walks, and hence decided a longstanding controversy whether or not one should have already some nanocrystalline order in polymer melts (typically there is no such order).

It is of course evident from Fig. 3.4 that the power-law behavior from Eqs. (3.36) and Eq. (3.37) holds only for an intermediate range of q, or – equivalently – for an intermediate range of length scales. Only for length scales much larger than the distance ℓ between subsequent effective monomers along the chain, and much smaller than the gyration radius $\langle R_g^2\rangle^{1/2}$, is the structure of the polymer coil "self-similar", and it is precisely this self-similarity, absence of a characteristic length scale, that is reflected in the power law, Eqs. (3.36) and (3.37). Self-similarity means, that a picture on a large scale and a magnification of a picture of the object on a small scale look just the same, at least qualitatively. Polymer coils are a good example of such self-similar objects, characterized by a fractal dimensionality d_f, and further examples (such as percolation clusters at the percolation threshold) will be encountered later in this chapter.

It must also be emphasized that no remnant of this fractal structure of a single polymer is seen when we study the coherent structure factor $S(q)$

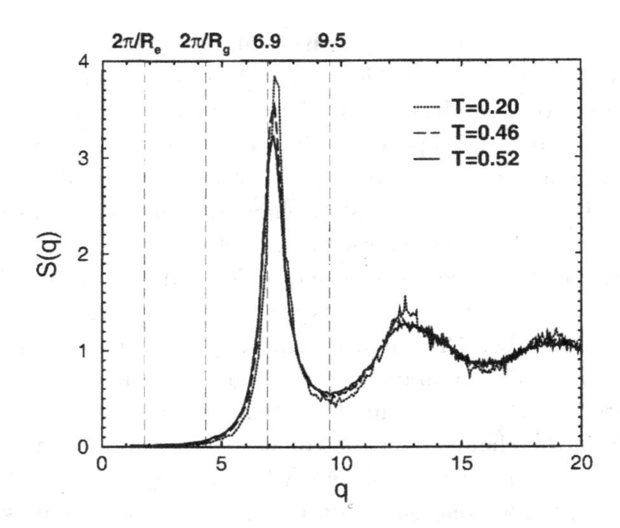

Fig. 3.5 Plot of the collective static structure factor $S(q)$ for an off-lattice bead-spring model of polymers with $n = 10$ versus q at three temperatures. Beads interact with a Lennard-Jones potential. Temperature and length are measured in units of the Lennard-Jones parameters. From Baschnagel *et al.* (2000b).

of the total polymer melt or glass (obtained from neutron scattering of a melt where all chains are deuterated, Fig. 3.5, or from X-ray scattering. Indeed, since the density of monomers of a chain inside its own volume in $d = 3$ is very small, $\rho \propto n^{-1/2}$, as argued above, it is clear that in a melt the coils interpenetrate each other strongly. The computer simulation snapshot picture (Fig. 3.6) gives a visual impression of this fact. Since the coherent scattering from all the monomers does not distinguish whether monomers belong to the same chain or to different chains, $S(q)$ for a dense melt looks qualitatively very similar to the structure factor of any other dense fluid.

3.2 Percolation: A First Example of a Fractal Structure

3.2.1 *The Percolation Probability and Percolation Threshold*

As already mentioned in Chap. 1, percolation theory deals with the geometry of disorder on lattices (Deutscher *et al.* 1983, Kirkpatrick 1973, Stauffer and Aharony 1992, Bollobás and Riordan 2006). The concepts developed here can also be carried over to off-lattice systems, the so-called "continuum percolation" (Shante and Kirkpatrick 1971, Scher and Zallen

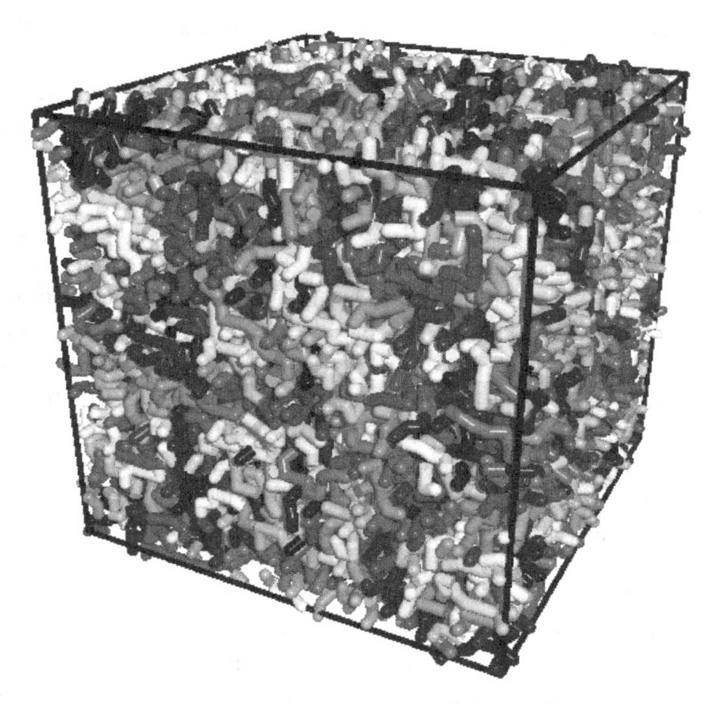

Fig. 3.6 Snapshot picture of a configuration of polymers in a melt, using the so-called bond fluctuation model of polymers on the simple cubic lattice, for $n = 20$, and a $L \times L \times L$ simulation box (indicated by the frame) with $L = 40$ lattice spacings, and periodic boundary conditions. Effective monomers (not shown) block all 8 sites of an elementary cube of the lattice. Bonds are indicated as thin cylinders.

1970), although here this extension will not be considered in detail. As we will see later, concepts from percolation theory find various applications in the theory of disordered solids and the theory of the glass transition. Specific applications to the physics of disordered matter include dilute ferromagnets, granular amorphous metal films on surfaces and granular superconductors, viscosity of gelling fluids and elasticity of networks, fluid flow through porous rocks, etc. (Sahimi 1994). However, one of the main reasons for the great popularity of this very active field is not just the large number of these applications, but the fact that the percolation problem has the character of a model for which various methods for the statistical mechanics of disordered matter can be developed and tested. In addition, the percolation transition is a very simple and pedagogically useful example for a phase transition, which in this context is described purely geometrically.

Concepts from percolation theory turn out to be very useful also for the interpretation of the thermal phase transitions (Fisher 1967, Kasteleyn and Fortuin 1969, Binder 1976, Coniglio and Klein 1980). Finally, the incipient percolating cluster at the percolation threshold is an excellent example for a statistical fractal (Mandelbrot 1982, Feder 1988, Stanley and Ostrowsky 1986, 1990, Bunde and Havlin 1991).

A key concept in this field is the "percolation cluster", since the size distribution of these clusters near the percolation threshold (only clusters of finite size occur on one side of this threshold, while on the other side a finite fraction of all sites or bonds of the lattice belong to the percolating cluster) also allow us to introduce the notion of critical exponents. Figure 1.9 of Chap. 1 can already serve as an illustration of "site percolation": Consider a regular lattice of infinite size and occupy the sites of this lattice at random with probability p. Unlike the problem of "correlated percolation" (Coniglio and Klein 1980), the occupation probability of a lattice site is completely independent of the occupation probability of its neighbors. Analogously, "bond percolation" means that one randomly puts bonds (with probability p) on the links of the lattice that connect nearest neighbor sites. In both variants it is useful to introduce "clusters", which are defined in terms of a neighborhood criterion: In the site percolation problem, every occupied site in a particular cluster must have an occupied neighbor site that belongs to the same cluster, and no site in the cluster can be a neighbor of an occupied site belonging to a different cluster. Figure 3.7 shows a few examples of clusters on the square lattice.

In the example shown in Fig. 3.7a, only 13 out of 64 sites are occupied and this means one is far below the percolation threshold p_c ($p_c \approx 0.592746$ for the site percolation on the square lattice; $p_c = 1/2$ holds exactly for bond percolation, as one can prove via the so-called "duality transformations", see Kirkpatrick 1973). Thus, when we consider the total number $N_s(p)$ of clusters of size s, we find that (on average) $N_s(p)$ decreases rapidly for increasing s as long as $p \leq p_c$. In order to allow to take a sensible thermodynamic limit, $N \to \infty$, where N is the total number of lattice sites, it makes sense to define a cluster concentration $n_s(p) = N_s(p)/N$. We refer to a cluster containing s occupied sites as a "$s-$cluster".

Near p_c a typical realization of a large but finite lattice will contain a large cluster that contains a connected path ranging from one boundary of the lattice to the opposite one. We call this path a "percolating path" and the cluster that contains this path a "percolating cluster". Obviously, in the limit $N \to \infty$ percolating clusters must have $s \to \infty$ occupied sites. If

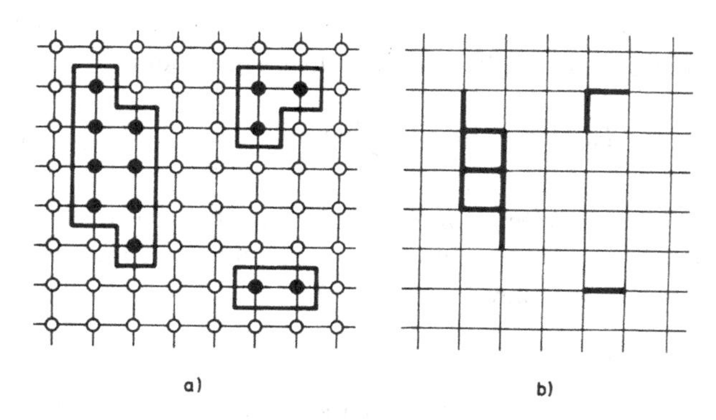

Fig. 3.7 a) Clusters on the square lattice for the site percolation problem. Lattice sites are occupied with the probability p (filled dots) or empty with probability $1-p$ (open circles). Neighboring occupied sites form clusters. The 8×8 lattice shown here contains three clusters, with $s = 2, 3$, and 8 occupied sites. b) Clusters on the square lattice for the bond percolation problem. Bonds are occupied with probability p (thick lines) or empty with probability $1-p$ (thin lines). Occupied bonds which are connected to each other form clusters. The lattice shown contains three clusters, with $s = 1$, $s = 2$ and $s = 9$ occupied bonds. From Binder and Heermann (1988).

such a cluster is present, we say that "percolation has occurred".

Let us suppose that $p < p_c$, so percolation has not yet occurred. Then a lattice site is either empty (with probability $1-p$) or it is one of the s sites of a $s-$cluster (with probability $sn_s(p)$). Making now use of the principle of probability theory that the probabilities of mutually exclusive events are additive, we find that the total probability, which must of course add up to unity, is $1 = 1 - p + \sum_{s=1}^{\infty} sn_s(p)$, i.e.

$$\sum_{s=1}^{\infty} sn_s(p) = p, \qquad p < p_c \,. \tag{3.38}$$

This sum rule simply expresses the fact that every occupied site must belong to some cluster, if we count sites that have no occupied neighbors as "unimers" ($s = 1$).

Note that Eq. (3.38) no longer holds if a percolating cluster is present. Instead one can define the percolation probability $P_\infty(p)$ as the ratio of the number of occupied sites in the percolation cluster relative to the number of all occupied sites ($= Np$) and conclude that the probability that any lattice site is part of a percolating cluster is $pP_\infty(p)$. Hence for $p > p_c$ we have instead of Eq. (3.38) that $1 = 1-p+\sum' sn_s(p)+pP_\infty(p)$, where $\sum'$ means that the percolating cluster (also referred to as "percolating network") is not counted in the summation. This sum rule hence yields a relation between the percolation probability and the cluster numbers,

$$P_\infty(p) = 1 - \frac{1}{p} \sum_{s=1}^{\infty}{}' s n_s(p) \,. \tag{3.39}$$

Of course, $P_\infty(p) \equiv 0$ for $p < p_c$. It turns out that the transition to nonzero $P_\infty(p)$ for $p > p_c$ is a continuous one, and $P_\infty(p)$ vanishes for $p \to p_c^+$ according to a power-law,

$$P_\infty(p) = \widehat{P}_\infty (p - p_c)^{\beta_p}, \quad p \to p_c^+ \,, \tag{3.40}$$

where $\widehat{P}_\infty$ is the critical amplitude of the percolation probability and β_p its critical exponent.

3.2.2 *Diluted Magnets and Critical Exponents*

The notation used in the following is similar to the one used for thermally driven second order phase transitions (Stanley 1971). Actually we can make this analogy even more explicit, if we return to the application of the site percolation problem to diluted ferromagnets (Fig. 1.9 of Chap. 1). Thus, we assume a mixed crystal where the lattice sites are randomly occupied by two atomic species, A and B, and only the B-atoms carry a magnetic moment. We also assume that exchange interactions between the B-atoms are ferromagnetic and restricted to nearest neighbors. Obviously, the percolation problem describes the magnetic ground state of this system in which all magnetic moments in a cluster are aligned parallel to each other, while the resulting total magnetic moments of different clusters orient independently of each other. (In reality, this is not quite true due to the occurrence of long range magnetostatic dipole-dipole interactions among the clusters, but this complication will here be ignored.) Hence for $p < p_c$ the system behaves like a "super-paramagnet" in which instead of single spins (carrying a magnetic moment μ) we have clusters of spins (carrying a magnetic moment $s\mu$), which can be oriented by applying an external magnetic field H.

For simplicity, we assume here that the system is magnetically uniaxial and hence can be described by the Ising model in which spins can only point up or down. Thus an isolated spin in a magnetic field has only two states with magnetic moments $\pm\mu$. The Zeeman energies are therefore $\pm\mu H$ and hence the partition function is $Z \equiv \exp(\mu H/k_B T) + \exp(-\mu H/k_B T)$. One thus obtains immediately that the average magnetization is $m =$

$[\mu \exp(\mu H/k_BT) - \mu \exp(-\mu H/k_BT)]/Z = \mu \tanh(\mu H/k_BT)$. Analogously also a s-cluster has only two states (with magnetic moments $\pm\mu s$) in the limit $T \to 0$, and hence $Z_s = \exp(s\mu H/k_BT) + \exp(-s\mu H/k_BT)$, $m_s = s\mu \tanh(s\mu H/k_BT)$. Normalizing the total magnetization $M(T, H)$ by its saturation value, which is $Np\mu$, we find for $p \le p_c$

$$m(T, H) = \sum_{s=1}^{\infty} N_s(p) s\mu \tanh(s\mu H/k_BT)/Np\mu \tag{3.41}$$

$$= \frac{1}{p} \sum_{s=1}^{\infty} n_s \, s \, \tanh(s\mu H/k_BT) \,. \tag{3.42}$$

For $p > p_c$ a percolating net occurs which has a spontaneous magnetization,

$$m(T, H) = \pm P_\infty + \frac{1}{p} {\sum_{s=1}^{\infty}}' n_s s \tanh(s\mu H/k_BT) \,, \tag{3.43}$$

where the sign in Eq. (3.43) has to be chosen corresponding to the sign of H. Thus, we immediately recognize that P_∞ plays the role of the spontaneous magnetization m_s,

$$m_s \equiv \lim_{H \to 0^+} m(T, H) = P_\infty \,. \tag{3.44}$$

The partition function of the super-paramagnetic system for $p < p_c$ is simply

$$Z = \prod_{s=1}^{\infty} \left[\exp\left(\frac{s\mu H}{k_BT}\right) + \exp\left(-\frac{s\mu H}{k_BT}\right) \right]^{N_s(p)} , \tag{3.45}$$

and hence the free energy f per lattice site becomes

$$f(T, H) = -k_BTN^{-1} \ln Z$$

$$= -k_BT \sum_{s=1}^{\infty} n_s(p) \ln[\exp(s\mu H/k_BT) + \exp(-s\mu H/k_BT)] \,. \tag{3.46}$$

Of course Eqs. (3.41) and (3.45) are consistent with the usual relation $M(T, H) = -N(\partial f/\partial H)_T$.

Without magnetic field, for $p < p_c$ we see that for the free energy of the superparamagnet we have a generalized ideal gas law,

$$F(T, 0) = -k_BT \ln 2 \sum_{s=1}^{\infty} n_s(p) \,. \tag{3.47}$$

Of particular interest is the susceptibility of the diluted ferromagnet. Remember that the susceptibility in zero field for the corresponding ideal paramagnet is $\chi_0 = \mu^2/k_BT$. Hence we obtain, for $p > p_c$ as well as for $p < p_c$,

$$\chi/\chi_0 = (1/\chi_0)(\partial\mu m/\partial H)_T = \frac{1}{p}\sum_{s=1}^{\infty}{}' s^2 n_s[1 - \tanh^2(s\mu H/k_BT)] . \quad (3.48)$$

We define the limiting value of this result for $H \to 0$ as the "percolation susceptibility χ_p", so that

$$\chi_p = \frac{1}{p}\sum_{s=1}^{\infty}{}' s^2 n_s . \quad (3.49)$$

At this point, it is useful to recall what is known about the critical behavior of standard ferromagnets (Stanley 1971). Both the spontaneous magnetization $m_s(T)$, the zero-field susceptibility $\chi(T, H = 0)$ as well as the singular part $f_{\text{sing}}(T, H = 0)$ of the free energy are characterized by power-laws near the critical temperature T_c, which define the critical exponents α, β, γ:

$$f(T, H) = f_{\text{reg}}(T, H) + f_{\text{sing}}(T, H), \quad f_{\text{sing}}(T, 0) \propto |T - T_c|^{2-\alpha}, \quad (3.50)$$

$$m_s(T) \propto (1 - T/T_c)^\beta , \quad (3.51)$$

$$\chi(T, H = 0) \propto |T - T_c|^{-\gamma} . \quad (3.52)$$

Further quantities of interest are the critical exponent defined by the variation of $m(T, H)$ along the critical isotherm,

$$m(T = T_c, H) \propto H^{1/\delta} , \quad (3.53)$$

and the spin pair correlation function

$$G_{\text{spin}}(R) = \langle S_0 S_{\vec{R}} \rangle , \quad (3.54)$$

which gives the conditional probability that the spin $S_{\vec{R}}$ at site $\vec{R}$ points up, if the spin S_0 at the origin points up (for simplicity this definition is given here for Ising spins $S_j = \pm 1$ but a generalization is straightforward). The behavior expected for the correlation function is, in d dimensions,

$$G_{\text{spin}}(R) \propto \begin{cases} \exp(-R/\xi), & T > T_c, \quad R \gg \xi \\ R^{-(d-2+\eta)}, & T = T_c \end{cases} . \quad (3.55)$$

Here ξ is the correlation length, while η is a critical exponent that describes the decay of the correlations at the critical point. Close to T_c the correlation length ξ diverges like a power-law with an exponent ν:

$$\xi \propto |T - T_c|^{-\nu} . \tag{3.56}$$

Equations (3.50)- (3.56) describe a phase transition of a ferromagnet driven by thermal disorder.

After this reminder of the properties of standard phase transitions, we return to the case of diluted magnets. Let us imagine to work with a sample of a diluted ferromagnet in which we can change the concentration p. For $T \to 0$ the system is then described by Eqs. (3.41)-(3.49), i.e. we encounter a magnetic transition between a ferromagnetic and a paramagnetic phase driven by a variation of the frozen-in geometrical disorder. The percolation probability P_∞ corresponds then to the spontaneous magnetization, Eqs. (3.39) and (3.44), and in analogy to the pure ferromagnetic system we expect that the various other quantities defined for the diluted ferromagnet exhibit a singular behavior similar to the behavior of the corresponding quantities at the thermal phase transitions:

$$f_{\text{sing}}(T,0) = -k_B T \ln 2 \Big[\sum_{s=1}^{\infty} n_s(p) \Big]_{\text{sing}} \propto |p - p_c|^{2-\alpha_p}, \qquad p \to p_c\,, \tag{3.57}$$

$$\chi_p(T, H=0) = \frac{1}{p} \sum_{s=1}^{\infty}{}' s^2 n_s \propto |p - p_c|^{-\gamma_p}, \qquad p \to p_c\,, \tag{3.58}$$

$$\Big[\sum_{s=1}^{\infty} s n_s(p = p_c) \exp(-hs) \Big]_{\text{sing}} \propto h^{1/\delta_p}, \quad h \to 0\,. \tag{3.59}$$

Note that in Eq. (3.59) we have used a simple $\exp(-hs)$ function instead of the hyperbolic tangent function that appears in Eq. (3.41), but of course for the character of the singular part in terms of a power law this difference is irrelevant. We also mention that often one uses percolation theory in a context that has nothing to do with diluted magnets, and therefore we also no longer imply that the quantity h has anything to do with a magnetic field. As a consequence, in the context of percolation, h is often denoted as the "ghost field", without any direct physical interpretation.

Finally one wishes to introduce also for the percolation problem a pair correlation function that plays an analogous role to the spin pair correlation, that we have considered in Eq. (3.54). Thus the correlation function that we are interested in is the conditional probability that a site at a distance $\vec{R}$ belongs to the same cluster as the site at the origin. This function is called the "pair connectedness function". In a diluted ferromagnet for $H = 0$ and $T \to 0$ the spin pair correlation function gives exactly the same result as the pair connectedness functions, since spins in the same cluster are aligned parallel with each other, while spins belonging to different clusters are uncorrelated. Thus we define, in complete analogy to Eqs. (3.54) -(3.56),

$$G(R) \propto \begin{cases} \exp(-R/\xi_p), & p < p_c, \quad R \gg \xi_p \\ R^{-(d-2+\eta_p)}, & p = p_c, \quad R \to \infty \end{cases}, \tag{3.60}$$

where ξ_p, the "correlation length" in the percolation problem, has an obvious geometrical interpretation: It describes the typical size of a percolation cluster. The exponent η_p characterizes the decay of $G(R)$ at p_c for an incipient percolating cluster (which dominates the behavior for $R \to \infty$ in comparison with all the small clusters that are also present). Again ξ_p shows for $p \to p_c^-$ a power-law

$$\xi_p \propto |p_c - p|^{-\nu_p}, \quad p < p_c, \quad p \to p_c^- . \tag{3.61}$$

Due to the very similar behavior of the percolation transition of diluted magnets and the thermal transition of pure magnets these two systems can be combined in a common phase diagram, see Fig. 3.8. We have a line of transitions with exponents $\alpha, \beta, \gamma, \cdots$ for the thermal transition ($p_c < p < 1$) and *different* exponents $\alpha_p, \beta_p, \gamma_p, \cdots$ for $p = p_c$, $T_c(p = p_c) = 0$. At the percolation transition there occur very interesting and nontrivial thermal excitations in the incipient percolating cluster since in fact this point can be considered as a special multicritical point. Note that the percolation exponents can only be observed by varying p at $T = 0$. Of course, one can cross the line $T_c(p)$ by varying p also at $T > 0$, but then one finds the critical exponents of the thermal transition (Stinchcombe 1983).

For $p > p_c$ Eq. (3.60) does of course not hold since even lattice sites that are arbitrarily far apart from each other are part of the same cluster, provided they belong to the percolating cluster. (Here we anticipate the result that in the limit of infinitely large lattices for $p > p_c$ a single percolating cluster occurs and not multiple ones.) The probability that an occupied site at $\vec{r}_1$ belongs to the percolating network is just the percolation probability $P_\infty(p)$, and the probability that another occupied site at $\vec{r}_2$ belongs to the percolating cluster is also $P_\infty(p)$. If $R = |\vec{r}_1 - \vec{r}_2| \gg 1$ these two events are statistically independent, and hence the probability that both occupied sites $\vec{r}_1$, $\vec{r}_2$ are part of the percolating cluster is simply

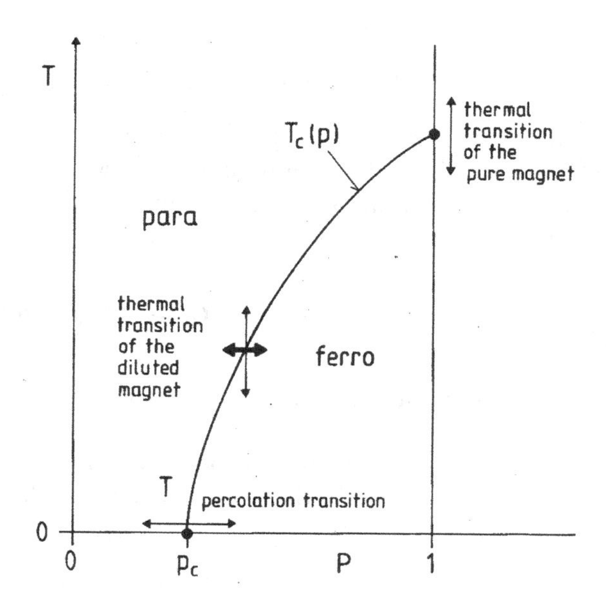

Fig. 3.8 Schematic phase diagram of a diluted ferromagnet, using temperature T and the concentration of ferromagnetic bonds p as variables. A line of critical points $T_c(p)$ separates the disordered paramagnetic phase (para) from the phase with ferromagnetic long range order (ferro). This line begins at the critical point of the pure system, $T_c(p = 1)$, and ends at the percolation threshold, $T_c(p = p_c) = 0$. If one crosses the line $T_c(p)$ at $T > 0$ one observes a thermal transition, with critical exponents that, for three-dimensional Ising systems, are believed to be universal along the whole line $T_c(p)$ as long as $p_c < p < 1$ (i.e., excluding the end-points since the diluted Ising ferromagnet belongs to a different "universality class" than the pure one). The geometric percolation transition is observed by varying p at $T = 0$. The various transitions are indicated by double arrows.

the product $P_\infty(p) \cdot P_\infty(p)$. Thus, we conclude

$$G(R) = [P_\infty(p)]^2, \quad R \to \infty, \quad p > p_c\,. \tag{3.62}$$

This result suggests that it makes sense to study for $p > p_c$ a reduced correlation function $G'(R)$ defined as

$$G'(R) \equiv G(R) - [P_\infty(p)]^2 \propto \exp(-R/\xi_p), \quad R \to \infty\,, \tag{3.63}$$

and one finds then that also the correlation length ξ_p satisfies Eq. (3.61). Of course, a similar reasoning applies to the thermal phase transition as well, where

$$G_{\text{spin}}(R) = \left[m_s(T)\right]^2, \quad R \to \infty, \quad T < T_c\,, \tag{3.64}$$

and

$$G'_{\rm spin}(R) \equiv \langle S_0 S_{\vec{R}} \rangle - [m_s(T)]^2 \propto \exp(-R/\xi), \quad R \to \infty, \qquad (3.65)$$

and now ξ in Eq. (3.65) also satisfies Eq. (3.56). Note that our treatment of correlation functions is still simplified since in Eqs. (3.55), (3.60), (3.63) and (3.65) we have only written down the leading exponential variation, ignoring the power-law prefactors in R that are present.

3.2.3 *The Fractal Dimensionality and the Concept of Finite Size Scaling*

We now discuss in more detail the physical content of the power-law decay of the correlation function at $p = p_c$, see Eq. (3.60). Let us consider the number n of sites within a (hyper-)sphere of radius R around a site of the largest cluster (i.e., an incipient percolating cluster) that also belongs to this cluster:

$$n = \int d\vec{r}\, [G(\vec{r})/P_R] \propto P_R^{-1} \int_0^R dr r^{d-1} G(r)\,, \qquad (3.66)$$

or, using Eq. (3.60),

$$n \propto P_R^{-1} \int_0^R r^{1-\eta_p} dr \propto P_R^{-1} R^{2-\eta_p}, \quad p = p_c\,. \qquad (3.67)$$

Here P_R is the probability that an occupied site within the hypersphere indeed belongs to the largest cluster (and not to any of the many smaller clusters that also are present and also are counted in $G(\vec{r})$). Anticipating the result, shown below, that P_R also depends on R via a power-law, namely $P_R \propto R^{-\beta_p/\nu_p}$, we thus obtain

$$n \propto R^{2-\eta_p+\beta_p/\nu_p} = R^{d_f}\,, \qquad (3.68)$$

where d_f is the fractal dimensionality of the incipient percolating cluster. For a compact object, we obviously would have $n \propto R^d$, so it is clear that we must have $d_f < d$ in order to have a fractal structure.

The justification for the relation $P_R \propto R^{-\beta_p/\nu_p}$ can be given, for instance, by a "finite size scaling" argument (Fisher 1971, Barber 1983, Privman 1990, Binder 1992). First we emphasize that there cannot be a sharp and well-defined percolation threshold if we consider a finite hypercubic lattice of linear dimension L since obviously the percolation probability is nonzero for all values of $p > 0$. (E.g., consider simply a one-dimensional

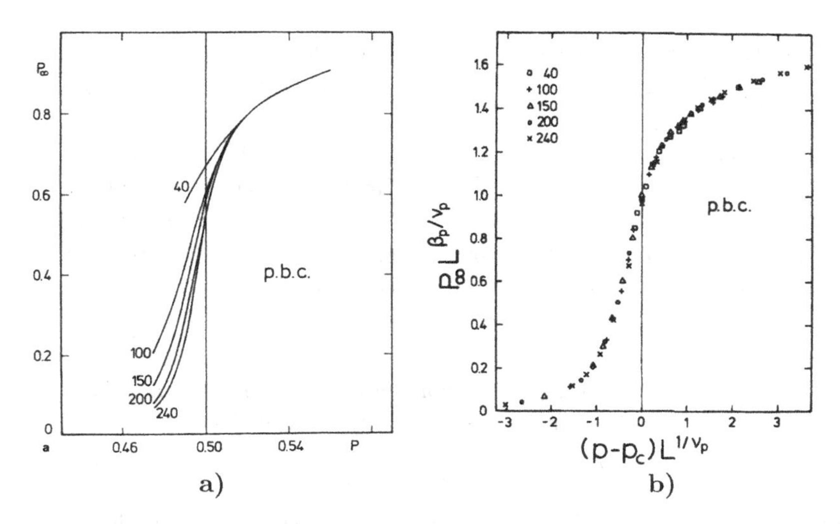

Fig. 3.9 a) Percolation probability $P_\infty(p)$ plotted vs. p, for the bond percolation problem on $L \times L$ square lattices, using periodic boundary conditions (p.b.c.). The curves correspond to different lattice sizes L (indicated in the figures). The vertical straight line indicates the percolation threshold p_c (=1/2) that shows up via the singular behavior described by Eqs. (3.40), (3.57- 3.61) only for $L \to \infty$. b) Same data as shown in a), but plotted in the rescaled form $L^{\beta_p/\nu_p} P_\infty$ vs. $(p - p_c)L^{1/\nu_p}$, with the following choice of exponents: $\beta_p = 0.139$, $\nu_p = 4/3$. From Heermann and Stauffer (1980).

chain of occupied sites spanning from one boundary to the opposite one. Such an object occurs with probability $P_{\text{chain}} \propto p^L = \exp(-L \ln p)$. Thus, the percolation probability for $0 < p \ll p_c$ is exponentially small but nonzero). This is also immediately seen from numerical simulations, see e.g. Fig. 3.9 (Heermann and Stauffer 1980): P_∞ decreases strongly as p approaches p_c from above, but varies completely smoothly through p_c. The only difference between $p < p_c$ and $p > p_c$ is that for $p < p_c$ the so-called "finite size tails" approach zero as L increases, while for $p > p_c$ they converge to a unique and nontrivial function, which ultimately exhibits the singular behavior of Eq. (3.40).

This rounding of the percolation transition by finite size effects can immediately be interpreted geometrically – recall that the diverging correlation length ξ_p, Eq. (3.61), sets the scale for the size of those percolation clusters that dominate the physical properties of interest (P_∞, χ_p, etc.). In a finite lattice of linear dimension L the growth of ξ_p is obviously bounded by L and thus we expect that the quantities P_∞, χ_p, etc. should be functions of the ratio L/ξ_p, that converge to their asymptotic behavior, Eqs. (3.40) and (3.58), only in the limit where $L/\xi_p \to \infty$. Thus, we write

$$P_\infty = (p - p_c)^{\beta_p}\, \widetilde{P}(L/\xi_p), \qquad \chi_p = (p - p_c)^{-\gamma_p}\, \widetilde{\chi}(L/\xi_p)\,, \tag{3.69}$$

where $\widetilde{P}(\zeta)$, $\widetilde{\chi}(\zeta)$ are so-called "scaling functions" with the behavior $\widetilde{P}(\zeta \to \infty) \to$ const., $\widetilde{\chi}(\zeta \to \infty) \to$ const.. In order that Eq. (3.69) is compatible with a smooth variation through p_c at finite L, it is clearly necessary that the singular prefactors in Eq. (3.69) are canceled by the small ζ behavior of the scaling functions. This is achieved by requiring

$$\widetilde{P}(\zeta) \propto \zeta^{-\beta_p/\nu_p}, \qquad \widetilde{\chi}(\zeta) \propto \zeta^{\gamma_p/\nu_p}\,. \tag{3.70}$$

From Eq. (3.69) we thus immediately conclude that for $p = p_c$ in a finite lattice singular power-laws result if we consider the variation with L,

$$P_\infty(p = p_c) \propto L^{-\beta_p/\nu_p}, \qquad \chi_p(p = p_c) \propto L^{\gamma_p/\nu_p}\,. \tag{3.71}$$

Of course, the Ansatz that P_∞ and χ_p depend on the two variables p, L in the special form involving this "homogeneity assumption" is clearly no more than a heuristic argument, but it is supported both by Monte Carlo simulation results (e.g. Fig. 3.9b) and by more rigorous theories, such as the renormalization group theory of critical phenomena (Fisher 1974, Cardy 2000). Eq. (3.71) then immediately yields our above statement that $P_R \propto R^{-\beta_p/\nu_p}$, since the exponents controlling these size effects should neither depend on the shape of the considered finite region (as long as it is compact), nor on the boundary conditions.

3.2.4 *Scaling of the Cluster Size Distribution*

A very illuminating way to discuss the critical phenomena associated with percolation relates them to the scaling behavior of percolation clusters (Stauffer 1979). For this one makes the assumption that the cluster concentration $n_s(p)$ depends for $p \to p_c$ and $s \to \infty$ on the two variables $s, p - p_c$ through a homogeneous function $F(\zeta)$,

$$n_s(p) = s^{-\tau}\; F[s^\sigma (p - p_c)]\,, \tag{3.72}$$

where σ, τ, are two critical exponents from which all other ones can be derived, as we will see below. Again, one can verify Eq. (3.72) very nicely by Monte Carlo simulations, and finite size effects – which in principle require to include, as in Eq. (3.69), the variable L/ξ_p – can be avoided by working with much larger lattices in the simulations than were used for Fig. 3.9, such as $L = 95000$ (Stauffer and Aharony 1992). Such large scale simulations have allowed to verify the power-law at $p = p_c$, $n_s(p) \propto s^{-\tau}$, over 5 decades in s $(10 \le s \le 10^6)$.

Inserting now Eq. (3.72) into Eq. (3.49) yields, for p close to p_c

$$\chi_p = \frac{1}{p}\sum_{s=1}^{\infty}{}' s^2 n_s(p) \approx \frac{1}{p}\int_0^{\infty} s^{2-\tau} F[s^\sigma(p-p_c)]ds\,. \tag{3.73}$$

Here we have replaced the sum by an integral, since the divergence of χ_p must be due to the fact that the sum in Eq. (3.73) for $p = p_c$ no longer is convergent – each individual term in the sum is in fact small, and one can show that $\tau > 2$. Now it is convenient to redefine the scaling function $F(s^\sigma(p-p_c)) \equiv f(x)$ with $x = s|p-p_c|^{1/\sigma}$, which thus gives, by changing the integration variable from s to x.

$$\chi_p = \frac{1}{p}|p-p_c|^{-(3-\tau)/\sigma}\int_0^{\infty} dx x^{2-\tau} f(x)\,. \tag{3.74}$$

The integral in Eq. (3.74) is just a finite nonzero constant, and hence $\chi_p \propto |p-p_c|^{-(3-\tau)/\sigma}$. Comparing with Eq. (3.58) we hence conclude that $\gamma_p = (3-\tau)/\sigma$.

We now want to discuss $P_\infty(p)$ with the help of the scaling assumption for the cluster size distribution. For this we start with Eq. (3.40) and extract the singular part by taking the derivative with respect to p:

$$\frac{d}{dp}[pP_\infty(p)] = p_c\beta_p\widehat{P}_\infty(p-p_c)^{\beta_p-1}, \quad p \to p_c^+\,. \tag{3.75}$$

On the other hand we obtain from Eqs. (3.39) and (3.72)

$$\frac{d}{dp}[pP_\infty(p)] = 1 - \sum_{s=1}^{\infty}{}' s\frac{dn_s(p)}{dp} = 1 - \sum_{s=1}^{\infty}{}' s^{1-\tau+\sigma}F'[s^\sigma(p-p_c)]\,, \tag{3.76}$$

where $F'(\zeta) = dF(\zeta)/d\zeta$ (note that $F' < 0$). Again we redefine the scaling function, $F'(s^\sigma(p-p_c)) \equiv g(s(p-p_c)^{1/\sigma})$, to find

$$\begin{aligned}\frac{d}{dp}[pP_\infty(p)] &= 1 - \int_0^{\infty} ds s^{1-\tau+\sigma} g[s(p-p_c)^{1/\sigma}] \\ &= 1 - (p-p_c)^{-1-\frac{2-\tau}{\sigma}}\int_0^{\infty} dx x^{1-\tau+\sigma} g(x)\,.\end{aligned} \tag{3.77}$$

The comparison of the singular part of Eq. (3.75) with (3.77) yields $\beta_p = (\tau-2)/\sigma$. With similar arguments one can also extract the singular parts of $f_{\rm sing}(T,0)$, Eq. 3.57, and of the response to the "ghost field", Eq. (3.59), and show that $2-\alpha_p = (\tau-1)/\sigma$, $1/\delta_p = \tau-2$. Consequently, we can write $\tau = 2+1/\delta_p$, $\sigma = 1/(\beta_p\,\delta_p)$ and express the other exponents in terms of β_p and δ_p. This yields the well-known "scaling laws" for the critical exponents, namely (recall that $\gamma_p = (3-\tau)/\sigma$, as stated after Eq. (3.74)

$$2-\alpha_p = \gamma_p + 2\beta_p = \beta_p(\delta_p+1)\,. \tag{3.78}$$

With a different reasoning (which is skipped here) one can also derive the so-called "hyper-scaling relation" involving the correlation length exponent (Stauffer and Aharony 1992)

$$d\nu_p = 2-\alpha_p\,. \tag{3.79}$$

Now we wish to establish more directly a relation with the fractal dimension, considering again a hypercubic lattice of linear dimension L. In order to have a cluster spanning from one boundary to the opposite one, it must have at least the size $s(L) \propto L^{d_f}$. The probability to find such a cluster in the considered systems is

$$L^d \int\limits_{s(L)}^{\infty} n_s ds \propto L^d \int\limits_{s(L)}^{\infty} s^{-\tau} ds \propto L^d[s(L)]^{1-\tau} \propto L^{d+d_f(1-\tau)}\,. \tag{3.80}$$

Now we must have a finite nonzero probability for such a spanning cluster when $L \to \infty$. Obviously this requires that the exponent in Eq. (3.80) vanishes, i.e.

$$d_f = d/(\tau-1) = d/[\sigma(2-\alpha_p)] = \beta_p\,\delta_p/\nu_p = \frac{\gamma_p+\beta_p}{\nu_p} = d - \frac{\beta_p}{\nu_p}\,, \tag{3.81}$$

where the scaling laws given by Eqs. (3.78) and (3.79) have been used. Using then a further scaling assumption for the pair connectedness function

$$G(\vec{R}) = R^{-(d-2+\eta_p)}\,\widetilde{G}(R/\xi_p), \qquad p \to p_c, \quad |\vec{R}| = R \to \infty \tag{3.82}$$

which contains the power-law of Eq. (3.60) as well as the exponential decay as special limiting cases, one can use the sum rule (note the analogy to $\chi = \sum\limits_R \langle S_0\, S_{\vec{R}}\rangle$)

$$\chi_p = \sum_{\vec{R}} G(\vec{R}) \underset{p\to p_c}{\longrightarrow} \int d\vec{R}\, G(\vec{R}) \propto \int dR\, R^{1-\eta_p}\widetilde{G}(R/\xi_p) \propto \xi_p^{2-\eta_p} \tag{3.83}$$

to show with Eqs. (3.58) and (3.61) that $\gamma_p = \nu_p(2-\eta_p)$. Using this scaling relation together with $d_f = 2-\eta_p+\beta_p/\nu_p$, Eq. (3.68), we recover Eq. (3.81), which demonstrates the self-consistency of the scaling description.

An alternative discussion of the fractal dimensionality starts from the following scaling Ansatz for the typical linear dimension for a s-cluster:

$$L_s = s^{1/d_f}\, \widetilde{L}[s^\sigma(p - p_c)]\,, \tag{3.84}$$

where $\widetilde{L}$ is a suitable scaling function. Now we can construct an effective correlation length not from an exponential decay, but from the second moment of the cluster size distribution function,

$$\xi_p^2 = \sum_{s=1}^{\infty} L_s^2 s^2 n_s(p) / \sum_{s=1}^{\infty} s^2 n_s(p)\,. \tag{3.85}$$

Converting once more sums into integrals, as it is appropriate for p near p_c, we find, using Eq. (3.72),

$$\xi_p^2 = \int_0^\infty ds\; s^{2-\tau+2/d_f}\, \widetilde{L}^2[s^\sigma|p - p_c|] F[s^\sigma|p - p_c|] / \int_0^\infty ds\; s^{2-\tau} F[s^\sigma|p - p_c|] \tag{3.86}$$

which yields

$$\xi_p^2 \propto |p - p_c|^{-2/(d_f\sigma)} = |p - p_c|^{-2\nu_p}\,. \tag{3.87}$$

Thus we see that Eq. (3.87) implies that $d_f = 1/(\nu_p\sigma)$, which is equivalent to Eq. (3.81).

3.2.5 *Percolation for Low and High Lattice Dimensions*

By now the reader should have developed a feeling for the derivation and use of scaling arguments – a type of reasoning that will be used in other contexts as well. It is, however, also of interest to consider simple limiting cases which can be treated exactly. The case of percolation in $d = 1$ is really trivial, since $p_c = 1$ for the linear chain. Each s-cluster has two empty sites next to its borders, and hence we have

$$n_s(p) = p^s(1 - p)^2, \tag{3.88}$$

while for $d > 1$ we have a nontrivial relation with the statistical problem of the so-called "lattice animals",

$$n_s(p) = \sum_t g_{\mathrm{s,t}} p^s (1 - p)^t, \tag{3.89}$$

where $g_{\mathrm{s,t}}$ is the number of configurations of clusters containing s occupied sites and t empty perimeter sites (separating the cluster from the remaining lattice sites, which can either be occupied or empty). No full analytic

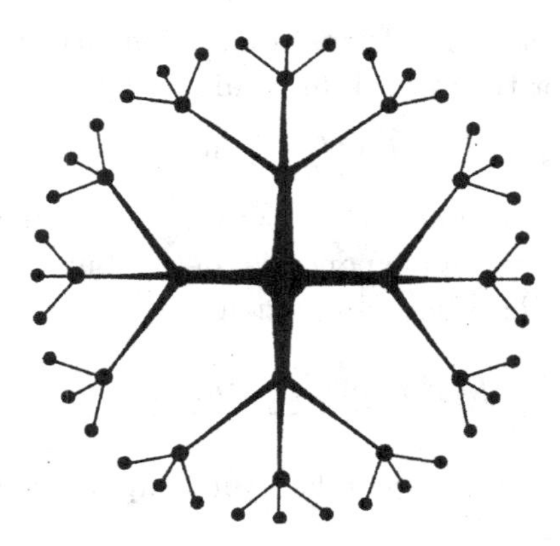

Fig. 3.10 Example of a Bethe lattice, or Cayley tree, of coordination number $z = 4$. From Ziman (1979).

formula for $g_{s,t}$ in Eq. (3.89) can be given, and hence all one can do are either exact enumerations of all geometric possibilities to calculate $g_{s,t}$ (which in practice is feasible only for rather small s) or Monte Carlo sampling.
In $d = 1$, however, Eq. (3.88), allows to readily extract the critical behavior by noting that $p \equiv \exp(\ln p) \underset{p\to 1}{\approx} \exp(p-1) = \exp(p-p_c)$, so that the scaling structure of Eq. (3.88) is more clearly revealed,

$$n_s(p) = (p_c - p)^2 \exp[s(p - p_c)] = s^{-2}F(\zeta), \quad \zeta \equiv (p - p_c)s\,. \tag{3.90}$$

Eq. (3.90) has precisely the form of Eq. (3.72), with $\tau = 2$, $\sigma \equiv 1$, and $F(\zeta) = \zeta^2 \exp(\zeta)$. The exponents α_p, β_p, γ_p, δ_p and ν_p are found to be

$$\alpha_p = 1, \quad \beta_p = 0, \quad \gamma_p = 1, \quad \delta_p = \infty, \quad \nu_p = 1\,. \tag{3.91}$$

The result $\beta_p = 0$ means that the percolation probability $P_\infty(p)$ has a discontinuous jump from zero to unity as $p \to p_c = 1$, see Eq. (3.40). This somewhat pathological behavior is related to the fact that $d = 1$ is the lower critical dimension for the percolation problem.

Another exactly soluble case, which is clearly more interesting, deals with percolation on the "Bethe lattice", or "Cayley tree", an example of which is shown in Fig. 3.10. The problem is analytically tractable because such lattices lack any closed loops. In a sense, the results are equivalent to results for ordinary lattices for $d \to \infty$ dimensions.

Let us consider *bond* percolation on the Cayley tree and discuss the probability $R(p)$ that a branch that is attached to an arbitrary cluster, via a bond that we will call "root", is in fact a "dead end" of this cluster, where by "dead end" we mean that the total number of bonds in the branch is finite. There are two mutually exclusive cases: Either the bond "root" is empty (this happens with probability $1-p$) or all the $z-1$ bonds that are connected to the bond "root" and lead away from the cluster lead to dead ends. The probability for this case is $p[R(p)]^{z-1}$. Since these two cases are mutually exclusive their probabilities are additive and thus the probability that "root" starts a dead end is given by

$$R(p) = 1 - p + p[R(p)]^{z-1}\,. \tag{3.92}$$

The solutions of this self-consistency equation for $R(p)$ are easily found. For $p < \frac{1}{z-1}$ we have only the solution $R(p) = 1$, which means there is no percolation. Conversely, for $p > 1/(z-1)$ one finds a solution $R < 1$, i.e. a percolating network exists. In order to obtain $P_\infty(p)$, one needs to calculate the probability that a randomly chosen bond belongs to the infinite percolating cluster. This bond must be occupied (probability p), and in addition at least one of the branches that are attached at the two vertices of the bond must continue on indefinitely. Since there are $2(z-1)$ branches on the lattice continuing at the 2 vertices of this bond, the probability that at least one branch continues on indefinitely is $1 - [R(p)]^{2(z-1)}$. Hence

$$P_\infty(p) = p\{1 - [R(p)]^{2(z-1)}\}\,. \tag{3.93}$$

As an example, consider $z = 3$: then $R(p) = 1 - p + pR^2(p)$, i.e. $R(p) = 1/p - 1$ (for $p > p_c = 1/2$), and $P_\infty(p) = p[1-(1/p-1)^4] \approx 8p(1-p_c/p)$ for p near p_c. Thus, we find that, close to p_c, $P_\infty(p)$ vanishes linearly and hence $\beta_p = 1$ This result also holds for other values of z. With similar reasonings (and a lot of algebra) one can work out exactly all the exponents that we have introduced above, to obtain (Stauffer and Aharony 1992):

$$\tau = \frac{5}{2},\ \sigma = \frac{1}{2},\ \alpha_p = -1,\ \beta_p = 1,\ \gamma_p = 1,\ \delta_p = 2, \nu_p = 1/2,\ \eta_p = 0\,. \tag{3.94}$$

While these exponents satisfy all the standard "thermodynamic" scaling relations, Eqs. (3.78), the hyper-scaling relation, Eq. (3.79), is only satisfied for the "marginal dimensionality" $d^* = 6$. For this dimensionality also all the scaling relations involving the fractal dimensionality d_f, Eq. (3.81), hold, with $d_f = 4$.

Renormalization group arguments (Cardy 2000) and various other studies (including also extrapolations of exact enumerations, Monte Carlo studies of lattices in $d = 6, 7, \cdots$, etc.) have indicated that in fact the exponents of the Bethe lattice, Eq. (3.94), are not only correct in the limit $d \to \infty$, but actually hold for dimensions $d \geq d^*$. (However, at $d = d^* = 6$ the power-laws in Eqs. (3.40), (3.57)-(3.61) are modified by logarithmic corrections.) In the most interesting cases of the physically relevant dimensions $d = 2$ and $d = 3$, however, the exponents deviate strongly from Eq. (3.94). Rather precise estimates for these exponents in $d = 3$ are available from Monte Carlo simulations from which one finds

$$\alpha_p \approx -0.62,\ \beta_p \approx 0.41,\ \gamma_p \approx 1.80,\ \nu_p \approx 0.88,\ d_f \approx 2.53\,. \tag{3.95}$$

For $d = 2$ one believes that the exponents are even known exactly from conformal field theory methods (Stauffer and Aharony 1992):

$$\alpha_p = -2/3,\ \beta_p = 5/36,\ \gamma_p = 43/18,\ \nu_p = 4/3,\ d_f = 91/48\,. \tag{3.96}$$

The concept developed in the context of percolation on the Cayley tree, that clusters contain branches that lead into "dead ends" and which do not contribute to percolation is, of course, also relevant for ordinary lattices. In particular, if one has in mind the application of bond percolation to random resistor networks, i.e. a network with connections that are conducting with probability p and isolating with probability $1 - p$, it is clear that only the "backbone" of the percolating cluster matters for the conductivity of the network. This backbone is obtained if all the "dangling ends" that do not contribute to conductivity are removed. In fact, if one defines for the number of sites (or bonds, respectively) that belong to the backbone of an incipient percolating cluster a relation that is analogous to Eq. (3.68),

$$n_B \propto R^{d_{fB}}, \quad p = p_c, \quad R \to \infty\,, \tag{3.97}$$

one finds for the fractal dimensions $d_{fB} \approx 1.6$ $(d = 2)$, $d_{fB} \approx 1.7$ $(d = 3)$, and $d_{fB} = 2$ for the Bethe lattice, i.e. in all cases we have $d_{fB} < d_f$. In fact, the structure of an incipient percolating cluster can be envisaged as a network which is composed of "blobs" (containing multiple connections) that are connected by one-dimensional structures, which are pieces of self-avoiding random walks. If one asks what fraction of sites (or bonds) is in these one-dimensional connections, one finds for their number (Stauffer and Aharony 1992)

$$n_1 \propto R^{d_{f1}}, \quad p = p_c, \quad R \to \infty\,, \tag{3.98}$$

with $d_{f1} = 1/\nu_p$. Since $d_{fB} > d_{f1}$ for $1 < d < 6$, asymptotically all the mass of the backbone is concentrated in the blobs of the backbone, and only a vanishing fraction in the one-dimensional connections. Only for $d = 6$, the number of sites (bonds) in the one-dimensional links and in the blobs scales with the same exponent. For $d > 6$ the links dominate, as expected (recall that no blobs can occur on the Bethe lattice). The typical number of sites contained in these one-dimensional paths between two blobs scales with the end-to-end distance of such a path L as $n_{\rm SAW} \propto L^{d_{f,\rm SAW}}$ where $d_{f,\rm SAW} = 1/\nu_{\rm SAW}$, with $\nu_{\rm SAW}$ the exponent characterizing the statistics of self-avoiding walks (see Sec. 3.1). As was discussed there, already for $d \geq 4$ one has $\nu_{\rm SAW} = 1/2$ and hence $d_{f,\rm SAW} = 2$, while for $d = 2$ $d_{f,\rm SAW} = 4/3$ and for $d = 3$ one has $d_{f,\rm SAW} \approx 1.70$. While for $d < 6$ one expects that

$$1/\nu_p < 1/\nu_{\rm SAW} < d_{fB} < d_f = d - \beta_p/\nu_p \tag{3.99}$$

holds, see Eq. (3.81), we have for the Bethe lattice $1/\nu_p = 1/\nu_{\rm SAW} = d_{fB} = 2$ while $d_f = 4$.

3.2.6 *Rigidity Percolation*

The details of the percolation phenomena depend certainly on what physical properties we have in mind. E.g, for the magnetic properties it is natural to consider site percolation while for electrical conductivity (which will be discussed in detail in Chap. 4) it is natural to start with a bond percolation model. However, both site and bond percolation belong to the same "universality class", i.e. the exponents defined above are the same, while the actual percolation thresholds are, of course, different.

This statement is no longer true, when we consider the so-called "rigidity percolation", where one asks whether or not a network can be mechanically deformed without energy cost, since this problem belongs to a different universality class. "Rigidity percolation" is a rather natural concept to study the mechanical stability of network glasses (Phillips 1979, 1981, Thorpe 1983, Phillips and Thorpe 1985, Guyon *et al.* 1990) and finds experimental applications for glasses such as amorphous $\rm Ge_xSe_{1-x}$ alloys (Bresser *et al.* 1986, Thorpe 1995, Feng *et al.* 1997, Kerner and Micoulaut 1997, Micoulaut and Phillips 2003, Mauro and Varshneya 2007), where Ge-ions are fourfold coordinated while Se-ions are only two-fold coordinated. Thus, this topic is of particular interest here, although it is much less well understood than the standard "connectivity percolation" treated so far.

The generic model of central-force rigidity percolation is a random network of Hookean springs, i.e. one assumes a potential energy of the form

$$V = \frac{1}{2} \sum_{\langle ij \rangle} \alpha_{ij} p_{ij} (\ell_{ij} - \ell_{ij}^o)^2 , \tag{3.100}$$

where ℓ_{ij} is the length of the bond connecting sites i and j in the network, $p_{ij} = 1$ with probability p and zero else, the sum $\langle ij \rangle$ is over all bonds in the network, $\alpha_{ij} > 0$ are the spring constants, and ℓ_{ij}^o the equilibrium bond lengths. The location of the sites are arbitrary, because the network is generic. Note that rigidity is a static concept, involving only infinitesimal virtual displacements of sites from their minimum energy positions. Therefore it does not matter whether or not the real potential is anharmonic.

Now a collection of sites is said to form a "rigid cluster" if no relative motion within that cluster is possible without a cost of energy. In contrast to this "floppy modes" are motions which do not cost energy. (Of course, any mechanical system has trivial floppy modes because of momentum and angular momentum conservation, namely, in d dimensions, d global translations and $d(d-1)/2$ global rotations, but those are of no interest here.) If we have N sites, we are interested in the number F of floppy modes in the network normalized per degree of freedom:

$$f = F/dN . \tag{3.101}$$

One possibility to solve this problem is to study the dynamical matrix of the potential energy given by Eq. (3.100) and to determine the eigenmodes and eigenfrequencies. Eigenfrequencies with $\omega^2 > 0$ have a vibrational character and are hence of no interest here, while any mode with $\omega^2 = 0$ is a floppy mode.

Of course, in the general case it is not possible to find an analytic solution of this problem. But there is a rather good approximate mean field type approach, called "constraint counting" (Phillips 1979), which goes back to ideas of Maxwell (1864). We associate with each bond a constraint, and assign $r/2$ constraints to each r-coordinated atom. The total number of constraints is hence $\frac{1}{2} \sum_r n_r r$, where n_r is the number of r-coordinated atoms in the network and thus $N = \sum_r n_r$. Thus F is approximated by the total number of degrees of freedom minus the total number of constraints, an approximation that corresponds to mean-field, and hence the fraction f becomes

$$f = \left[dN - \sum n_r r/2 \right] / dN = \left[dN - \frac{1}{2} N z p \right] / dN , \tag{3.102}$$

where $pz = \sum n_r r / \sum n_r$, z being the coordination number of the underlying lattice. Thus, we see that f can be written as $1 - p/p^*$ with $p^* \equiv 2d/z$. For a triangular lattice one thus obtains $p^* = 2/3$, while numerical estimates for the actual percolation threshold, i.e. $f = 1 - p_c/p^* = 0$, give $p_c = 0.6602 \pm 0.0003$ (Jacobs and Thorpe 1996).

Thus, the simple Maxwell estimate for the rigidity percolation threshold is surprisingly accurate. Note that the percolation threshold p_c for rigidity percolation is very different from that for simple bond percolation which is $p_c \approx 0.34729$ for the triangular lattice, see Stauffer and Aharony (1992). If one analyzes simulation results for rigidity percolation on the triangular lattice with linear dimensions L up to $L = 1150$ and applies finite size scaling methods, one finds (Jacobs and Thorpe 1996)

$$\begin{aligned} \alpha_{rp} &= -0.48 \pm 0.05, \quad \beta_{rp} = 0.175 \pm 0.02\,, \\ \nu_{rp} &= 1.21 \pm 0.06, \quad d_f^{(rp)} = 1.86 \pm 0.02\,. \end{aligned} \tag{3.103}$$

Thus, although the estimate for the percolation threshold is so close to the mean field result, the exponent β_{rp} is very different from the Maxwell prediction, $f \propto (1-p/p^*)^{\beta_{rp}}$ with $\beta_{rp} = 1$, implied by Eq. (3.102). Furthermore we emphasize that the simple estimate given by Eq. (3.102), although mean field in nature, is not the (exact) result for a Bethe lattice, for which one finds that the transition at p_c is *first* order (Moukarzel *et al.* 1997). In that case neither the concepts of critical exponents nor of a fractal dimensionality have any meaning.

Finally we mention that in $d = 3$ dimensions it is necessary to go beyond the simple model of Eq. (3.100) and to include also three-body bond angle potentials, that yield further constraints (Phillips and Thorpe 1985), but this extension will not be discussed here. We only mention a physical interpretation of why rigidity percolation yields a universality class that is different from standard connectivity percolation: The general question asked by percolation theory is the following one: "Is it possible to transmit some quantities (i.e., a scalar, or a vector) across a graph" (here we have considered lattices with some missing bonds or sites as a "graph"). It turns out that connectivity percolation amounts to the transmission of scalar quantities (e.g. for conduction electrical charges are transported over the network) while rigidity percolation amounts to the transmission of forces, i.e. vectors. Therefore "rigidity percolation" is often also referred to as "vector percolation" (Phillips 1979, 1981). In connectivity percolation the removal of a single bond of the percolating cluster can at most cause the splitting of this into two subclusters. In contrast to this the first order transition found for the Bethe lattice then implies that at the percolation threshold there exists a "critical bond" whose removal causes the percolating cluster to split into infinitely many subclusters, i.e. removing one constraint generates many floppy modes, thus vividly illustrating the non-local character of the rigidity concept.

3.3 Other Fractals (Diffusion-Limited Aggregation, Random Surfaces, etc.)

3.3.1 *General Concepts on Fractal Geometry*

In the discussion of simple random walks and self-avoiding walks, as well as in the discussion of the incipient percolating cluster at the percolation threshold, the concept of the "fractal dimension" d_f has already been introduced. In the present section, we reconsider this concept from a slightly more general point of view, and briefly discuss also other fractal objects which are of interest in the context of this book. More extensive discussions can be found in many books (Mandelbrot 1982, Pynn and Skjeltorp 1985, Stanley and Ostrowsky 1986, 1990, Jullien and Botet 1987, Feder 1988, Meakin 1988, 1998, Vicsek 1989, Bunde and Havlin 1991, Barabasi and Stanley 1995).

Consider an object in a d-dimensional space formed by N subunits, or building blocks, in the limit $N \to \infty$. Take a volume of size L^d centered at the center of mass of the object, and count the number n of subunits that fall into this volume. If we find a power-law relation $n \propto L^{d_f}$ with $d_f < d$, we term the object a "fractal", while for $d_f = d$ the object is compact, space-filling, and not of interest for the present section. Of course, it is required that L is much larger than the linear dimensions of the subunits, but otherwise L is arbitrary.

The fractals considered so far are all "random fractals" or "statistical fractals", but it is also possible to form non-random fractals by regular constructions, typically defined in terms of a recursive construction, such as the well known "Sierpinski gasket" which can be defined as follows: Start with an equilateral triangle in $d = 2$ and remove from its interior the equilateral triangle formed by the three lines that connect the midpoints of the lines forming the first triangle. Repeat this process for the remaining three triangles, and continue this process indefinitely. In this way one creates a self-similar structure that has on all length scales holes of triangular shape.

Obviously the number of subunits $n(2L)$ on scale $2L$ is simply $3n(L)$ and thus we have immediately that $n(L) \propto L^{d_f}$ with $d_f = \ln 3/\ln 2 \approx 1.585$. The density on scale $2L$ obeys the equation $\rho(2L) = (3/4)\rho(L) = 2^{d_f-d}\rho(L)$. Iterating this equation from the scale $L = 1$ to the scale L one readily finds $\rho(L) \propto L^{d_f-d}$ which goes to zero for $L \to \infty$. The difference $d_f - d$ is often called the "co-dimension". This particular regular fractal not only has the merit that d_f can be calculated exactly, but also that these power-laws hold on all scales, i.e. there are no "correction to scaling" terms, which are expected to be present for the case of statistical fractals. However, for glassy systems regular fractals are not useful models since their properties differ too much from the ones of glass-forming systems and hence they will not be considered further.

A very simple and useful example of a statistical fractal is the so-called "fractional Brownian motion" (Mandelbrot 1982), which is a simple generalization of ordinary random walks, or standard Brownian motion, and where $x(t)$ is a single valued function of a variable t (usually time), whose increments $x(t_2) - x(t_1)$ are gaussian distributed with variance

$$\langle [x(t_2) - x(t_1)]^2 \rangle \propto |t_2 - t_1|^{2/z} . \tag{3.104}$$

For $z = 2$ this gives of course the familiar Brownian motion discussed already in Chap. 2. Equation (3.104) shows a statistical scaling behavior: If the time difference $t = t_2 - t_1$ is increased by a factor λ, the increments $\Delta x = x(t_2) - x(t_1)$ change by a factor $\lambda^{1/z}$. Of course the curve $x(t)$ can also be interpreted in a completely different way. E.g. one could describe the coast line of Norway, taking x and t as the two coordinates on the (locally flat) surface of the Earth. In such an interpretation the two variables x and t must be equivalent since they are just spatial coordinates. There are, however, other situations in which one has an anisotropic scaling, i.e. if one coordinate, t, is scaled by a factor λ the other coordinate, x, is scaled by a factor $\lambda^{1/z}$ with exponent $z \neq 1$. Such structures, which are called "self-affine" instead of "self-similar", do have physical significance, for instance, if we consider the random growth of a surface when particles are randomly deposited on a substrate (Vicsek 1989, Barabasi and Stanley 1995).

Consider now a topologically one-dimensional fractal "curve", e.g. one which is generated by a fractional Brownian walk in space (x, t) as described by Eq. (3.104). Then a quantity of interest is the "length" ℓ of the curve. Surprisingly this (apparent) length varies with the size L of the yardstick used to measure this length as can be seen as follows: Imagine that we have determined $\ell(L_1)$. If we now use a shorter yardstick $L_2 = \lambda L_1$, with $\lambda < 1$, a segment of the curve that had (with the ruler L_1) a length L_1 will now, due to the self-similarity of the curve, consist of $n = \lambda^{-d_f}$ segments of length L_2. Then the length $\ell(L_2)$ is given by $\ell(L_2) = L_2 \cdot n = \lambda L_1 \cdot \lambda^{-d_f} = \lambda^{1-d_f}\ell(L_1)$, which is larger than $\ell(L_1)$ since $d_f > 1$.

The fractal dimension d_f is sometimes also called the "box dimension" since it characterizes the covering of the (fractal) set S by d-dimensional boxes of linear size L. If the entire set S is contained within one box of size $L = L_{\rm max}$, one can use the same line of reasoning used in the previous paragraph to see that the number $n_{\rm box}(L)$ of boxes of size L needed to cover S is given by $n_{\rm box}(L) = (L_{\rm max}/L)^{d_f}$. Since these are the boxes that contain the "mass" that is contained in the fractal object, if one considers a fractal formed by aggregation of particles, this is also called the "mass dimension".

In a self-affine transformation each point $\vec{x}$ of a set S is transformed into a point $\vec{x}' = (\lambda_1 x_1, \ldots, \lambda_d x_d)$ of a set S' with scaling factors λ_i that depend on the direction in space. As an example, let us consider the relation between d_f and z for the self-affine fractional Brownian motion of Eq. (3.104) (Voss 1985). Consider one particular path $x(t)$ spanning a time $\Delta t = 1$ and a range $\Delta x = 1$. We divide the time span into N intervals $\delta t = 1/N$. Each of these intervals will contain a portion of $x(t)$ with a range $\delta x = (\delta t)^{1/z} = N^{-1/z}$. The occupied portion of each interval can be covered by $\delta x/\delta t = N^{1-1/z}$ square boxes of linear scale $L = 1/N$. The number of square boxes covering the whole path is therefore $n(L) = N \times N^{1-1/z} = N^{2-1/z} = L^{-(2-1/z)}$ which implies that $d_f = 2 - 1/z$ for the path $x(t)$.

It is also of interest to generalize Eq. (3.104) to higher dimensions, e.g. to consider a "fractal landscape" of height $h(x, y)$ as function of two variables x, y

$$\langle [h(x_2, y_2) - h(x_1, y_1)]^2 \rangle \propto [(x_2 - x_1)^2 + (y_2 - y_1)^2]^{1/z} \tag{3.105}$$

with a fractal dimension $d_f = 3 - 1/z$, while an intersection of a vertical plane with such a fractal surface would bring us back to the previous problem for which $d_f = 2 - 1/z$.

Suppose now that the fractal landscape we are talking about would simply be the surface of the earth. Then it would be natural to assume that the landscape up to a height $h = 0$ is filled by water, that there are well-separated regions of height $h > 0$ like islands in the sea, and that the contours of $h = 0$ which define these islands are fractal coastlines. In general one speaks of the "perimeter" of a fractal object, and one can associate a fractal dimensionality of the coastline d_f, as we have done for the fractional Brownian walk given by Eq. (3.104). If the island has a linear size $L_{\max}$, we have for the length P of the perimeter $P \propto L_{\max}^{d_f}$ or equivalently, if one measures the coastline by using a yardstick of length L one finds that the perimeter is $P \propto L^{d_f}$. Note however, that only the coastline of the island is a fractal since the island itself is "compact", i.e. its area A scales with the linear dimension $L_{\max}$ according to $A \propto L_{\max}^2$, and hence $P \propto A^{d_f/2}$. Of course, if we consider real islands on earth and their coastlines there is no proof (apart from empirical evidence, see Mandelbrot 1982) that they are actually fractal. But the concept of a fractal landscape is a nice idea to visualize the behavior of surfaces growing with time due to random adsorption of particles (Vicsek 1989).

It is of interest to ask whether for the geometrical characterization of a fractal object a *single* exponent, the fractal dimension d_f, is really enough for a complete description of the fractal structure. We have seen that the mass $M(L)$ of a statistical fractal scales like $M(L) \propto L^{d_f}$, but what was actually meant was a statement on the *average* mass

$$\langle M(L)\rangle \propto L^{d_f}, \quad \text{with } \langle M(L)\rangle = \sum_m mP(m,L) \tag{3.106}$$

where the mass distribution $P(m,L)$ is assumed to be normalized, i.e. $\sum\limits_m P(m,L) = 1$.

Instead of studying only this average mass it is also of interest to consider the fluctuation in mass,

$$\langle(\Delta M)^2\rangle = \langle[M(L) - \langle M(L)\rangle]^2\rangle \tag{3.107}$$

and to ask whether or not the L−dependence of this quantity involves another exponent. It turns out that for the mass of fractals such as random or self-avoiding walks, percolation etc. this is not the case since the so-called "lacunarity" ("lacuna" is the Latin word for "gap")

$$\Lambda(L) = \langle(\Delta M)^2\rangle / \langle M(L)\rangle^2 \tag{3.108}$$

simply tends to a (universal) constant for $L \to \infty$, and therefore there is no new exponent.

As we will see later, the situation is no longer that simple if we consider dynamic properties such as, e.g., the conductivity of a percolating cluster: Different moments of the distribution function of the electrical current scale with exponents which are not simply related to each other, so a quantity analogous to Eq. (3.108) for the current does involve a new exponent, and still other exponents occur in higher order moments. Such a property is termed "multifractality" (Aharony 1990).

3.3.2 *Diffusion-Limited Aggregation*

Diffusion limited aggregation (DLA) is an archetypal model of random irreversible growth processes (Witten and Sander 1981). Here we describe only the salient features and refer to Herrmann (1992), Kolb (1999), and Sander (2000) for a more extended discussion. Basically the model is defined in terms of a computer simulation algorithm: One starts with the first particle which is put on the origin of the lattice (note, however, that also off-lattice generalizations of DLA have been formulated and investigated, see e.g. Meakin 1988, 1998). From a randomly selected position on a hyperspherical surface of radius R_m centered at the origin, a particle is launched to start a random walk trajectory. If it arrives at one of the nearest neighbor sites of the particle at the origin, it sticks to this site and thereby the cluster grows by one unit. Then the next particle is launched, and if it touches the cluster, also it sticks irreversibly. If, however, a particle moves a distance R_f away from the cluster (e.g. $R_f = 2R_m$), then the random walk is stopped, and a new particle is launched. This process is iterated many times, and in this way structures are generated as it has already been shown in Fig. 1.8.

It is of interest to note that this process can also be viewed as a numerical implementation of the so-called "Laplacian growth" problem, where one seeks a solution of the Laplace equation but with a moving boundary condition (the surface of the cluster). A well-known physical example is the so-called "Stefan problem", i.e. the solidification of a supercooled melt where the latent heat has to diffuse toward the boundary (Langer 1980). The growth of the crystal is determined by the temperature field $T(\vec{r})$, which in equilibrium follows the Laplace equation, $\nabla^2 T(\vec{r}) = 0$, and one assumes that on the external boundary (e.g. at infinity) a temperature $T_0 < T_m$ (T_m is the melting temperature) is imposed. On the interface between the fluid and the solid, the Gibbs-Thomson equation implies a local temperature

$$T_i = T_m(1 - \gamma\kappa/Q), \tag{3.109}$$

where Q is the latent heat, γ the liquid-solid surface tension, and κ the curvature of the interface. Now motion of the interface is described through the normal growth

$$v_n = \frac{D_T C_p}{Q} \nabla_n T. \tag{3.110}$$

Here v_n is the (normal) velocity of the interface, D_T is the thermal diffusion coefficient, C_p the specific heat at constant pressure in the fluid at the melting temperature, and $\nabla_n T$ is the component of the gradient of T that is perpendicular to the surface. Equations. (3.109) and (3.110) together with Laplace's equation $\nabla^2 T(\vec{r}, t) = 0$ define what is called a "moving boundary problem": At any instant of time one has a well-defined Dirichlet boundary condition (fixed values of the temperature field) of an elliptic differential equation and thus a unique solution for the temperature field outside of the growing solid cluster. This solution determines, via Eq. (3.110), the motion of the phase boundary in a time interval δt. The resulting change in the shape of the boundary defines the new Dirichlet boundary condition and therefore the new solution for $T(\vec{r}, t)$, etc.

The reason that such physical situations lead to rather random irregular structures is the fact that moving interfaces resulting from the solution of such problems show instabilities as can be seen by making a linear stability analysis of the growth. Thus, in the Stefan problem one finds the Mullins-Sekerka (1963) instability, which states that long wavelength excitations of a flat surface are unstable if their wave-vector is less than $q_c \propto Q/\sqrt{D_T \gamma T_m C_p}$. These instabilities enhance any noise (thermal noise or disorder present in the initial condition) and cause ultimately the growth of statistical fractals rather than compact regular structures.

Replacing temperature by pressure in Eqs. (3.109) and (3.110) leads to the problem of "viscous fingering", i.e. the penetration of a fluid of low viscosity into a more viscous fluid. Related problems are dielectric breakdown, fracture, etc. Other applications of DLA include the explanation of patterns observed in electrodeposition, nanoscale surface pattern organization, cellular organization, etc. (see Kolb 1999 for more references).

Early simulation studies of DLA clusters on lattices concluded that the structure is a self-similar fractal (Meakin 1988). More recent work using a much larger number of particles showed, however, that this conclusion was premature, since on large scales lattice anisotropy starts to dominate the structure, and only for true off-lattice DLA (where one also has grown clusters which are very large, over 10^8 particles) does one find the mass radius relation for a true fractal, $M \propto R^{d_f}$ with $d_f \approx 1.71$ in $d = 2$ and $d_f \approx 2.485 \pm 0.005$ in $d = 3$ (Herrmann 1992), but even in this case serious doubts on the self-similarity of the cluster shape remain (Kolb 1999).

3.3.3 *Growth of Random Interfaces*

While in the DLA problem we have considered a random growth process of an object in d dimensions starting out from a zero-dimensional (i.e., point-like) seed, it also is of interest to consider the growth of objects in d dimensions by random addition of material on a $(d-1)$-dimensional substrate. Thus, rough surfaces may arise in crystal growth from the melt, vapor deposition, electroplating, spray painting and coating, and biological growth (Family 1990, Barabasi and Stanley 1995). Of course similar fractal surfaces may also be created if material is randomly removed, e.g. by corrosion processes.

Consider a flat, $(d-1)$–dimensional surface at time $t=0$, on which, by some processes that we will discuss later, particles are randomly deposited. We concentrate on a section of the surface with linear dimension L, perpendicular to the growth direction, and describe the growing interface by the function $h(\vec{x},t)$, the height of the interface at time t and position $\vec{x}$, above the substrate. We next define the average height $\langle h(t)\rangle$ and the interfacial width $w(L,t)$,

$$\langle h(t)\rangle = \frac{1}{L^{d-1}}\sum_{\vec{x}} h(\vec{x},t), \quad w(L,t) = (\langle h^2\rangle - \langle h\rangle^2)^{1/2}. \tag{3.111}$$

Usually one makes the following Ansatz for the scaling behavior (Vicsek 1989, Family 1990):

$$w(L\to\infty,t) \propto t^{\beta}, \quad w(L,t\to\infty) \propto L^{\alpha}, \tag{3.112}$$

where α,β are suitable exponents. In fact, one often assumes (Family and Vicsek 1985) that both power-laws can be combined into a "dynamic scaling" form, with a "scaling function" $\widetilde{w}$

$$w(L,t) = L^{\alpha}\widetilde{w}(t/L^{\alpha/\beta}), \quad \widetilde{w}(\zeta) \propto \zeta^{\beta}. \tag{3.113}$$

This type of "dynamic scaling" (Hohenberg and Halperin 1977) is familiar from dynamic critical phenomena at phase transitions in thermal equilibrium. Consider e.g. an Ising spin system, in which one starts with a completely random initial spin configuration, and then at time $t=0$ quenches the system to the critical temperature T_c. Then the susceptibility grows as (Sadiq and Binder 1984)

$$\chi(L,t) = L^{\gamma/\nu}\widetilde{\chi}(t/L^{z}), \quad \widetilde{\chi}(\zeta) \propto \zeta^{\gamma/(z\nu)}. \tag{3.114}$$

Equation (3.114) expresses the familiar "critical slowing down", i.e. at T_c the relaxation time scales as $\tau \propto L^z$. A similar interpretation can be given to Eq. (3.113), with a dynamic exponent $z = \alpha/\beta$.

The simplest model for growing interfaces is the "ballistic deposition model" (Vold 1959) in which particles simply "rain down" onto a substrate. For random ballistic deposition on a lattice the system develops an array of columns which grow independently of each other. Therefore the height of the columns follow a Poisson distribution and correspondingly $w \propto t^{1/2}$, i.e. $\beta = 1/2$, independent of dimension (Family 1990). A more realistic (and less trivial) model assumes random deposition with subsequent surface diffusion that mimics the local relaxation of the surface upon the deposition of new material. Depending on the detailed assumptions about the "rules" governing this hopping process of particles on the random interface different growth laws can be found (Shim and Landau 2001, Rikvold and Kolesik 2002).

A simple analytical description for the time evolution of a random interface was proposed by Edwards and Wilkinson (1982)

$$\frac{\partial h(\vec{x},t)}{\partial t} = \nu \nabla^2 h(\vec{x},t) + \eta(\vec{x},t) \, . \tag{3.115}$$

In this Langevin equation ν describes the surface diffusion, and $\eta(\vec{x},t)$ is a random noise term in the flux (with mean zero since we assume that in Eq. (3.115) the average height of the interface above the substrate has already been subtracted). Due to its linearity in h, Eq. (3.115) is easily solved, yielding (in $d \leq 3$ dimensions) $\beta = (3-d)/4$, $\alpha = (3-d)/2$, and $z = 2$. The result $\beta = 0$ in $d = 3$ means a logarithmic variation, $w^2 \propto \ln L$.

Another ballistic deposition model assumes that particles also "rain down" following straight-line trajectories but stick to the already existing aggregate on the substrate as soon as they become (on the lattice) a nearest neighbor of a particle that already is part of the aggregate. Thus, the growth rule is similar to DLA, apart from the fact that the "seed" for the growth is $(d-1)$ dimensional rather than zero dimensional, but the resulting structures locally look similar to local structures in a DLA cluster. It is believed that an appropriate theoretical description is provided by the equation due to Kardar, Parisi and Zhang (1986)

$$\frac{\partial h}{\partial t} = \nu \nabla^2 h + \frac{1}{2}\lambda (\nabla h)^2 + \eta(\vec{r},t) \, , \tag{3.116}$$

where a non-linear term proportional to $(\nabla h)^2$ has been added to Eq. (3.115), involving another parameter λ. Due to its nonlinear character,

Eq. (3.116) cannot be solved analytically in general, and the analysis by field-theoretic methods and numerical studies is a difficult problem (Lässig 1998, Miranda and Reis 2008). For one-dimensional interfaces ($d = 2$) one can show that $\alpha = 1/2$, $\beta = 1/3$ and $z = \alpha/\beta = 3/2$. These exponents satisfy also the scaling relation $\alpha + z = 2$, which can be derived more generally from Galilean invariance. For higher dimensions arguments have been presented (see Lässig 1998 for a review) to suggest that $\alpha = 2/5$, $z = 8/5$ in $d = 3$, and $\alpha = 2/7$, $z = 12/7$ in $d = 4$. However, the numerical data obtained for this problem does not yet seem to be fully conclusive.

3.4 Random Close Packing

In this section we discuss the modeling of amorphous structure for the simplifying case in which the attractive interactions between the atoms are completely neglected and the repulsive interactions are approximated by hard core potentials. In this case the problem reduces to a statistical dense packing of hard bodies. For the sake of simplicity we assume that there occurs only a single atomic species of spherical shape, and so the problem reduces to the packing of hard spheres that touch each other.[2] Despite this simplification, the geometrical properties of the resulting structure are at present understood only partially and thus remain in the focus of current research (Parisi and Zamponi 2010).

If we do the packing in $d = 2$ dimensions (packing of hard disks) the answer to this problem is really trivial: We simply build up a regular triangle (Fig. 3.11). In two dimensions there is no conflict between the tendency of the atoms to make a dense local packing and the formation of a long range ordered periodic structure that densely fills the available space.

In $d = 3$ dimensions the situation is completely different, however. Four spheres that touch each other form a regular tetrahedron, if we draw the lines that connect the centers of the spheres (Fig. 3.11). If we now join several tetrahedra together along a common edge, space is no longer completely filled as it is the case in $d = 2$ dimensions: We can pack at most 5 tetrahedra together, but there remains a small wedge of 7.4° into which no further tetrahedron fits in anymore (Fig. 3.11). Therefore packing of

[2] Note that presently very little is known on the packing of *non*-spherical objects (ellipsoids, rods, ...) although such systems show quite particular features, such as, for ellipsoids, a non-monotonic dependence of the maximum packing density as a function of the elongation (Donev *et al.* 2004).

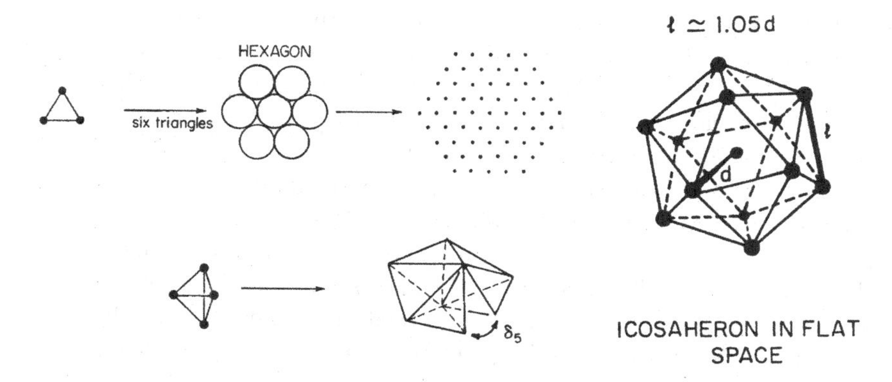

Fig. 3.11 The random close packing of disks yields a regular triangular lattice (left, upper part), while the packing of tetrahedra (connecting of centers of four hard spheres touching each other) does not result in a space-filling structure, due to a misfit angle of 7.4° (left, lower part). A small distortion of the tetrahedra that removes this misfit allows the formation of a regular icosahedron (right). From Nelson and Spaepen (1989).

regular tetrahedra does not allow for a periodic space-filling structure.

Thus this misfit in the packing of hard spheres requires a certain distortion of the ideal structure. Let us therefore assume that we pack tetrahedra together, but that we relax the condition that all edges must have equal length. Starting with an atom in the center, we may pack 12 atoms in a regular arrangement around it, to obtain a polyhedron which has only regular triangles of edge length ℓ at its surface. Each of these triangles at the surface forms a tetrahedron with the atom in the center, but the distance d between the atom in the center and the atoms forming the triangles is slightly less than ℓ ($\ell \approx 1.05\,d$), so the tetrahedra are no longer ideal equilateral tetrahedra. This regular polyhedron with 12 corners on its surface thus forms an icosahedron (Fig. 3.11).

For an isolated aggregate of only 13 atoms the icosahedron is not only the "best structure" if we consider the packing of hard spheres, but it is also the minimum energy configuration for typical pairwise potentials such as the Lennard-Jones potential.[3] For 13 atoms a cluster having the structure of hexagonal close packing (hcp) or face-centered cubic (fcc) ordering would have a higher energy. However, while the latter structures are suitable for forming a regular space-filling structure, it is not possible to fill space by an arrangement of icosahedra. Thus, in three dimensions there is a conflict

[3]Note that 13 is not the only number that gives rise to a locally compact structure with high symmetry. Other such "magic" numbers are 38, 98, ... (Wales 2004).

between the structures that optimize densely packed structures locally (such as the icosahedron) and globally (the hcp and fcc structures). This conflict has motivated the speculation that in $d = 3$ it is physically possible that a dense packing of hard spheres may also result in a completely random structure with no long range order.

The first systematic study of statistical packing of hard spheres, also called "random close packing" (RCP), was done by Bernal who, using ball bearings, concluded that it is difficult to pack spheres randomly beyond a volume fraction $\phi_{RCP} \approx 64\%$ (Bernal 1960). Later studies used computers to create such packings by, starting with a tetrahedron, adding sequentially spheres to an existing cluster, making sure that the new sphere touches three spheres of this cluster. Another possibility to generate a RCP structure is to slowly inflate in a fixed volume spheres that at time zero have been assigned a random velocity and to end the process when the structure cannot relax anymore (Lubachevsky and Stillinger 1990). Various studies showed that the notion of RCP is ill-defined in that the value that is obtained for the maximum packing density depends on the details of the procedure that is used to generate the packing (Torquato *et al.* 2000). However, later it has been conjectured that around ϕ_{RCP} the number of accessible states quickly goes to zero which would explain why in practice this value is nevertheless meaningful (Kamien and Liu 2007, Song *et al.* 2008). Despite these questions, simulation studies have revealed some of the interesting properties of the RCP structure. E.g. it has been found that the static structure factor $S(q)$ shows at small wave-vectors a *linear* dependence in q, instead the quadratic one that is observed in normal dense liquids (Donev *et al.* 2005), a feature that has been rationalized by the suppression of long range fluctuations in the system due to the "ideal" packing. Recently a number of studies have also been done to investigate the properties of RCP for multicomponent systems, or to connect these structures with the problem of "jamming", i.e. the state in which a system is under an external stress and, at high enough density, does not relax anymore (O'Hern *et al.* 2002, Song *et al.* 2008, Berthier and Witten 2009, Hermes and Dijkstra 2010, van Hecke 2010). However, since these results would take us away too much from the main focus of the book, we refer the interested reader to the cited papers.

Another interesting application of RCP is related to the fact that the statistical packing of hard spheres yields a radial density distribution that is in surprisingly good agreement with the experimental results for amorphous metallic glasses. This can be recognized, e.g., from Fig. 3.12, Finney (1970),

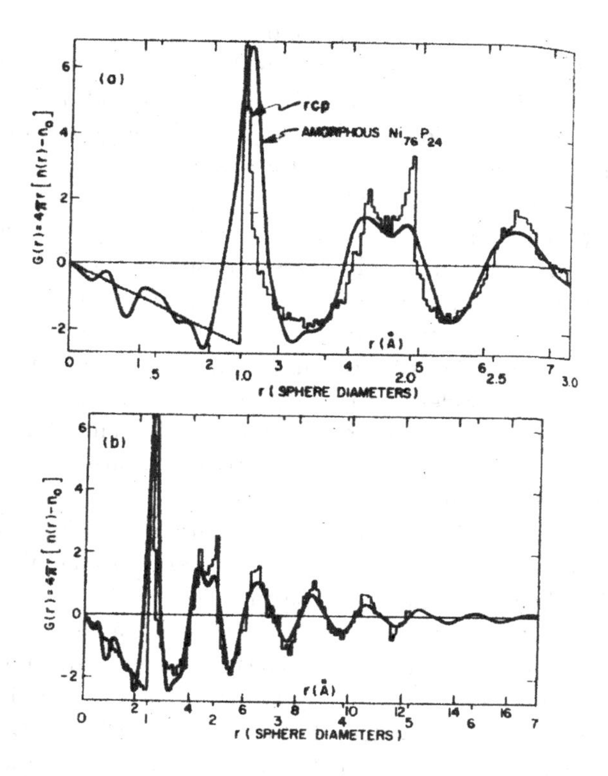

Fig. 3.12 Plot of the reduced radial distribution function $G(r) = 4\pi r[n(r) - n_0]$ versus distance r, n_0 being the average number density and $n(r) = n_0 g(r)$. The full curve is the experimental result for amorphous $Ni_{76}P_{24}$, while the histogram shows the result of the RCP structure. The upper panel shows the range up to 3 sphere diameters and the lower one the more extended range up to 7 sphere diameters. From Cargill (1975).

Cargill (1975), where we show $4\pi r(n_0 g(r) - n_0)$, where n_0 is the particle density. (Note that the term n_0 is subtracted in order to obtain a function that for $r \to \infty$ goes to zero.) It turns out that Ni and P have essentially the same effective atomic radius and by using the hard sphere radius of the RCP model as a fit parameter it is possible to make the first peak of the model coincide with the one of the data. Is is then found that also the heights and widths of these peaks agree quite well, and, in view of the simplicity of the model, the overall good agreement observed for all the further peaks is really remarkable.

In Fig. 3.12 we recognize also a well known characteristic feature of the RCP structure: The splitting of the next nearest neighbors peak. Such a feature is observed for most metallic glasses, but it is not seen for covalently bonded network glasses (amorphous Si, Ge, SiO_2 etc.). Roughly speaking this feature can be attributed to the fact that there are two typical second nearest neighbor distances. The first one is just two times the diameter of the hard spheres. The second one is the distance between two hard spheres that are placed on the line which is orthogonal to the connecting line between two touching hard spheres, i.e. the four particles form a cross. In molten metals and in other simple fluids this splitting is not seen. It is a unique characteristic of the amorphous RCP structure since at higher temperatures the difference between these two length scales is smeared out by thermal motion. From Fig. 3.12 we can also note that there is a slight discrepancy between the model and the experimental data in the heights of the two maxima of this split second-neighbor peak. This difference is attributed to the effect of the attractive interactions that are present in the real material but neglected in the model.

These packing considerations yield a plausible argument why it is possible to create amorphous structures from easily crystallizable metals such as iron by condensing them from the gas phase on a cold substrate: The atoms cluster together locally, forming tetrahedra and icosahedra, and, due to the low temperature, this local structure cannot be rearranged easily into a long range ordered structure. Thus the initially formed RCP structure is frozen in, even if it fills space less densely than the hcp or fcc structures. As discovered already by Kepler in 1611 (Zallen 1983), the maximum possible filling factor of space for a packing of hard spheres is indeed obtained in these latter structures, namely $P_f = \pi/(3\sqrt{2}) \approx 0.7405$. This has to be compared with the filling factor of the RCP structure which is only $P_f \approx 0.637$ (Cargill 1975, Bernal 1965), and hence only $0.637/0.7405 \approx 86\%$ of the density of a crystalline structure is reached.[4] The RCP structure is only kinetically favored because of the above packing considerations whereas from an energetic point of view it is disfavored. We also mention that if one considers melts of binary or ternary metallic glasses, for which it is possible to produce a glass from a continuous but rapid cooling from the melt, one also never discovers any sign of fcc-type or hcp-type or other crystalline short range order in the melt. Instead the short range order in the melt is

[4]While P_f for the crystalline structures is easily found in an exact way, for the RCP structure the filling factor is only known empirically. The currently best estimates come from computer simulations (Cargill 1975), while the first consideration of this problem actually was done almost 300 years ago by the biologist Hales (1727), who studied the filling of peas in a pot (in the context of their ability of water uptake).

described by a (softened and broadened) variant of the radial distribution function encountered in the RCP structure.

We now discuss some general geometrical features of random space-filling structures. Suppose we fill space by Voronoi polyhedra requiring that two neighboring polyhedra always share a single face, while each edge must be shared by three polyhedra, and each corner (or vertex) belongs to four polyhedra. These conditions are of course only bounds since it is also possible that an edge belongs to more than three polyhedra (remember, e.g., the packing of 5 regular tetrahedra with a common edge, Fig. 3.11). According to the Voronoi polyhedra construction principle, three polyhedra sharing an edge (or four polyhedra sharing a vertex) are minimal numbers, and if higher numbers of common edges or vertices occur, a special symmetry between the centers of the associated Voronoi polyhedra must exist.

Let us now combine this information with the Euler-Poincaré formula which gives rise to a topological constraint for the number of vertices (V), the number of edges (E) and faces (F) in a graph containing N cells (Coxeter 1958):

$$V - E + F - N = 1\,. \tag{3.117}$$

Let us consider the special case that all faces in the network of the Voronoi polyhedra are polygons with the same number (p) of edges (and vertices). Taking out an isolated polyhedron ($N = 1$) from the network, we must have for this object the equalities

$$pF = 3V = 2E\,, \tag{3.118}$$

since now only three faces (and edges) meet at each vertex, and two faces meet at each edge, while each face is in contact with p edges (and vertices). Substituting Eq. (3.118) in Eq. (3.117), one finds $F = 12/(6-p)$.

If one analyzes a RCP structure in terms of the Voronoi construction, one finds that the average number $\bar{F}$ of faces per polyhedron is about $\bar{F} \approx 14$. The corresponding value of p would be slightly larger than 5, indicating that many pentagonal faces occur in the structure. This occurrence of pentagonal faces hence simply emerges as a consequence of topological constraints.

Of course, it is necessary to take into account that the result $\bar{F} \approx 14$ applies to numerical studies of RCP structures with $N \gg 1$ polyhedra. Since in such a structure each vertex belongs to four edges, to six faces, and four cells, as simple geometrical considerations show, one finds (f being now the number of faces per cell):

$$(fp/2)N = pF = 3E = 6V\,. \tag{3.119}$$

Embedded in these equations is also the fact that each edge belongs to three faces and to three cells and to two vertices, while each face belongs to two cells and has both p vertices and p edges. Finally, each cell has f faces, and $fp/3$ vertices (each vertex belongs to six faces, there are p vertices per face, but only two faces belong to a cell).

Using Eq. (3.119) in the Euler-Poincaré-relation, by noting that for $N \to \infty$ the "1" on the right hand side can be neglected, one finds

$$f = \frac{12}{6-p}\,. \tag{3.120}$$

In a real Voronoi network constructed from a realization of the RCP structure, the polyhedra show a distribution with respect to the number of their faces and vertices, and thus one has to average Eq. (3.120) over this distribution. In order to do this one considers the number of faces F_p that have exactly p vertices, so that the Euler-Poincaré relation becomes

$$\sum_p (6-p)F_p = 6N\,. \tag{3.121}$$

Note that the factor 6 instead of 12 as in Eq. (3.120) results from the definition that F_p means the total number of faces, with p edges and vertices. It is no longer normalized per cell, and each face belongs to two cells, as noted above. Defining now

$$[p]_{\text{av}} = \sum_p pF_p / \sum_p F_p\,, \tag{3.122}$$

Eq. (3.121) can be rewritten as

$$\sum_p F_p(6 - [p]_{\text{av}}) = 6N\,, \tag{3.123}$$

and writing $\sum_p F_p = N[f]_{\text{av}}/2$ we finally obtain

$$[f]_{\text{av}} = \frac{12}{6 - [p]_{\text{av}}}\,. \tag{3.124}$$

While relations such as Eq. (3.124) are useful to interpret properly the numerical findings such as relating $[f]_{\text{av}} \approx 14$ to $[p]_{\text{av}}$, these topological constraints are not strong enough to determine $[f]_{\text{av}}$ itself uniquely. This is in contrast to $d = 2$ dimensions, where topological constraints are much stronger. There the Euler-Poincaré relation reads

$$V - E + F = \chi, \tag{3.125}$$

where χ is the so-called Euler-Poincaré constant which distinguishes a plane from closed curved surfaces. Each edge belongs to two faces, and (in the generic case in which one has no symmetries) three edges meet at a vertex. Also in this case each face, or plaquette, that has p vertices must have p edges. Thus, we conclude

$$E = \sum_p pF_p/2\,, \quad V = \sum_p pF_p/3\,, \tag{3.126}$$

and with Eq. (3.125) we thus obtain

$$\sum_p \left(\frac{p}{3} - \frac{p}{2} + 1\right) F_p = 0\,, \tag{3.127}$$

where we have again omitted χ in the limit $F_p \to \infty$. Equation (3.127) then implies

$$\sum_p F_p = \frac{1}{6} \sum_p pF_p \quad \text{and thus} \quad [p]_{\text{av}} = 6\,. \tag{3.128}$$

The average number of edges per face can now be obtained as

$$e = 2[E]/\sum_p F_p = \sum_p pF_p/\sum_p F_p = 6\,, \tag{3.129}$$

here the factor of 2 takes account the fact that each edge belongs to two faces.

So far, we have seen that topological considerations have an impact on the properties of the Voronoi networks that describe RCP structures. It is tempting to ask whether topological concepts are also useful to formulate the statistical mechanics of the glassy structure and its melting toward the fluid phase. An elegant approach of this type has been proposed by Nelson (1985). However, since this approach has not been followed up much in the recent literature, we will not attempt to go into the details of this (actually rather complicated) theory, but describe only qualitatively the salient features and ideas.

Starting point is the observation discussed above (Fig. 3.11) that 5 ideal tetrahedra cannot be packed perfectly since there is a missing angle of 7.4°. The idea now is to postulate a structure in which this ideal 5-fold coordination, which is evident in the icosahedral arrangement (Fig. 3.11), occurs as a prevailing element as much as possible, but to use also "defects" where the tetrahedra locally have a fourfold or sixfold coordination in order to achieve a dense filling of space. It turns out that these defects cannot be point defects but must be line defects, and they are closely related to the defects that were discussed in the context of two-dimensional melting, namely the disclinations (Fig. 2.10). In the perfect structure with fivefold coordination (which cannot exist in the standard euclidean three-dimensional space, but can be imagined to exist in a space with a specially adapted curvature) the tetrahedra configurations with 6-fold and 4-fold coordinations form disclination and antidisclination lines. In fact, if these lines form a regular lattice, a particular crystal structure results which actually is known in nature, the so-called Frank-Kasper phases, in which a periodic arrangement of sites with a icosahedral coordination alternates with sites with 6-fold coordination. According to Nelson (1985) one can view a glass as a disordered variant of such a phase, where the disclination (and antidisclination) network do not form a regular lattice but instead an irregular network. In the (supercooled) fluid this network is a dynamical object in that the motion of the atoms can deform it (but for topological reasons defect lines of the same type cannot cross, which gives rise to the very slow dynamics), while in the glass the defect line structure is frozen in. Some numerical simulations that follow up these ideas can be found in Sausset *et al.* (2008) and Sausset and Tarjus (2010).

One can use these concepts to develop a field-theoretic approach for the problem of glassy structures. The basic object is a triad of orthogonal unit vectors $\vec{n}_1(\vec{r})$, $\vec{n}_2(\vec{r})$, and $\vec{n}_3(\vec{r})$, describing the local orientation of the icosahedra. Without disclination defects the field theory would be a harmonic Hamiltonian similar to elasticity theory of liquid crystals, i.e. terms $(\nabla\vec{n}_i(\vec{r}))^2$ with suitable prefactors occur in the free energy functional. In order to introduce the defect lines, a vector potential is postulated (using some rather sophisticated symmetry considerations), where the special curvature of space enters that allows a tetrahedral packing without defects. The resulting statistical mechanical model leads to the concept of frustration, qualitatively similar to the "frustrated plaquettes" of Ising spin glasses mentioned in Chap. 1, since a globally satisfactory minimum of the resulting free energy functional in ordinary euclidean three-dimensional space does not exist. With some plausible assumptions this very original and rich theory can be used to even predict the structure factor of amorphous metallic glasses, in reasonable agreement to experiment (Sachdev and Nelson 1984), but involving a few adjustable parameters in the free energy functional. However, since it is neither easy to develop this theory further, nor is it clear how it explains other types of structural glasses, we will not follow up on it here.

Finally we mention that some of the concepts developed in the context of the packing of hard spheres and particles of other shapes can be carried over to the packing of particles in granular matter (see e.g. Duran 1999, Ono *et al.* 2002, Revuzhenko 2006, Keys *et al.* 2007, Metha 2007, Olsson and Teitel 2007), but granular materials will not be discussed further in the present textbook.

3.5 Continuous Random Networks

In the last section the amorphous structure was largely based on the use of the Voronoi construction (see also Chap. 2). The positions of the atoms in the structure were used as a starting point to construct the Voronoi cells, and we have seen that topological constraints are useful to help to understand some of the statistical properties of these cells.

We now continue to use the division of space into Voronoi cells, and consider a related concept, the so-called "Delaunay construction": We just connect the center of each Voronoi cell by straight lines to all the centers of Voronoi cells that share a common face with the first one. If there are no special symmetries in the structure, the resulting network of lines is a "simplicial graph", i.e. it contains only "simplex polyhedra" (tetrahedra in $d = 3$) and "simplex polygons" (triangles in $d = 2$).

If the original structure was a completely random arrangement of points in space (or in a plane, respectively), this construction yields the so-called "continuous random network" (CRN). Such a structure is an idealized starting point for a discussion of the structure of network forming glasses as occur for covalently bonded materials such as amorphous semiconductors, silicate glasses, etc. The geometrical bonds of the graph are then identified with the chemical bonds of the real system.

The three-dimensional "continuous random network" has a tetrahedral coordination, i.e. the coordination number is $z = 4$. This is compatible with the covalent bonds occurring in Si and Ge. Of course, chemistry also allows other possibilities. E.g., in Se we have $z = 2$ covalent bonds, so the material in the crystalline structure forms regular chains, oriented in a preferred direction in the quasi – one dimensional trigonal crystal structure. In the amorphous variant, Se forms chains resembling random walks like amorphous polymers, or closed rings. Branched structures can only form if one introduces suitable defects with a higher coordination number, such as As ($z = 3$) or Ge ($z = 4$), which lead to the formation of an irregular network, reminiscent of the structure of a rubber formed from gaussian coil-like polymer chains interconnected by chemical crosslinks. In the case of sulfur also rings occur, but typically the rings are very small, containing 8 atoms only. In molten Se and molten S one can choose conditions where the covalent bonds can open and close again, a situation analogous to the so-called "living polymers" where chains can break and reconnect or also form branched structures.

It also is of interest to consider the two-dimensional case for which $z = 3$ allows the formation of networks. An example is As, whose crystal structure has planes with atoms forming a honeycomb lattice (actually in the real material these planes are stacked onto each other, hold together by weak van der Waals forces). Related materials are As_2Se_3, or As_2S_3, where the planes exhibit the structure of a decorated honeycomb lattice (see Fig. 1.1).

For all these structures an associated random network structure is conceivable, and it is useful to incorporate the microscopic properties of the covalent bonds of the particular material that is considered into the "rules" according to which the network is built. For instance, for the elementary two-dimensional network with a single atomic species we may require:

(i) $z = 3$, each atom is threefold coordinated.
(ii) The distances between nearest neighbors, i.e. the lengths of the covalent bonds, are held strictly constant. This condition takes into account, that the energy scale for covalent bonds is in the eV range, thus largely exceeding thermal energies (recall 1eV = 11600K) and hence it would be energetically very unfavorable, and in fact does not occur in nature, to have bond lengths that deviate strongly from their optimal distance. Note, however, that we allow for a distribution of the bond angles, while in the crystal structure also the bond angles are precisely fixed.
(iii) The network structure is assumed to be "ideal", in the sense that no loose ends, so-called "dangling bonds", occur in the network.

A network formed according to rules that are motivated by the chemistry differs markedly from an ideal CRN where only the earlier described geometrical construction is invoked. In particular, the short-range order is different: While here we have strictly $z = 3$, in the Delaunay network we

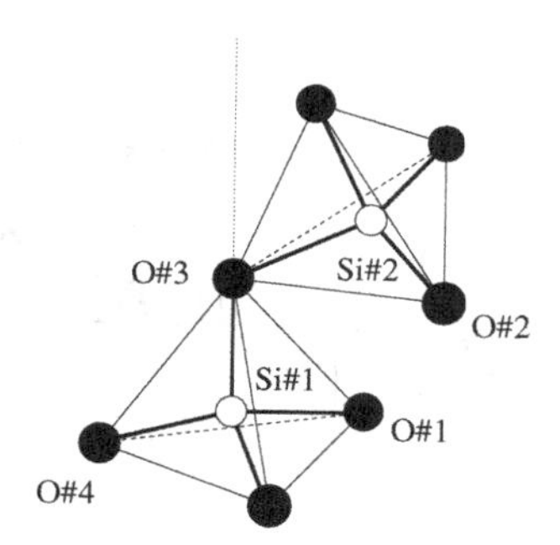

Fig. 3.13 Local arrangement of atoms found in amorphous silica (SiO_2). The silicon atom sits in the center of a tetrahedron with oxygen atoms on its corners. Most of these corners are shared by two tetrahedra and the common oxygen atom is denoted as "bridging". Note that even if all the oxygen atoms are bridging, there is still disorder in the structure since the bond length Si#1-O#1 or the angle Si#1-O#3-Si#2 will vary. From Vollmayr *et al.* (1996).

have a distribution of z, and the average value of z is $[z]_{\rm av} = 6$, as follows from the Euler-Poincaré formula in $d = 2$ mentioned in the previous section. Also the bond lengths in the ideal geometric CRN are not at all constant but instead broadly distributed. The advantage of the CRN constructed according to the chemical rules is, of course, a reasonable description of the short-range order of the real material. Unlike the crystal, it has, however, no long range positional and bond orientational order.

These rules can be generalized to amorphous structures of the type A_2B_3, where one now considers an additional degree of freedom, namely the bond angle of the atom in the bridging position with $z = 2$. Of course it is very natural to assume that this bond angle can vary much more than the bond angles of three-fold coordinated atoms. Therefore one can devise rules for a network construction such that the only disorder that is allowed concerns these bond angles for the bridging atoms, while all bond lengths as well as the bond angles of the three-fold coordinated atoms remain fixed, at their values known from the crystal. Actually, these ideas have already been formulated in the early work by Zachariasen (1932) in which he proposed the first statistical network model for A_2B_3 glasses.

Similar rules apply to three-dimensional networks with $z = 4$, appropriate in the monoatomic case for Si or Ge , and to amorphous silica (SiO_2) in the diatomic case, see Fig. 3.13. For the monoatomic case one requires

(i) $z = 4$, each atom is fourfold coordinated.
(ii) Bond lengths are strictly constant.

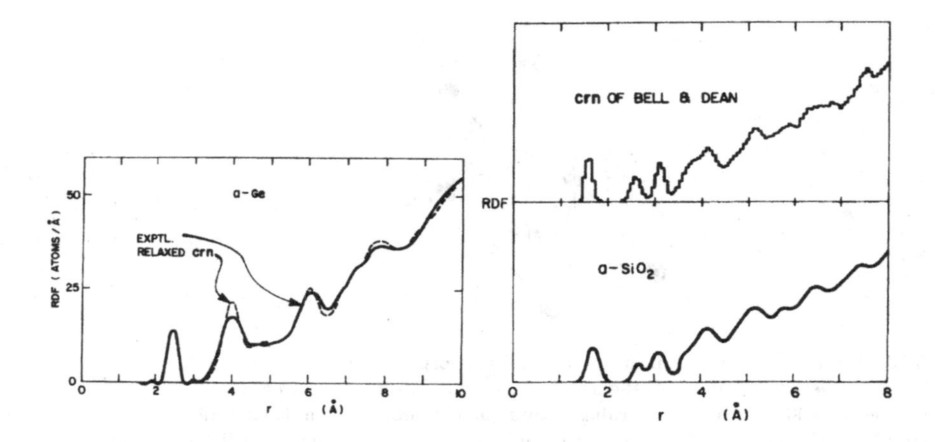

Fig. 3.14 Left: Comparison of the radial distribution function of the refined version of the four-coordinated continuous random network with that observed for amorphous germanium. From Steinhardt *et al.* (1974). Right: Comparison between calculated (upper part) and observed (lower part) radial distribution function (RDF) of amorphous silica (a - SiO_2). From Bell and Dean (1972).

(iii) The network is ideal (no dangling bonds).

Again the CRN built according to these rules is distinguished from the diamond lattice of the corresponding crystal by the lack of long range positional and bond orientational order and by a distribution of bond angles. The left panel of Fig. 3.14 shows that a network built according to such rules reproduces nicely the experimentally observed radial distribution function, and it certainly is quite remarkable that the agreement is so good. Actually, for a long time it was controversial whether or not the rules (i)-(iii) above would be too restrictive to allow the construction of a macroscopically large network at all (one could imagine that the constraints actually are too strong). This controversy was settled, however, when Polk (1971) built a hand-made ball-stick structural model containing 440 atoms according to these rules.

While in the ideal diamond lattice the tetrahedral angle has the value $\Theta = \arccos(1/3) \approx 109°$, in the CRN built according to the above rules one observes fluctuations of the bond angle of about $\Delta\Theta \approx 9°$. It now is possible to refine the structure somewhat by considering the energetics of both small bond angle and bond length variations, which in the harmonic approximation can be written as $\Delta E_\Theta = \frac{1}{2}C_\Theta(\Delta\Theta)^2$ and $\Delta E_\ell = \frac{1}{2}C_\ell(\Delta\ell/\ell)^2$, where the constant C_ℓ for the bond length variations is much larger, $C_\ell \gg C_\Theta$

($C_\ell \to \infty$ according to the above rule). If the constants C_Θ and C_ℓ are known (either from empirical fits or from electronic structure calculations), one can relax the strict bond length constraint and modify slightly the structure by minimizing its total energy. (The RDF shown in Fig. 3.14 has actually been obtained by this approach.)

Bell and Dean (1972) generalized these chemical rules of modeling structures to amorphous SiO_2. One requires perfect chemical order for the coordination numbers and a perfect tetrahedral angle,

(i) z(Si)$= 4$, z(O) $= 2$.
(ii) Constant length of the Si-O bond, perfect tetrahedral bond angle at the Si-atom as in the crystal.
(iii) No dangling bonds.

Unlike crystalline SiO_2, we thus have here disorder in the Si-O-Si angle (which varies by about $\Delta\Theta \approx 15°$ with an average value of about $150°$). Thus the structure consists of regular tetrahedra, with oxygens at the corners, Si-atoms in the centers, and each vertex of a tetrahedra is shared with a second one, so the stoichiometry is correct. Again, this model gives a quite good prediction for the radial distribution function (RDF), right panel of Fig. 3.14.

We now briefly discuss some useful extensions that help to understand the physical effects of impurities incorporated into the network. In fact, it turns out that many properties of amorphous materials are much less sensitive to the presence of impurities than their crystalline counterparts. Consider for instance As impurities in Ge: As belongs to the 5th group of the periodic table of elements ($n = 5$ s-and p-electrons occur in the outermost shell, three electrons hence are unpaired in this shell), and thus it wants to form 3 covalent bonds. If we bring As in a single crystal of Ge, which has $n = 4$ and forms a diamond lattice where each site is fourfold coordinated ($z = 4$), it does not really "fit" in from the chemical point of view since there is one excess electron in its outermost shell. Therefore As acts as "donor" as far as the electrical conductivity of the semiconductor Ge is concerned, and consequently the remaining As-ion has $z = 4$ in the lattice. In amorphous Ge, however, an ideal network can be formed even if As-impurities are present, simply by forming nodes with a different z in the network. In such an "ideal glass" each Ge-atom has $z = 4$, each As has $z = 3$, each Se has $z = 2$, etc. This is the so-called "$z = 8 - n$-rule" (Mott 1969). It explains why electrical properties of amorphous semiconductors are rather insensitive to the presence of impurities that can be incorporated by covalent bonds into the network. Of course, in a real network also defects may occur that violate the $z = 8-n$ rule, such as Se-atoms with $z = 3$ or $z = 1$ (dangling bonds). The situation is similar as for organic networks formed from crosslinked macromolecules, where one also uses the concepts of "ideal networks" (which do not have dangling ends) in a simplified discussion of rubber elasticity, see Rubinstein and Colby 2003).

Particularly important defects in the inorganic network glasses are defects consisting of pairs that respect charge neutrality: e.g. a pair of Se-atoms with $z = 1$ and $z = 3$. Such defects are called "valence alternation pairs" (VAP), and are of central importance to discuss the optical and dielectric properties of amorphous semiconductors. However, such properties are outside the scope of the present treatment.

Of course, these concepts of structural modeling using chemical rules (and possibly refinement according to some assumed potentials for bond lengths, bond angles, etc.) have been generalized to many other systems, including glasses with three kinds of atoms (see, e.g., Wang *et al.* 2000). Nevertheless we will not dwell on these developments here, since the approach is rather heuristic. Also, it should be remembered that the properties of actual amorphous materials are dependent on the "history" of their production process: E.g. if a glass is produced from the melt by a slow cooling process, then physical properties such as the density, but also structural details such as the distribution of bond angles do depend on the cooling velocity. In Fig. 3.15 we reproduce an example from a molecular dynamics (MD) simulation for SiO_2 (Vollmayr *et al.* 1996) showing this for the bond angles. One can see a rather clear dependence of the angular distribution on the cooling rate. Even more sensitive to the cooling rate are the properties relating to intermediate range order, such as the ring size distribution (see next section). Hence it is not clear to what physical production protocol a glass structure produced from "chemical rules" (with or without energetic refinement) actually corresponds to. Since the RDF is actually a rather insensitive probe of the detailed glass structure (this fact is shown in the next section, where it is pointed out that relatively different potential models for MD work lead to very similar RDF's, whereas other physical properties come out rather differently), the conclusion is that the construction procedures discussed here all involve rather uncontrollable errors, and MD simulations are preferable, whenever applicable. The "chemical rule" construction approach of Polk (1971), Bell and Dean (1972), Steinhardt *et al.* (1974), Cargill (1975), and others clearly played an important role in the historical development of a qualitative understanding of the

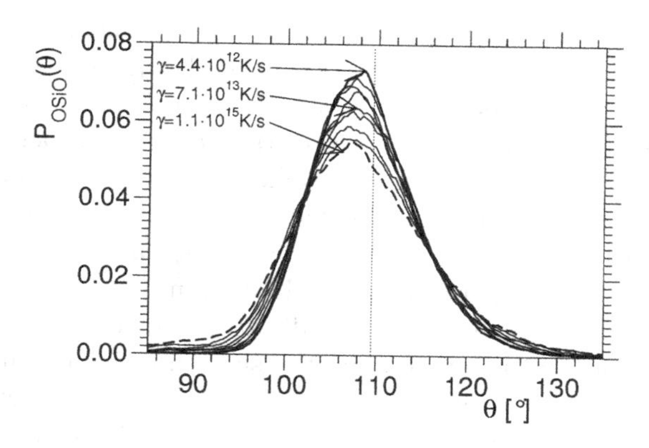

Fig. 3.15 Cooling rate dependence of angular distributions $P_{\mathrm{OSiO}}(\theta)$ of the tetrahedral bond angle θ, as obtained from molecular dynamics simulations using the BKS model discussed in Sec. 3.6, for various cooling rates γ as shown in the figure. The thin vertical line indicates the ideal tetrahedral bond angle of perfect tetrahedra. From Vollmayr *et al.* (1996).

random character of the structure of amorphous materials, but is no longer a competitive research tool in present times.

When we discussed the RCP models (Sec. 3.4) we have argued that a key concept for the theoretical description of glassy structures is the concept of "topological disorder", and we have mentioned the idea of "disclination lines" in a structure with an icosahedral type order that would be preferable on local scales because of "packing" considerations. Although the CRN's based on chemical rules as discussed here have a fixed coordination number z, it would be wrong to conclude that there is no topological disorder. Instead the topological disorder that is present has a somewhat different character than discussed in the previous section. Consider the closed loops ("rings") of covalent bonds formed in the structure (in a crystal such as As the smallest rings all involve $n = 6$ bonds, see Fig. 1.1). The corresponding disordered structure (Fig. 1.2) exhibits disorder of two types:

(i) Instead of regular hexagons we have irregular hexagons, with disorder in bond length and bond angles.
(ii) Instead of hexagons we occasionally also find pentagons and heptagons. These polygons with $n = 5$ and $n = 7$ also occur in the structural model for amorphous semiconductors by Polk (1971), and can be identified in structural models for amorphous SiO_2 as well (e.g. Vollmayr *et al.* 1996).

Now one can show that in three-dimensional CRN's the mentioned pentagons and heptagons cannot occur as localized isolated objects, but instead they must form line-like defects (Rivier 1983): If we connect the center of each pentagon (heptagon) with the centers of the neighboring pentagons (heptagons) we obtain lines that run throughout the random network! While the disorder present in the irregular shape of the hexagons can be removed by continuously deforming them back into a regular hexagon, there is clearly no way to remove the pentagons and hexagons without breaking some bonds and therefore this (topological) disorder is much more fundamental. From a theoretical point of view, the smooth deformations of a polygon can be treated analogously to gauge transformations in electrodynamics, and since the disorder present in the irregularity of hexagons can be eliminated completely by such smooth deformations (as the disorder present in the "hot crystal", Fig. 2.3), one says that this disorder is not "gauge invariant", while topological disorder cannot be removed in this way and is called "gauge invariant". Such considerations have led Rivier (1983) to attempt to formulate a "gauge theory of glasses", in which the "odd lines" connecting pentagons or hexagons are the basic objects of the theoretical description. Similar to the approach by Nelson (1985), one postulates a free energy functional based on a continuum elasticity description with topological defects, representing the "odd lines" and the gauge invariance that is needed is that of the symmetry group SO(3). Although this description (which is similar to the Yang-Mills gauge theory known in the theory of elementary particles) is very elegant formally, it has not yet led to any specific predictions and hence will also not be discussed further.

3.6 Chemically Realistic Models of Structural Glasses

In the previous two sections we have discussed two generic concepts for the modeling of amorphous solids, namely random close packing of hard spheres, and continuous random networks. While such models clearly capture some general qualitative features of the structure of amorphous solids, it is clear that *a priori* these models do not differentiate between the properties of chemically different materials since, at least in their simplest version, system-specific forces do not enter at all. Thus it is not possible to predict any physical properties that depend on these forces (elastic constants, thermal expansion, specific heat, or dynamical properties).

To overcome this drawback it is thus necessary to take into account these system specific forces and to express them in terms of the coordinates of the particles. For the simplest cases these forces can be described by pairwise interactions between the different types of atoms but often it is necessary to include also three-body (or even higher order) terms. However, due to the resulting complexity of the problem, simple analytical treatments are no longer feasible and therefore the standard methods to investigate these systems are Monte Carlo and molecular dynamics computer simulation techniques.

In the Monte Carlo techniques (Binder and Heermann 1988, Binder and Ciccotti 1996, Landau and Binder 2000) one generates a stochastic trajectory through the configuration space of the system. Each proposed individual steps of this random walk through configuration space must satisfy the principle of detailed balance and are accepted or rejected according to the Metropolis *et al.* (1953) acceptance criterion. In the molecular dynamics method, one simply solves numerically Newton's equations of motion for all the atoms in the system, calculating the forces acting on them from the interatomic potentials (Allen and Tildesley 1987, Frenkel and Smit 2001, Binder and Ciccotti 1996). Regarding the physical properties of the resulting models, we note that both methods disregard completely the electrons and treat the atoms (or ions, respectively) in the framework of classical statistical mechanics, hence ignoring all quantum effects. To some extent, the latter can be taken into account by an extension known as "ab initio Molecular Dynamics" (Car and Parrinello 1985, Marx and Hutter 2009) as will be discussed below.

If the interaction potential between the atoms is given by a Lennard-Jones potential with a length scale σ and strength ε, and if we measure the distances in units of σ, one readily finds that the natural time unit of a molecular dynamics simulation is

$$\tau_0 = (m\sigma^2/48\varepsilon)^{1/2}, \tag{3.130}$$

where m is the mass of the particles.[5] For most systems τ_0 is of the order of 10^{-13} s, and since the time step δt of the numerical integration scheme of Newton's equations of motion has to be significantly smaller than τ_0 (e.g. $\delta t = 0.01\,\tau_0$), one immediately can conclude that δt is of the order of 1 femtosecond (10^{-15} sec). From this consideration it becomes obvious that it is intrinsically difficult to mimic by computer simulations, e.g., the production process of real glasses, such as the slow cooling from the melt

[5]Following standard practice we have included here a factor 48. In some papers this factor is, however, not present.

until a temperature is reached at which the structural relaxation time equals the time constant of the cooling and the structure is frozen-in. Even after a million integration steps δt we have only reached times of the order of nanoseconds (10^{-9}sec) and thus are 10 orders of magnitude below the time scale of cooling in a real experiment.

In principle, the difficulty that only very short time scales are accessible to equilibrate the (supercooled) fluid from which the glass is produced can be avoided by using Monte Carlo rather than Molecular Dynamics methods, since for Monte Carlo methods it is possible to use "unphysical" moves which propagate the system through its phase space much faster than more "physical" moves do – provided one can find such moves which satisfy detailed balance and have a reasonable acceptance rate, of course. Unfortunately, for most systems such moves are not known, however. A notable exception is a continuous random network model of amorphous silicon, for which Tu *et al.* (1998) have proposed the potential

$$E_{\rm pot}(\psi,\{\vec{r}_i\}) = \sum_{\ell,j\in\psi} \frac{1}{2} K_\Theta (\cos\Theta_{\ell j} - \cos\Theta_0)^2 + \sum_{j\in\psi} \frac{1}{2} K_b (b_j - b_0)^2 \,. \quad (3.131)$$

This potential depends both on the positions $\{r_i\}$ of the atoms and on the set ψ of bonds connecting pairs (i,k) of atoms. Here j represents the j'th bond and b_j its length, $\Theta_{\ell j}$ is the angle between bonds ℓ and j connected to a common atom. The preferred bond length is b_0, the preferred bond angle is Θ_0, and K_Θ and K_b are "spring constants". If one wishes to focus on the role of network structure, one considers the energy as a function solely of bond topology, minimizing $E_{\rm pot}$ with respect to the geometrical coordinates $\{\vec{r}_i\}$, $E(\psi) = \min_{\{\vec{r}_i\}} E_{\rm pot}\,(\psi,\{\vec{r}_i\})$. Using $E(\psi)$, the system can be studied via Monte Carlo simulations using the so-called "bond switching" moves proposed by Wooten *et al.* (1985): From an initial configuration ψ_1, a bond is chosen randomly (call it BC), and one further bond connected to each terminus is also chosen randomly (bonds BA and CD). The only constraint is that all four atoms A,B,C,D must be distinct. The switching move is then given by cutting the bonds BA and CD and by forming new bonds AC and BD. The so obtained new configuration ψ_2 is then relaxed and used as a trial configuration in the Metropolis acceptance criterion. In this way the system can sample topologically distinct configurations without introducing dangling bonds or changing the number of bonds to any atom. Using this efficient algorithm, Tu *et al.* (1998) investigated the interface between amorphous and crystalline Si.

Note that if one studies a suitable model for Si at temperatures below the melting temperature by means of molecular dynamics, most of the simulation time is spent simulating the vibrations of the network, and only very rarely does one simulate the bond-breaking events. This is the reason why the Monte Carlo method discussed above is advantageous for the problems treated by Tu *et al.* (1998). On the other hand, such Monte Carlo methods do not yield information on the dynamical properties of the glass, and moreover they are not advantageous at very high temperatures, where the coordination number of the atoms (which according to the procedure described above is strictly fixed) also fluctuates. It is clear that then molecular dynamics algorithms are the method of choice, since they do not imply any unphysical constraints on the topology of the system, and yield static and dynamic properties simultaneously.

As an example, we focus here on another network-glass-forming material, molten SiO_2, for which, due to the importance of the material, numerous molecular dynamics simulations have been done for various potentials (Woodcock *et al.* 1976, Soules 1979, Mitra *et al.* 1981, Garofalini 1982, Mitra 1982, Erikson and Hostetler 1987, Tsuneyuki *et al.* 1988, Vashishta *et al.* 1990, Rustad *et al.* 1991, Vessal *et al.* 1993, Rino *et al.* 1993, Vollmayr *et al.* 1996, Wilson *et al.* 1996, Poole *et al.* 1997, Badro *et al.* 1997, Taraskin and Elliott 1997, Horbach and Kob 1999a, Saika-Voivod 2001, Shell *et al.* 2002). Here we only mention the so-called BKS-potential (van Beest *et al.* 1990), which includes a pseudo-Coulomb term and a Buckingham-type potential,

$$E_{\text{pot}}(\{\vec{r}_i\}) = \sum_{i<j} \phi_{\alpha\beta}(|\vec{r}_i - \vec{r}_j|)\,, \tag{3.132}$$

$$\phi_{\alpha\beta}(r) = \frac{q_\alpha q_\beta e^2}{r} + A_{\alpha\beta}\exp(-B_{\alpha\beta}r) - \frac{C_{\alpha\beta}}{r^6}\,, \tag{3.133}$$

where $\alpha, \beta \in$ Si, O, and the effective charges are $q_{\text{O}} = -1.2$ and $q_{\text{Si}} = 2.4$, while the parameters $A_{\alpha\beta}$, $B_{\alpha\beta}$ and $C_{\alpha\beta}$ can be found in the original paper (van Beest *et al.* 1990). This potential succeeds to mimic the directional covalent bonding between the Si-atom and the oxygen atoms through a clever competition between the Si-O pair potential and the O-O pair potential. Of course, it should be kept in mind that classical pair potentials such as Eq. (3.132) cannot properly describe physical effects such as changes of the polarization of electron clouds through varying atomic environments, charge transfer between ions, etc. Thus, one must expect limitations of potentials such as Eq. (3.132), when one discusses the precise character of localized atomic vibrations, etc. A more reliable description of structure and dynamical properties can be obtained by the so-called Car-Parrinello (1985) "*ab initio* molecular dynamics" method, which has been applied to SiO_2 by Pasquarello and Car (1997), Benoit *et al.* (2000, 2001), Tangney and Scandolo (2002), van Ginhoven *et al.* (2005). The disadvantage, however, is that the *ab initio* MD is several orders of magnitude slower than classical MD, and hence only time scales on the order of 10 picoseconds are accessible, and also the length scales that can be studied are rather small (since only a few hundered atoms can be simulated). For systems like SiO_2 it is, however, known that the dynamics and elastic properties show finite size effects if one uses systems with less than a few thousand atoms (Horbach *et al.* 1996 and Leonforte *et al.* 2006). Thus while ab initio MD clearly is useful both in amorphous Si (Car and Parrinello 1988) and SiO_2 to clarify structure and dynamics on the scales of covalent bonds and vibrations of their lengths and angles between them, one has to resort to classical molecular dynamics methods for a study of slower motions and the intermediate range order that we will discuss in the following. However, recent extensions in the efficiency of *ab initio* Molecular Dynamics (Kühne *et al.* 2007, Marx and Hutter 2009) should allow more extensive applications of this technique to amorphous materials in the future.

Despite the computational advantage of the classical simulations, the preparation of a well-equilibrated glassy structure is a severe problem since the time scales accessible are still short (typically 10-100 nanoseconds and thus the cooling rate is around 10^{12} K/s). Early studies (Angell *et al.* 1981) have claimed that the effect of the ultrarapid cooling rate on a simulated structure can be neglected, i.e. that one obtains essentially the same structure as in a laboratory experiment in which the cooling rate is around $10^2 - 10^{-2}$ K/s. However, one now knows that this claim was overly optimistic since in these early simulations system sizes were used that were very small (e.g., Soules (1979) used only 128 oxygen atoms and 64 silicon atoms) and thus it was not possible to achieve the necessary statistical accuracy to obtain meaningful results on this issue. More recent work (Vollmayr *et al.* 1996) has instead given clear evidence that both local properties like the distribution of the tetrahedral angle (Figs. 3.13 and 3.15) or the "ring statistics" (Figs. 3.16 and 3.17) depend rather strongly on the cooling rate. In these simulations, the samples were cooled to zero temperature by reducing $T(t)$ linearly in time,

$$T(t) = T_i - \gamma t, \quad \text{with}\, 4.44 \times 10^{12}\ \text{K/s} < \gamma < 1.14 \times 10^{15}\ \text{K/s}\,, \tag{3.134}$$

where T_i=7000 K is the starting temperature of the quench. One can see

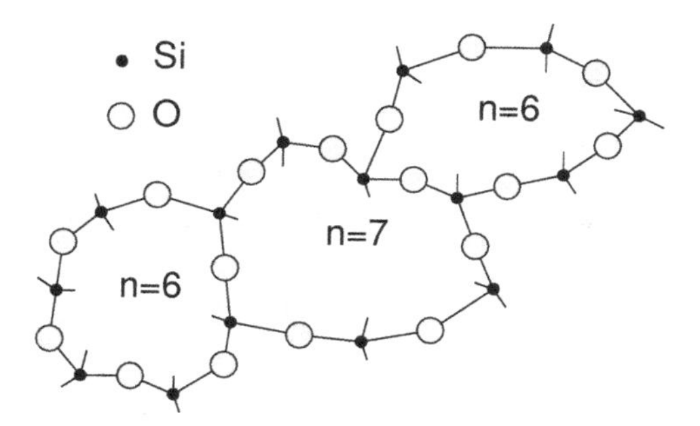

Fig. 3.16 Schematic example of rings identified in the Si-O-network. To define a ring one starts at a Si atom (Si#1) and takes two arbitrary nearest oxygen neighbors that are connected to this Si atom (O#1 and O#2). The shortest path of Si-O bonds that leads from O#1 to O#2, avoiding Si#1, is called the ring associated with O#1,Si#1,O#2. The number of Si atoms in this path defines the length of the ring.

from Fig. 3.15 that with decreasing γ the width of the tetrahedral angular distribution gets systematically narrower, which means that the local order in the tetrahedral network increases. Also with respect to the ring statistics, which characterizes the structure on intermediate length scales (Fig. 3.16), clear cooling rate effects are seen, Fig. 3.17. While many rings with $n = 4$ and $n = 8$ are present for very fast cooling, the number of such rings, $P(n)$, decreases systematically with decreasing γ, while $P(n = 6)$, $P(n = 5)$ and $P(n = 7)$ either somewhat increase or stay constant. Thus this is again evidence that with decreasing cooling rate the system becomes, at least locally, more homogeneous, i.e. ordered.

Similar cooling rate effects are also detected if one studies the density of the system in the framework of a NpT ensemble simulation (the simulated system is then coupled to a "barostat"). Since it is known that molten SiO_2 has a density anomaly at high temperatures, around $T = 1820$ K, caused by a change of sign of the thermal expansion coefficient, it is not easy at all to obtain the correct value of the density at room temperature from cooling runs such as described by Eq. (3.134), since in the simulation the system falls out of equilibrium at a much higher temperature than in the experiment. (In the experiment T_g is around 1450 K whereas the kinetic glass transition temperature accessible in present day simulations is above 2500 K.) Therefore it is better to fix the density at an experimentally reasonable value, rather than trying to predict it, and hence to carry out the simulation in the NVT ensemble (Horbach and Kob 1999a). For pure SiO_2 (described by the BKS potential given by Eq. (3.132)), it is presently

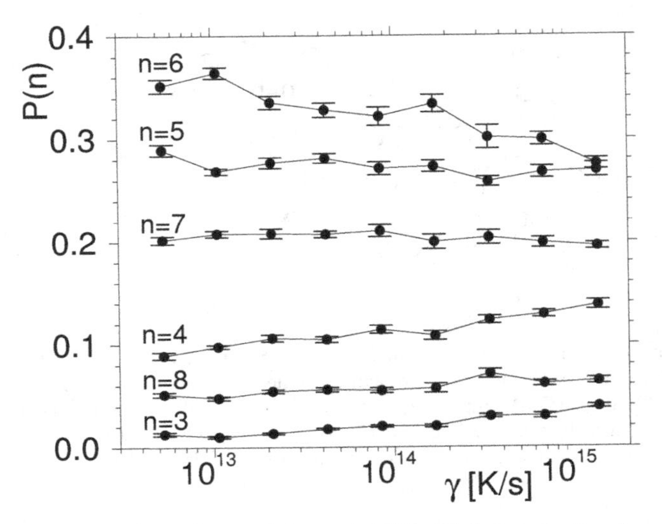

Fig. 3.17 Cooling rate dependence of the probability $P(n)$ that, in the glass at $T = 0$, a silicon atom is member of an n-membered ring. See Fig. 3.16 for the definition of a $n-$membered ring. From Vollmayr *et al.* (1996).

possible to equilibrate the melt at a temperature $T = 2750$ K, a temperature at which the structural relaxation time is of the order of a few nanoseconds. Using such well-equilibrated configurations as the starting state of a cooling run, with a cooling rate of the order of 10^{12} K/s as described above, one then can produce a glass whose structure can be analyzed and compared with experimental data. With this procedure, the neutron structure factor $S^{\rm neu}(q)$ obtained from the model is in very good agreement with the corresponding experimental data, cf. Fig. 2.5. As discussed in Chap. 2, the scattering experiments obtain only one linear combination of the partial structure factors $S_{\alpha\beta}(q)$, while the molecular dynamics modeling readily yields both the partial radial distribution functions $g_{\alpha\beta}(r)$, Fig. 2.8, as well as the partial structure factors, Fig. 3.18. Note that the nearest neighbor distance for Si-O is around 1.6 Å, see Fig. 2.8b, and that therefore this distance is seen in the partial structure factors as a pronounced peak at 2.8Å^{-1}. At smaller wave vectors, $q \approx 1.6\text{Å}^{-1}$, an additional peak can be seen, the so-called "first sharp diffraction peak". The microscopic reason for this peak is the local chemical ordering of the ions in tetrahedron-like structures and the position of the peak corresponds to the distance between neighboring tetrahedra. It is worthwhile to mention that this peak is observed already at temperatures as high as $T = 4000$ K thus showing that

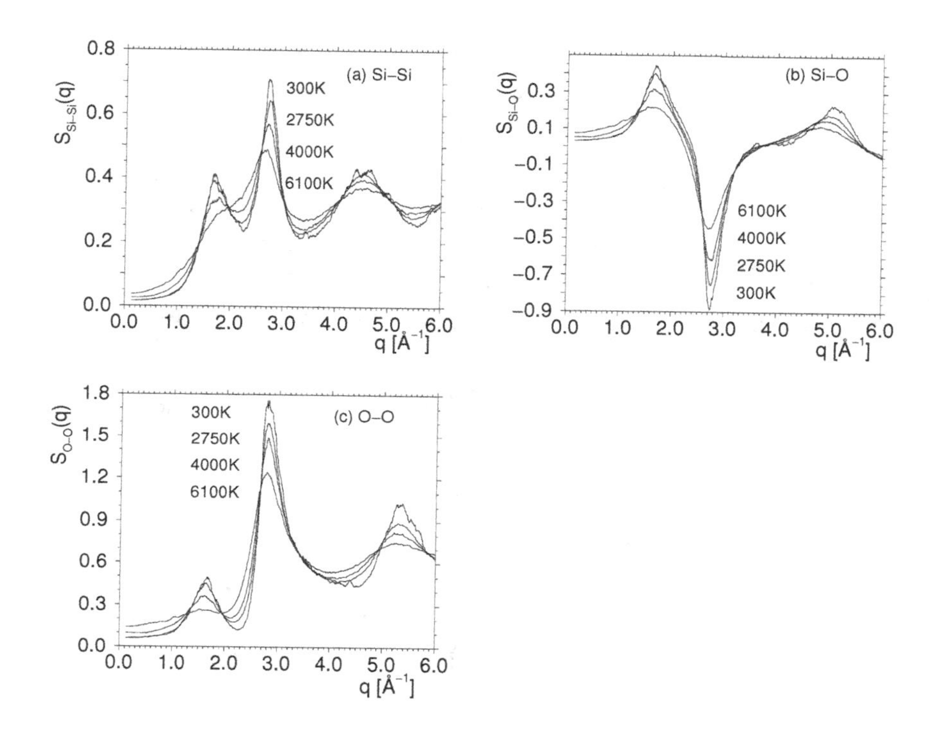

Fig. 3.18 Partial structure factors $S_{\alpha\beta}(q)$ as predicted from simulations using the BKS potential for molten and amorphous SiO_2. The curves for $T \geq 2750$ K are for equilibrated melts whereas the curves for $T = 300$ K have been obtained by quenching the system from $T = 2750$ K to $T = 300$ K and therefore are not in equilibrium. a) Si-Si, b) Si-O, and c) O-O. From Horbach and Kob (1999a).

even at such high temperatures the local structure of the system is given by a corner-shared tetrahedral network.

The same conclusion is reached by investigating the coordination numbers. Since the $g_{\alpha\beta}(r)$ show well-defined minima between the first and second peak (cf. Fig. 2.8), it is possible to identify for each ion its nearest neighbors by requiring that they are within the first neighbor shell, defined by the location of this minimum, as described in Chap. 2. For the Si-Si, Si-O and O-O correlations these minima occur at 3.64 Å, 2.35 Å, and 3.21 Å, respectively. Figure 3.19 shows the probability $P_{\alpha\beta}(z)$ that an ion of type α has exactly z nearest neighbors of type β, for all relevant values of z. One recognizes that even at high temperatures more than 85% of the silicon and oxygen atoms are four and two fold coordinated, respectively, as expected

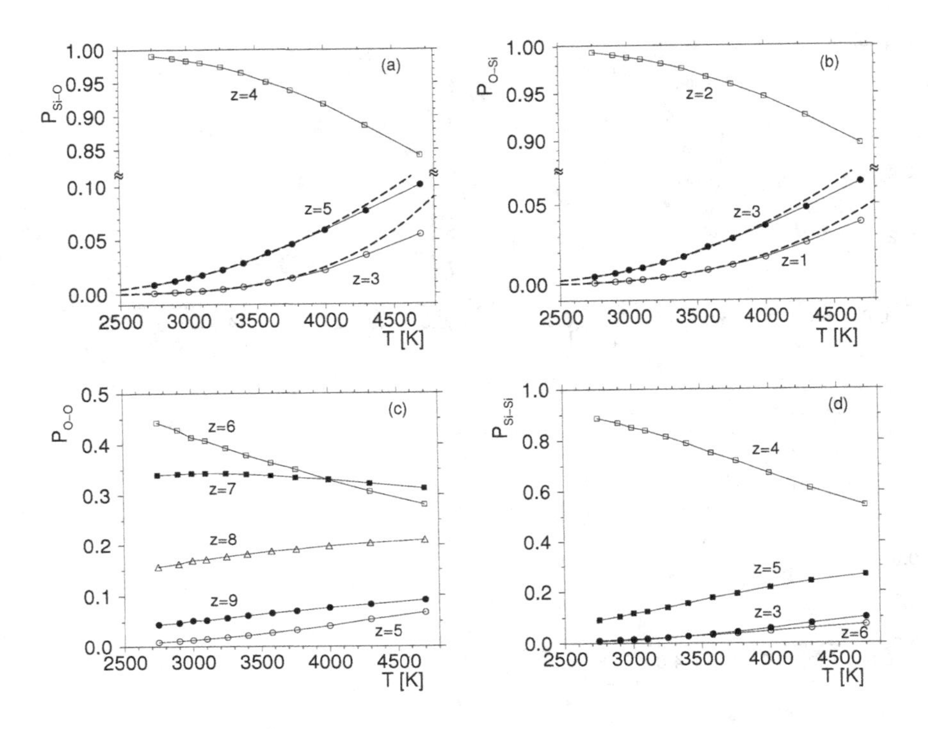

Fig. 3.19 Temperature dependence of the probability $P_{\alpha\beta}(z)$ that an ion of type α in SiO_2 has exactly z nearest neighbors of type β. The dashed lines in a) and b) are fits with an Arrhenius law. a) $P_{\text{Si-O}}$, b) $P_{\text{O-Si}}$, c)$P_{\text{O-O}}$, d)$P_{\text{Si-Si}}$. From Horbach and Kob (1999a).

for the local environment of a corner-shared tetrahedral network. At the lowest temperature investigated this fraction is higher than 99%. The remaining defects are silicon atoms that are three and five fold coordinated and oxygen atoms that are three and one fold coordinated, the latter thus forming "dangling bonds". For $T < 3700$ K the probability of these defects is compatible with Arrhenius laws, $P_{\alpha\beta}(T) = \pi_{\alpha\beta}\exp(-e_{\alpha\beta}/T)$, with parameters $\pi_{\text{SiO}} = 4.4$, $e_{\text{SiO}} = 17130$ K for $z = 5$, $\pi_{\text{SiO}} = 58.6$, $e_{\text{SiO}} = 31100$ K for $z = 3$, for instance (see dashed lines in Fig. 3.19). At low T the most frequent defects are Si-atoms coordinated with 5 oxygens and oxygens bound to 3 silicons. When the system reaches the glass transition temperature, these defects are frozen in. Using the so obtained activation energies and extrapolating the Arrhenius law to the experimental value $T_g = 1450$ K for SiO_2 (Brückner 1970) one can estimate that five-fold coordinated Si occurs

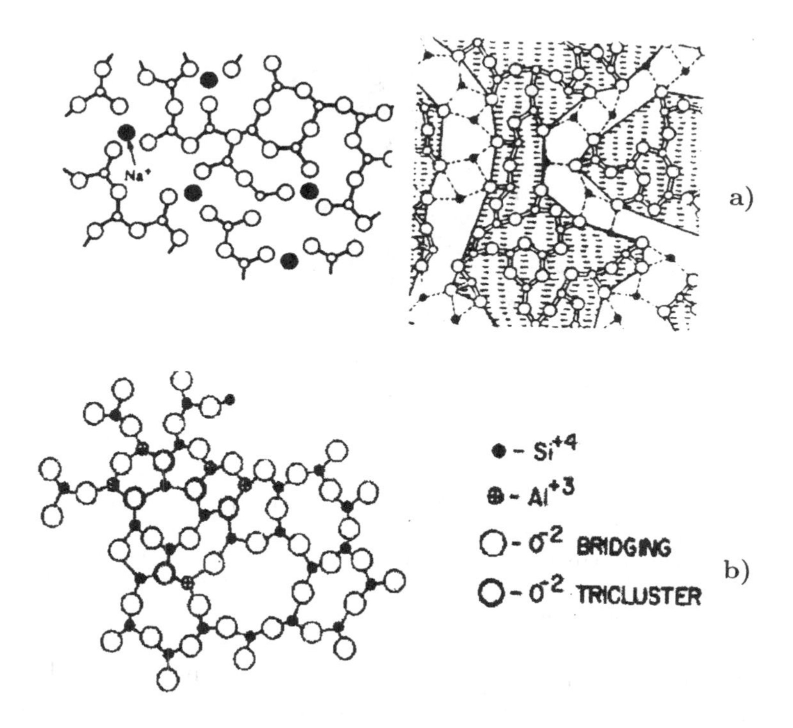

Fig. 3.20 a) Na_2O act in molten and glassy SiO_2 as a network modifier, creating dangling oxygen bonds. Zachariasen (1932) assumed a random distribution of the Na^+ ions in the network (left part), while Greaves (1985) postulated the existence of sodium-rich channels (right part). b) Al-ions in mixtures of Al_2O_3 and SiO_2 can be incorporated into the covalent network, via formation of 3-fold coordinated oxygen ions ("triclusters").

with probability $3.2 \cdot 10^{-5}$ in the glass, and three-fold coordinated oxygen with probability $1.5 \cdot 10^{-5}$. Finally we mention that no such simple temperature dependence is found for $P_{\mathrm{O-O}}(z)$ and $P_{\mathrm{Si-Si}}(z)$, see Fig. 3.19c and d, which shows that for defects occurring on somewhat larger distances a simple minded view of activated states is incorrect.

Very interesting changes in the structure and the dynamics are found if one mixes SiO_2 with other oxides, such as Na_2O (Horbach *et al.* 2001, Horbach *et al.* 2002a) or Al_2O_3 (Horbach *et al.* 2002b). Na^+ ions have the effect to break up the network and thus to create dangling bonds, Fig. 3.20a (Zachariasen 1932). The resulting tetrahedral structure in the melt is much less rigid than the one of pure molten SiO_2 and hence the glass transition temperature is substantially lower in the Na_2O-SiO_2 mixtures than in pure

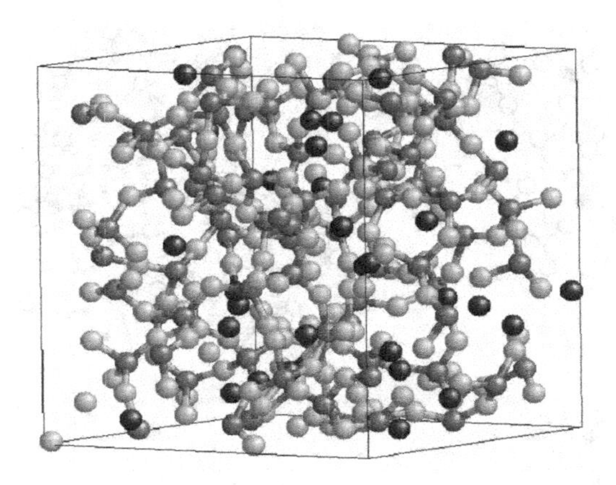

Fig. 3.21 Snapshot of $(Na_2O)\cdot(5\ SiO_2)$ at $T = 2700$ K, showing a cube of linear dimension 16.19 Å (the linear dimension of the total simulated box is about 48 Å, containing 8064 atoms). Silicon atoms are in light gray, oxygen atoms in dark gray, connected by covalent bonds, while sodium atoms are shown as spheres without any bonds attached to them. From Binder *et al.* (2003).

silica, a fact that is important for glass industry (Rawson 1980). While Zachariasen (1932) assumed a random distribution of the Na^+ ions in the network, Greaves (1985) postulated the existence of Na-rich channels in order to explain the fact that the self-diffusion constants of Na^+-ions are orders of magnitude larger than those of O or Si.

In contrast to this, the introduction of Al_2O_3 into the network does not lead to such a formation of channel and anomalous enhancement of the mobility of Al^{3+} ions, and therefore one believes that the Al^{3+} is incorporated into the network, replacing the silicon. However, since the valence of Al is different from the one of Si, such a replacement is not directly possible without changing the local topology of the network (see the discussion on this in Sec. 3.5 in the context of continuous random networks) and therefore it is believed that so-called "triclusters" are formed (MacDowell and Beall 1969), i.e. configurations in which an oxygen atom is connected to one Al and two Si atoms, see Fig. 3.20b.

We now illustrate these statements by briefly reviewing some of the results of Horbach *et al.* (2001, 2002a). Figure 3.21 shows a snapshot of $(Na_2O)(5\ SiO_2)$ at $T = 2700$ K (Binder *et al.* 2003). Already from this snapshot one gets a hint that some clustering of the Na^+-ions is present, although not in the form of compact Na-rich drops, but instead in string-

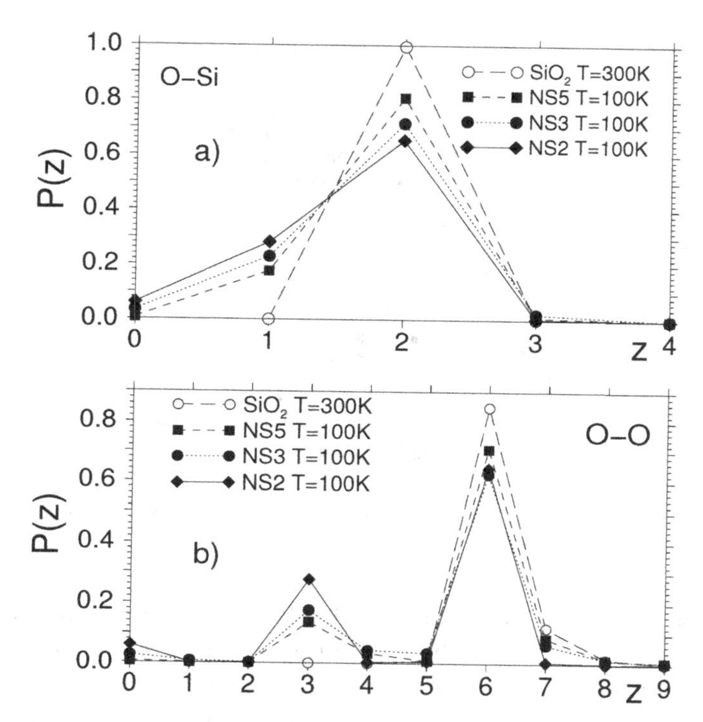

Fig. 3.22 Distribution $P(z)$ of the coordination number z of silicon neighbors of an oxygen atom (upper part) and of oxygen neighbors of an oxygen atom (lower part). The systems with compositions $(Na_2O)\cdot(x\ SiO_2)$ are abbreviated as NSx. The data for SiO_2 is for $T = 300$ K, and the one for NSx are for $T = 100$ K. From Winkler (2002).

like structures, and, as expected, there are indeed dangling bonds (O-atoms covalently bonded to a single Si-atom only), see Fig. 3.20a. These observations can be quantified if we use Eq. (2.36) to determine from the partial radial distribution functions (Fig. 2.8) the number of nearest Si neighbors of all oxygen atoms, $z_{\text{O-Si}}$, and the number of nearest O-neighbors of all oxygen atoms. The resulting distributions $P(z)$ at low temperatures are shown in Fig. 3.22 for the compositions $(Na_2O)\cdot(x\ SiO_2)$, denoted by NSx, with $x = 2, 3$ and 5 as well as the corresponding results for pure SiO_2 (Winkler 2002). (Note that it is irrelevant that the data for SiO_2 is at 300 K whereas the one for NSx is at 100 K since at these low temperatures the topology of the network is independent of T.) While for SiO_2 the data for the O-Si coordination is indistinguishable from perfect coordination $\{P(z = 2) = 1,\ P(z \neq 2) = 0\}$, for NSx we have quite a few oxygen atoms that are connected to only one Si atom, $P(z = 1) > 0$, thus forming

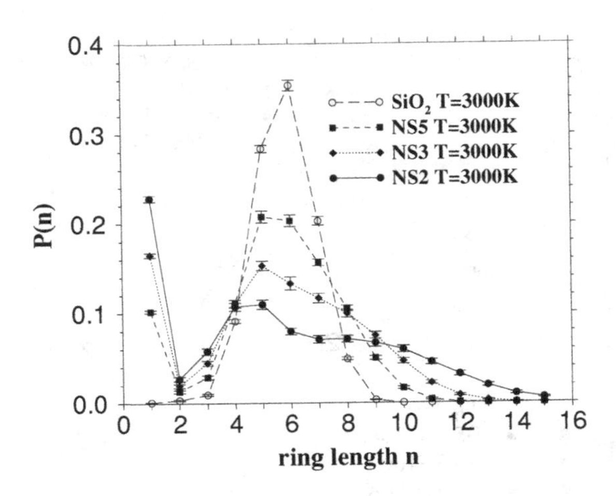

Fig. 3.23 Probability distribution $P(n)$ to find a ring of length n for SiO_2 and various sodium silicate melts at $T = 3000$ K. From Winkler (2002).

dangling bonds. For the O-O coordination, the situation is a bit more complex, since already in the "chemically ideal" SiO_2-network this coordination is not fixed by any simple rule. Thus we find that although the maximum at $z = 6$ is very pronounced, there are also contributions for $z = 5$ and $z = 7$. The peak at $z = 6$ can be simply understood from the fact that each oxygen, which is at a corner shared by two tetrahedra, has $2 \times 3 = 6$ other corners of these two tetrahedra in its neighborhood. Interestingly the $P(z)$ for NSx shows a second peak for $z = 3$, while $P(z = 4)$ and $P(z = 5)$ remain rather small, which is due to configurations in which the oxygen atom considered forms a dangling bond. These dangling bonds are also responsible for the presence of large loops in the network in contrast to the situation in pure silica for which the length of the rings is relatively short (Winkler *et al.* 2004), see Fig. 3.23 for data at high temperatures in the fluid region.

Particularly interesting is the fact that the distribution of sodium ions in the network is not completely random, but instead the sodium ions are concentrated in a (percolating) network of "channels" (Greaves 1985, Ingram 1989, Oviedo and Sanz 1998, Jund *et al.* 2001, Horbach et al. 2002a, Meyer *et al.* 2002, Meyer *et al.* 2004). This irregular microphase separation is recognized rather clearly in the partial structure factors (Fig. 3.24), which exhibit a pronounced "prepeak" at $q \approx 0.9$ Å^{-1}. However, this prepeak is hardly seen in the neutron structure factor $S^{\rm neu}(q)$, which is just the

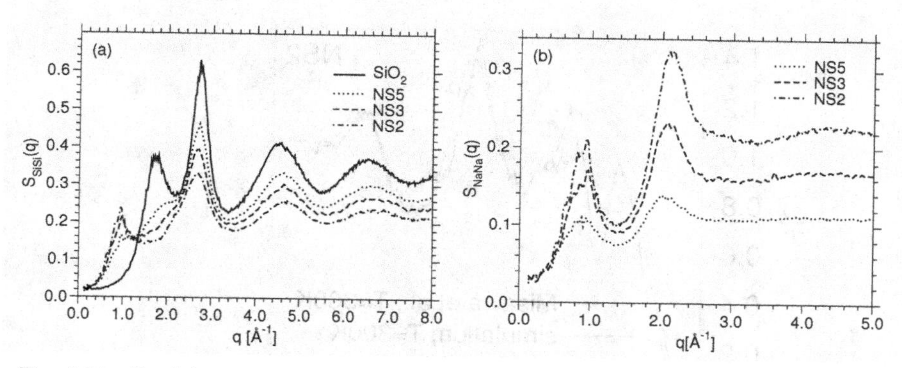

Fig. 3.24 Partial structure factors of Na_2O-$xSiO_2$ (=NSx) at T = 300 K showing a prepeak at around 0.9 $Å^{-1}$. a) $S_{SiSi}(q)$. Also included is the result for pure SiO_2 in order to show that there the prepeak is completely absent. b) $S_{NaNa}(q)$. From Winkler (2002).

weighted sum of the partial structure factors (see Eq. (2.39)), Fig 3.25. The reason for this is that the partial structure factors for $S_{NaSi}(q)$ and $S_{NaO}(q)$ contain a pronounced minimum at $q \approx 0.9$ $Å^{-1}$, which almost cancels the contribution from $S_{SiSi}(q)$ and $S_{NaNa}(q)$ (Horbach and Kob 1999b). Only very recently Meyer *et al.* (2002) were able to show that at sufficiently high temperatures the predicted prepeak can indeed be seen in neutron scattering experiments, since, due to the thermal expansion of the sample, at high T the deconstructive interference effects of the partial structure factors is less pronounced (see also Meyer *et al.* 2004).

As we have seen above, amorphous SiO_2 at low temperatures is an almost perfect tetrahedral network, and the introduction of Na-ions breaks up this network. The addition of Al-atoms has instead a different effect since the Al-ions are incorporated into the network. However, the resulting network is far from ideal, as can be seen immediately from the coordination number distributions $P_{\alpha\beta}(z)$, Fig. 3.26. In this figure O(Si, Al) means that one does not distinguish between Si and Al in the oxygen neighborhood. As one can infer from P_{SiO} and P_{AlO}, at low temperatures most of the Si- and Al-atoms have four O-neighbors, i.e. Al-ions are incorporated into the tetrahedral network structure. However, the relative arrangement of the AlO_4 tetrahedra is somewhat particular in that, see Fig. 3.26c, only 70% of the O-atoms are two-fold coordinated, whereas around 30% of the O-atoms are threefold coordinated by Si and Al-atoms, forming the so-called triclusters, in agreement with the results from NMR experiments (Schmücker *et al.* 1997). It is remarkable that, according to the simulation,

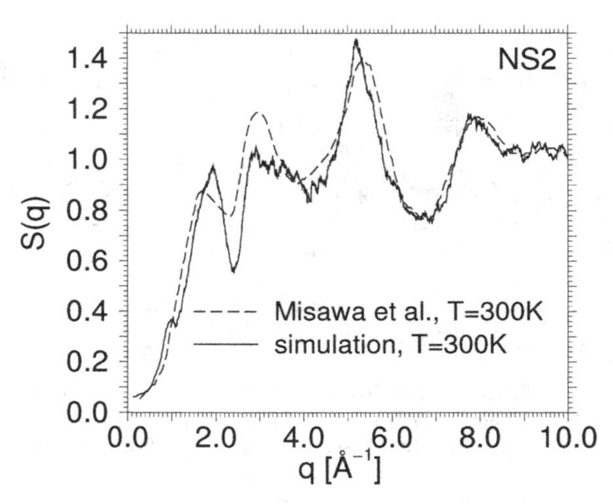

Fig. 3.25 Neutron static structure factor of sodium disicilicate as observed in a neutron scattering experiment and in a computer simulation. The full curve is the molecular dynamics simulation of Horbach and Kob (1999b), where the experimental neutron scattering lengths for Si, O and Na atoms were used, so there is no adjustable parameter whatsoever. The broken curve represents the corresponding experimental data of Misawa *et al.* (1980). From Horbach and Kob (1999b).

the percentage of these triclusters is nearly independent of temperature, Fig. 3.26c. The distributions P_{OSi} and P_{OAl} show that the probability for an O-atom to be three-fold coordinated by Si-atoms is very low (essentially zero at $T = 300$ K), whereas the probability that an oxygen atom is surrounded by three Al-atoms is not negligible (Fig. 3.26d). Most triclusters contain at least one Al-atom. These triclusters also lead to specific features in the angular distribution of the O-Al-O bond angle and the distribution of rings $P(n)$, Fig. 3.27 (Winkler *et al.* 2004). For the definition of rings in this figure no distinction between Si- and Al-atoms has been made. It is seen that there occurs a significant probability for $n = 2$ and $n = 3$, unlike the behavior of pure SiO_2 (Fig. 3.17). It is found that the rings with $n = 2$ almost always contain two Al-ions and one can correlate clearly the existence of triclusters with the presence of these small-membered rings (Winkler *et al.* 2004).

Also in this model for aluminium-silicate glass a prepeak is found in the partial structure factors $S_{\alpha\beta}(q)$ at $q \approx 0.5$ Å^{-1}, indicating a microphase separation between SiO_2 and Al_2O_3 on the scale of nanometers. Since the experimental phase diagram of the metastable fluid (if crystallization is avoided) shows at lower temperature a miscibility gap between a melt rich in

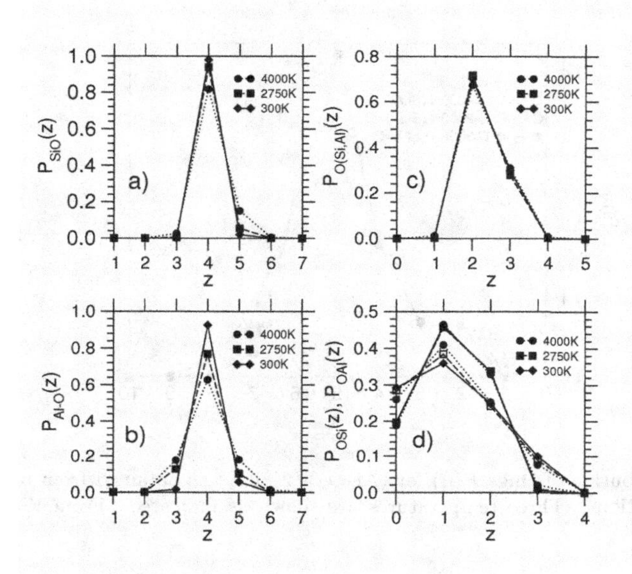

Fig. 3.26 Distributions $P_{\alpha\beta}(z)$ of the coordination number z for the temperature $T =$ 4000, 2300, and 300 K as found in a molecular dynamics simulation of a model for $(Al_2O_3)\cdot(2SiO_2)$. a) $P_{\mathrm{SiO}}(z)$, b) $P_{\mathrm{AlO}}(z)$, c) $P_{\mathrm{O(SiAl)}}(z)$, and d) $P_{\mathrm{OSi}}(z)$ and $P_{\mathrm{OAl}}(z)$. The three curves that are significantly different from zero at $z = 3$ correspond to $P_{\mathrm{OAl}}(z)$. From Winkler *et al.* (2004).

Al_2O_2 and a melt poor in Al_2O_3, one can attribute this prepeak to precursor effects of a macroscopic phase separation. In fact, many amorphous systems are rather inhomogeneous in their chemical constitution on the nanoscale, and these structures are an active topic of research.

Of course, molten SiO_2 and its mixtures with Na_2O and Al_2O_3 are by no means the only systems for which chemically realistic models of structural glasses have been studied. Another example for a continuous random network, amorphous Si, has already been mentioned (Tu *et al.* 1998, Car and Parrinello 1988). Another prominent example of an oxidic glass-forming melt is B_2O_3, which also has been studied by molecular dynamics methods extensively (e.g. Fernández-Perea *et al.* 1996). A particularly interesting glassformer is GeO_2, since for this material the existence of different Ge isotopes has enabled the neutron scattering study of partial structure factors $S_{\mathrm{GeGe}}(q)$, $S_{\mathrm{OO}}(q)$ and $S_{\mathrm{GeO}}(q)$ in the melt (Salmon et al. 2007). Since also in this case an effective potential similar to Eq. 3.132 suitable for Molecular Dynamics work is available (Oeffner and Elliott 1998),

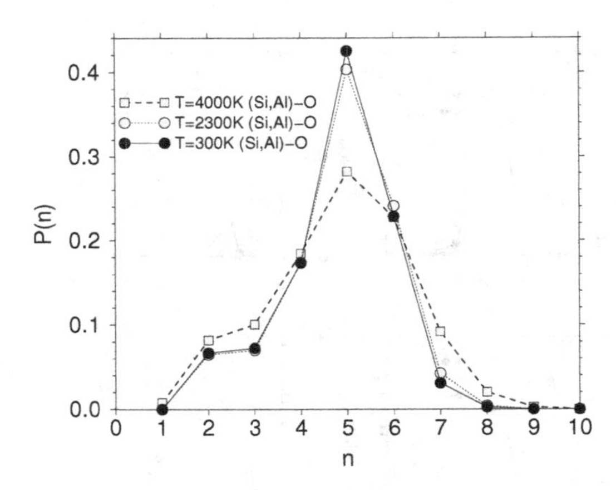

Fig. 3.27 Distribution of rings $P(n)$ for $(Al_2O_3)\cdot(2\ SiO_2)$ as obtained from molecular dynamics simulations. Three temperatures are shown as indicated. From Winkler *et al.* (2004).

extensive simulations have been performed (Hawlitzky *et al.* 2008) and very good agreement with experiment was found. The structural details are qualitatively very similar to those of SiO_2, and hence will not be reviewed here. We only mention that Hawlitzky *et al.* (2008) have further tested the accuracy of their model by a comparison to (presumably more accurate) results from *ab initio* Molecular Dynamics methods. Also glass-forming melts of $CaAl_2Si_2O_8$ have been studied (Nevins and Spera 1998, Ganster *et al.* 2004), as well as the mixture of SiO_2 and $NaAlSiO_4$ (Stein and Spera 1995, 1996). Recent work on other materials includes BeF_2 (Agarwal *et al.* 2007, Agarwal and Chakravarty 2007, 2009) and hydrous silica melts (Pöhlmann *et al.* 2004, Benoit *et al.* 2008); in the latter case, *ab initio* Molecular Dynamics methods had to be used in order to take into account the hydrogen atoms. No attempt is made to give a complete bibliography on the rich work attempting the modeling of network glasses here but instead we refer the reader to the review article by Poole *et al.* (1995). In addition, very extensive work exists also on chemically realistic models for amorphous polymers, such as polyethylene (e.g. Paul *et al.* 1997, Smith *et al.* 2000), polybutadiene (Smith *et al.* 2001, 2002, Paul and Smith 2004, Paul 2007), etc. Since the primary focus of these studies is on dynamic properties rather than structure, we will not discuss them here. A very large activity has also been devoted to the structural modeling of systems which are related to the random close packing of hard spheres (or mixtures of spheres with different sizes), etc. The physical systems that come closest to these types of models of amorphous structures are metallic glasses, e.g. NiZr (Teichler 1996, 1999) and Ni_{25} Zr_{60} Al_{15} alloys (Guerdane and Teichler 2001). Also a model for $Al_{80}Ni_{20}$ molten alloy exhibits properties typical for glassformers (Das *et al.* 2008). More schematic models of this type (not addressing a particular material) are binary mixtures of soft spheres (see e.g. Bernu *et al.* 1985, Roux *et al.* 1989, Barrat *et al.* 1990, Hiwatari *et al.* 1994) or of Lennard-Jones particles (Kob and Andersen 1995a 1995b, Kob 1999). Again, the most interesting aspect of these models is their slow dynamics, and hence we will not dwell further on the details of their structure here.

At the end of this section, we emphasize that we deliberately have discussed a single example for chemically realistic models of structural glasses (namely SiO_2 and its mixtures with other oxides) in detail, as a prototype of glassforming materials. In addition, we have focused on research results from the group of the authors, for simplicity: of course, there exists a wealth of studies from other groups, both on SiO_2 and on other materials. But it would be beyond the scope of this book to give an encyclopedic survey of all the work that has been done. Rather we have tried to give a pedagogic introduction to the basic concepts only; interesting topics of ongoing research (such as the possibility to observe multiple different amorphous structures in the same material and possible transitions between them, see Loerting *et al.* 2009) hence could not be covered.

3.7 Liquid and Solid Foams: Random Cellular Structures

Cellular structures occur under many different circumstances and have many remarkable properties. A common feature of such structures is that one has small compartments enclosed by walls packed together such that they fill space without gaps. In many cases, the walls enclosing the cells are solid, and the contents of the cell may be a fluid. Examples of such "solid foams" or "cellular solids" (Gibson and Ashby 1997) are wood, cork, sponges, corals, occurring in nature, but due to remarkable properties also artificial cellular solid materials are of great technological importance: polymeric solid foams are used for disposable coffee cups, as materials for packaging, for thermal isolation purposes in house constructions, etc.; metal foams find use as car bumpers, and other applications where mechanical strength and light weight are simultaneously needed properties.

Liquid foams are even more common since we find them in properly brewed beer, whipped cream, bath and shaving foams to fire extinguishers and many other applications. Liquid foams consist of two fluids that have unmixed and typically one fluid forms the walls of the structure, while the less dense fluid (often vapor, e.g. simply air) fills the cells - so one can consider the system also as a network of gas bubbles. Typically the thickness of a wall between two neighboring bubbles is much smaller than the linear dimension of the bubbles, and thus a coarse-grained view of the structure just treats the walls as curved surfaces. Two walls meet on lines, and three walls met at vertices - so the idealized view of a foam is just a problem in geometry: one speaks of a "dry foam" (Weaire and Hutzler 1999, Aste and Weaire 2008, Schliecker 2002). In real foams, the walls take a finite volume fraction of the total structure, and then the foam is called "wet foam". Despite their fragility and limited lifetime (e.g. of the order of seconds or minutes), the mechanical response of fluid foams on microscopic timescales is solid-like. The reasons for the limited lifetime of fluid foams on macroscopic timescales even under ideal conditions (absence of drainage caused by gravity, absence of film rupture) is their coarsening behavior: Consider two neighboring bubbles which are separated by a thin liquid film with interface tension 2γ and constant average curvature $H = (1/R_1 + 1/R_2)/2$, where R_1, R_2 are the two principal radii of curvature. The Young-Laplace law implies then a pressure difference $\Delta p = 2\gamma H$. Thus in any irregular structure of bubbles the bubbles with smaller radii of curvature have a larger internal pressure than the larger ones. These pressure differences act as driving force that makes the larger bubbles grow on the expense of the smaller ones, and in the late stages there is a power law describing the increase of the average bubble volume $\langle V(t)\rangle$ with time, $\langle V(t)\rangle \propto t^{3/2}$, and the structure of the foam coarsens in a self-similar, scale-invariant manner (Weaire and Hutzler 1999).

In the limit of dry foams simple geometric laws apply, which are known since a long time ago (Plateau 1873). E.g., when three walls between bubbles meet in a straight line, the angles between the walls must be equal, namely 120°. Those lines are called the "Plateau borders". If four Plateau borders meet in a vertex, they also build a symmetrical tetrahedral structure: i.e., they form the tetrahedral angle 109.6^o. The film curvatures sum to zero at a vertex, and furthermore it is found that no more than 4 Plateau borders meet at a vertex. In fact, these conditions have been proven as necessary conditions for stability (Taylor 1976).

On short enough timescales, where the diffusion of gas between neighboring cells (as is necessary for coarsening) is still negligible, a dry foam can be interpreted as a structure with a partition of space into cells of given volumes with a minimal cost of surface (free) energy. When the cells would have all the same volume, this is known as the "Kelvin problem" (Kelvin suggested as a solution of this problem a packing of identical 14-sided truncated octahedra, and it took more than 100 years to find a slightly better structure, filling space with cells of two different shapes but equal volume, Weaire and Phelan 1994). However, it is still not proven that this structure is really the lowest energy structure of a three dimensional foam.

An interesting variant of cellular structures occurs in two dimensions. In disordered cell structures, one just has Voronoi tesselations, as discussed in another context in previous sections of this chapter, and each Voronoi cell may have $n = 3, 4, 5, \cdots$ neighbors, where one again may apply Euler's relation to conclude that the average number of sides n of a polygon must be six, $\langle n \rangle = \sum_n np(n) = 6$, with $\sum_n p(n) = 1$, $p(n \leq 2) = 0$. The second moment of this distribution, $\mu_2 = \langle (n - \langle n \rangle)^2 \rangle$, is then very useful to characterize the distribution. To a very good approximation, "Lemaitre's law" implies a unique relation between μ_2 and $p(6)$ for a broad variety of real disordered cellular structures in $d = 2$ dimensions (Lemaitre et al. 1993, Schliecker 2002). Lack of space (and expertise) prevents us from giving a more extensive account, and hence we refer the interested reader to books dealing with the physics of foams (Gibson and Ashly 1997, Weaire and Hutzler 1999, Aste and Weaire 2008).

Last but not least we mention that foams can also be sheared and it is found that they show a behavior that is surprisingly similar to the one found in granular materials or glass-forming liquids in that one has a yield stress, the formation of shear bands, etc. (Durian 1995, Höhler and Cohen-Addad 2005, Dennin 2008, Katgert *et al.* 2008). Therefore they are often used as simple model systems that allow to study these complex phenomena on a scale of individual particles and hence to increase our understanding of the disordered systems in which the constituent particles are so small that they cannot be tracked individually.

References

Agarwal, M., and Chakravarty, C. (2007) *J. Phys. Chem. B* **111**, 13294.

Agarwal, M., and Chakravarty, C. (2009) *Phys. Rev. E* **79**, 030202 (R).

Agarwal, M., Sharma, R., and Chakravarty, C. (2007) *J. Chem. Phys.* **127**, 164502.

Aharony, A. (1990) *Physica* **A168**, 479.

Allen, M.P., and Tildesley, D.J. (1987) *Computer Simulation of Liquids* (Clarendon Press, Oxford).

Angell, C.A., Clarke, J.H.R., and Woodcook, L.V. (1981) *Adv. Chem. Phys.* **48**, 397.

Aste, T., and Weaire, D. (2008) *The Pursuit of Perfect Packing, 2nd ed.* (Taylor and Francis, London).

Badro, J., Teter, D.M., Downs, R.T., Gillet, P., Hemley, R.J., and Barrat, J.-L. (1997) *Phys. Rev.* **B56**, 5797.

Baeurle, S.A. (2009) *J. Math. Chem.* **46**, 363.

Barabasi, A.L., and Stanley H.E. (1995) *Fractal concepts in surface growth* (Cambridge University Press, Cambridge).

Barber, M.N. (1983) in *Phase Transitions and Critical Phenomena, Vol. 8* (C. Domb and J.L. Lebowitz, eds.) p. 145 (Academic Press, New York).

Barrat, J.-L., Roux, J.-N., and Hansen, J.-P. (1990) *Chem. Phys.* **149**, 197.

Baschnagel, J., Binder, K., Doruker, P., Gusev, A.A., Hahn, O., Kremer, K., Mattice, W.L., Müller-Plathe, F., Murat, M., Paul, W., Santos, S., Suter, U.W., and Tries, V. (2000a) *Adv. Polymer Sci.* **152**, 41.

Baschnagel, J., Bennemann, C., Paul, W., and Binder, K. (2000b) *J. Phys.: Cond. Mat.* **12**, 6365.

Bell, R.J., and Dean, P. (1972) *J. Appl. Phys.* **43**, 2727.

Benoit, M., Ispas, S., Jund, P., and Jullien, R. (2000) *Eur. Phys. J.* **B13**, 631.

Benoit, M., Ispas, S., and Tuckerman, M.E. (2001) *Phys. Rev.* **B64**, 224205.

Benoit, M., Pöhlmann, M., and Kob, W. (2008) *Europhys. Lett.* **82**, 57004.

Bernal, J.D. (1960) *Nature* **185**, 68.

Bernal, J.D. (1965) in *Liquids: Structure, Properties, Solid Interactions* (T. J. Hughes, ed.) p. 25 (Elsevier, Amsterdam).

Bernu, B., Hiwatari, Y., and Hansen, J.P. (1985) *J. Phys. C. Solid State Phys.* **18**, L371.

Berthier, L., and Witten, T.A. (2009) *Phys. Rev. Lett.* **80**, 021502.

Binder, K. (1976) *Ann. Phys.* **98**, 390.

Binder, K. (1992) in *Computational Methods in Field Theory* (H. Gausterer and C.B. Lang, eds.) p. 59, (Springer, Berlin).

Binder, K., and Ciccotti, G. (1996) (eds.) *Monte Carlo and Molecular Dynamics of Condensed Matter Systems* (Società Italiana di Fisica, Bologna).

Binder, K., and Heermann, D.W. (1988) *Monte Carlo Simulation in Statistical Physics. An Introduction* (Springer, Berlin).

Binder, K., Horbach, J., Kob, W., and Winkler, A. (2003) *Computing in Science and Engineering* **5**, 60.

Bollobás, B., and Riordan, O. (2006) *Percolation* (Cambridge University Press, New York).

Bresser, W., Boolchand, P., and Suranyi, P. (1986) *Phys. Rev. Lett.* **56**, 2493.

Brückner, R. (1970) *J. Non-Cryst. Solids* **5**, 123.

Bunde, A., and Havlin, S. (1991) *Fractal and Disordered Systems* (Springer, Berlin).

Car, R., and Parrinello, M. (1985) *Phys. Rev. Lett.* **55**, 2471.

Car, R., and Parrinello, M. (1988) *Phys. Rev. Lett.* **60**, 204.

Cardy, J. (2000) *Scaling and Renormalization in Statistical Physics* (Cambridge University Press, Cambridge).

Cargill, G. S. (1975) *Solid State Phys.* **30**, 227.

Coniglio, A., and Klein, W. (1980) *J. Phys.* **A13**, 2775.

Coxeter, H.S.M. (1958) *Ill. J. Math.* **2**, 746.

Das, S.K., Kerrache, A., Horbach, J., and Binder, K. (2008) in *Phase Transformations in Multicomponent Melts* (Herlach, D.M., ed.) p. 141 (Wiley-VCH, Weinheim).

De Gennes, P.G. (1979) *Scaling Concepts in Polymer Physics* (Cornell University Press, Ithaca).

De Gennes, P.G. (1984) *J. Phys. Chem.* **88**, 6469.

Dennin, M. (2008) *J. Phys.: Condens. Matter* **20**, 283103.

Des Cloizeaux, J., and Jannink, G. (1990) *Polymers in Solution: Their Modelling and Structure* (Oxford University Press, Oxford).

Deutscher, G., Zallen, R., and Adler, J. (1983) *Percolation Structures and Processes* (Adam Hilger, Bristol).

Donev, A., Cisse I., Sachs D., Variano E.A., Stillinger, F.H., Connelly R., Torquato S., and Chaikin P.M. (2004) *Science* **303**, 990.

Donev, A., Stillinger, F.H., and Torquato, S. (2005) *Phys. Rev. Lett.* **95**, 090604.

Duran, D.J. (1999) *Sands, Powders, and Grains: An Introduction of the Physics of Granular Materials* (Springer, Berlin).

Durian, D.J. (1995) *Phys. Rev. Lett.* **75**, 4780.

Edwards, S.F., and Wilkinson, D.R. (1982) *Proc. R. Soc. London* **A381**, 17.

Erikson, R.L., and Hostetler, C.J. (1987) *Geochimica et Cosmochimica Acta* **51**, 1209.

Family, F. (1990) *Physica* **A168**, 561.

Family, F., and Vicsek, T. (1985) *J. Phys.* **A18**, L75.

Feder, J. (1988) *Fractals* (Plenum Press, New York).

Feng, X., Bresser, W., and Boolchand, P. (1997) *Phys. Rev. Lett.* **78**, 4422.

Fernández-Perea, R., Bermejo, F.J., and Enciso, E. (1996) *Phys. Rev.* **B53**, 6215.

Finney, J.L. (1970) *Proc. Roy. Soc. London A* **319**, 479.

Fisher, M.E. (1967) *Physics* **3**, 267.

Fisher, M.E. (1969) *J. Phys. Soc. Japan Suppl.* **26**, 44.

Fisher, M.E. (1971) in *Critical Phenomena* (M.S. Green, ed.) p. 1 (Academic Press, New York).

Fisher, M.E. (1974) *Rev. Mod. Phys.* **46**, 587.

Flory, P.J. (1953) *Principles of Polymer Chemistry* (Cornell University Press, Ithaca).

Flory, P.J. (1969) *Statistical Mechanics of Chain Molecules* (Interscience, New York).

Frenkel, D., and Smit, B. (2001) *Understanding Molecular Simulation — From Algorithms to Applications* (Academic Press, San Diego).

Ganster, P., Benoit, M., Kob, W., and Delaye, J.-M. (2004) *J. Chem. Phys.* **120**, 10172.

Garofalini, S.H. (1982) *J. Chem. Phys.* **76**, 3189.

Gibson, L.G., and Ashby, M.F. (1997) *Cellular Solids, 2nd ed.* (Cambridge University Press, Cambridge).

Graessley, W.W. (2008) *Polymeric Liquids and Networks: Dynamics and Rheology* (Garland Science, London).

Greaves, G.N. (1985) *J. Non-Cryst. Solids* **71**, 203.

Grosberg, A.Yu., and Khokhlov, A.R. (1994) *Statistical Physics of Macromolecules* (American Inst. of Physics, Woodbury).

Guerdane, M., and Teichler, H. (2001) *Phys. Rev.* **B65**, 014203.

Guyon, E., Roux, S., Hansen, A., Bideau, D., Trodec, J.-P., and Crapo, H. (1990) *Rep. Progr. Phys.* **53**, 373.

Hales, S. (1727) *Vegetable Staticks* (Innys and Woodward, London).

Hawlitzky, M., Horbach, J., Ispas, S., Krack, M., and Binder, K. (2008) *J. Phys.: Condens. Matter* **20**, 285106.

Heermann, D.W., and Stauffer, D. (1980) *Z. Phys.*, **B40**, 133.

Hermes, M., and Dijkstra, M. (2010) *Europhys. Lett.* **89**, 38005.

Herrmann, H. J. (1992) in *The Monte Carlo Method in Condensed Matter Physics* (K. Binder, ed.) p. 93 (Springer, Berlin).

Hiwatari, Y., Matsui, J., Uehara, K., Muranakat, T., Miyagawa, H., Takasu, M., and Odagaki, T. (1994) *Physica* **A204**, 306.

Hohenberg, P.C., and Halperin, B.I. (1977) *Rev. Mod. Phys.* **49**, 435.

Höhler, R., and Cohen-Addad, S. (2005) *J. Phys.: Condens. Matter* **17**, R1041.

Horbach, J., and Kob, W. (1999a) *Phys. Rev.* **B60**, 3169.

Horbach, J., and Kob, W. (1999b) *Phil. Mag.* **B79**, 1981.

Horbach, J., Kob, W., Binder, K., and Angell, C.A. (1996) *Phys. Rev.* **E54**, R5897.

Horbach, J., Kob, W., and Binder, K. (2001) *Chem. Geol.* **174**, 87.

Horbach, J., Kob, W., and Binder, K. (2002a) *Phys. Rev. Lett.* **88**, 125502.

Horbach, J., Winkler, A., Kob, W., and Binder, K. (2002b) in *High Performance Computing in Science and Engineering '02* (E. Krause, W. Jäger, eds.) p. 109 (Springer, Berlin).

Ingram, M.D. (1989) *Phil. Mag.* **B60**, 729.

Jacobs, D.J., and Thorpe, M.F. (1996) *Phys. Rev.* **E53**, 3682.

Jullien, R., and Botet, R. (1987) *Aggregation and Fractal Aggregates* (World Scientific, Singapore).

Jund, P., Kob, W., and Jullien, R. (2001) *Phys. Rev.* **B64**, 134303.

Kamien, R.D., and Liu, A.J. (2007) *Phys. Rev. Lett.* **99**, 155501.

Kardar, M., Parisi, G., and Zhang, Y. (1986) *Phys. Rev. Lett.* **56**, 889.

Kasteleyn, P.W., and Fortuin, C.M. (1969) *J. Phys. Soc. Japan* (Suppl.) **26**, 11.

Katgert, G., Möbius, M.E., and van Hecke, M. (2008) *Phys. Rev. Lett.* **101**, 058301.

Kerner, R., and Micoulaut, M. (1997) *J. Non-Cryst. Solids* **2-3**, 298.

Keys, A.S., Abate,. A.R., Glotzer,. S.C., Durian, D.J. (2007) *Nature Phys.* **3**, 260.

Kirkpatrick, S. (1973) *Rev. Mod. Phys.* **45**, 574.

Kirste, R., Kruse, W.A., and Ibel, K. (1975) *Polymer* **16**, 120.

Kob, W. (1999) *J. Phys.: Condens. Matter* **11**, R85.

Kob, W., and Andersen, H.C. (1995a) *Phys. Rev.* **E51**, 4626.

Kob, W., and Andersen, H.C. (1995b) *Phys. Rev.* **E52**, 4134.

Kolb, M. (1999) in *Anomalous Diffusion. From Basics to Application* (R. Kutner, A. Pekalsi, and K. Sznajd-Weron, eds.) p. 253 (Springer, Berlin).

Kühne, T.D., Krack, M., Mohamed, F.R., and Parrinello, M. (2007) *Phys. Rev. Lett.* **98**, 066401.

Landau, D.P., and Binder, K. (2000) *A Guide to Monte Carlo Simulations in Statistical Physics* (Cambridge University Press, Cambridge).

Langer, J.S. (1980) *Rev. Mod. Phys.* **52**, 1.

Lässig, M. (1998) *J. Phys.: Condens. Matter* **10**, 9905.

Lemaitre, J., Gervois, A., Troadec, J.P., Rivier, N., Ammi, M., Oger, L., and Bideau, D. (1993) *Phil. Mag.* **B67**, 347.

Leonforte, F., Tanguy, A., Wittmer, J.P., and Barrat, J.L. (2006) *Phys. Rev. Lett.* **97**, 05501.

Loerting, T., Brazhkin, V.V., and Morishita, T. (2009) *Adv. Chem. Phys.* **143**, 29.

Lubachevsky, B.D., and Stillinger, F.H. (1990) *J. Stat. Phys.* **60**, 561.

MacDowell, J.F., and Beall, G.H. (1969) *J. Amer. Ceram. Soc.* **52**, 17.

Mandelbrot, B.B. (1982) *The fractal geometry of nature* (Freeman, San Francisco).

Marx, D., and Hutter, J. (2009) *Ab Initio Molecular Dynamics* (Cambridge University Press, Cambridge).

Mauro, J.C., and Varshneya, A.K. (2007) *J. Am. Ceram. Soc.* **90**, 192.

Maxwell, J.C. (1864) *Philos. Mag.* **27**, 294.

Meakin, P. (1988) in *Phase Transitions and Critical Phenomena*, Vol. 12 (C. Domb and J.L. Lebowitz, eds.) p. 336 (Academic Press, New York).

Meakin, P. (1998) *Fractals, Scaling and Growth far from Equilibrium* (Cambridge University Press, Cambridge).

Metha, A. (2007) *Granular Physics* (Cambridge University Press, Cambridge).

Metropolis, N., Rosenbluth, A.W., Rosenbluth, M.N., Teller, A.H., and Teller, E. (1953) *J. Chem. Phys.* **21**, 1087.

Meyer, A., Schober, H., and Dingwell, D.B. (2002) *Europhys. Lett.* **59**, 708.

Meyer, A., Horbach, J., Kob, W., Kargl, F., and Schober, H. (2004) *Phys. Rev. Lett.* **93**, 027801.

Micoulaut, M., and Phillips, J.C. (2003) *Phys. Rev. B* **67**, 104204.

Miranda, V.G., and Reis, F.D.A.A. (2008) *Phys. Rev. E* **77**, 031134.

Misawa, M., Price, D.L., and Suzuki, K. (1980) *J. Non-Cryst. Solids* **37**, 85.

Mitra, S.K. (1982) *Phil. Mag.* **B45**, 529.

Mitra, S.K., Amini, M., Fincham, D., and Hockney, R.W. (1981) *Phil. Mag.* **B43**, 365.

Mott, N.F. (1969) *Phil. Mag.* **19**, 835.

Moukarzel, C., Duxbury, P.M., and Leath, P.L. (1997) *Phys. Rev.* **E55**, 5800.

Mullins, W.W., and Sekerka, R.F. (1963) *J. Appl. Phys.* **34**, 323.

Nelson, D.R. (1985) in *Applications of Field Theory to Statistical Mechanics* (L. Garrido, ed.) p. 13, (Springer, Berlin).

Nelson, D.R., and Spaepen, F. (1989) *Solid State Phys.* **42**, 1.

Nevins, D., and Spera, F.J. (1998) *Am. Min.* **83**, 1220.

Oeffner, R.D., and Elliott, S.R. (1998) *Phys. Rev.* **B58**, 14791.

O'Hern, C.S., Langer, S.A., Liu, A.J., and Nagel, S.R. (2002) *Phys. Rev. Lett.* **88**, 075507.

Olsson, P., and Teitel, S. (2007) *Phys. Rev. Lett.* **99**, 178001.

Ono, I.K., O'Hern, C.S., Durian, D.J., Langer, S.A., Liu A.J., and Nagel, S.R. (2002) *Phys. Rev. Lett.* **89**, 095703.

Oviedo, J., and Sanz, J.F. (1998) *Phys. Rev.* **B58**, 9047.

Parisi, G., and Zamponi, F. (2010) *Rev. Mod. Phys.* **82**, 789.

Pasquarello, A., and Car, R. (1997) *Phys. Rev. Lett.* **79**, 1766.

Paul, W. (2007) *Revs. Comput. Chem.* **26**, 1.

Paul, W., and Smith, G.D. (2004) *Rep. Progr. Phys.* **67**, 1117.

Paul, W., Smith, G.D., and Yoon, D.Y. (1997) *Macromol.* **30**, 7772.

Phillips, J.C. (1979) *J. Non-Cryst. Solids* **34**, 153.

Phillips, J.C. (1981) *J. Non-Cryst. Solids* **43**, 37.

Phillips, J.C., and Thorpe, M.F. (1985) *Solid State Commun.* **53**, 699.

Plateau, J.A.F. (1883) *Statique experimentale et theorique des liquides soumis aux seules forces moleculaires* (Gauthier-Villars, Paris).

Pöhlmann, M., Benoit, M., and Kob, W. (2004) *Phys. Rev. B* **70**, 184209.

Polk, D.E. (1971) *J. Non-Cryst. Solids* **5**, 365.

Poole, P.H., McMillan, P.F., and Wolf G.H. (1995) in *Structure, Dynamics and Properties of Silicate Melts* (J.F. Stebbins, P.F. McMillan, and D.B. Dingwell, eds.) p. 563 (Mineralogical Society of America, Washington).

Poole, P.H., Hemmati, M., and Angell, C.A. (1997) *Phys. Rev. Lett.* **79**, 2281.

Praprotnik M., Delle Site L., and Kremer K. (2008) *Ann. Rev. Phys. Chem.* **59**, 545.

Privman, V.P. (1990) (ed.) *Finite Size Scaling and Numerical Simulation of Statistical Systems* (World Scientific, Singapore).

Pynn, R., and Skjeltorp, A. (1985) (eds.) *Scaling Phenomena in Disordered Systems* (Plenum Press, New York).

Rawson, H. (1980) *Properties and Applications of Glass* (Elsevier, Amsterdam).

Revuzhenko, A.F. (2006) *Mechanics of Granular Media* (Springer, Berlin).

Rikvold, P.A, and Kolesik, M. (2002) *J. Phys. A* **35**, L117.

Rino, J.P., Ebbssö, I., Kalia, R.K., Nakano, A., and Vashishta P. (1993) *Phys. Rev.* **B47**, 3053.

Rivier, N. (1983) in *Topological Disorder in Condensed Matter* (F. Yonezawa and T. Ninomiya, eds.) p. 13 (Springer, Berlin).

Roux, J.-N., Barrat, J.-L., and Hansen, J.-P. (1989) *J. Phys.: Condens. Matter* **1**, 7171.

Rubinstein, M., and Colby, R. (2003) *Polymer Physics* (Oxford University Press, Oxford).

Rustad, J.R., Yuen, D.A., and Spera, F.R. (1991) *Phys. Rev.* **B44**, 2108.

Sachdev, and Nelson D. R. (1984) *Phys. Rev.* **B53**, 1947.

Sadiq, A., and Binder, K. (1984) *J. Stat. Phys.* **35**, 517.

Sahimi, M. (1994) *Applications of Percolation Theory* (Taylor & Francis, London).

Saika-Voivod, I., Sciortino, F., and Poole, P.H. (2001) *Phys. Rev.* **E63**, 011202.

Salmon, P.S., Barnes, A.C., Martin, R.A., and Cuello, G.J. (2007) *J. Phys. Condens. Matter* **19**, 415110.

Sander, L.M. (2000) *Contem. Phys.* **41**, 203 (2000).

Sausset, F., Tarjus, G., and Viot P. (2008) *Phys. Rev. Lett.* **101**, 155701.

Sausset, F., and Tarjus, G. (2010) *Phys. Rev. Lett.* **104**, 065701.

Schäfer, L. (1999) *Excluded Volume Effects in Polymer Solutions* (Springer, Berlin).

Scher, H., and Zallen, R. (1970) *J. Chem. Phys.* **53**, 3759.

Schliecker, G. (2002) *Adv. Phys.* **51**, 1319.

Schlesinger, M.F., Zaslavsky, G.M., and Frisch, U. (1995) *Lévy Flights and Related Topics in Physics* (Springer, Berlin).

Schmücker, M., Mackenzie, K.J.D., Schneider, H., and Meinhold, R. (1997) *J. Non-Cryst. Solids* **217**, 99.

Shante, V.K.S., and Kirkpatrick, S. (1971) *Adv. Phys.* **20**, 325.

Shell, M.S., Debenedetti, P.G., and Panagiotopoulos, A.Z. (2002) *Phys. Rev.* **E66**, 011202.

Shim, Y., and Landau, D.P. (2001) *Phys. Rev.* **E64**, 036110.

Smith, G.D., Paul, W., Monkenbusch, M., and Richter, D. (2000) *Chem. Phys.* **261**, 61.

Smith, G.D., Borodin, O., Bedrov, D., Paul, W., Qui, X., and Ediger, M.D. (2001) *Macromol.* **34**, 5192.

Smith, G.D., Borodin, O., and Paul, W. (2002) *J. Chem. Phys.* **117**, 10350.

Sokal, A.D. (1995) in *Monte Carlo and Molecular Dynamics Simulations in Polymer Science* (K. Binder, ed.) p. 47. (Oxford University Press, Oxford).

Song, C., Wang, P., and Makse, H.A. (2008) *Nature* **453**, 629.

Soules, T.F. (1979) *J. Chem. Phys.* **71**, 4570.

Stanley, H.E. (1971) *Introduction to Phase Transitions and Critical Phenomena* (Oxford University Press, Oxford).

Stanley, H.E., and Ostrowsky, N. (Eds.) (1986) *On Growth and Form* (Martinus Nijhoff, Dordrecht).

Stanley, H.E., and Ostrowsky, N. (Eds.) (1990) *Correlations and Connectivity* (Kluwer, Dordrecht).

Stauffer, D. (1979) *Phys. Rep.* **54**, 1.

Stauffer, D., and Aharony, A. (1992) *Introduction to Percolation Theory* (Taylor & Francis, London).

Stein, D.S., and Spera, F.J. (1995) *Am. Min.* **80**, 417.

Stein, D.J., and Spera, F.J. (1996) *Am. Min.* **81**, 284.

Steinhardt, P., Alben, R., and Weaire, D. (1974) *J. Non-Cryst. Solids* **15**, 199.

Stinchombe, R.B. (1983) in *Phase Transitions and Critical Phenomena*, Vol. 7 (C. Domb and J.L. Lebowitz, eds.) p. 150 (Academic Press, New York).

Strobl, G. (1996) *The Physics of Polymers* (Springer, Berlin).

Tangney, P., and Scandolo, S. (2002) *J. Chem. Phys.* **117**, 1.

Taraskin, S.N., and Elliott, S.R. (1997) *Phys. Rev.* **B56**, 8605.

Taylor, J.E. (1976) *Ann. Math.* **123**, 489.

Teichler, H. (1996) *Phys. Rev. Lett.* **76**, 62.

Teichler, H. (1999) *Phys. Rev.* **B59**, 8473.

Thorpe, M.F. (1983) *J. Non-Cryst. Solids* **57**, 355.

Thorpe, M.F. (1995) *J. Non-Cryst. Solids* **182**, 355.

Torquato, S., Truskett, T.M., and Debenedetti, D.B. (2000) *Phys. Rev. Lett.* **84**, 2064.

Tsuneyuki, S., Tsukada, M., Aoki, H., and Matsui, Y. (1988) *Phys. Rev. Lett.* **61**, 869.

Tu, Y., Tersoff, J., Grinstein, G., and Vanderbilt, D. (1998) *Phys. Rev. Lett.* **81**, 4899.

van Beest, B.W., Kramer, G.J., and van Santen, R.A. (1990) *Phys. Rev. Lett.* **64**, 1955.

van Ginhoven, R.M., Jonsson, H., and Corrales, L.R. (2005) *Phys. Rev. B* **71**, 024208.

van Hecke, H. (2010) *J. Phys.: Condens. Matter* **22**, 033101.

Vashishta, P., Kalia, R.K., and Rino, J.P. (1990) *Phys. Rev.* **B41**, 1297.

Vessal, B., Amini, M., and Catlow, C.R.A. (1993) *J. Non-Cryst. Solids* **159**, 184.

Vicsek, T. (1989) *Fractal Growth Phenomena* (World Scientific, Singapore).

Vold, M.J. (1959) *J. Colloid Sci.* **14**, 168.

Vollmayr, K., Kob, W., and Binder, K. (1996) *Phys. Rev.* **B54**, 15808.

Voss, R.F. (1985) in *Scaling Phenomena in Disordered Systems* (R. Pynn and A. Skjeltorp, eds.) p. 1, (Plenum Press, New York).

Wang, Y., Boolchand, P., and Micoulaut, M. (2000) *Europhys. Lett.* **52**, 633.

Wales D. (2004) *Energy Landscapes* (Cambridge University Press, Cambridge).

Weaire, D., and Hutzler, S. (1999) *Physics of Foams* (Oxford University Press, Oxford).

Weaire, D., and Phelan, R. (1994) *Phil. Mag. Lett.* **70**, 345.

Wilson, M., Madden, P.A., Hemmati, M., and Angell, C.A. (1996) *Phys. Rev. Lett.* **77**, 4023.

Winkler, A. (2002) *Ph.D. Thesis* (Johannes Gutenberg Universität Mainz).

Winkler, A., Horbach, J., Kob, W., and Binder, K. (2004) *J. Chem. Phys.* **120**, 384.

Witten, T.A., and Sander, L.M. (1981) *Phys. Rev. Lett.* **47**, 1400.

Woodcock, L.V., Angell, C.A., and Cheeseman, P.A. (1976) *J. Chem. Phys.* **65**, 1565.

Wooten, F., Winer, K., and Weaire, D. (1985) *Phys. Rev. Lett.* **54**, 1392.

Yamakov, V., Milchev, A., and Binder, K. (1997), *J. Phys. II France* **7**, 1123.

Zachariasen, W.H. (1932) *J. Am. Chem. Soc.* **54**, 3841.

Zallen, R. (1983) *The Physics of Amorphous Solids* (Wiley, New York).

Ziman, J.M. (1979) *Models of disorder. The theoretical physics of homogeneously disordered systems* (Cambridge University Press, Cambridge).

Chapter 4

General Concepts and Physical Properties of Disordered Matter

In the preceding chapter we have seen that many different concepts exist to develop geometrical models that describe the static structure of disordered matter: random walks, random occupation of lattices and percolation, random close packing of spheres, random aggregation models and other fractals, random networks, and last but not least chemically realistic structures defined by means of model Hamiltonians for particular materials. It turns out that these models for the structures are connected quite naturally to various physical properties of interest, such as the thermodynamic properties of the system as well as its relaxation dynamics. In the present chapter, we will first explore these connections, and then discuss some models (such as models for magnetic spin glasses and related physical systems) which are already too complex to be understood in full detail, but the simple concepts (such as percolation and its consequences) do nevertheless help to obtain at least a basic understanding. All these concepts and ideas are also used, at least in some form, by most of the (many!) competing theories that have been proposed to explain the transition from the supercooled fluid to the glass, a topic that will be discussed in Chaps. 5 and 6.

4.1 The Rouse Model for Polymer Dynamics: A Simple Example for the Consequences of the Random Walk Picture

It is obvious that for a discussion of the physical properties of a polymer solution or melt the intra-chain interactions considered in Sec. 3.1 are generally not sufficient, and the interaction between monomers belonging to different chains and with the molecules of the solvent have to be taken into account. In fact, the details of these interactions are indispensable if we want to describe the equation of state of a polymer solution or a polymer melt, and cooperative phenomena such as the crystallization of polymers or the phase separation of polymer solutions or mixtures between different kinds of polymers. Even more complicated order-disorder transitions can occur for semi-flexible polymers (which may show liquid crystalline order) or polymers composed of several chemically distinct blocks (which may show long range ordered meso-phases based on microphase separation), etc. However, all these phenomena are outside of the scope of this textbook but a detailed discussion can be found in the excellent books of Flory (1953), de Gennes (1979a), des Cloizeaux and Jannink (1990), Grosberg and Khokhlov (1994), Strobl (1996), Rubinstein and Colby (2003), and Graessley (2008).

Here we will only treat the dynamical properties of polymers (with a chain length that is not too large) in the dense melt, for which the so-called "Rouse model" (Rouse 1953) is a rather good approximation. In this very simple model the chain is described as a fully flexible bead-spring model, and the interactions of the effective monomers (the "beads") of a chain with other chains in the melt are simply described in terms of friction and random forces. Alternatively, the Rouse model can be interpreted as a description of an ideal chain, i.e. it has gaussian random walk statistics, in a dilute solution with an immobile solvent.[1] The additional physical constraint mentioned above, that the chains should not be too long, is required because we will ignore completely the fact that chains cannot cross each other in the course of their random motions, i.e. they are not entangled. In fact, for very long chains the condition that the dynamics of the chains must respect the constraints imposed by their mutual entanglement does give rise to a very different behavior, which one attempts to describe by the so-called "tube model" or by the concept of "reptation" (Doi and Edwards 1986, de Gennes 1979a). Ignoring this non-crossability constraint can also be interpreted as dealing with "phantom chains" which do not interact with each other directly.

Thus, in the Rouse model each bead is subjected to *i*) the forces $\vec{F}_{\mathrm{ch}}$ from the neighboring beads along the chain, *ii*) the friction force $\vec{F}_{\mathrm{fr}}$ due to the solvent, and *iii*) to the random force $\vec{F}_{\mathrm{r}}$ which appears when the considered bead collides with solvent molecules. As discussed already in Chap. 2 in the context of Brownian motion, the random forces and the friction are not

[1]Note that in a physically realistic description of a dilute polymer solution, the mobility of the solvent particles actually gives rise to long range hydrodynamic interactions that affect the dynamics of polymers substantially, and that are described by the model of Zimm (1956).

independent of each other, but instead linked by the fluctuation-dissipation theorem. Therefore, the equation of motion for the j'th bead is (Grosberg and Khokhlov 1994)

$$m\frac{d^2\vec{x}_j}{dt^2} = \vec{F}_{\mathrm{ch},j} + \vec{F}_{\mathrm{fr},j} + \vec{F}_{\mathrm{r},j}\,, \tag{4.1}$$

where m is the mass of the bead and $\vec{x}_i$ its position in space. Instead of identifying a bead with a monomer of the polymer, one also can do a coarse-graining of the chain, i.e. a bead is then describing an effective Kuhn segment, i.e. a sub-block of the backbone that can be considered as rigid. In this case one does no longer have to deal with the torsional and bending angles between the monomers of the polymer (since they are all lumped together in the Kuhn segment) and thus the interaction between neighboring beads can be described to be essentially of entropic nature. The effective interaction energy due to chain connectivity becomes thus

$$U = \sum_{j=1}^{n-1} U_{j,j+1} = \sum_{j=1}^{n-1} \frac{3k_BT}{2a^2}\left(\vec{x}_{j+1} - \vec{x}_j - a\right)^2, \tag{4.2}$$

where we have ignored constant terms that are independent of the conformation of the macromolecules. Here n is the number of effective Kuhn segments which have an average length a. Hence the force $\vec{F}_{\mathrm{ch},j}$ becomes

$$\vec{F}_{\mathrm{ch},j} = -\frac{\partial U}{\partial \vec{x}_j} = \frac{3k_BT}{a^2}(\vec{x}_{j+1} - 2\vec{x}_j + \vec{x}_{j-1}), \quad j \neq 1, n\,. \tag{4.3}$$

A further simplification arises from the fact that in Eq. (4.1) the inertial term can be neglected, i.e. we assume the friction to be large. Making the Ansatz for the friction force $\vec{F}_{\mathrm{fr},j} = -\zeta d\vec{x}_j/dt$, where ζ is the friction coefficient, we find from Eqs. (4.1) and (4.3) that

$$\zeta\frac{d\vec{x}_j}{dt} = \frac{3k_BT}{a^2}\left(\vec{x}_{j+1} - 2\vec{x}_j + \vec{x}_{j-1}\right) + \vec{F}_{\mathrm{r},j}\,. \tag{4.4}$$

The random force has zero mean value and is assumed to be delta-correlated gaussian noise,

$$\langle\vec{F}_{\mathrm{r},j}(t)\rangle = 0, \quad \langle F^{\alpha}_{\mathrm{r},j}(t)F^{\beta}_{\mathrm{r},j'}(t')\rangle = 2\zeta k_BT\,\delta_{jj'}\,\delta_{\alpha\beta}\,\delta(t-t')\,, \tag{4.5}$$

α, β denoting cartesian components. Equations (4.5) state that the random force (components) acting on different beads (or at different time) are statistically independent of each other. Thus Eq. (4.4) is a linear Langevin equation. It is convenient to consider it in the continuum limit $n \gg 1$ and hence to make the approximation $\vec{x}_{j\pm1} - \vec{x}_j = \pm\frac{\partial \vec{x}}{\partial j} + \frac{1}{2}\frac{\partial^2 \vec{x}}{\partial j^2}$. In this way Eq. (4.4) becomes a diffusion equation in the coordinate system $\{j\}$ along the contour of the chain:

$$\zeta \frac{\partial \vec{x}(t,j)}{\partial t} = \frac{3k_BT}{a^2} \frac{\partial^2 \vec{x}(t,j)}{\partial j^2} + \vec{F}_\mathrm{r}(t,j), \tag{4.6}$$

with

$$\langle F_\mathrm{r}^\alpha(t,j) F_\mathrm{r}^\beta(t',j')\rangle = 2\zeta k_B T \delta_{\alpha\beta}\delta(j-j')\delta(t-t'). \tag{4.7}$$

Of course, like any partial differential equation, Eq. (4.6) must be supplemented by boundary conditions at $j = 0$ and $j = n$. These boundary conditions can easily be derived by considering explicitly the forces on the first and last bead (which differ from Eq. (4.3) due to the "missing neighbors"). This yields

$$\left.\frac{\partial \vec{x}}{\partial j}\right|_{j=0} = 0, \quad \left.\frac{\partial \vec{x}}{\partial j}\right|_{j=n} = 0. \tag{4.8}$$

A solution of Eq. (4.6) subject to these boundary conditions is immediately found by performing the Fourier transformation with respect to the variable j, i.e.

$$\vec{x}(t,j) = \vec{y}_0(t) + 2\sum_{p=1}^{\infty} \vec{y}_p(t)\cos(\pi p j/n), \quad 0 \le j \le n. \tag{4.9}$$

The "normal coordinates" $\vec{y}_p(t)$ are called the "Rouse modes" and are expressed in terms of $\vec{x}(t,j)$ by an inverse Fourier transformation:

$$\vec{y}_p(t) = \frac{1}{n}\int_0^n dj\, \cos(\pi p j/n)\vec{x}(t,j), \quad p = 0,1,2,\ldots. \tag{4.10}$$

Differentiating Eq. (4.10) with respect to t, and using Eq. (4.6) one readily finds

$$\zeta \frac{\partial \vec{y}_p(t)}{\partial t} = -\frac{\zeta}{\tau_p}\vec{y}_p(t) + \vec{f}_p(t), \quad p \neq 0, \tag{4.11}$$

where the Rouse spectrum is

$$\tau_p = \frac{n^2 a^2 \zeta}{3k_B T \pi^2 p^2}, \quad p \neq 0 \tag{4.12}$$

while

$$\zeta \frac{\partial \vec{y}_0(t)}{\partial t} = \vec{f}_0(t)\,, \tag{4.13}$$

and the random forces $\vec{f}_p(t)$ acting on the Rouse modes are

$$\vec{f}_p(t) = \frac{1}{n}\int_0^n dj\, \cos(\pi p j/n)\vec{F}_{\rm r}(t,j)\,. \tag{4.14}$$

Equations (4.5) and (4.7) yield then

$$\langle \vec{f}_p(t)\rangle = 0, \quad \langle f_p^{\alpha}(t) f_q^{\beta}(t')\rangle = \frac{\zeta k_B T}{n}(1+\delta_{p0})\delta_{pq}\delta_{\alpha\beta}\delta(t-t')\,. \tag{4.15}$$

Thus, the Rouse equation Eq. (4.4), or equivalently Eq. (4.6), has been decomposed into a set of independent ordinary differential equations for the coordinates $\vec{y}_p(t)$, the Rouse modes. Thus, in the framework of the Rouse model the motion of a polymer chain is represented by a superposition of independent Rouse modes. The solution of Eq. (4.11) is

$$\vec{y}_p(t) = \frac{1}{\zeta}\int_{-\infty}^{t} dt' \exp\left(-\frac{t-t'}{\tau_p}\right)\vec{f}_p(t')\,, \tag{4.16}$$

and using Eq. (4.15) one easily derives a simple exponential decay for the correlation function of the Rouse modes,

$$\langle \vec{y}_p(t)\cdot\vec{y}_p(0)\rangle = \frac{3k_B T\tau_p}{\zeta}\exp(-t/\tau_p)\,. \tag{4.17}$$

Combining this result with Eq. (4.12) gives an expression for the mean squared amplitude of the Rouse mode

$$\langle |\vec{y}_p(0)|^2\rangle = \frac{na^2}{\pi^2 p^2}\,. \tag{4.18}$$

For $\vec{y}_0(t)$ we have instead from Eqs. (4.13) and (4.10)

$$\vec{y}_0(t) = \frac{1}{\zeta}\int_{-\infty}^{t} dt' \vec{f}_0(t') = \frac{1}{n}\int_0^n dj\,\vec{x}(t,j)\,, \tag{4.19}$$

and hence we note that $\vec{y}_0(t)$ is nothing else but the location of the center of mass of the coil. Therefore the mean squared displacement of the center of mass can be expressed as, using Eq. (4.15),

$$\langle[\vec{y}_0(t) - y_0(0)]^2\rangle = \frac{1}{\zeta^2}\int_0^t dt' \int_0^t dt'' \langle \vec{f}_0(t')\vec{f}_0(t'')\rangle = \frac{6k_BTt}{n\zeta} . \tag{4.20}$$

Thus the diffusion constant of the coil scales with n as $\mathcal{D} = k_BT/(n\zeta)$, while the largest relaxation time τ_1, the "Rouse time", scales as $\tau_1 = n^2a^2\zeta/(3k_BT\pi^2)$.

It is also of interest to calculate the mean squared displacements of individual beads, which become for times $t \ll \tau_1$

$$\langle[\vec{x}(t,j) - \vec{x}(0,j)]^2\rangle \approx \frac{4na^2}{\pi^2}\int_0^\infty \frac{dp}{2p^2}\left[1 - \exp\left(-\frac{tp^2}{\tau_1}\right)\right] = \left(\frac{12}{\pi}\frac{k_BTa^2}{\zeta}t\right)^{1/2} . \tag{4.21}$$

Thus we see that here the dependence on the chain length n has disappeared and therefore one often denotes the mean squared displacement by $g_1(t)$, i.e. with no reference to the $n-$dependence. Furthermore we notice that this function increases only as $t^{1/2}$, instead of the standard Einstein relation, Eq. (4.20), i.e. it shows an anomalous diffusion. At the time $t \approx \tau_1$, there is a smooth crossover between Eqs. (4.20) and (4.21), a time at which the mean squared displacement is comparable to the mean squared radius of gyration of the coil ($R_g^2 = na^2/6$), see Eq. (3.34).

Equation (4.21) also leads to the so-called Kohlrausch-Williams-Watts, or stretched exponential, behavior for the intermediate scattering function of the chain, i.e. a time dependence proportional to $\exp(-(t/\tau)^\beta)$, with a Kohlrausch exponent $\beta \leq 1$. This is most easily seen for the incoherent function, which is defined as, see Eq. (2.75),

$$\begin{aligned} F_s(\vec{q},t) &= \frac{1}{n}\sum_{j=1}^{n}\langle \exp\{i\vec{q}\cdot[\vec{x}_j(t) - \vec{x}_j(0)]\}\rangle \\ &= \frac{1}{n}\sum_{j=1}^{n}\exp\left\{-\frac{1}{6}q^2\langle[\vec{x}_j(t) - \vec{x}_j(0)]^2\rangle\right\} , \end{aligned} \tag{4.22}$$

where the second equality follows from the approximation that $\vec{x}_j(t) - \vec{x}_j(0)$ is a gaussian distributed random variable. Inserting Eq. (4.21) into

Eq. (4.22) we immediately find that

$$F_s(\vec{q},t) = \exp\left\{-\frac{q^2}{6}\left(\frac{12}{\pi}\frac{k_BTa^2}{\zeta}t\right)^{1/2}\right\}, \quad t \ll \tau_1 , \tag{4.23}$$

which thus shows indeed the announced stretched exponential relaxation. The physical interpretation of this result is evident: Due to the random walk structure of the coil it is p, the coordinate conjugated to the coordinate j along the chain, rather than $\vec{q}$ which is the "good quantum number". While the correlation functions $\langle \vec{y}_p(t) \cdot \vec{y}_p(0)\rangle$ show a simple exponential decay, see Eq. (4.17), the decay of $F_s(\vec{q},t)$ is governed by the full spectrum of Rouse modes and the integration over this spectrum, cf. Eq. (4.21), then gives rise to the anomalous diffusion and the stretching of the correlator, Eq. (4.23).

Without derivation we simply mention that if one includes the long range hydrodynamic forces mediated by the solvent, which then leads to the so-called "Zimm model" (Zimm 1956), one finds that (Doi and Edwards 1986)

$$\mathcal{D} \propto k_BT/(6\pi\eta_s R_g), \quad \langle[\vec{x}(t,j) - \vec{x}(t,0)]^2\rangle \propto t^{2/3}, \tag{4.24}$$

where η_s is the viscosity of the solvent. Since Eq. (4.22) is still valid, also in this case $F_s(q,t)$ shows stretched relaxation: $F_s(q,t) \propto \exp(-\text{const}\ q^2t^{2/3})$.

Still a different behavior results if we abandon the assumption of gaussian chains and include excluded volume interactions, so that $R_g^2 \propto a^2n^{2\nu}$ with an exponent $\nu > 1/2$. Keeping still the assumption of an immobile solvent (this is also called the "free draining limit"), one finds that Eq. (4.20) still holds, but the Rouse time scales now like

$$\tau_1 \propto \frac{a^2\zeta}{k_BT} n^{2\nu+1} . \tag{4.25}$$

Again Eq. (4.25) can easily be understood from the fact that the largest relaxation time corresponds to a diffusive motion of the coil over a distance comparable to the coil size, since $\langle[\vec{y}_0(t) - \vec{y}_0(0)]^2\rangle = 6\mathcal{D}\tau_1 \propto a^2n^{2\nu} \propto R_g^2$. For the monomer mean squared displacement the scaling argument that $g_1(t) \equiv \langle[\vec{x}(t,j) - \vec{x}(t,0)]^2\rangle$ should be independent of n for $t \ll \tau_1$ and $g_1(t = \tau_1) \propto R_g^2$ implies

$$g_1(t) \propto a^2(k_BTt/a^2\zeta)^{\frac{2\nu}{2\nu+1}} . \tag{4.26}$$

For $d = 2$ spatial dimensions where $\nu = 3/4$, see Eq. (3.19), this yields $g_1(t) \propto t^{3/5}$ while for $d = 3$ where $\nu \approx 0.59$ this yields $g_1(t) \propto t^{0.54}$ (Kremer and Binder 1984). These power-laws have been verified by numerical simulations both for $d = 3$ (Paul *et al.* 1991) and $d = 2$ (Milchev and Binder 1996).

It is also of interest to discuss the dynamic structure factor $S(q, \omega)$ for single chains (de Gennes 1967). It can be shown that this structure factor satisfies a dynamic scaling hypothesis (Hohenberg and Halperin 1977)

$$S(q, \omega)/S(q) = [\omega_c(q)]^{-1} \widetilde{S}\{\omega/\omega_c(q)\}\,. \tag{4.27}$$

Here $S(q)$ is the static single-chain structure factor discussed in Sec. 3.1, $\widetilde{S}(X)$ is a scaling function, and the characteristic frequency $\omega_c(q)$ scales like

$$\omega_c(q) \propto q^z\,, \tag{4.28}$$

where z is the dynamic exponent, which is $z = 2 + 1/\nu$ in the case of the Rouse model, while $z = 3$ in the case of the Zimm model. From Eq. (4.28) it follows immediately that the mean squared displacement scales like $g_1(t) \propto t^{2/z}$. Later we will introduce the so-called "spectral dimension" d_s which is related to the length scale measured by $g_1(t)$ as $g_1(t) \propto t^{d_s/d_f} = t^{\nu d_s}$ (remember that we have seen in Sec 3.1 that polymers have the fractal dimension $d_f = 1/\nu$). This relation then implies for the Rouse model

$$d_s = 2/(\nu z) = 2/(2\nu + 1)\,. \tag{4.29}$$

Note that this expression gives $d_s = 1$ if we set $\nu = 1/2$.

To conclude this section, we describe some tests of these concepts by computer simulation.

The first investigation was done by Baumgärtner and Binder (1981) using Monte Carlo simulations of a melt of freely jointed chains of $N = 16$ beads, that interacted with each other with a Lennard-Jones potential, keeping the length ℓ of the links fixed. Monte Carlo moves consisted of choosing a bead (i) at random and attempting a rotation around an axis joining the two neighboring beads $i - 1$, $i + 1$ by a randomly chosen angle $\varphi \in [0, \pi]$ (Fig. 4.1a).

Indeed the resulting incoherent intermediate scattering function $F_s(q, t)$ of single chains complies nicely with the predicted stretched exponential behavior from Eq. (4.23), i.e. $\ln F_s(q, t) \propto t^{1/2}$, see Fig. 4.1b. However, this early work did not investigate the properties of the individual Rouse modes, Eqs. (4.12) and (4.17). A more extensive test of another model, the so-called bond-fluctuation model (Carmesin and Kremer 1988, Deutsch and Binder 1991) on the simple cubic lattice, was performed later by Okun

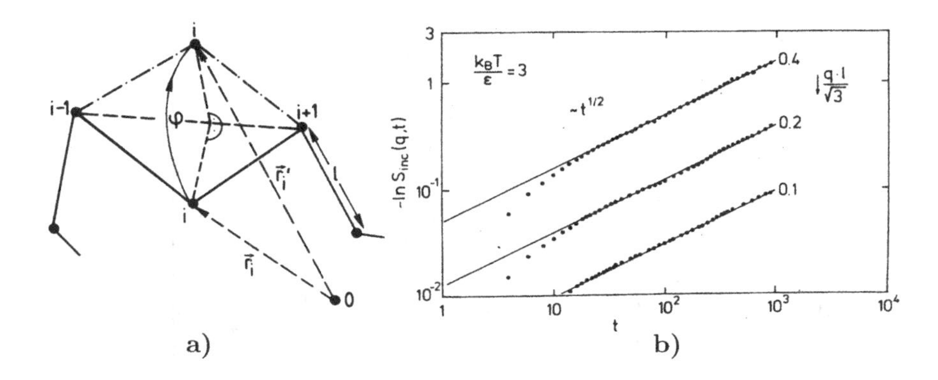

Fig. 4.1 a) Off-lattice model for a polymer chain. In this freely jointed chain rigid links of length ℓ are jointed at beads, marked by dots, forming arbitrary angles with each other. The stochastic chain conformational changes are modeled by random rotations of bonds around the axis connecting the nearest-neighbor beads along the chains, as indicated. Beads interact with a Lennard-Jones potential, using $\sigma = 0.4\ell$. b) Log-log plot of the logarithm of the incoherent scattering function $F_s(q,t)$, here denoted by S_{inc}, versus time (in units of Monte Carlo steps per bead) and $k_BT/\varepsilon = 3$. Three wave-vectors q are shown for a single off-lattice freely jointed chain with $N = 16$ beads. From Baumgärtner and Binder (1981).

et al. (1997) using $n = 10$. Fig. 4.2 shows that the equal-time mean square amplitude of the Rouse modes is in reasonable agreement with the prediction from Eq. (4.18), $\langle|\vec{y}_p(0)|^2\rangle = n^2a^2/(\pi^2p^2)$.

Of course, the smallness of n in the simulation (which is necessary to avoid the onset of reptation effects) renders the continuum approximation, done to replace Eq. (4.4) by Eq. (4.6), somewhat inaccurate, in particular if p becomes comparable to n. In fact, one can solve also Eq. (4.4) directly to find $\tau_p = [\zeta a^2/12k_BT]/[\sin(p\pi/2n)]^2$, $p = 1, \cdots, n-1$, and this result agrees with the simulation even better. However, there are more important systematic deviations of the simulation results from the Rouse predictions than these effects related to the finite length of the chain: Fig. 4.3 shows that the decay of the time correlation function is not the simple exponential predicted by Eq. (4.17), but instead it is somewhat stretched, and that this stretching becomes more pronounced with increasing mode index. These deviations must be due to many-body effects and thus are beyond the approximation done within the Rouse model that the interaction of a single chain with all the other monomers are simply described by a friction coefficient.

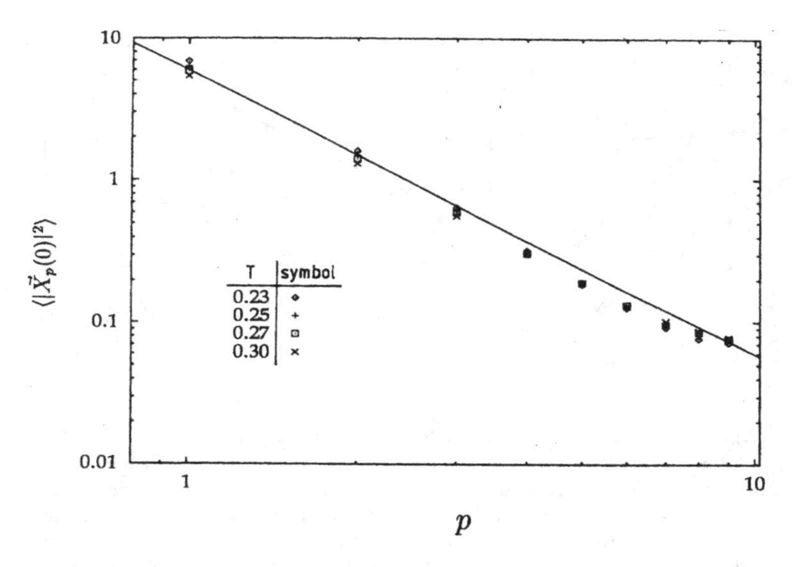

Fig. 4.2 Log-log plot of the mean squared amplitude of the Rouse modes, denoted here as $\langle|\vec{X}_p(0)|^2\rangle$, versus the mode index p, for a chain containing $n = 10$ for the bond fluctuation model on the simple cubic lattice at a volume fraction of $\phi = 0.5$ occupied sites. In this model temperature is introduced by choosing an energy $\varepsilon = 0$ to the bond $(3, 0, 0)$ a_0, where a_0 is the lattice spacing, or permutations thereof, while all other bonds correspond to an energy $\varepsilon = 1$. The straight line shows the relation $\langle|\vec{X}_p(0)|^2\rangle = na^2/(\pi^2 p^2)$, see Eq. (4.18), where the effective length a is taken from the mean squared end-to-end distance as $a^2 = \langle R^2\rangle/(n-1)$. Four temperatures are shown as indicated. In this model, the critical temperature of mode coupling theory turns out to be $T_c \approx 0.15$, while the Vogel-Fulcher temperature T_0 is $T_0 \approx 0.12$. From Okun *et al.* (1997).

Also the gaussian approximation used in Eq. (4.22) is slightly inaccurate since one finds that $F_s(q,t)$ decays somewhat slower than expected from Eqs. (4.22) and (4.23), see Fig. 4.4. Note that the graph shown in Fig. 4.4 is completely independent of the Rouse model - both $F_s(q,t)$ and $g_1(t)$ are directly observed in the simulation, and no adjustable parameters of any kind are involved.

At high temperatures, $T \geq 0.30$, the decay of $F_s(q,t)$ with increasing time (note that $g_1(t)$ increases uniformly with time, so t is an implicit parameter of the curves in Fig. 4.4) is very close to the exponential decay $\exp(-g_1 q^2/6)$ and there is only a weak stretching. In contrast to this one notes for the two lowest temperatures, $T = 0.21$ and $T = 0.23$, that the correlator develops a shoulder at about $F_s(q,t) \approx 0.6$. This shoulder turns out to be a precursor of the plateau that is seen for $T \to T_c$, where T_c is

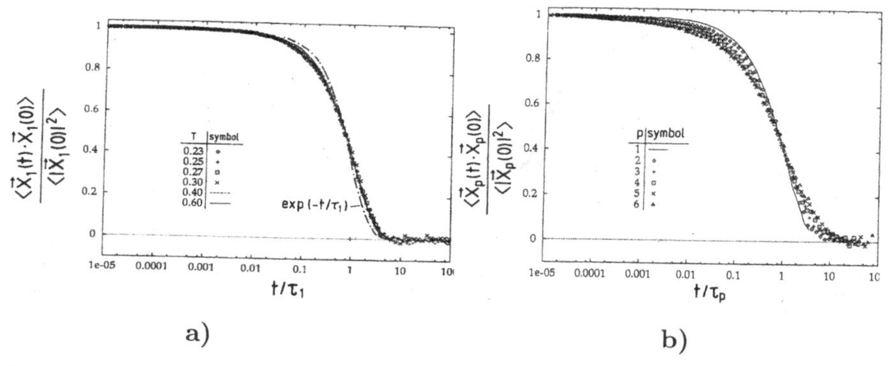

Fig. 4.3 a) Semi-log plot of the normalized autocorrelation function of the first Rouse mode versus scaled time t/τ_1, for the same model as Fig. 4.2, and compared with the theoretical prediction, $\exp(-t/\tau_1)$, Eq. (4.17). Here τ_1 has been estimated from the simulation from the condition that at $t = \tau_1$ this correlator has decreased to $1/e$. Various temperatures are included, as indicated in the figure. From Okun i*et al.* (1997). b) Semi-log plot of the normalized autocorrelation functions of the first six Rouse modes versus scaled time t/τ_p, for the same model as Fig. 4.2, at $T = 0.23$. From Okun *et al.* (1997).

the critical temperature of mode coupling theory, see Chap. 5, and which for this model is $T_c \approx 0.15$ and which is associated with the so-called "cage effect". This prominent feature of glassy relaxation will be discussed in more detail in Chap. 5. But we see already here that stretched exponential relaxation and non-gaussian behavior of dynamic correlators are fairly general features of the behavior of polymeric liquids over a wide temperature regime, and occur not only close to the glass transition.

In the Monte Carlo simulations shown in Figs. 4.2-4.4 the stochastic character of the relaxation is built in by construction, and hence one does not test at all the first step of the treatment, where in the passage from Eq. (4.1) to Eq. (4.4) the inertial term $md^2\vec{x}_j/dt^2$ was neglected. In order to test for possible effects of that term, it is useful to use molecular dynamics simulations of a bead-spring model. Figure 4.5 shows the mean squared displacement from such a simulation. It is seen that for very short times (times shorter than the characteristic time unit t_0 of the molecular dynamics model, corresponding to displacements that are significantly smaller than the distance between monomers) a "free flight" ballistic motion of the monomers is found, which is not described by Eqs. (4.4) or (4.6). Only for displacements of the order of the mean distance between monomers (or larger), which in this model is around 1.0, the Rouse description becomes applicable.

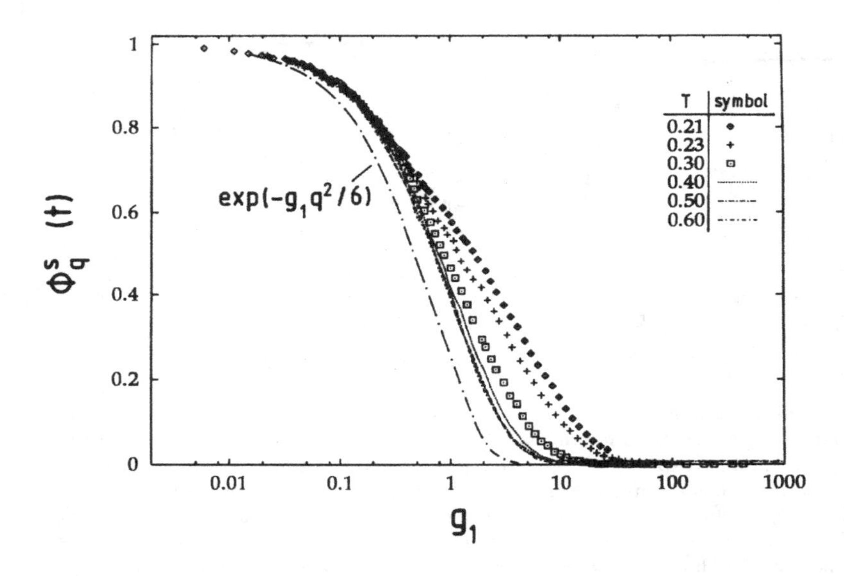

Fig. 4.4 Plot of the intermediate incoherent scattering function $F_s(q,t)$ [denoted here as $\phi_q^s(t)$] versus the mean squared displacement $g_1 = \langle[\vec{x}_j(t) - \vec{x}_j(0)]^2\rangle$ of the monomers, for the same model as in Fig. 4.2, and various temperatures as indicated in the figure. The result of Eq. (4.22), $F_s(q,t) = \exp(-g_1q^2/6)$, is shown by a dash-dotted line. All data refer to $q = 2.92/a_0$, where the static structure factor of the melt exhibits its "first sharp diffraction peak" also called "amorphous halo". From Okun *et al.* (1997).

Figure 4.5 also demonstrates further limitations of the Rouse model: While Eq. (4.20) predicts for this model an Einstein relation that is valid for all times and not only in the asymptotic limit $t \to \infty$, the figure shows that in the simulation the Einstein relation $g_3(t) = 6\mathcal{D}t$ holds only if $g_3(t) > R_g^2$, whereas for $a^2(=1) < g_3(t) < R_g^2$ one observes anomalous diffusion, $g_3(t) \propto t^{0.8}$. This behavior is not an artifact of this particular model, but it is found in a wide variety of models, and is hence believed to be a general property of the dynamics of dense polymer melts. Also the mean squared displacements of the monomers deviates from Eq. (4.21) systematically, since one observes $g_1(t) \propto t^{0.63}$ instead of the predicted $g_1(t) \propto t^{0.5}$. Nevertheless the decomposition of the mean squared displacements into the Rouse correlators $\langle\vec{y}_p(t)\cdot\vec{y}_p(0)\rangle$ works very well, as Fig. 4.5 shows, and hence interference between different Rouse modes (correlators $\langle\vec{y}_p(t)\cdot\vec{y}_q(0)\rangle$ with $p \neq q$) are indeed negligible.

We have presented here this detailed assessment of the Rouse model of polymer melts, since polymer melts are important model materials for

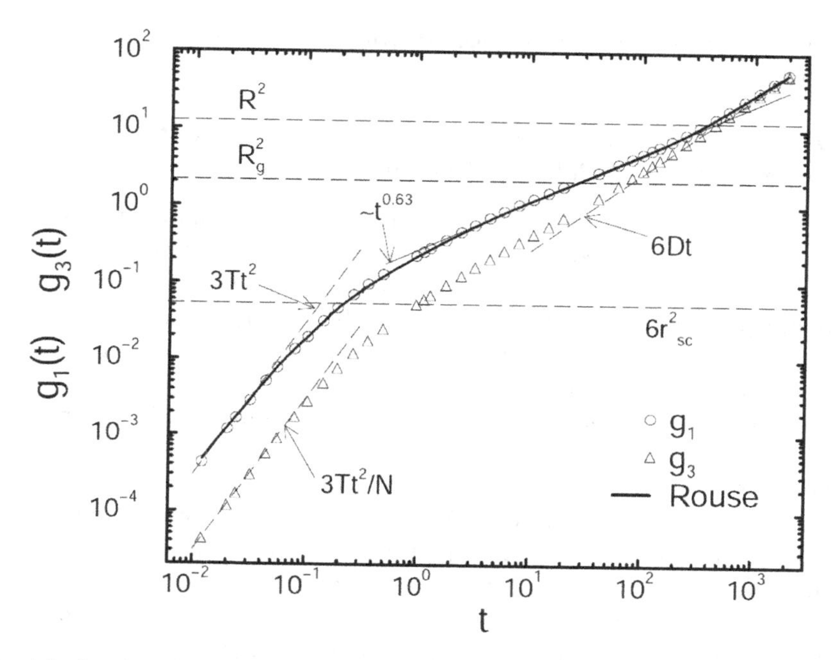

Fig. 4.5 Log-log plot of $g_1(t) = \langle[\vec{x}_j(t) - \vec{x}_j(0)]^2\rangle$, the mean square displacement of an inner monomer, and of $g_3(t)$, the chain's center of mass, versus time for a melt of polymers interacting via a Lennard-Jones potential. The chain length is $N = 10$, the monomer density is $\rho = 1$, and the temperature is $T = 1$. Distance and time is measured in Lennard-Jones units σ and $t_0 = \sigma(m/\varepsilon)^{1/2}$ with a monomer mass $m = 1$. The thick solid line through $g_1(t)$ is the prediction of the Rouse model, using as input the correlation functions $\langle y_p(t)\, y_p(0)\rangle$ of the Rouse modes for $p = 1, \cdots, 9$. Broken straight lines labeled $3Tt^2$ and $3Tt^2/N$ show the ballistic behavior of the monomers and the center of mass, respectively, for very short times, where the interaction of a monomer with neighboring monomers is not yet felt. Characteristic lengths shown are the mean squared gyration radius R_g^2 and the mean square end-to-end distance R^2. Also, the diffusion of the center of mass, $g_3(t) = 6\mathcal{D}t$, is indicated, as well as an effective power-law $g_1(t) \propto t^{0.63}$ observed at intermediate times for the monomers. $6r_{sc}^2$ is the size of the next nearest neighbor case. From Baschnagel *et al.* (2000).

studying the slow relaxation near the glass transition, and hence have been investigated experimentally in great detail (Donth 2001). Moreover, in order to understand the relaxation dynamics close to the glass transition, it is necessary to clarify their dynamical properties at temperatures far above the glass transition as well, and this was the purpose of the present discussion.

4.2 Application of the Percolation Problem to Physical Systems

4.2.1 *Percolation Conductivity and a Naive Treatment of the Elasticity of Polymer Networks*

In Sec. 3.2, we have already discussed the percolation problem on lattices, and have seen that site percolation is a rather natural concept to discuss the physical properties of dilute magnets. We also have considered bond percolation, where on a lattice the nearest neighbor bonds connecting the lattice sites are present with probability p and absent with probability $1-p$. In fact, considering this problem on the Bethe lattice (or "Cayley tree", respectively, see Fig. 3.10) has allowed us to formulate a mean field theory for percolation.

In the present section, we will extend these considerations and we explore further applications of concepts from percolation theory to various physical problems. In fact, the first application of these concepts dates back long before the term "percolation" was coined: This is the theory of Flory (1941a-c) and Stockmayer (1943) of the gelation transition of branched polymers. Consider a solution of branched polymers, the so-called "sol phase", in which chemical reactions occur such that the size of these polymers grows with time, since more and more clusters aggregate together. In this way one images the formation of large, statistically branched, macromolecules, until at some time t_p the gelation transition occurs, i.e. an infinite polymer network has formed, the so-called "gel phase". The chemical details of his process can be very complicated since in practice one may start the dendritic polymerization reaction by putting initiators into a solution containing only monomers, and the reaction forms chemical bonds between the monomers, including crosslinks between the macromolecules. To a first approximation one can assume that the time t after the polymerization was started is proportional to the fraction p of bonds on a lattice, and then the time t_p corresponds to the percolation threshold p_c (de Gennes 1976a, 1976b, Stauffer 1976, Stauffer *et al.* 1982). Then the gel fraction $G_\infty(t)$ corresponds to the percolation probability $P_\infty(p)$, etc. The coordination number z of the lattice sites corresponds to the functionality of the crosslinks in the network. Of course, for this simple mapping between gelation and percolation one has several caveats: The branched macromolecules in the sol phase carry out a complicated Brownian motion, and hence it is not clear whether the formation of additional crosslinks is a completely random uncorrelated process, like the random addition of bonds to clusters on a rigid lattice. The dynamics of the complicated inhomogeneous fluid that we encounter in the sol phase near the gelation transition may influence rather distinctly the structure of the gel that is formed. On a local scale, the macromolecular dynamics (similar to the Rouse and Zimm models discussed in the previous section) may also need consideration. However, if one accepts this percolation interpretation of a gel, the elastic shear modulus of such a network becomes analogous to the electrical conductivity of a random resistor network, which we will consider now.

The physical situation that is envisaged is a random mixture of materials with different conductivities. Suppose that we have a lattice in which the links of the lattice are occupied with probability p with a conducting material of conductivity σ_0, while the remaining bonds are occupied by a non-conducting material. If a voltage is applied to the system, we will obtain a nonzero current only if an infinitely large percolating cluster formed from the conducting materials is present. Since for $p \to p_c^+$ the percolation probability $P_\infty(p)$ vanishes continuously with a power-law (Sec. 3.3), it is tempting to assume a power-law for the percolation conductivity near p_c as well,

$$\Sigma(p)/\sigma_0 = \widehat{\sigma}(p - p_c)^\mu, \quad p \gtrsim p_c \,, \tag{4.30}$$

where $\Sigma(p)$ is the conductivity of the sample, $\hat{\sigma}$ is a (dimensionless) critical amplitude, and μ a critical exponent.

Of course, it is now natural to ask whether there is a scaling relation by which μ is related to some of the other exponents that have been discussed in Sec. 3.2. Clearly, only the backbone of the percolating cluster can maintain the current, while the dead ends contribute nothing to the conductivity. It turns out that the precise numerical estimation of the exponent μ is a difficult problem. Actually the first estimation of μ has been obtained from a real experiment and not a simulation: Last and Thouless (1971) used a conducting foil in which they randomly punched holes, measuring then the conductivity as a function of the density of the holes. The Monte Carlo estimation of μ remained controversial for some time. Normand *et al.* (1988) even constructed a special purpose computer to determine μ in $d = 2$ dimensions, and obtained from a finite size scaling analysis that $\mu/\nu_p = 0.9745 \pm 0.0015$. Remembering $\nu_p = 4/3$ {cf. Eq. (3.96) from Chap. 3} we hence obtain $\mu \approx 1.30$ in $d = 2$, while $\mu \approx 2.0$ in $d = 3$, the result for a Bethe lattice being $\mu = 3$ (Stauffer and Aharony 1992).

A related problem is the statistical mixture of a material with finite conductivity σ_0 and a superconducting material which can be used as an approximate description for granular superconducting films (in $d = 2$ dimensions). Considering the regime $p < p_c$, where the superconducting regions do not yet percolate, one finds a conductivity that is still finite, but diverges if the percolation threshold is approached:

$$\Sigma(p)/\sigma_0 = \widehat{\sigma}'(p_c - p)^{-s}, \quad p \leq p_c \,. \tag{4.31}$$

Numerical estimates give $s \approx 1.3$ in $d = 2$, while $s \approx 0.73$ in $d = 3$ and $s = 0$ for the Bethe lattice.

We now discuss the phenomenological theory of random resistor networks. A convenient starting point is the well known Kirchhoff laws of electrodynamics. Consider a large d-dimensional lattice of linear dimensions L, and let us introduce the local electrical potential V_i at lattice site i. Between neighboring lattice sites i, j we have the conductivity $\sigma_{ij} = \sigma_0$ if a bond is present and $\sigma_{ij} = 0$ if it is not. For the case that a stationary electrical current flows through the network, at each lattice site the current that enters on one side must leave on the other side, i.e. the sum of all currents at each site (in the interior of the lattice, we do not consider the free boundaries of the system here) must vanish:

$$\sum_j \sigma_{ij}(V_i - V_j) = 0 \,. \tag{4.32}$$

Here we have simply used Ohm's law to express the current along the bond (ij) in terms of the potential difference $V_i - V_j$ as $I_{ij} = \sigma_{ij}(V_i - V_j)$.

We consider a situation in which we have free boundary conditions in one lattice direction, but periodic boundary conditions in all other directions, we choose $V_i = 0$ on all sites of the bottom free boundary and $V_i = V$ on all sites of the top free boundary. We want to compute the resulting steady-state current density (current per unit area) through the system which is given by

$$I = \sum_{i \in \mathcal{B}} \sum_j I_{ij}/L^{d-1} = \sum_{i \in \mathcal{B}} \sum_j \sigma_{ij}(V_i - V_j)/L^{d-1} \,. \tag{4.33}$$

Here $\mathcal{B}$ means the bottom free boundary of the system. For $p < p_c$ we have of course $I = 0$, i.e. there is no current through the system. For $p > p_c$ we can define the macroscopic conductivity Σ from Ohm's law as follows,

$$I = \Sigma V/L\,, \tag{4.34}$$

the resistance being $R = L/\Sigma$. From Eqs. (4.33) and (4.34) we thus conclude that Σ is a solution of the eigenvalue equation

$$\Sigma V - \frac{1}{L^{d-2}} \sum_{i \in \mathcal{B}} \sum_{j} \sigma_{ij}(V_i - V_j) = 0\,. \tag{4.35}$$

Before discussing further the approaches to calculate Σ explicitly, we consider the connection of this problem to the elasticity of gels. We assume that the gel can be described as a harmonic solid of crosslink points connected with spring constants k_{ij} that are independent of the cartesian components α of the displacement vectors $\vec{u}_i$ of the crosslinks:

$$U_{\text{elastic}} = U_0 + \frac{1}{2} \sum_{i \neq j} k_{ij} (\vec{u}_i - \vec{u}_j)^2\,. \tag{4.36}$$

Note that this Ansatz is in contrast to the one one would use in a crystal treated in the harmonic approximation for which we would instead have a tensorial expression:

$$U_{\text{harmonic crystal}} = U_0 + \frac{1}{2} \sum_{i \neq j} \sum_{\alpha,\beta} k_{ij}^{\alpha,\beta} (u_i^\alpha - u_j^\alpha)(u_i^\beta - u_j^\beta)\,. \tag{4.37}$$

The assumption that a gel can be described by Eq. (4.36) instead of Eq. (4.37) is known as "scalar elasticity". While at first sight this assumption seems to be plausible (unlike a crystal a gel does not single out any lattice directions and still exhibits the full rotation symmetry), one now knows that despite the isotropy of amorphous materials tensorial expressions of the elastic energy, as written in Eq. (4.37), are needed. Actually, if we use Eq. (4.36) for a square lattice with harmonic bonds between nearest neighbors, we have an instability against shear forces even if all bonds are present ($p = 1$)! Nevertheless, for the moment we ignore this problem, and continue with the simpler Eq. (4.36).

Now the condition of mechanical equilibrium requires that at each crosslink point of the network the components of the forces acting on these points vanish:

$$\frac{\partial U_{\text{elastic}}}{\partial u_i^\alpha} = \sum_j k_{ij} (u_i^\alpha - u_j^\alpha) = 0\,. \tag{4.38}$$

If we now describe the (disordered) gel as a lattice in which springs $k_{ij} = k_0$ are present with probability p and no springs occur ($k_{ij} = 0$) with probability $1 - p$, we immediately recognize that the problem is isomorphic with the conductivity problem given by Eq. (4.32): We only need to identify V_i with u_i^α and σ_0 with k_0! Within this naive analogy, the elastic constant that then corresponds to the conductivity Σ is simply the shear modulus. If we approach the gelation transition from the gel side, the gel becomes "soft" against shear deformations. In the fluid sol phase, the shear modulus is zero by definition, of course. However, in a later section we will discuss how this treatment needs to be refined.

It should also be noted that amorphous solids (including gels) have a further elastic constant, in addition to the shear modulus: This is the bulk modulus B, the inverse of the compressibility:

$$B = -V(\partial p/\partial V)_T \,. \tag{4.39}$$

Of course also in the fluid sol phase we have $B > 0$. Actually, for dense fluids it is often a good approximation to consider them to be incompressible, i.e. $B = 0$. This approximation is in fact made, if we treat percolation conductivity on a rigid lattice and simply translate the result to gel elasticity. Recalling that only the backbone of the percolating cluster carries the current while dead ends do not contribute to the conductivity, we immediately can state the (seemingly obvious) fact that the resistance against shear deformations is only due to the backbone, while dead ends in the percolating cluster are irrelevant for the elastic properties. (Note that this conclusion is not quite correct, since also steric hindrance effects can contribute to an increase of B.)

Before we discuss further analogies between percolation conductivity and gel elasticity, we make a brief comment on the chemistry of gels. So far we have implicitly assumed that the crosslinking process of branched polymers in solution is the generic mechanism for the sol-gel transition. However, we emphasize that actually sol-gel-transitions occur in a wide variety of physical systems. A particularly important gel is the inorganic "silica gel" (or the related "silica hydrogel", which is a silica gel that is swollen due to the water that is incorporated into the gel structure), Brinker and Scherer (1990). Although also silica gel also is a random network structure with covalent bonds, it differs from the continuous random network structure of amorphous SiO_2 (where in the ideal case all oxygens occur in the bridging position connecting two Si-atoms, Si-O-Si, which are the centers of neighboring tetrahedra, see Sec. 3.5) due to the presence of many "dead ends" in the network, where hydroxy groups (Si-OH) occur. One can produce silica gel with a much smaller density than regular amorphous SiO_2. The porosity of silica gel is crucial for its use as catalyst (typical pores have diameters in the range from 20 to 100 Å, and thus huge "internal surfaces" are created, offering adsorption sites suitable for reactants involved in the desired chemical reaction). If we, very roughly, interpret the structure of silica gel in the framework of percolation theory, the bond probability p corresponds to the percentage of O-atoms in the Si-O-Si bridging position relative to the total number of oxygen atoms in the material.

The gelation process of silica gel is started in the sol phase, starting with $Si(OH)_4$ molecules. Thus clusters grow from small molecules until at p_c, the gel point, a percolating network is formed: Only exactly at the gel point does the network have a fractal dimension $d_f < d$. Many sol-gel transitions of this type are known also for organic molecules, leading then to polymer network formation. A somewhat different polymer network, however, is obtained in the "vulcanization transition" in which one already starts with a solution or melt of flexible linear macromolecules (with $n \gg 1$ effective monomers), and which are then transformed into a network via the introduction of chemical crosslinks. The fact that between the nodes of the resulting network one has rather long flexible chains which also have $n_c \gg 1$ subunits (although typically $n_c \ll n$ needs to be considered) leads to the particular features of rubber elasticity. However, this subject is beyond the scope of the present book and thus we refer here only to other literature (Treloar 1975, Erman and Mark 1997, Rubinstein and Colby 2003, Graessley 2008).

4.2.2 *Excitations of Diluted Magnets Near the Percolation Threshold*

Another interesting analogy exists between percolation conductivity (or scalar elasticity of gels, respectively) and the dynamics of diluted Heisenberg ferromagnets. To see this we consider a model of the type discussed in the introduction (see Sec. 1.2), namely

$$\mathcal{H} = -\sum_{\langle i,j \rangle} J_{ij}(\vec{S}_i \cdot \vec{S}_j), \quad P(J_{ij}) = p\,\delta(J_{ij} - J) + (1-p)\delta(J_{ij}) \,. \tag{4.40}$$

Introducing the operators $S_j^{\pm} = S_j^x \pm i\, S_j^y$ we obtain

$$\begin{aligned}\vec{S}_i \cdot \vec{S}_j &= S_i^z S_j^z + \frac{1}{2}(S_i^+ S_j^- + S_i^- S_j^+) \\ &\approx S^2 + \frac{1}{2}(S_i^+ S_j^- + S_i^- S_j^+) - \frac{S}{2}(S_i^- S_i^+ + S_j^- S_j^+)\,, \end{aligned} \tag{4.41}$$

where we have used the expansion, valid at low enough temperatures,

$$S_i^z = \sqrt{S^2 - S_i^- S_i^+} \approx S - \frac{1}{2} S_i^- S_i^+ , \quad S \to \infty\,. \tag{4.42}$$

Inserting Eq. (4.41) into Eq. (4.40) we obtain the quantum-mechanical equations of motion

$$i\hbar \frac{dS_i^-}{dt} = [\mathcal{H}, S_i^-(t)] \tag{4.43}$$

and thus we find a linearized system of equations,

$$i\hbar \frac{dS_i^-}{dt} = 2S \sum_j J_{ij} \{S_i^-(t) - S_j^-(t)\}\,. \tag{4.44}$$

This equation has the same structure as the generalization of Eq. (4.32) to the case of non-steady state time evolution of the electrical current and voltage distribution,

$$C \frac{dV_i(t)}{dt} = -\sum_j \sigma_{ij} [V_i(t) - V_j(t)]\,. \tag{4.45}$$

(Here we have attributed to each lattice site i an electrical capacitance C.)

In order to discuss the dynamic behavior of $S_j^-(t)$ and $V_i(t)$, governed by Eqs. (4.44) and (4.45), respectively, it is convenient to make an Ansatz of the form

$$S_j^-(t) \propto \exp(i\hbar\, \omega_\alpha t)\,, \tag{4.46}$$

where the eigenvalue ω_α follows from the condition that Eq. (4.46) must be a nontrivial solution of Eq. (4.44). In order to write the eigenvalue equation in a compact notation, we define $\hbar\omega_\alpha = 2SJ\lambda_\alpha$ and the associated eigenfunction $|i, \alpha\rangle$. Then Eqs. (4.44) and (4.46) reduce to

$$\sum_j \underline{\Delta}_{ij} |j, \alpha\rangle = \lambda_\alpha |i, \alpha\rangle\,, \tag{4.47}$$

where the operator $\underline{\Delta}_{ij} \equiv |i\rangle\langle i| - |j\rangle\langle j|$ characterizes the differences between the values of an elementary excitation on neighboring connected sites.

Of course, if we have a highly diluted lattice well below the percolation threshold, only finite clusters exist and long wavelength excitations (with wavelengths larger than the size of the cluster) cannot occur, because the clusters are statistically independent of each other, and no magnetic interactions occur between disjunct clusters. Therefore, we again focus our interest on the case $p > p_c$, in which an infinitely large percolating network exists, which hence can carry long wavelength spin waves.

However, due to the disorder present in the system for $p < 1$, the states $|j, \alpha\rangle$ on the percolating network are not eigenstates of the momentum operator $\widehat{\vec{p}}$ (unlike spin waves on a perfect periodic lattice). Therefore we have to define an effective momentum $\vec{q}_\alpha$ belonging to the eigenstate $|\alpha\rangle$ as

$$q_\alpha = \langle \alpha, j|\frac{\hbar}{i}\nabla|j, \alpha\rangle\,, \tag{4.48}$$

and if we assume that the state $|j, \alpha\rangle$ is written in position space representation Eq. (4.48) can be evaluated. Now for $q_\alpha \to 0$ we expect that the operator $\underline{\Delta} \equiv \sum_{i,j} \underline{\Delta}_{ij}$ is just the discrete analogy of the Laplace operator $-a^2\nabla^2$ on the percolating network, a being the lattice constant. Therefore, we expect that λ_α in Eq. (4.47) for $q_\alpha \to 0$ should be quadratic in q_α, since $\underline{\Delta}$ is quadratic in the gradient operator ∇. Thus we define

$$\lambda_\alpha = a^2 q_\alpha^2 \rho(p)\,, \tag{4.49}$$

where the coefficient $\rho(p)$ is called the "spin wave stiffness constant". For an undiluted lattice with no disorder the translation invariance resulting for Eq. (4.44) allows to solve it directly by Fourier transformation. This approach yields exactly Eq. (4.49), and $\rho(p = 1) = z_\ell$, the coordination number of the lattice (note that in Eq. (4.40) the exchange was restricted to nearest neighbors only). For $p < 1$ a rigorous analytic calculation of $\rho(p)$ is, however, not possible and therefore one has to resort either to numerical solutions of the eigenvalue problem Eq. (4.47), or to use approximate treatments, such as the effective medium approximation that will be discussed in the next subsection.

A detailed analysis of Eq. (4.45), that is however beyond the scope of our present discussion, yields then the following relation between the electrical conductivity $\Sigma(p)$ and the spin wave stiffness $\rho(p)$,

$$\Sigma(p) = \sigma_0 P_\infty(p)\,\rho(p)\,. \tag{4.50}$$

Combining this relation with Eq. (4.30), $\Sigma(p) \propto (p - p_c)^\mu$, and Eq. (3.40),

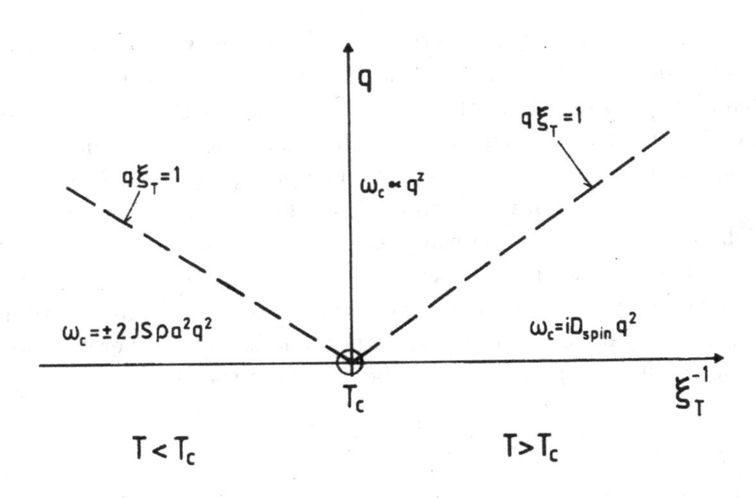

Fig. 4.6 Long wavelength modes in the plane of variables wavenumber q and inverse correlation length ξ_T^{-1} (measuring the distance from the critical point $T_c(p)$) for a (diluted) Heisenberg ferromagnet. For $T < T_c$ and $q\xi_T < 1$ the response function has poles on the real axis, describing spin wave excitations $\omega = \pm 2JS\rho(T)a^2q^2$. For $T > T_c$ and $q\xi_T < 1$ there occurs a pole on the imaginary axis $\omega = i\mathcal{D}_{\text{spin}}q^2$, describing spin diffusion. At $T = T_c$ one has a non-hydrodynamic law, $\omega_c \sim q^z$.

$P_\infty(p) \propto \infty(p - p_c)^{\beta_p}$, implies that the magnon frequencies vanish when $p \to p_c$, i.e. one has "soft magnons":

$$\rho(p) \propto (p - p_c)^{\mu - \beta_p} . \tag{4.51}$$

Such soft magnons are encountered both at $T = 0$ for $p \to p_c^+$ and for $T > 0$ when the transition line $T_c(p > p_c) > 0$ between the ferromagnetic state and the paramagnetic state of the dilute magnet is crossed. The corresponding vanishing of the magnon frequency is called "critical slowing down" (Hohenberg and Halperin 1977) and is simply controlled by the vanishing of the magnetization,

$$\hbar\omega(q) \approx a^2q^2\, 2J\, S\rho(T), \quad \rho(T) \propto m_s(T) \propto (1 - T/T_c(p))^\beta . \tag{4.52}$$

Note that, however, the statements on the spectrum of the long wavelength modes in a (diluted) Heisenberg ferromagnet near the percolation threshold or near the Curie temperature of the thermally driven transition to the paramagnetic phase hold only in the hydrodynamic regime (Fig. 4.6), $q\xi_T < 1$ (Hohenberg and Halperin 1977), or $q\xi_p < 1$ (if one crosses the percolation threshold at $T = 0$). This means that Eqs. (4.51) and (4.52) hold only if the wavelength $2\pi/q$ of the excitation is distinctly larger than the correlation length ξ_T (or ξ_p, respectively). By the "dynamic scaling hypothesis" (Hohenberg and Halperin 1977) one can extend the description also in the opposite regime, $q\xi_T \gg 1$. Defining a characteristic frequency $\omega_c(q)$ from the condition

$$\int_{-\omega_c(q)}^{+\omega_c(q)} S(q,\omega)\, d\omega = \frac{1}{2}\int_{-\infty}^{+\infty} S(q,\omega) d\omega\,, \tag{4.53}$$

where $S(q,\omega)$ is the dynamic structure factor, Eq. (2.86), which describes the intensity of inelastic magnetic scattering of neutrons, Eq. (2.80), dynamic scaling asserts that

$$\omega_c(q) = q^2(1 - T/T_c)^{\beta}\widetilde{\omega}(q\xi_T)\,. \tag{4.54}$$

Just as for the finite size scaling, see Sec. 3.2.3, where a scaling assumption involving the ratio of the correlation length and the linear dimension of the system was made, here the ratio of the correlation length and the wavelength of the excitation enters. But in both cases the basic idea is the comparison of characteristic lengths.

At T_c we can find the behavior of $\omega_c(q)$ from the condition that the scaling function $\widetilde{\omega}(\zeta)$ in Eq. (4.54) must exhibit a power-law behavior, since only then it is possible that the temperature dependence of the spin wave stiffness constant $\{\propto (1 - T/T_c)^{\beta}$, see Eq. (4.52)$\}$ is canceled by the temperature dependence of the correlation length. Thus, we conclude that $\widetilde{\omega}(\zeta) \propto \zeta^{\beta/\nu}$, and, using the scaling relation (Fisher 1974),

$$\beta/\nu = \frac{1}{2}(d - \gamma/\nu) = \frac{1}{2}(d - 2 + \eta) \tag{4.55}$$

one predicts (Hohenberg and Halperin 1977)

$$\omega_c(q) \propto q^z\,,\ T = T_c,\ z = 2 + \frac{\beta}{\nu} = \frac{1}{2}(d + 2 + \eta) = (5 + \eta)/2\,, \tag{4.56}$$

where in the last step we have assumed the dimension to be $d = 3$.

A completely analogous reasoning applies to the behavior of the characteristic frequency at the percolation transition: With Eq. (4.51) and (4.52) one obtains

$$\omega_c = q^2\rho(p)\widetilde{\omega}_p\{q\xi_p\} \propto q^2(p - p_c)^{\mu-\beta_p}\,\widetilde{\omega}_p\{q(p - p_c)^{-\nu_p}\}\,. \tag{4.57}$$

The scaling function $\widetilde{\omega}_p(\zeta)$ must behave as $\widetilde{\omega}_p(\zeta) \propto \zeta^{(\mu-\beta_p)/\nu_p}$, and hence the characteristic frequency scales like

$$\omega_c(q) \propto q^{2+(\mu-\beta_p)/\nu_p}, \quad p = p_c . \tag{4.58}$$

This result can also be interpreted as the dispersion relation of a long wavelength excitation on a fractal object, since right at $p = p_c$ the percolating cluster is a fractal. In the next section, we will discuss excitations of fractals in more detail.

4.2.3 *Effective Medium Theory*

In Sec. 3.2 we have seen, that the qualitative aspect of static properties near a percolation threshold can be accounted for by a self-consistent analytic treatment via the theory of Flory and Stockmayer. In this subsection, we try to formulate a counterpart of such a treatment for dynamic phenomena, considering the electrical conductivity of a disordered medium via phenomenological equations from electrostatics as a simple example.

The idea is to assume that small regions with known properties are embedded in an "effective medium" whose properties we determine in a self consistent way. For this we take the electric field inside such a region, $\vec{E}_{\text{in}}$, as a constant, and relate it to the electric field $\vec{E}_{\text{m}}$ of the medium as

$$\vec{E}_{\text{in}} = \vec{E}_{\text{m}} - \frac{4\pi}{d}\vec{P} , \tag{4.59}$$

where d is the dimensionality of the system, and $\vec{P}$ the dielectric polarization. The latter can be attributed to the charge density ρ_n on the surface of the embedded region,

$$\rho_n = \widehat{\vec{n}} \cdot \vec{P} , \tag{4.60}$$

where $\widehat{\vec{n}}$ is a unit vector perpendicular to the surface of the embedded region. Outside of the embedded region we describe the field with the help of a depolarization tensor $\underline{L}$

$$\vec{E}_{\text{out}} = \vec{E}_{\text{m}} + 4\pi\vec{P} \cdot \underline{L} . \tag{4.61}$$

For spheres (in $d = 3$) or circles ($d = 2$) the depolarizing field is easily calculated explicitly, see standard texts on electrodynamics, and the result can be summarized as

$$\underline{L} = \widehat{\vec{n}} \otimes \widehat{\vec{n}} - \frac{1}{d}\underline{1} . \tag{4.62}$$

In order to derive a relation between $\vec{P}$ and $\vec{E}_{\rm m}$, we require that at the surface of the embedded region the component of the density of the electrical current in the direction normal to the surface is continuous:

$$\hat{\vec{n}} \cdot (\underline{\sigma}_{\rm in} \cdot \vec{E}_{\rm in}) = \hat{\vec{n}} \cdot (\underline{\sigma}_{\rm m} \cdot \vec{E}_{\rm out}) \,. \tag{4.63}$$

Here $\underline{\sigma}_{\rm in}$ is the tensor of electrical conductivity of the embedded region, and $\underline{\sigma}_{\rm m}$ the corresponding quantity of the effective medium. For simplicity we assume perfect isotropy, i.e.

$$\underline{\sigma}_{\rm in} = \sigma_{\rm in} \begin{pmatrix} 1 & 0 & 0 \\ 0 & 1 & 0 \\ 0 & 0 & 1 \end{pmatrix} , \quad \underline{\sigma}_{\rm m} = \sigma_{\rm m} \begin{pmatrix} 1 & 0 & 0 \\ 0 & 1 & 0 \\ 0 & 0 & 1 \end{pmatrix} , \quad d = 3 \,. \tag{4.64}$$

Choosing the direction of $\hat{\vec{n}}$ as the z-axis, and discussing only the z-components of the fields, we obtain from Eqs. (4.59)-(4.64)

$$\left(E^z_{\rm m} - \frac{4\pi}{d} P^z \right) \sigma_{\rm in} = \left[E^z_{\rm m} + 4\pi P^z \frac{d-1}{d} \right] \sigma_{\rm m} \,, \tag{4.65}$$

which yields

$$\frac{4\pi P^z}{d} = E^z_{\rm m} \frac{\sigma_{\rm in} - \sigma_{\rm m}}{\sigma_{\rm in} + \sigma_{\rm m}(d-1)} \,. \tag{4.66}$$

Self consistency implies now that σ_m is defined such that the average polarization $\langle P^z \rangle$ vanishes if we average over all values of the conductivity $\sigma_{\rm in}$ of the embedded region appropriately, i.e.

$$\left\langle \frac{\sigma_{\rm in} - \sigma_{\rm m}}{(d-1)\sigma_{\rm m} + \sigma_{\rm in}} \right\rangle_{\rm in} = 0 \,. \tag{4.67}$$

For the problem of percolation conductivity, we take $\sigma_{\rm in} = \sigma_0$ with probability p and $\sigma_{\rm in} = 0$ with probability $1 - p$, and therefore we find

$$p \frac{\sigma_0 - \sigma_{\rm m}}{(d-1)\sigma_{\rm m} + \sigma_0} - (1-p) \frac{\sigma_{\rm m}}{(d-1)\sigma_{\rm m}} = 0 \,. \tag{4.68}$$

This relation tells us how the conductivity $\sigma_{\rm m}$ of the effective medium depends on p:

$$\sigma_{\rm m} = \sigma_0 \frac{pd-1}{d-1} , \quad p > \frac{1}{d}; \quad \sigma_{\rm m} \equiv 0, \quad p \le \frac{1}{d} \,. \tag{4.69}$$

According to this treatment one thus has $p_c = 1/d$ and hence $\sigma_m = [\sigma_0/(d-1)](1 - p/p_c)$, i.e. the exponent $\mu = 1$.

Although also the effective medium theory has the character of a self-consistent effective field approximation, it is not quantitatively compatible with the Flory-Stockmayer theory, as we already see from the different result for p_c, $p_c = 1/(z-1)$ in the Flory-Stockmayer theory, where z is the coordination number of the lattice. Of course, the assumption of spherical symmetry of embedded regions, Eq.(4.62), is inconsistent with a lattice description.

4.3 Elementary Excitations of Fractal Structures

4.3.1 *Diffusion on a Percolation Cluster: The "Ant in the Labyrinth"*

We now consider the problem of percolation conductivity from a different point of view, namely we treat the stochastic hopping of charged particles from sites to sites in a randomly disordered structure. We are interested in analyzing how the mean squared displacement of such particles grows with time. De Gennes (1976b) introduced this problem as the "motion of an ant in the labyrinth" (see also Stauffer and Aharony, 1992). Let us denote with $P_i(t)$ the conditional probability that the ant is at site i at time t after it started at the origin (0) at time $t = 0$ $\{P_0 = 0$ for $i \neq 0\}$. This probability satisfies a master equation

$$P_i(t+1) - P_i(t) = \sum_j [\sigma_{ji} P_j(t) - \sigma_{ij} P_i(t)] , \tag{4.70}$$

where we have assumed discrete time increments $\Delta t = 1$ at which jumps of the ant from each site to a neighboring site can occur, and σ_{ji} is the probability that the jump leads from the site j to its neighbor i. The first term on the right hand side of Eq. (4.70) describes all jumps which lead to the considered site i (and hence lead to an increase of the probability P_i), while the second term describes all jumps which lead away from the considered site (decrease of the probability P_i).

We suppose the choice of the jump direction is completely random, i.e. each of the z neighbors is tried with the same *a priori* probability ($1/z$ if there are z neighbors), but the jump can take place only if the site i is occupied, so $\sigma_{ji} = 1/z$ if i is occupied and $\sigma_{ji} = 0$ if i is empty. Instead of this so-called "blind ant" one can also treat a diffusion process in which jumps are only tried to occupied sites (then $\sigma_{ji} = 1/z_j$, z_j being the number of occupied neighbor sites of site j). However, it is believed that this latter problem ("shortsighted ants" rather than "blind ants") leads to the same universal exponents for $p \to p_c$, (Harris *et al.* 1987), and hence will not be considered further.

If the probabilities P_i change only weakly with time, which is the case for large t, Eq. (4.70) is equivalent to

$$\frac{dP_i(t)}{dt} = \sum_j [\sigma_{ji} P_j(t) - \sigma_{ij} P_i(t)]. \tag{4.71}$$

For any percolation cluster of finite size a stationary solution P_i^{stat} of this equation is reached, which is independent of time, and simply given by $P_i^{\text{stat}} = 1/s$ for a cluster containing s sites. The mean squared displacement reached then is nothing else than the mean squared cluster radius R_s^2. Averaging this result (for $p < p_c$) over the cluster size distribution, n_s yields

$$R^2(t \to \infty, p < p_c) = \sum_s n_s s R_s^2 \ , \tag{4.72}$$

since the probability that the diffusion of the ant (which may choose any occupied site with equal probability as origin at $t = 0$) occurs on a cluster with s sites is sn_s. For p near p_c we may use the scaling assumption for n_s {cf. Sec.3.2} to obtain

$$R^2(t \to \infty, p < p_c) \propto (p_c - p)^{\beta_p - 2\nu_p} \ . \tag{4.73}$$

Note that since the weighting of clusters in Eq. (4.72) differs from the weighting that yields the correlation length squared, see Eq. (3.87), $\xi_p^2 \propto (p_c - p)^{-2\nu_p}$, one has in Eq. (4.73) a different critical exponent.

For $p > p_c$ an infinite percolating network exists, and then we expect that the asymptotic behavior of the (Euclidean) distance that the ant has diffused follows the Einstein relation,

$$R^2(t \to \infty) = 2dDt \, , \tag{4.74}$$

D being the diffusion constant. Using further the relation between diffusion constant and conductivity Σ,

$$\Sigma = e^2 n D / k_B T \, , \tag{4.75}$$

where e is the charge of the diffusing particles and n their concentration with, Eq. (3.40), $n \propto P_\infty(p) \propto (p - p_c)^{\beta_p}$. Using Eqs. (4.50) and (4.51) we thus conclude

$$D \propto \Sigma(p)/n \propto (p - p_c)^{\mu - \beta_p} \propto \xi_p^{-(\mu - \beta_p)/\nu_p} \, . \tag{4.76}$$

Note that Eq. (4.74) is no longer true right at the percolation threshold $p = p_c$, since then $D = 0$, although the mean squared displacement still diverges to infinity, but with a subdiffusive behavior. We can obtain the power-law for this anomalous diffusion by noting that Eq. (4.74) holds only if the distances that are traveled are much larger than the percolation correlation length ξ_p. Therefore, it is tempting to generalize Eq. (4.76) by defining a diffusion constant $\mathcal{D}(L)$ that depends on the Euclidean distance L considered: For $L \gg \xi_p$ Eq. (4.76) must result, while for L of order ξ_p we make the Ansatz:

$$\mathcal{D}(L) = \xi_p^{-(\mu-\beta_p)/\nu_p} \widetilde{D}(L/\xi_p) \underset{\xi_p \to \infty}{\longrightarrow} L^{-(\mu-\beta_p)/\nu_p}, \tag{4.77}$$

where we have assumed that the scaling function $\widetilde{D}(\xi \to 0)$ behaves as a power-law such that the exponents of ξ_p cancel. Using then a relation analogous to Eq. (4.74) but with a diffusion constant $D(L)$ given by Eq. (4.77) and which hence depends on the scale L that the ant has traveled,

$$L^2 \propto L^{-(\mu-\beta_p)/\nu_p} t \quad \text{or} \quad L^{2+(\mu-\beta_p)/\nu_p} \propto t\,, \tag{4.78}$$

we predict for the anomalous diffusion of a particle traveling on an incipient percolating cluster

$$L(t) \propto t^{\nu_p/[2\nu_p+\mu-\beta_p]}, \quad t \to \infty\,. \tag{4.79}$$

Note that Eq. (4.79) does not show a smooth crossover to Eq. (4.73), since in Eqs. (4.72) and (4.73) we have allowed to move the ant on any cluster, while in Eq. (4.76), we have constrained the ant to move on the percolating cluster, and disregarded the contribution from ants moving on finite clusters. If the latter are included, the diffusion constant simply scales like the conductivity, $D' \propto \Sigma(p)$. We now make a scaling Ansatz of the form

$$R(t) = t^k \widetilde{R}[(p - p_c)t^x] \tag{4.80}$$

and require that for $p > p_c$ we have $R(t) \propto (D't)^{1/2}$ while for $p < p_c$ we demand to recover Eq. (4.73). For $p < p_c$ we assume $\widetilde{R}(y) \propto y^z$, with an unknown exponent z. From this and Eqs. (4.80) and (4.73) we thus find that $z = \beta_p/2 - \nu_p$ as well the relation $k = -zx$. For $p > p_c$ we assume $\widetilde{R}(y)$ to be of the form $y^{z'}$, with a new exponent z'. With Eq. (4.80) and $R^2 \propto Dt$ we find $z' = \mu/2$ and $2k + 2z'x = 1$. Putting these results together one concludes that the exponents k and x must have the values

$$k = \frac{\nu_p - \beta_p/2}{2\nu_p + \mu - \beta_p}, \quad x = [2\nu_p + \mu - \beta_p]^{-1}\,. \tag{4.81}$$

This result is consistent with Eq. (4.79) since from the argument of the scaling function in Eq. (4.80) we see that the characteristic time scales like $(p-p_c)^{-1/x}$ which is proportional to $\xi_p^{1/(x\nu_p)}$. A comparison with Eq. (4.79) shows thus that the exponent x is indeed given by the expression (4.81).

Many numerical studies have been devoted to an estimation of the exponent y in the relation $L(t) \propto t^y$ {Eq. (4.79} for diffusion on an incipient percolation cluster. Recent Monte Carlo estimates (Blavatska and Janke 2009) are $y = 0.353(3)$ in $d = 2$ and $y = 0.273(3)$ in $d = 3$. Systematically larger exponents are obtained if the diffusion takes place only on the backbone of the percolation cluster, namely $y = 0.372(2)$ in $d = 2$ and $y = 0.306(2)$ in $d = 3$.

4.3.2 *The Spectral Dimension and Fracton Excitations*

It is also interesting to consider the number of sites $N(t)$ that have been visited by the ant in the course of its random walk on the percolating cluster. One can expect that

$$\begin{aligned} N(t) \propto [L(t)]^{d_f} &= t^{d_f \nu_p/[2\nu_p+\mu-\beta_p]} = t^{(\gamma_p+\beta_p)/[2\nu_p+\mu-\beta_p]} \\ &= t^{d_s/2}, \quad t \to \infty, \quad d_s \equiv d_f/[1+(\mu-\beta_p)/2\nu_p]\,, \end{aligned} \tag{4.82}$$

where we have made use of the relation $d_f \nu_p = \gamma_p + \beta_p$ from Eq. (3.81). In the last step we have introduced the so-called "spectral dimension" d_s (Alexander and Orbach 1982, Rammal and Toulouse 1983). For $d \geq 6$ this exponent has the value 4/3 since there $\mu = 3$, $\beta_p = 1$, $\gamma_p = 1$ and $\nu_p = 1/2$. Since initial estimates for the numerical values in $d = 2$ and $d = 3$ were compatible with $d_s = 4/3$ as well, Alexander and Orbach (1982) proposed the conjecture that $d_s = 4/3$ holds exactly in all dimensions. However, more recent numerical evidence suggests that this Alexander-Orbach conjecture is not correct (Stauffer and Aharony 1992).

The probability that the diffusing ant returns to the starting point of its random walk at time t is proportional to the inverse of $N(t)$,

$$P_0(t) \propto 1/N(t) \propto t^{-d_s/2}, \quad t \to \infty\,. \tag{4.83}$$

It is useful to consider also the Laplace transform $\widetilde{P}_0(s)$ of $P_0(t)$,

$$\widetilde{P}_0(s) \equiv \int_0^\infty dt e^{-st} P_0(t) \tag{4.84}$$

which can be used to define a density of states $n(\varepsilon)$ as follows:

$$n(\varepsilon) = -\frac{1}{\pi} \lim_{\delta \to 0^+} \mathrm{Im} \widetilde{P}_0(-\varepsilon + i\delta) , \tag{4.85}$$

which yields

$$n(\varepsilon) \propto \varepsilon^{d_s/2-1} . \tag{4.86}$$

As a consequence of the close analogy between the equations describing percolation conductivity {Eq. (4.35)} and elasticity of random harmonic networks {Eqs. (4.36)-(4.38)} one can derive for the phonon frequencies $\{\omega\}$, that are the eigenfrequencies of the Hamiltonian, Eq. (4.36), a result that corresponds to Eq. (4.86). The density of states of these phonons is

$$\zeta(\omega) d\omega \propto \omega^{d_s - 1} d\omega \ , \quad \text{percolation cluster,} \tag{4.87}$$

where the spectral dimension d_s is the same as for the ant in the labyrinth problem, as will be shown later. (Below we will also comment on the fact that here we have an exponent $d_s - 1$ instead of $d_s/2 - 1$, as in Eq. (4.86).) Recall at this point the analogy of Eq. (4.87) to the density of states resulting from acoustic phonons in a harmonic crystal,

$$\zeta(\omega) d\omega = \omega^{d-1} d\omega , \quad \text{crystal,} \tag{4.88}$$

where d is the dimensionality of space. We thus find that with respect to the spectrum of vibration modes, the euclidean dimension of space for a fractal object gets replaced by the spectral dimension d_s and not by the fractal dimension d_f. (This difference between d_s and d_f is in agreement with our result for the Rouse model for a gaussian polymer chain, Sec. 4.1, for which we have found the result $d_s = 1$ whereas $d_f = 2$, see Eq. (4.29)). Since Eq. (4.87) is a rather general result we can conclude that the concept of the "spectral dimension" is applicable to all fractal structures and, as we will see below, it is indeed very useful for the characterization of the excitations of fractal aggregates (such as those shown in Fig. 1.6 in the introduction).

For acoustic phonons the relation $\zeta(\omega) \propto \omega^{d-1}$ is intimately connected to the phonon dispersion relation $\omega = cq$, where c is the velocity of sound, and the fact that in reciprocal space the density of possible wave-vectors $\vec{q}$ is uniform. It is thus of interest to ask what is the dispersion relation of the "fracton excitations", i.e. the relation between the eigenfrequency ω and the associated wavelength λ of the (propagating) mode. Recall that we have answered this question already for the related problem of magnons on a percolation cluster in a diluted ferromagnet at the percolation threshold, Eq. (4.58). Setting $q = 2\pi/\lambda$ we conclude for magnons

$$\omega \propto \lambda^{-[2+(\mu-\beta_p)/\nu_p]} = \lambda^{-2d_f/d_s}\,, \tag{4.89}$$

where we have made use of Eq. (4.82). Just as ferromagnetic magnons in a crystal obey $\omega \propto q^2$ while acoustic phonons obey $\omega \propto q$, acoustic modes on a percolation cluster (or other fractal object, respectively) satisfy the dispersion relation

$$\omega \propto \lambda^{-d_f/d_s}\,. \tag{4.90}$$

The reason that for acoustic vibrations we find d_s instead of $d_s/2$ as it is the case for diffusion on fractals or for magnons is the simple fact that the corresponding differential equation which yields the eigenvalues is now a *second* order differential equation in time:

$$m\frac{d^2u_i^\alpha}{dt^2} = -\sum_j k_{ij}(u_i^\alpha - u_j^\alpha), \quad \alpha = x, y, z\,. \tag{4.91}$$

Here we have assumed that the particles whose displacements are described by u_i^α have mass m, and hence their (classical) kinetic energy is $m\dot{\vec{u}}_i^2/2$, so Eq. (4.91) is simply Newton's equation of motion, using the potential given by Eq. (4.36). In contrast to this the equations describing conductivity (or diffusion) and magnetic excitations are first order differential equations in time, cf. Eqs. (4.44) and (4.45). Thus, the eigenvalue λ_α in Eq. (4.47) corresponds to ω^2 in the case of acoustic vibrations, while it corresponds to the spin wave energy $\varepsilon = \hbar\omega^{(\mathrm{sw})}$. We can transform from the spin wave density of states $n(\varepsilon)d\varepsilon$ to the vibrational density of states $\zeta(\omega)d\omega$ by making the identification $\varepsilon = \omega^2$,

$$\zeta(\omega) = n(\omega^2)\frac{d\varepsilon}{d\omega} \propto (\omega^2)^{d_s/2-1}\omega = \omega^{d_s-1}\,, \tag{4.92}$$

which hence justifies Eq. (4.87). Finally we note that such fracton excitations with a density of states as given by Eq. (4.87) would also give rise to an anomalous specific heat at low temperatures T, $C_v \propto T^{d_s}$ rather than the Debye law, $C_v \propto T^d$. However, it is wrong to attribute the anomalous specific heat of glasses, which at low temperatures is approximately linear in T, to such excitations, as will be discussed in Sec. 4.4.

An experimental verification of the fracton concept is difficult, since objects that are really fractal must be expected to be rather fragile if they are in free space, which would be necessary if one wishes to study their

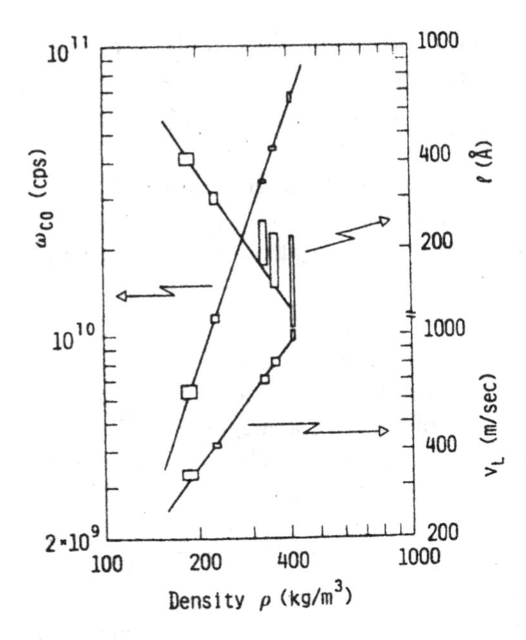

Fig. 4.7 Log-log plot of the crossover frequency ω_{co} (left ordinate), the crossover length ℓ (upper right ordinate) and the longitudinal sound velocity v_L (lower right ordinate) plotted vs. density ρ for silica aerogels. Straight lines are power-law fits with exponent values quoted in the text. From Courtens *et al.* (1987).

intrinsic vibrations, i.e. their motion unaffected by surrounding matter. However, one can, e.g., study fractons for porous silica aerogels, which are fractal up to some length scale $\ell(\rho)$, where ρ is their density, while for scales much larger than $\ell(\rho)$ these aerogels are essentially homogeneous. Vibrational modes of such silica aerogels have been probed by Brillouin scattering (Courtens *et al.* 1987) over a range of densities from 0.103 g/cm^3 to 0.407 g/cm^3, while the density of pure silica glass is about 2.2 g/cm^3. Of course, for wavelengths that significantly exceed ℓ, one observes ordinary acoustic phonons, $\omega \propto q$, while a crossover to fracton excitations sets in at $\lambda \approx \ell$. From the observed dispersion relations ω vs. q the crossover length $\ell(\rho)$ can be measured as function of density, and using the result $\rho \propto \ell^{d_f - d}$ (see Chap. 2) the fractal dimension d_f can be extracted. This yields $d_f = 2.36 \pm 0.02$ for this material (Courtens *et al.* 1987).

Now it is clear that the crossover in the dispersion relation at $\lambda \approx \ell$ from $\omega_q = v_L(\ell)q$, where v_L is the longitudinal sound velocity, to $\omega_q = \text{const} \cdot q^{d_f/d_s}$ from Eq. (4.90) must be smooth. Setting hence $q = 2\pi/\ell$ into these dispersion relations and equating them we obtain a result for $v_L(\ell)$

$$v_L(\ell) \propto \ell^{-(d_f/d_s-1)} \tag{4.93}$$

which can be measured, cf. Fig. 4.7 (Courtens *et al.* 1987), yielding the estimate $d_s \approx 1.252 \pm 0.061$. As a caveat, however, we emphasize that the analysis based on Eqs. (4.36), (4.91), and (4.92) relies on the assumption that for the longitudinal acoustic modes scalar elasticity (rather than tensorial elasticity) gives quantitatively correct results.

Fig. 4.7 also includes a crossover frequency ω_{co} that refers to the damping of these sound waves. Recall that in a homogeneous elastic medium there occurs a scattering of sound waves caused by density fluctuations, the so-called Rayleigh scattering, giving rise to the fact that the Brillouin scattering is not described by a delta function $\delta(\omega - \omega_q)$, but instead the Brillouin line has an intrinsic line width Γ_q that scales with the fourth power of ω_q,

$$\Gamma_q = \omega_q^4/\omega_{\text{co}}^3 \,. \tag{4.94}$$

Here we have defined ω_{co} in terms of the proportionality constant in the relation $\Gamma \propto \omega^4$. Clearly, for $\omega_q \ll \omega_{\text{co}}$ the scattering is very weak and hence the sound modes are damped only weakly. However, for $\omega_q > \omega_{\text{co}}$ a transition to a regime of strong scattering occurs, since ω_q and Γ_q are now of the same order, and thus we no longer have an ordinary propagating wave. A detailed analysis of this situation shows that localized modes occur, the excitation is not infinitely extended in the system but instead exists only on the scale of the so-called "localization length". Localization of states in disordered media is a very general phenomenon, and will be considered in Sec. 4.4 in more detail. Here we only mention that for fractons one finds that the localization length is in general of the same order as the wavelength $\lambda = 2\pi/q$ of the fracton itself, and hence one concludes $\omega_{\text{co}} \propto \ell^{-d_f/d_s} \propto \omega_q$, cf. Eq. (4.90). This conclusion is also compatible with the data of Fig. 4.7.

Finally let us see how the sound velocity for longitudinal sound would behave on a percolating network close to p_c, putting the length ℓ in Eq. (4.93) equal to the percolation correlation length $\xi \propto (p - p_c)^{-\nu_p}$. This yields

$$v_L(p) \propto (p - p_c)^{(d_f/d_s-1)\nu_p} = (p - p_c)^{(\mu-\beta_p)/2} \,, \tag{4.95}$$

i.e. the sound velocity vanishes if we approach p_c since the mechanical coupling between different parts of the percolating cluster goes to zero for $p \to p_c^+$.

A dynamical behavior at the percolation threshold that is very different from the standard critical slowing down embodied in the vanishing of sound velocity {Eq. (4.95)}, spin wave stiffness {Eq. (4.51)} or diffusion constant {Eq. (4.76)} and conductivity {Eq. (4.30)} occurs if we consider the critical dynamics of diluted anisotropic ferromagnets, e.g. the Ising model at the percolation threshold (see Fig. 1.9 of Chap. 1) instead of the isotropic Heisenberg model assumed in Sec. 4.2.2. In a percolating cluster Ising spins are most easily overturned relative to their neighbors in that part of the network which are just linear chains since this involves simply an energy cost $2J$, and hence the rate at which this occurs in the framework of a Glauber kinetic Ising model (Glauber 1963) is $\exp(-2J/k_BT)$ as $T \to 0$. Once such a defect (relative to a uniform ferromagnetic alignment of the spins) is created, the domain walls between spin down and spin up in such one-dimensional parts of the network can diffuse freely forward and backward. However, additional activation is required if the moving domain wall reaches a branching point of the network, at which point further motion of the wall requires the breaking of an additional bond. Since a percolating cluster at the percolation threshold contains loops on all length scales, the treatment of the kinetic Ising model on such a randomly branched structure is highly nontrivial. Stinchcombe (1985) investigated this problem with a real space recursive renormalization group method, to find that the relaxation time increases more strongly than the simple Arrhenius relaxation $\tau \propto \exp(2J/k_BT)$ that would hold for the strictly one-dimensional Ising chain, namely

$$\tau \propto \exp\left\{A\left(\frac{2J}{k_BT}\right)^2 + B\left(\frac{2J}{k_BT}\right)\right\}, \tag{4.96}$$

where A and B are constants that are known only approximately. Since for the Ising model the thermal correlation length ξ_T, i.e. the typical size of the domains, at p_c behaves simply as $\xi_T \propto \varepsilon^{-\nu_p}$ with $\varepsilon = \exp(-2J/k_BT)$, Eq. (4.96) implies that in this case the standard dynamical scaling relation $\tau \propto \xi_T^z$, where z is a dynamical exponent, is not true.

4.3.3 *The Sol-Gel Transition Revisited*

In Chap. 3, we have already discussed the geometrical aspects and the analogy to percolation of the sol-gel transition, where more and more crosslinks are formed between branched polymers in solution until at the gel point an infinitely extended polymer network is formed (Stauffer 1976, Stauffer *et al.* 1982, de Gennes 1976a). The Flory (1941a-c)– Stockmayer (1943)

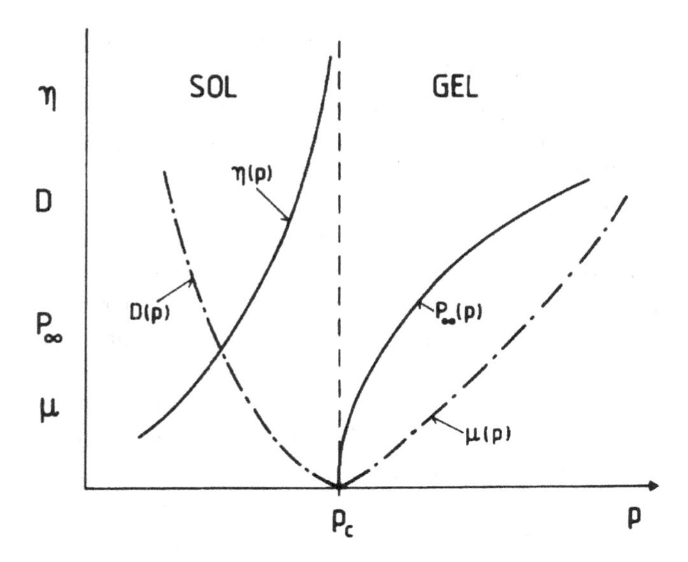

Fig. 4.8 Schematic variation of viscosity $\eta(p)$, diffusion constant $D(p)$, gel fraction $P_\infty(p)$, and shear modulus $\mu(p)$ with the concentration p of crosslinks; at p_c the sol-gel transition takes place.

theory of the gelation transition is in fact the simplest mean field approach to the percolation problem. In Sec. 4.2.1, we have addressed the problem of the elasticity of the gel in a preliminary way. Now we return to this transition, which is of particular interest since it can be viewed as a continuous thermodynamic phase transition from a fluid to an amorphous solid (Goldbart and Goldenfeld 1987, 1989). Of interest for this transition are not only static quantities, such as the structure or the gel fraction (fraction of monomers in the system belonging to the gel phase, local topology,...), but also dynamical properties: If we approach the transition from the sol side, the viscosity of the solution diverges and the effective self-diffusion constant of the macromolecules vanishes (Fig. 4.8), see Broderix *et al.* (2000) for a discussion of the theoretical aspects and Martin and Adolf (1991) for a review of experiments.

Approaching the transition from the gel side, the shear modulus μ vanishes, and on both sides of the transition the complex frequency-dependent shear relaxation modulus $G(\omega)$ is expected to exhibit an interesting singular behavior (Goldenfeld and Goldbart 1992). An important aspect in the gel phase is also the tensorial character of elasticity (Bergman and Kantor 1984, Kantor and Webman 1984, Benguigui 1984, Kantor 1985). However, we will not discuss here in great detail all these interesting aspects of the sol-gel transition, which has been a subject of intense research (e.g. Farago and Kantor 2000a, 2000b, 2002), de Arcangelis *et al.* 2002, del Gado *et al.* 2002), but summarize only some salient features.

First of all we turn to the dynamics in the sol phase, and wish to discuss the critical behavior of the transport coefficients (Fig. 4.8). These anomalies are due to the formation of larger and larger branched macromolecules as the gel point p_c is approached. The description of these branched macromolecules in terms of percolation clusters is of course insufficient to specify the dynamical properties of such a polymer solution. The simplest possibility to associate an intrinsic dynamics to these branched macromolecules is again provided by the use of the Rouse model, which we have seen to be very useful for the description of linear chains (Sec. 4.1). This model neglects all interactions in the system except those that describe the connectivity of the cluster, treating all bonds between the beads in the branched polymer as harmonic springs. Of course this model is terribly crude–but one should recall that for melts of short (i.e., non-entangled) polymers it does work rather well.

De Gennes (1978) suggested that the viscosity η can be related to the Rouse relaxation times τ_n of the polymers of size n as

$$\eta \propto \sum_s s\, n_s \tau_s \ , \tag{4.97}$$

with n_s being the probability that a "cluster" (i.e., a branched polymer) has size s. Assuming that τ_s scales like R_s^2, the squared gyration radius of the cluster, and noting $R_s \propto s^{1/d_f}$, where $d_f = d - \beta_p/\nu_p$ is the fractal dimension of the percolation problem, see Eq. (3.81), one can evaluate Eq. (4.97) similarly as we have done for similar sums in Chap. 3, using the same scaling assumption for the cluster number n_s as used there. This yields a power-law

$$\eta(p) \propto (p_c - p)^{-k} \quad \text{with } k = 2\nu_p - \beta_p \, . \tag{4.98}$$

However, the validity of this approach is doubtful, since it ignores the complicated internal structure of percolation clusters, and it would be rather surprising if in this case a dynamic exponent (k) could be entirely expressed in terms of static ones (ν_p, β_p). Despite this, in recent years different groups have used the framework of mode coupling theory, which will be discussed in Chap. 5, to predict the relaxation dynamics of gel-forming systems using as input only static quantities (essentially the static structure factor) (Bergenholtz and Fuchs 1999, Cates *et al.* 2004, Kroy *et al.* 2004, Viehman and Schweizer 2008). The obtained results show that this approach does indeed have its merits and should therefore not be discarded too quickly. Therefore we will come back to this point in Chap. 6 when we discuss the connection between gels and glasses.

An alternative connection (de Gennes 1979b) relates the viscosity at the sol-gel transition to the conductivity of a random resistor network made up of a fraction p of superconducting and a fraction $1-p$ of normal conducting links. Approaching the percolation transition from below, the conductivity diverges as $\sigma \propto |p-p_c|^{-s}$, and de Gennes (1979b) predicted $s = k$. However, Broderix *et. al.* (2000) derived a different scaling relation implying

$$k = \phi - \beta_p \,, \tag{4.99}$$

where ϕ is the so-called crossover exponent (Harris and Lubensky 1987). Within the framework of the Rouse model, and neglecting the correlation between the crosslinks, it is also possible to derive an exact correspondence between the static shear viscosity and the resistance of a random resistor network (Broderix *et al.* 2000). However, one should recall that in reality the Rouse model does not give a reliable description of the dynamics of polymers in (dilute) solution, since it neglects the hydrodynamic interactions among the monomers of the chains. For a single chain in dilute solution, it is known that the Zimm (1956) model, which takes these hydrodynamic interactions into account, gives rise to a significantly different behavior from the Rouse model (Sec. 4.1 and Eq. (4.24)). Furthermore, even if the long range hydrodynamic interactions are included, we must realize that near the gelation threshold it is questionable whether the clusters in the solution can indeed be treated as independent of each other. Thus, it is presently not clear whether the theory of Broderix *et. al.* (1997, 2000) is experimentally relevant. For the intermediate scattering function $S(q,t)$, this theory makes for long times an Ansatz where again a decomposition in terms of clusters of different sizes s is made, see Chap. 3:

$$S(q,t) \underset{t\to\infty}{\longrightarrow} \sum_{s=0}^{\infty} s n_s \exp(-D_0 q^2 t/s) \,. \tag{4.100}$$

Here the result of the Rouse model enters that for a cluster of size s the diffusion constant D_s scales inversely with its mass, and hence $D_s = D_0/s$, where D_0 is a constant. Using again the scaling description

$n_s(p) = s^{-\tau}F(s^\sigma(p-p_c))$], see Sec. 3.2.4, Eq. (3.72), and then transforming the sum in Eq. (4.100) into an integral, one obtains a generalized stretched exponential relaxation (Broderix *et al.* 2000)

$$S(q,t\to\infty) \propto (p_c-p)^{(2\tau-5)/4\sigma}\,[D_0q^2t]^{-(2\tau-3)/4} \times \exp\{-2[D_0q^2(p_c-p)^{1/\sigma}\,t]^{1/2}\} \tag{4.101}$$

if $p < p_c$, while for $p = p_c$ one finds

$$S(q,t\to\infty) \propto (D_0q^2t)^{\tau-2}\,. \tag{4.102}$$

These results imply that the effective diffusion constant D_{eff}, defined from $S(q,t\to\infty) \propto \exp[-2(D_{\text{eff}}q^2t)^{1/2}]$, vanishes as $(p_c-p)^{1/\sigma}$, and at p_c there is a power-law decay of $S(q,t\to\infty)$ with a very small exponent, since $\tau = 2+1/\delta$ (cf. Sec. 3.2.4). These results are at least in qualitative accord with experiments using quasi-elastic light scattering on silica gels (Martin and Adolf 1991). However, instead of the result from Eq. (4.101) one finds in experiments $S(q,t\to\infty) \propto \exp[-2(D_{\text{eff}}q^2t)^{2/3}]$, suggestive of a Zimm model type description. In addition, the experimental values of k do not support Eq. (4.99) but are systematically larger.

Also the critical exponent f describing the vanishing of the shear modulus of the gel $\{\mu \propto (p-p_c)^f\}$ is a subject of much discussion (Farago and Kantor 2000a, 2000b, 2002, del Gado *et al.* 2002). Farago and Kantor (2000a) proved that for a phantom network (i.e. chains without excluded volume interactions) of gaussian springs each having the energy $U = (1/2)Kr^2$, where r is the spring length, the exponent f equals the exponent describing the conductivity of random resistor networks near p_c. However, more recently the same authors have pointed out (Farago and Kantor 2002) that also the topology of the connectivity of the network matters: If the network has the connectivity of a cubic lattice, two distinct shear moduli μ_1 and μ_2 need to be distinguished, whose difference is characterized by a new exponent, $\mu_1-\mu_2 \propto (p-p_c)^h$, with $h \approx 4$. Furthermore they find in their model $f \approx 2$, which is in contradiction to a result of Daoud (2000) and Daoud and Coniglio (1981) who, using a scaling argument, suggest that the elastic moduli of a gel scale as k_BT/ξ_p^d, where $\xi_p \propto (p-p_c)^{-\nu_p}$ is the percolation correlation length, which hence would yield, making use of Eq. (3.95), $f = d\nu_p \approx 2.64$ in $d = 3$ dimensions. If instead the main contribution to the elasticity comes from a bond bending energy, one would predict $f = d\nu_p + 1 \approx 3.64$ (Feng and Sen 1984, Kantor and Webman 1984). In fact, as discussed by del Gado *et al.* (2002), for each of these predictions one seems to find some experimental confirmation, so the question arises whether the gel elasticity is really characterized by universal behavior at all. The simulations of del Gado *et al.* (2002) confirm the result $f = d\nu_p$ and hence suggest that (at least) two different universality classes for critical behavior of gel elasticity exist. However, a deeper theoretical understanding of what properties distinguish these classes is still lacking.

4.4 Physical Properties of Amorphous Solids

The physical properties of amorphous solids are of course of utmost importance for the applications of these materials. E.g., it is obvious that the optical isotropy of glasses used in lenses is just a simple consequence of the fact that for any amorphous structures no direction in space is singled out, in contrast to the case of crystals. However, there are a large number of optical properties of glasses for which our understanding of the relation between the structure and the chemical composition of the glass is still very incomplete (Scholze 1990, Bach and Neuroth 1998, Bach and Krause 1999). These properties include, e.g., the (frequency-dependent) refractive index, and also properties related to the transmission, scattering and absorption of light, from the infrared to the ultraviolet. Specific point impurities (such as Fe) may give glasses a specific color, and it is clear that an understanding of optical properties such as optical absorption or frequency-dependent reflectivity need to be discussed in conjunction with the electronic states and band structure of the material (Bach and Neuroth 1998, Zallen 1983). The latter topic is also crucial for an understanding of electrical conductivity and all kinds of electronic excitations, a topic of central importance for amorphous semiconductors (Brodsky 1979, Mott and Davis 1971, Hamakawa 1982, Singh and Shimakawa 2003) but also for metallic glasses (Güntherodt and Beck 1981, Beck and Güntherodt 1983, 1994, Kovalenko *et al.* 2001, Miller and Liaw 2008, Suryanarayana and Inoue 2010) for which also magnetic properties may be of interest. However, we emphasize from the outset that all these properties are completely outside of the scope of the present book, as well as properties that are closely related to chemistry (chemical resistance and corrosion of glasses, ionic processes between glasses and their environment, see e.g. Bach and Krause 1999) or to the crystallization of glasses (Gutzow and Schmelzer 1995). The large number of books that we have quoted here on these subjects already indicated the richness of these topics, and the fact that a very broad empirical knowledge on these subjects has already been accumulated. However, here we focus only on some physical properties that are of central interest from the point of view of statistical thermodynamics, such as specific heat, thermal expansion, and certain transport properties (diffusion and ionic conductivity, thermal conductivity, and related probes like the inelastic scattering of light and neutrons). In addition, the goal is not on exhaustive discussion of properties of various materials, but instead to focus on the extent to which these properties can be understood in terms of simple models.

4.4.1 *Two-Level Systems*

Already in the first chapter we have pointed out that one of the most striking hallmarks of glassy solids is the anomalous behavior of the specific heat and the thermal conductivity at low temperatures (Zeller and Pohl 1971, Phillips 1981, Buchenau 2001), see Fig. 1.3. It is of course tempting to assume that these anomalies are related to the picture of a glass as "frozen liquid" (Fig. 1.2), with a large number of metastable states, a fact that is also plausible from the large residual zero point entropy, see Fig. 1.17. When the glass is frozen into one of these states which corresponds to a local minimum of the energy landscape, most other states are no longer accessible from that state if the temperature is far below the glass transition temperature, since they are separated from the considered state by very high energy barriers. However, if some states are accessible (on the timescales relevant for the experiment), they will contribute to the specific heat.

This view is the key idea of the tunneling model proposed by Phillips (1972) and by Anderson *et al.* (1972). Of course, for the low temperatures of interest (1 K and below, see Fig. 1.3) thermally *activated* processes between these states are highly improbable, and hence one is left with the idea of a *tunneling* process between two states that correspond to the two local minima of particle configurations. Quantum mechanics tells us that such a process can only occur if the mass of the degree of freedom that shows the tunneling motion is sufficiently small and if the tunneling barrier is not too high. Hence we can expect that the dominating contribution must come from localized excitations, such as small rearrangements of single atoms or small groups of atoms as hypothesized in Fig. 1.5. However, the precise nature of these tunneling objects that are held responsible for the low temperature anomalies of glasses is still controversial. Nevertheless, this ambiguity does not prevent a phenomenological description of this tunneling in terms of asymmetric double-well potentials (Fig. 1.4). Here x is the "reaction coordinate" that connects the two states over a low-lying saddle point in phase space. Tunneling can occur if the height V and the width δ of this energy barrier (Fig. 1.4) are small enough, and also the energy difference Δ between the two states must be not too large. In the following, we shall explore the most salient consequences of this tunneling picture, following von Löhneysen (1981) and Kovalenko *et al.* (2001).

The task is to solve the Schrödinger equation

$$\widehat{\mathcal{H}}\varphi(x) = \Big[-\frac{h^2}{2m}\frac{d^2}{dx^2} + U(x)\Big]\varphi(x) = E\varphi(x)\,, \tag{4.103}$$

where m is the mass of the degree of freedom whose spatial coordinate is x, $U(x)$ is the double-well potential of Fig. 1.4, and E is the enrgy. We assume that, unlike the situation shown in that figure, both wells have the same characteristic frequency ω_0, i.e., placing the origin at the left minimum, $d^2U/dx^2\,\big|_{x=0} = d^2U/dx^2\,\big|_{x=\delta} = m\omega_0^2$. If the two wells could be treated as harmonic potentials independent of each other, we would have the following ground state wave functions for the right (R) and left (L) minimum:

$$\varphi_L(x) = (x_0\sqrt{\pi})^{-1/2}\exp(-x^2/2x_0^2)\,, \quad x_0 = (\hbar/m\omega_0)^{1/2} \tag{4.104}$$

$$\varphi_R(x) = (x_0\sqrt{\pi})^{-1/2}\exp[-(x-\delta)^2/2x_0^2]\,. \tag{4.105}$$

By adding an appropriate constant to $U(x)$ we can make that $U(0) = \Delta/2$, $U(\delta) = -\Delta/2$, and hence the energies belonging to these solutions become $E_L = (\hbar\omega_0 + \Delta)/2$, $E_R = (\hbar\omega_0 - \Delta)/2$.

Recall that the distance from the ground state to the first excited state in a single well is $\hbar\omega_0$, an energy that is of the order of the Debye energy $\hbar\omega_D$ of the respective crystal. Since the latter energy is significantly larger than Δ, such excited states can safely be ignored.

Now we want to take into account that in reality there is only a finite barrier V between the right and the left well, and hence due to tunneling neither $\varphi_L(x)$ nor $\varphi_R(x)$ are true eigenstates of $\widehat{\mathcal{H}}$ in Eq. (4.103). If V is not too small the true solution will resemble very much a trial solution of the form

$$\varphi(x) = C_L\varphi_L(x) + C_R\varphi_R(x)\,, \tag{4.106}$$

where a condition for the coefficients C_L, C_R follows from the normalization

$\int_{-\infty}^{+\infty} |\varphi(x)|^2 dx = 1$ as

$$C_L^2 + C_R^2 + 2C_L C_R S = 1\,, \tag{4.107}$$

where S is the overlap integral

$$S = \int_{-\infty}^{+\infty} \varphi_L(x)\varphi_R(x)dx = \exp(-\lambda), \quad \lambda = \delta^2/4x_0^2 = \delta^2 m\omega_0/4\hbar\,. \tag{4.108}$$

Thus we find that, as expected, tunneling is negligible when either δ or m is too large.

The energy can be written as

$$E = \int \varphi^*(x)\widehat{\mathcal{H}}\varphi(x)dx = C_L^2 \int \varphi_L(x)\widehat{\mathcal{H}}\varphi_L(x)dx + C_R^2 \int \varphi_R(x)\widehat{\mathcal{H}}\varphi_R(x)dx$$
$$+ 2C_L C_R \int \varphi_R(x)\widehat{\mathcal{H}}\varphi_L(x)dx\,. \tag{4.109}$$

We now make use of the variational principle, i.e. minimize E with respect to C_L and C_R which yields, noting the constraint Eq. (4.107)

$$C_L \int \varphi_L \widehat{\mathcal{H}}\varphi_L dx + C_R \int \varphi_R \widehat{H}\varphi_L dx = E(C_L + C_R S), \tag{4.110}$$

$$C_R \int \varphi_R \widehat{H}\varphi_R dx + C_L \int \varphi_R \widehat{H}\varphi_L dx = E(C_R + C_L S)\,. \tag{4.111}$$

With a little algebra one finds $\int \varphi_L \widehat{H}\varphi_L dx \approx \hbar\omega_0/2 + \Delta/2 \approx \int \varphi_R \widehat{H}\varphi_R dx$, while

$$\int \varphi_R \widehat{\mathcal{H}}\varphi_L dx \approx e^{-\lambda}\Big\{\frac{1}{4}\hbar\omega_0 - \frac{\delta^2}{8x_0^2}\hbar\omega_0 + U(\delta/2) + \frac{\hbar}{4m\omega_0}\frac{d^2U}{dx^2}\Big|_{x=\frac{\delta}{2}}\Big\}\,. \tag{4.112}$$

For the sake of simplicity we choose the function $U(x)$ such that $d^2U/dx^2|_{x=\frac{\delta}{2}} = -m\omega_0^2$, and $V \approx U(\delta/2) = \hbar\omega_0\delta^2/8x_0^2$. Note that, with Eqs. (4.108) and (4.112), this implies that $\lambda = 2mVx_0^2/\hbar^2$ and $\int \varphi_R \widehat{H}\varphi_L dx \approx 0$. Defining new constants $C_1 = C_L + C_R S$ and $C_2 = C_R + C_L S$ and making use of the fact that $\Delta \ll \hbar\omega_0$ and $S \ll 1$, Eqs. (4.110) and (4.111) take the form of a simple matrix equation

$$\frac{1}{2}\begin{pmatrix} \Delta & -\Delta_0 \\ -\Delta_0 & -\Delta \end{pmatrix}\begin{pmatrix} C_1 \\ C_2 \end{pmatrix} = \Big(E - \frac{1}{2}\hbar\omega_0\Big)\begin{pmatrix} C_1 \\ C_2 \end{pmatrix} \tag{4.113}$$

where

$$\Delta_0 = \hbar\omega_0 \exp\Big(-\frac{1}{2}\frac{\delta}{\hbar}\sqrt{2mV}\Big) . \tag{4.114}$$

Equation (4.113) is readily diagonalized and one finds the two eigenvalues and eigenvectors as

$$E_{\rm gr} = \frac{1}{2}\Big(\hbar\omega_0 - \sqrt{\Delta^2 + \Delta_0^2}\Big), \quad C_1 = -(\varepsilon - \Delta)C_2/\Delta_0 \tag{4.115}$$

$$E_{\rm exc} = \frac{1}{2}\Big(\hbar\omega_0 + \sqrt{\Delta^2 + \Delta_0^2}\Big), \quad C_1 = -\Big(\varepsilon + \Delta\Big)C_2/\Delta_0 , \tag{4.116}$$

which describe the ground state and the first excited state, and hence are separated by $\varepsilon = \sqrt{\Delta^2 + \Delta_0^2}$. Note that in the degenerate case, $\Delta = 0$, this splitting can be very small.

Due to the fact that in this problem there are essentially two relevant energy levels, Eqs. (4.115) and (4.116), one calls the system described by the potential $V(x)$ a "two-level system (TLS)".

It is also of interest to consider the interaction of the two-level-systems with phonons. One can expect that an elastic deformation of the surroundings of a TLS will change its asymmetry, and hence one introduces a perturbation linear in the strain tensor $\underline{u}$,

$$\widehat{\mathcal{H}}_1 = -\begin{pmatrix} \gamma_{ik} & 0 \\ 0 & -\gamma_{ik} \end{pmatrix} u_{ik} , \tag{4.117}$$

where $\gamma_{ik} = d\Delta/du_{ik}$ is called the deformation, usually taken to be diagonal, and we set $\gamma_{ik} = \gamma\delta_{ik}$, due to the isotropy of amorphous solids. (Note that in Eq. (4.117) we have summed over repeated indices.) In a first approximation the change of Δ_0 due to $\underline{u}$ can be neglected. The transformation which diagonalizes $\widehat{\mathcal{H}}_0$ (i.e. $\widehat{\mathcal{H}}$) to $\widehat{\mathcal{H}}'_0$ will, due to the presence of $\widehat{\mathcal{H}}_1$, produce off-diagonal terms so that an effective perturbation results

$$\widehat{\mathcal{H}}'_1 = -\begin{pmatrix} \Delta/\varepsilon & 2\Delta_0/\varepsilon \\ 2\Delta_0/\varepsilon & -\Delta/\varepsilon \end{pmatrix} \gamma u_{ii} , \tag{4.118}$$

and thus the perturbed Hamiltonian can be rewritten as a pseudospin Hamiltonian $\widehat{\mathcal{H}}' = \widehat{\mathcal{H}}'_0 + \widehat{\mathcal{H}}'_1$:

$$\widehat{\mathcal{H}}' = \varepsilon\widehat{S}_z + (2M\widehat{S}_x + D\widehat{S}_z)u_{ii} \tag{4.119}$$

where $M = (\Delta_0/E)\gamma$, $D = (2\Delta/E)\gamma$, and

$$\widehat{S}_x = \frac{1}{2}\begin{pmatrix} 0 & 1 \\ 1 & 0 \end{pmatrix} \quad \text{and} \quad \widehat{S}_z = \frac{1}{2}\begin{pmatrix} 1 & 0 \\ 0 & -1 \end{pmatrix}. \tag{4.120}$$

Thus, one can interpret the diagonal term $D\widehat{S}_z u_{ii}$ in terms of the change of the energy splitting of the TLS through the strain field. The off-diagonal term $2\,MS_x u_{ii}$ describes resonant transitions from $\varepsilon/2$ to $-\varepsilon/2$. The problem is fully analogous to a spin $\frac{1}{2}$ particle in a magnetic field $\vec{B}$ (note that for a sound wave that oscillates with the sound frequency the term relating to ε_{ii} corresponds to a frequency-dependent magnetic field), $\widehat{\mathcal{H}} = \hbar\gamma_G(\vec{B}\cdot\widehat{\vec{S}})$, γ_G being the "gyromagnetic ratio". Therefore it is not really a surprise that one can deduce equations of motion for the pseudospin $\vec{S}$ in Eq. (4.119) that are fully analogous to the well known Bloch equations of magnetism (Golding *et al.* 1973, Black and Halperin 1977). Some of the most direct and hence most convincing evidence for the actual existence of TLS, namely the resonant interaction between TLS and phonons enabling the possibility to saturate ultrasound absorption (Golding *et al.* 1978, Doussineau *et al.* 1978) is based on this result.

Here we are mostly interested in the consequences that result from this description for the specific heat. For a given atomic configuration and no elastic deformation (i.e. $u_{ik} \equiv 0$) the Hamiltonian of the two-level system in diagonal form is simply $\widehat{\mathcal{H}}_0 = \begin{pmatrix} E_{\text{gr}} & 0 \\ 0 & E_{\text{exc}} \end{pmatrix}$, and hence its thermal expectation value is

$$\langle\widehat{\mathcal{H}}_0\rangle = \frac{1}{Z}(E_{\text{gr}}e^{-E_{\text{gr}}/k_BT} + E_{\text{exc}}e^{-E_{\text{exc}}/k_BT}), \tag{4.121}$$

with the partition function $Z = \exp(-E_{\text{gr}}/k_BT) + \exp(-E_{\text{exc}}/k_BT)$. Hence, the contribution of one TLS to the specific heat is a simple Schottky anomaly, described by

$$C_V^{(1\text{TLS})} = \left(\partial\langle\mathcal{H}_0\rangle/\partial T\right)_V = \frac{\varepsilon^2}{k_BT^2}\,\frac{1}{4}\text{sech}^2\left(\frac{\varepsilon}{2k_BT}\right), \tag{4.122}$$

where we have made use of Eqs. (4.115) and (4.116).

In order to obtain the observed contribution of all the TLS to the specific heat we need to know the distribution of the parameters Δ and Δ_0, or alternatively of ε and $r \equiv \Delta_0^2/(\Delta_0^2+\Delta^2)$. The latter quantity is restricted by definition to $0 \le r \le 1$: $r = 1$ corresponds to the special case of symmetric double-well potentials, whereas $r = 0$ corresponds to a diverging barrier height. Due to the structural disorder of the glass, one expects indeed a broad distribution $p(\varepsilon, r)$ so that

$$C_V = \overline{C_V^{(1\mathrm{TLS})}} = k_B \int_0^\infty d\varepsilon \int_{r_0}^1 p(\varepsilon, r) \Big(\frac{\varepsilon}{2k_B T} \, \mathrm{sech} \frac{\varepsilon}{2k_B T} \Big)^2 dr \,, \qquad (4.123)$$

where we have introduced r_0 as a maximal cut-off parameter of the barrier height, which can be a function of the characteristic time of the experiment (Black 1978): the larger this time-scale, the larger barriers will contribute due to rare tunneling events. The fact that one can indeed identify a contribution in the specific heat that depends logarithmically on the observation time (Loponen *et al.* 1980, Meißner and Spitzmann 1981, Zimmermann and Weber 1981) is another nice piece of evidence that supports the picture of the TLS.

Since even today the precise microscopic nature of the TLS's is uncertain, the precise form of $p(\varepsilon, r)$ is not known. Most important at low T is of course the form of $p(\varepsilon, r)$ for $\varepsilon \to 0$, as is obvious from the integrand in Eq. (4.123). Already Anderson *et al.* (1972) suggested to take $p(\varepsilon, r)$ simply equal to a constant. Then Eq. (4.123) simply implies

$$C_V = \mathrm{const} \cdot T \,, \qquad (4.124)$$

i.e. a linear variation of the specific heat at low temperatures. If the same approximations are used in the Debye model for the thermal conductivity κ (Berman 1976), one finds (Anderson *et al.* 1972, Hunklinger and Piché 1975)

$$\kappa \propto T^2 \,, \qquad (4.125)$$

again in good agreement with experimental data (e.g. Zeller and Pohl 1971, von Löhneysen 1981).

One can extend the theoretical modeling of the TLS to include ultrasonic absorption (Anderson *et al.* 1972, Phillips 1972, Hunklinger and Arnold 1976), the related logarithmic contribution in the temperature dependence of the sound velocity (Piché *et al.* 1974), etc. For more details on this we refer the reader to the beautiful reviews of von Löhneysen (1981) and Kovalenko *et al.* (2001). It seems that a consistent description of many properties of a broad variety of different glassy materials in the temperature range from a few hundreds of a K to a few K has been obtained. However, other experiments at very low temperatures around 1 mK present unexpected phenomena (Rogge *et al.* 1996, Stechlow *et al.* 1998, Ludwig *et al.* 2002) which, to our knowledge, are not yet fully explained.

Another extension of the tunneling model is needed if one considers higher temperatures. While the analysis presented above, leading to the effective Hamiltonian given by Eq. (4.119), implicitly assumes, following Sussmann (1964), that the basic coupling of a TLS to phonons is a single-phonon-process (via the bilinear term $\widehat{S}_x u_{ii}$), it has been suggested that many-phonon processes may give rise to fluctuations of the double-well potential which cause a reduction of its height (Fleurov and Trakhtenberg 1982, 1983, 1986). Also TLS with very small barriers in which intrinsically anharmonic motions need to be considered may be important (Karpov *et al.* 1982, 1983, Klinger and Karpov 1983). Extensions of this model are called the "soft potential model" (see Buchenau *et al.* 1991, 1992 and Parshin 1994 for a review) or asymmetric double-well potential model (where one still uses the same picture for the potential, Fig. 1.4, but describes the dynamics entirely classically, see Gilroy and Phillips (1981). We will not review these theoretical developments here, but instead restrict us to a brief discussion of anomalous properties of glasses at intermediate temperatures in the next subsection.

4.4.2 *Anomalies of Glasses at Intermediate Temperatures: Excess Specific Heat, Thermal Conductivity Plateau, and Boson Peak*

By "intermediate temperatures" we mean here temperatures typically in the regime 1 K$< T <$ 100 K, i.e. well below the glass transition temperature T_g of typical amorphous materials. Although this region of temperature is clearly of greater practical importance than the one discussed in the context of the two-level systems of the previous section, it is nevertheless much less well understood. Here we will give only a cursory review of the main facts and ideas, and will neither attempt to review exhaustively the experiments (for reviews on experiments see Courtens *et al.* (2001, 2003), Nakayama (2002), and Buchenau 2001) nor the corresponding theory and simulations (Karpov *et al.* 1983, Buchenau *et al.* 1991, 1992, Sokolov 1999, Sokolov *et al.* 1993, Götze and Mayr 2000, Theenhaus *et al.* 2001, Taraskin and Elliot 1997, 1999a, 1999b, 2000a, 2000b, Guillot and Guissani 1997, Horbach *et al.* 2001, Gurevich *et al.* 2003, Wyart *et al.* 2005, Shintani and Tanaka 2008, Ilyin *et al.* 2009). Note that most of this work is concerned mostly with anomalies in vitreous silica, since in this material and similar "strong glass-

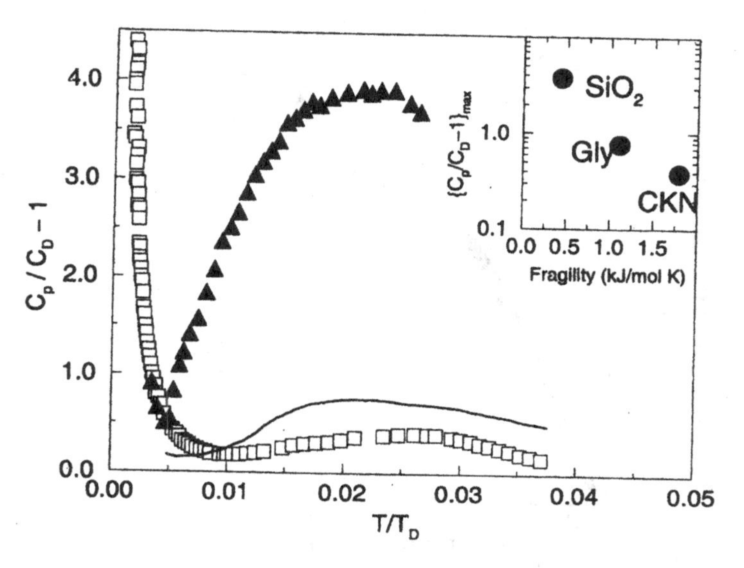

Fig. 4.9 The excess specific heat $C_p/C_D - 1$ versus the reduced temperature T/T_D, where T_D is the Debye temperature. Both T_D and the Debye approximation for the specific heat C_D are calculated from the known acoustic sound velocity of the respective materials. Full triangles refer to vitreous silica, the line refers to glycerol, and the open squares to CKN {= the mixed nitrate $Ca_2\,K_3\,(NO_3)_7$}. Note that the glass temperatures of these three materials are $T_g = 186$ K (glycerol), $T_g = 333$ K (CKN) and T_g=1450 K (SiO_2). The inset shows the maximum excess as function of the fragility parameter which is defined here from the slope of the $\log \eta$ vs. T_g/T plot at T_g (defined by $\eta(T_g) = 10^{13}$ poise), $F = R[\partial \ln \eta / \partial(T_g/T)]|_{T=T_g}$, R being Avogadro's constant. From Sokolov *et al.* (1997).

formers" (Angell 1991) these anomalies are much more pronounced than in "fragile glass-formers" (Angell 1991, Sokolov *et al.* 1993). As an example we show in Fig. 4.9 the excess specific heat for vitreous silica, glycerol and CKN, normalized to the Debye value C_D (Sokolov *et al.* 1997). Of course, relative to the Debye approximation, C_p/C_D diverges as T^{-2} for $T \to 0$, due to the linear specific heat at very low temperatures. This happens for all glasses, and is of no interest in the present context. Here we rather focus on the hump that occurs in the region $0.02 \leq T/T_D \leq 0.03$, T_D being the Debye temperature. From the figure we recognize that this excess is very pronounced for "strong glasses" such as SiO_2, while for the "fragile glass" CKN there is almost no excess at all.

In the same regime of temperatures (i.e., for T near 10 K in the case of SiO_2) the thermal conductivity has a pronounced plateau (Fig. 4.10)

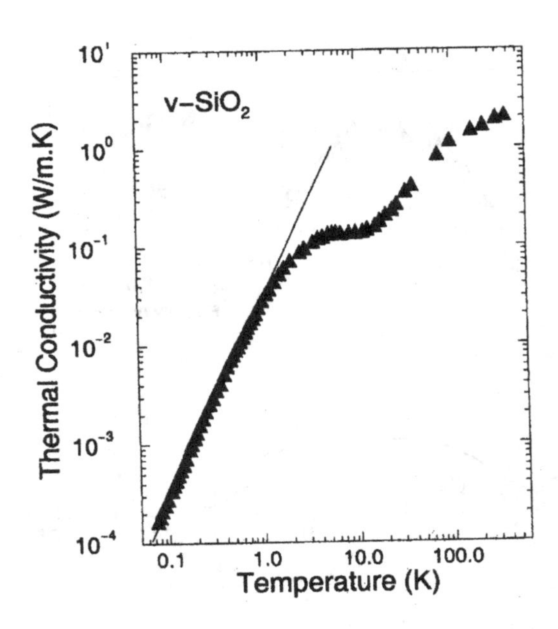

Fig. 4.10 The thermal conductivity of vitreous SiO_2 (v-SiO_2) plotted versus temperature on a log-log-plot. The straight line illustrates the $\kappa \propto T^2$ law, Eq. (4.125). From Raychaudhuri (1989).

(Raychaudhuri 1989). The occurrence of this plateau implies that at high frequency Ω the mean free path of the acoustic phonons must decrease strongly, e.g. $\ell \propto \Omega^{-4}$, or the excitations must even become localized (Feldman *et al.* 1999).

However, the feature that has caused the most controversy on its interpretation is the excess scattering seen in inelastic scattering of neutrons, X-rays or light, the so-called "boson peak" in the THz frequency range (Courtens *et al.* 2001, Nakayama 2002). Figures 4.11 and 4.12 show typical experimental data of this peak (Hehlen *et al.* 2000, Rufflé *et al.* 2003) and Fig. 4.13 shows a corresponding simulation (Horbach *et al.* 2001) for the BKS model (van Beest *et al.* 1990) of liquid and vitreous silica (see Sec. 3.6). Note that the intensity of the incoherent inelastic neutron scattering, where an energy $\hbar\omega$ is transferred from the neutron to the solid (or liquid), simply is proportional to the density of states $\zeta(\omega)$ of the system[2]

[2]This is true for one-component liquids for which the neutron scattering cross section is independent of the atoms. For multi-component systems this is no longer true in general. A similar caveat exists for the Raman intensity.

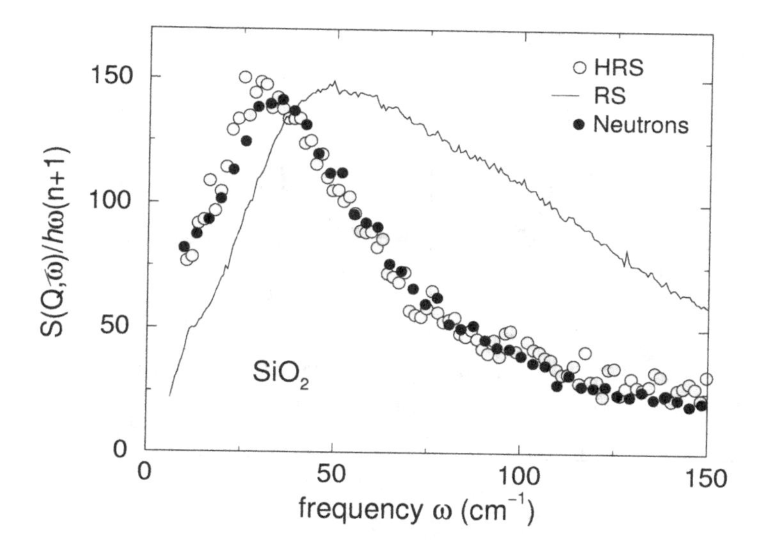

Fig. 4.11 Inelastic scattering intensity $S(Q, \omega)$ for silica from incoherent inelastic neutron scattering (filled circles), scaled hyper-Raman intensity $I(\omega)$ (open circles), and Raman intensity (solid line). Note that in the main text, the scattering wavenumber is denoted q instead than Q as in this figure. Adapted from Hehlen *et al.* (2000).

$$S_{\text{inc}}(q, \omega) \propto [n(\omega) + 1]\zeta(\omega)/\omega\,, \tag{4.126}$$

while the intensity of hyper-Raman experiments (Hehlen *et al.* 2000) can be expressed as

$$I(\omega) = [n(\omega) + 1]\zeta(\omega)c(\omega)/\omega\,. \tag{4.127}$$

Here $n(\omega)$ is the thermal occupation number of the states, $n(\omega) = [\exp(\hbar\omega/k_B T) - 1]^{-1}$, which takes into account the bosonic character of the excitations of the system (which can be either delocalized vibrations such as phonons or localized excitations and give rise to the name "boson peak"), and $c(\omega)$ is the hyper-Raman coupling coefficient which *a priori* is not known. Since, however, Fig. 4.11 implies that in the frequency range of interest $c(\omega)$ is essentially constant, it will here not be considered further. As is well known, the contribution of vibrational excitations of a (harmonic) solid to the specific heat can be written as

$$C_V(T) = \frac{\partial}{\partial T}\int_0^\infty \hbar\omega n(\omega)\zeta(\omega)d\omega\,. \tag{4.128}$$

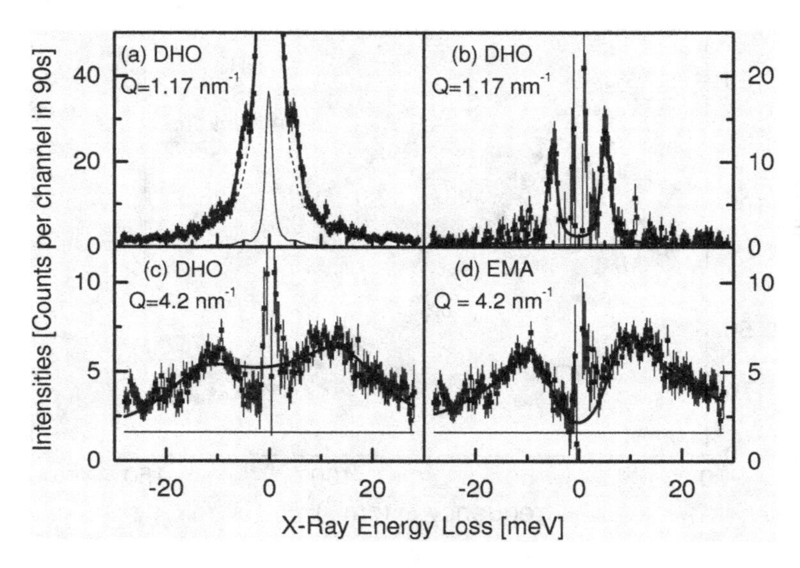

Fig. 4.12 Brillouin X-ray scattering signals obtained on densified silica glass, $d\mathrm{-SiO_2}$, for different wave-vectors. a) $Q = 1.17\ \mathrm{nm}^{-1}$. The bold line is the measured signal. The thin line is the same data divided by 20 in order to allow to see the whole peak. The dashed line is the estimated elastic contribution. In b)-d) this elastic contribution has been subtracted. b): Data for $Q = 1.17\ \mathrm{nm}^{-1}$ including a fit with a damped harmonic oscillator. c) and d): $Q = 4.2\ \mathrm{nm}^{-1}$ with a damped harmonic oscillator and a fit from the effective medium approximation, respectively. Adapted from Rufflé *et al.* (2003).

In a standard (crystalline) solid, the density of vibrational states in the Terahertz regime (i.e. frequencies $\omega \ll \omega_D$, the Debye frequency) simply scales as ω^2,

$$\zeta(\omega) = 6\pi\omega^2/\omega_D^3\,, \tag{4.129}$$

and ω_D is related to the velocities of transverse (v_t) and longitudinal (v_ℓ) sound by

$$\omega_D^3 = \frac{9\rho}{32\pi^4}\left(\frac{2}{v_t^3} + \frac{1}{v_\ell^3}\right)^{-1}, \tag{4.130}$$

where ρ is the density of the crystal. Using the experimental values $v_t = 3.767 \cdot 10^3$ m/s and $v_\ell = 5.970 \cdot 10^3$ m/s for vitreous silica, one can estimate ω_D to be 10.4 THz, which corresponds to a Debye temperature $T_D = \hbar\omega_D/k_B \approx 500$ K.

Obviously, using Eq. (4.129) in Eqs. (4.126) and (4.127), and noting that $n(\omega) + 1 \approx k_B T/\hbar\omega$ for small ω, one predicts $S_{\mathrm{inc}}(\omega) \propto I(\omega) \propto \mathrm{const}$

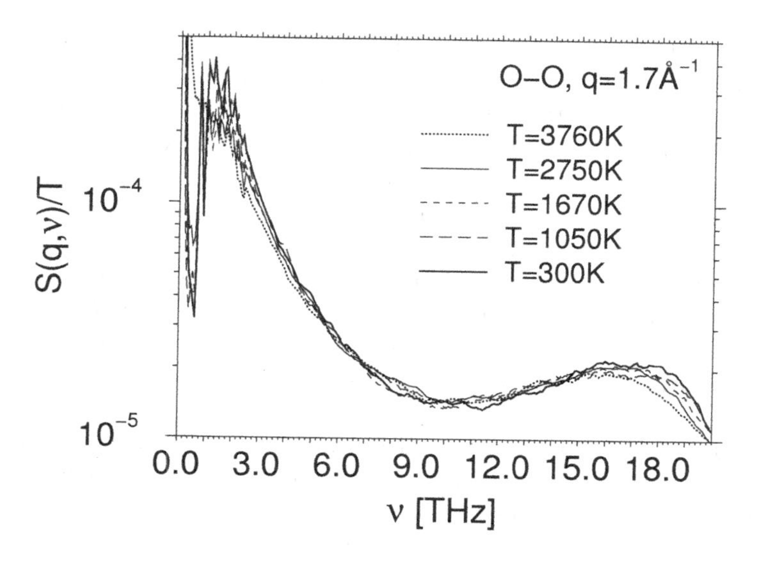

Fig. 4.13 $S(q,\nu)/T$ for the O-O correlations in SiO_2 at $q = 1.7$ Å^{-1} for different temperatures, as obtained from molecular dynamics simulations in the framework of the BKS model of SiO_2. Note that in this figure the energy transfer of the scattering is denoted as $h\nu$ rather than $\hbar\omega$ as in the text. From Horbach *et al.* (2001).

for $\omega \ll \omega_D$, i.e. from Eq. (4.129) it follow that there is no peak in this range of frequencies.

Of course, the result $\zeta(\omega) \propto \omega^2$ results for phonon excitations, i.e. one has propagating eigenmodes $\omega(\vec{k}) = v_\ell k$ and $\omega(\vec{k}) = v_t k$, respectively, and $\vec{k}$ is a "good quantum number". The latter is true for crystals as a consequence of the perfect periodicity of the crystalline lattice, but questionable for amorphous solids. Due to their disordered structure we expect the phonon modes to be damped at finite k since they are well-defined eigenmodes only in the limit $k \to 0$, because on the scale of a wavelength $\lambda = 2\pi/k$ the system looks homogeneous, i.e. density fluctuations are averaged out. The fact that at frequencies around 300 GHz propagating sound modes that are only weakly damped still occur can be shown by pulse-echo experiments (Zhu *et al.* 1991). However, the situation becomes considerably more complex at significantly higher values of k, i.e. for length scales that correspond to $O(10)$ nearest neighbor distances. One must expect that plane waves (characterized by ω and k) exist as well-defined excitations (i.e only weak attenuation occurs) only up to some crossover frequency crossover ω_{co} (and associated wavenumber k_{co}), at which the disorder (lack of periodicity) of the structure becomes important. One question is, what happens to acoustic-like excitations at $\omega > \omega_{co}$, and whether there is a situation comparable to strong localization of electronic states (Anderson 1958, Mott 1966, Mott and Davis 1971, Alexander 1989), for which the eigenmodes decay in real space exponentially with distance? The other question is, whether this crossover is related to the boson peak (Figs. 4.11-4.13) and to the anomalies of thermal conductivity and specific heat (Figs. 4.9-4.10). So far, the answers to all these questions are still controversially discussed in the literature (see Nakayama 2002 for a review and Courtens *et al.* 2001, 2003, Schirmacher 2006, Schirmacher *et al.* 2007, Rufflé *et al.* 2008, and Schmid and Schirmacher 2008 for examples on the ongoing discussion on the subject).

First of all, in inelastic coherent neutron scattering experiments as well as in simulations (Horbach *et al.* 1998, 2001), one observes phonon-like peaks in $S_{\rm coh}(q,\omega)$ over a surprising wide range of q, even in the liquid state (Fig. 4.14). Of course, no liquid can support shear stress, and hence the transverse modes at $T = 2750$ K in Fig. 4.14 should become overdamped at very small q. However, at this temperature the damping of the transverse modes is not yet significant even for the smallest wave-vectors accessible in present day simulations. Therefore the dispersion relation at $T = 2750$ K in the liquid state, Fig. 4.14a, and in the glass state, Fig. 4.14b, are very similar. The observation that at rather high frequencies (as considered here) there is no essential difference between a viscous liquid and a glass is also implicit in the approximate description of the boson peak provided by mode coupling theory (Götze and Mayr 2000). Furthermore, the longitudinal branch is almost periodic with a minimum located around $q_m = 2.8$ Å^{-1}, which is the location of the second sharp diffraction peak in the static structure factor (Fig. 2.5), and which corresponds to length scales of intra-tetrahedral distances (Horbach and Kob 1999). Thus $q_m/2$ can be interpreted, in analogy to crystals, as a quasi Brillouin zone. The minimum in $\nu_\ell(q)$ at q_m can be easily understood since the particles tend to favor relative separations of $2\pi/q_m$, and therefore at these wavelengths it costs relatively little energy to excite a collective mode. Of course, in a crystal there is an exact periodicity, i.e. $\nu_\ell(q_m) = 0$, and only the dispersion of modes up to $q_m/2$ are of interest, while in the disordered material there is a nontrivial behavior in the entire q range.

Note that in contrast to what one would expect for simple liquids, the mentioned minimum in $\nu_\ell(q)$ is not observed at $q = 1.7$ Å^{-1}, the location of the first sharp diffraction peak (see Fig. 2.5) in the static structure factor

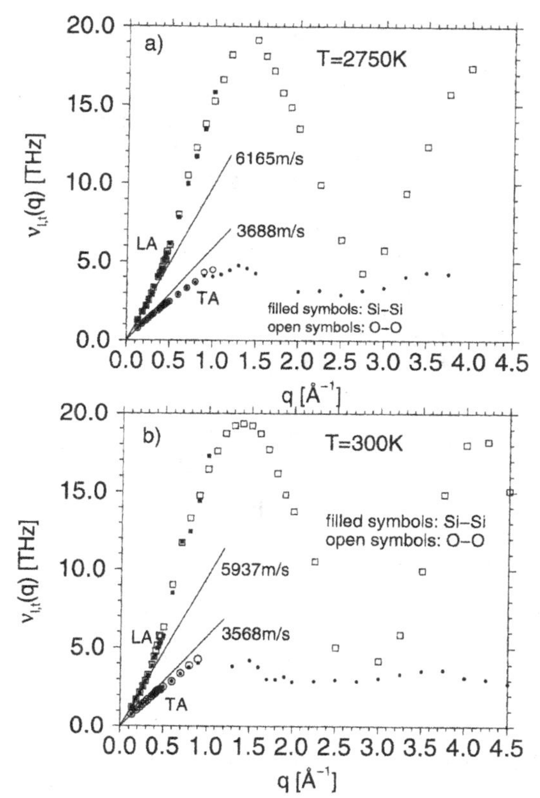

Fig. 4.14 Peak maximum positions $\nu_\ell(q)$ for the longitudinal acoustic (LA) and $\nu_t(q)$ for the transverse acoustic (TA) modes, for the Si-Si correlations (filled symbols) and the O-O correlations (open symbols) at (a) $T = 2750$ K and (b) $T = 300$ K, as obtained from molecular dynamics simulations. The bold lines are fits with linear laws $\nu_\alpha = v_\alpha q/(2\pi)$, for which the corresponding values of the sound velocities v_α are given in the figure. From Horbach *et al.* (2001).

$S(q)$. This is due to the fact that this q value corresponds to length scales of connected SiO_4 tetrahedra, a structural unit which is less rigid than one tetrahedron and hence the corresponding peak in the static structure factor is less pronounced. Also different from simple liquids is the linear dispersion of $\nu_\ell(q)$ for 0.4 Å$^{-1} < q < 1$ Å^{-1} with a slope that is higher than the linear dispersion found at smaller wave-vectors. This phenomenon, also known as "fast-sound", is probably related to the order on the length scale of the SiO_4-tetrahedra and is also found in other systems, such as water, that have a tetrahedron-like local structure (Sciortino and Sastry 1994).

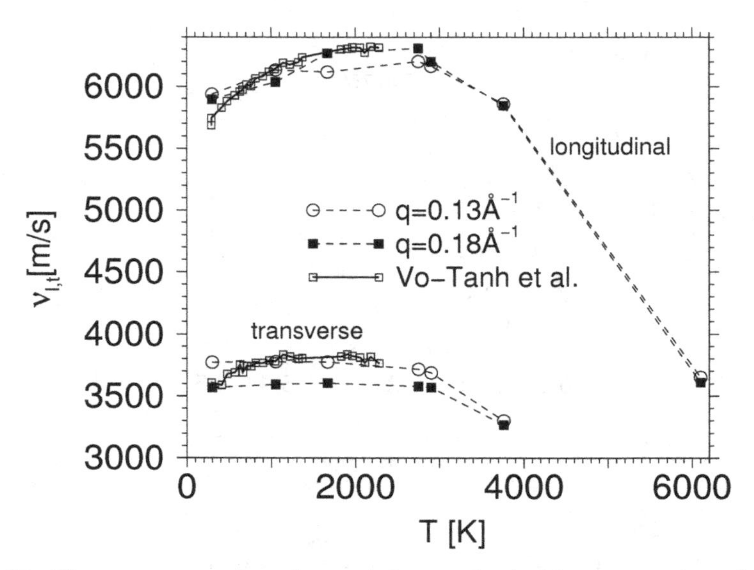

Fig. 4.15 The temperature dependence of the sound velocities v_t as estimated from $\nu_\ell(q)$, $\nu_t(q)$ at $q = 0.13$ Å^{-1} (open circles) and $q = 0.18$ Å^{-1} (filled squares) using molecular dynamics simulations as shown in Fig. 4.14. Also included are the experimental data of Polian *et al.* 2002 which were multiplied with the factor $(2.2/2.37)^{0.5}$ in order to account for the different densities in the simulation and the experiment. From Horbach *et al.* (2001).

The behavior of $\nu_\ell(q)$ is in agreement with findings from neutron scattering (Arai *et al.* 1999) and simulations using other potential models for SiO_2 (Taraskin and Elliott 1997). The transverse branch $\nu_t(q)$ becomes rather flat for $q > 0.9$ Å^{-1}, which is an indication of the overdamped character of the transverse acoustic excitations at these wave-vectors (Taraskin and Elliott 1997, Horbach *et al.* 2001).

Figure 4.15 shows that the temperature dependence of the effective sound velocities that one can extract from Fig. 4.14 is in good agreement with the experimentally determined sound velocities of vitreous silica. In addition, analyzing for 0.1 Å$^{-1} < q < 0.4$ Å^{-1} the line-widths $\Gamma(q)$ of the peaks (whose positions are shown in Fig. 4.14), one finds $\Gamma(q) \propto q^2$ for the LA modes, a result that one expects for propagating modes in an isotropic anharmonic elastic medium (Boon and Yip 1980). One also finds that in the range 0.6 Å$^{-1} < q < 2$ Å^{-1}, $\Gamma(q)$ decreases again, while $\nu_L(q)$ has a pronounced peak, and the line width $\Gamma(q)$ is in fact smaller than $\nu_L(q)$. It is clear, however, that it would be wrong to conclude from these observations that there are propagating modes in this q-range: Instead one has to realize that it is just not possible to infer the character of a mode from the shape of $S_{\text{coh}}(q, \omega)$. A much better method to decide on the nature of a mode is to diagonalize the dynamical matrix of the system after having cooled the system to low temperature and to determine numerically the curvature of the potential at the local minimum. This allows to obtain the eigenfrequencies and eigenvectors, and thus such an analysis would tell whether a mode is a propagating plane wave or not. Such an analysis has been attempted both for random regular lattices (Sheng *et al.* 1994, Schirmacher *et al.* 1998) and for vitreous silica (Taraskin and Elliott 2000b), and the conclusion is that propagating plane waves are not valid approximations beyond the crossover frequency ω_{co} mentioned above. For vitreous silica, the estimated value of $\omega_{co}/2\pi$ is around 1 THz (Taraskin and Elliott 2000b), and the modes above this so-called Ioffe-Regel (1960) crossover are not localized but diffusive in character. Actually, in the regime $\omega < \omega_{co}$ the line width of a propagating plane wave, which (in the absence of anharmonicity) is due to "Rayleigh scattering" by static fluctuations (Klemens 1955), should vary as $\Gamma \approx q^4$, and this has indeed been seen in simulations of random regular lattices (Schirmacher *et al.* 1998).

In the regime $\omega > \omega_{co}$ different types of modes can be distinguished: Localized vibrating modes, propagating or diffusive modes (the latter are modes that are quite delocalized in space but cannot be represented by a simple plane wave). The question whether or not a mode is localized or extended, can be answered by considering its so-called "participation ratio" (Bell and Dean 1972)

$$p_\alpha = \Big(\sum_{i=1}^{N} |\vec{u}_i^\alpha|^2\Big)^2 \Big/ \Big(N \sum_{i=1}^{N} |\vec{u}_i^\alpha|^4\Big), \tag{4.131}$$

where $\vec{u}_i^\alpha$ is the eigenvector at site i for mode α, N being the number of atoms. For delocalized excitations p_α is a fraction of unity that does not vanish as $N \to \infty$, unlike the case of localized excitations. It is found that in vitreous silica for $\omega > \omega_{co}$ modes are diffusive, but most of them stay delocalized (localized vibrations occur for relatively high frequencies, however). This issue is different from the case of electronic excitations in disordered metals, where localized states with an excitation energy $\varepsilon > \varepsilon_c$ (the "mobility edge") play a very important role. We also draw attention to the role of system dimensionality in this problem of "Anderson localization" (Anderson 1958) of electronic states since, e.g. in the presence of disorder, electronic states in $d = 1$ are always localized in the sense of $p_\alpha(N \to \infty) \to 0$, and no delocalized states can occur. The case of $d = 2$ is marginal, see

Abrahams *et al.* 1979, in that there is still not true metallic behavior, i.e. "quantum diffusion" of electrons is absent, unlike the behavior of a classical particle that moves in the presence of random fixed obstacles in $d = 2$ (see Höfling *et al.* 2007, who find anomalous diffusion only at the percolation threshold of this Lorentz model).

Finally we mention that the participation ratio is certainly not the best suited quantity to allow to decide whether or not a state is localized or not. In practice it turns out that two localized states can hybridize (i.e. become weakly coupled), giving thus rise to a new state that is quite extended and thus complicating severely the interpretation of p_α (Schober and Oligschleger 1996).

There has been some discussion in the literature on the quantitative description of the peak describing the excess intensity as a function of frequency in the spectrum (Figs. 4.11-4.13). Based on the idea that the boson peak is due to propagating modes that exhibit a homogeneous viscous damping ($\Gamma \propto q^2$), a damped harmonic oscillator (DHO) model has been proposed (Ruocco *et al.* 2000). Courtens *et al.* (2001) refute this idea, and point out that the DHO model at q values where the width of the peak exceeds the instrumental resolution distinctly (such as for $q = 4.2\text{nm}^{-1}$ in Fig. 4.12) the DHO model provides a poor fit, while a description based on the effective medium approximation (EMA) works much better. The latter description implies an inhomogeneous broadening, and is compatible with a diffusive rather than propagating nature of the excitations.

However, neither of these models really implies that the boson peak is due to an enhanced density of states of acoustic modes, and in fact clear evidence for such a speculative hypothesis is lacking. Instead there is growing evidence that the boson peak results from an excess in the density of states of optical modes, due to the rocking of small groups of SiO_4 tetrahedra. These coupled rotational motions of tetrahedra were already suggested in the early work by Buchenau *et al.* (1984), and since they lie in an appropriate frequency window, they strongly hybridize with the transverse acoustic phonons, leading to a strong attenuation of the latter (Nakayama 2002). This consideration is indeed compatible with the computer simulations of Horbach *et al.* (2001) in which it has been found that in the coherent scattering the peaks representing the acoustic peaks sit on top of a flat background which could be attributed to such effects. On the other hand, a similar behavior is also implied by the mode coupling theory of Götze and Mayr (2000), which refers to simple atomic glasses rather than vitreous silica. Still apparently different mechanisms for the boson peak have been proposed by Kantelhardt *et al.* (2001), Theenhaus *et al.* (2001), Bunde *et al.* (2002), and Taraskin *et al.* (2001, 2002). Taraskin *et al.* (2001, 2002) attribute the boson peak to disorder-induced level-repelling effects, whereas Theenhaus *et al.* point out the importance of a translation-rotation coupling of the tetrahedra or similar structural units. Kantelhardt *et al.* (2001) and Bunde *et al.* (2002) obtain the boson peak from a model with acoustic modes only, assuming a power-law distribution $P(f) \propto 1/f$ for the force constants between neighboring atoms, with suitable assumptions for the cutoffs $f_{\min}$, $f_{\max}$ of such a distribution. Clearly, this assumption is rather *ad hoc*, but a surprisingly large number of experimental observations can be accounted for.

A somewhat different treatment of the boson peak based on the idea of randomly fluctuating transverse elastic constants is due to Schirmacher (2006), see also Ilyin *et al.* (2009). Assuming Gaussian distributions, the configurational averaging of the approximate Green's function is done with the replica method familiar from the theory of spin glasses (see Sec. 4.5). The calculation of the density of states proceeds similarly as for the problem of electrons in a random potential. However, to keep the theory tractable, a mean field theory is introduced via the self-consistent Born approximation. It is found that the boson peak occurs when the mean free path becomes comparable to the wavelength of the excitations. This theory is intended to describe thermal conductivity, acoustic attenuation (Schirmacher *et al.* 2007) and Raman scattering (Schmid and Schirmacher 2008).

One problem that all theories have that attribute the boson peak to such rather general mechanisms is that they cannot really explain why the boson peak is only pronounced in the "strong glass-formers" and very weak (or even absent) for the fragile glass-formers. In any case, we feel that further research is needed to clarify these issues.

As a last point of this section, we consider now the specific heat of SiO_2 over a wide range of temperature (Fig. 4.16). As expected, the specific heat starts to reach the saturation value of the Dulong-Petit law if the temperature exceeds the Debye temperature, $T_D \approx 500$ K. The harmonic approximation stays close to the experimental data up to the glass transition temperature ($T_g \approx 1450$ K), where the experimental data show a step-like increase when the glass melts. Of course, the harmonic approximation can never describe the melting of the glass (or of any crystalline solid). In fact, the discrepancies between the simulation and the real data are largest for $T \approx 100-300$ K and not near T_g. This finding suggests that the main discrepancies between simulation and experiment are due to an

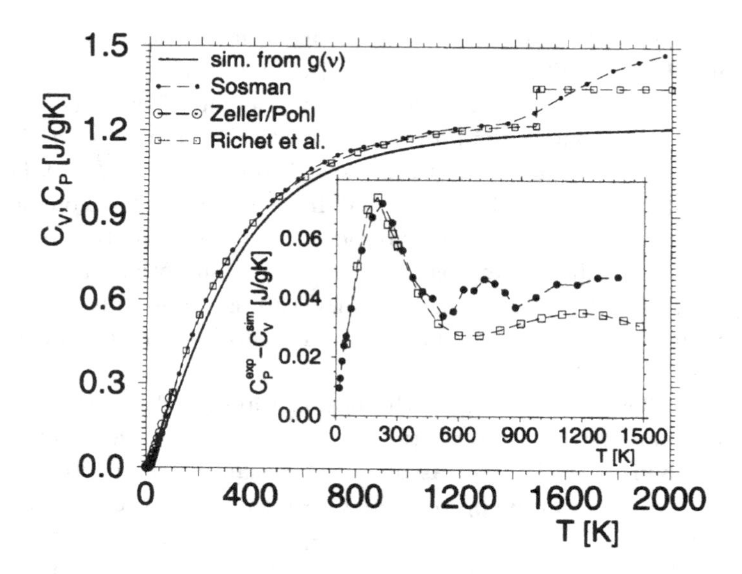

Fig. 4.16 Temperature dependence of the specific heat of SiO_2 as predicted by the harmonic approximation, using the density of states $\zeta(\omega)$ estimated from molecular dynamics simulations (solid line). The symbols are experimental data for the specific heat at constant pressure C_p from Sosman (1927), from Zeller and Pohl (1971) and from Richet *et al.* (1982). The inset shows the difference between the data of Richet *et al.* (1982) and of Sosman (1927) and the simulation data. From Horbach *et al.* (1999).

inaccurate estimation of the density of states $\zeta(\omega)$ by the simulation, and not due to anharmonic effects. Both the two-level systems (which make their important contribution to C_V at $T < 1$ K) and the anharmonicities described by the soft potential model (which make their important contribution to C_V at $T < 50$ K) yield only very small corrections to the specific heat when compared to the harmonic excitations that take over at temperatures of the order of T_D. Thus we can conclude that at the temperatures of interest for applications of glassy materials, a simple harmonic approximation is sufficiently reliable to account for physical properties such as the specific heat of silica glass.

4.5 Spin Glasses

In this section we deal with a prototype of a strongly disordered solid, for which the frozen-in randomness in the interactions leads to qualitatively new kinds of ordering phenomena. Already in Chap. 1 it has been pointed out that random dilution of a ferromagnet with competing ferro- and antiferromagnetic interactions (Figs. 1.10, 1.11) leads to a kind of "frustration effect" (Fig. 1.14) and therefore no uniform long range order of conventional ferro-or antiferromagnetic type can be found that is favorable for all exchange interactions. Therefore at a temperature T_f a freezing transition occurs, below which the spins seem to be frozen-in with seemingly random orientations but in fact the "spin glass" does have order, although this order is of an unconventional type (Binder and Young 1986, Fischer and Hertz 1991, de Dominicis and Giardina 2006, Boutet de Monvel and Bovier 2009). Despite enormous efforts, many aspects of spin glasses are still not yet well understood. Nevertheless, the statistical thermodynamics of simple models for spin glasses, such as the model of Edwards and Anderson (1975), could be worked out more completely and in much greater detail than has so far been possible for structural glasses. Thus, it has been a hope that the understanding of the phase transition to the spin glass state will promote a better understanding of structural glasses and the transition from the supercooled fluid to the amorphous solid as well. Although at this point it is not yet clear how far this expectation works out, it is true that many concepts developed first in the spin glass context have recently found interesting applications for the problem of the structural glass transition.

4.5.1 *Some Experimental Facts about Spin Glasses: Systems and Physical Properties*

It turns out that the main phenomena observed at the freezing transition of spin glasses are fairly universal, at least on a qualitative level, and are observed in a vast range of different systems (Binder and Young 1986, Young 1998). The archetypical spin glasses (Cannella and Mydosh 1972, Mulder *et al.* 1981) are noble metals (Au, Ag, Cu, Pt) weakly diluted with magnetic transition metals ions, such as Fe or Mn. The scattering of the conduction electrons of the host metal at the spins of these impurity ions leads to an indirect exchange interaction between the spins, which according to Ruderman and Kittel (1954), Kasuya (1956) and Yosida (1957), RKKY, oscillates at large distance $\vec{R}$, see Fig. 4.17 (Binder 1977), like

$$J(R) = J_0 \frac{\cos(2k_F R + \varphi_0)}{(k_F R)^3} . \tag{4.132}$$

Here J_0 and φ_0 are constants, and k_F is the Fermi wave-vector of the host

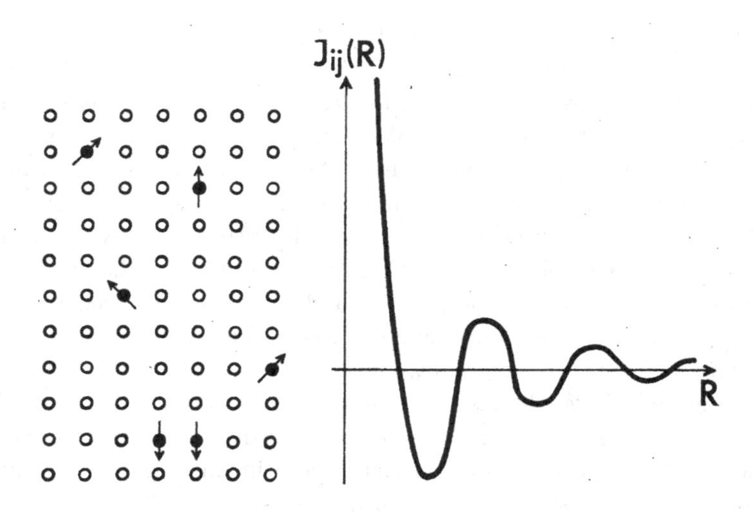

Fig. 4.17 Schematic sketch of magnetic moments that are randomly placed in a metallic matrix (left), and the resulting RKKY exchange integral plotted as a function of distance (right). From Binder (1977).

metal. Since the distances between the spins are random (Fig. 4.17), some interactions of a considered spin with other spins will be positive, favoring parallel alignment, and some will be negative, favoring anti-parallel alignment. Thus the "frustration effect" of the type as explained schematically in Fig. 1.14 is evident (Toulouse 1977).

Due to the long range of the RKKY interaction, Eq. (4.132), these metallic spin glasses can exist even at strong dilution. This is in contrast to the case of nonmetallic spin glasses, such as the randomly mixed crystal of ferromagnetic EuS with nonmagnetic SrS (Fig. 1.10), for which the competing interactions arise from super-exchange of the magnetic Eu ions via the electrons of sulfur: Nearest neighbor interactions are ferromagnetic ($J_{nn} > 0$), next nearest neighbor interactions are antiferromagnetic ($J_{nnn} < 0$), with $J_{nnn}/J_{nn} \approx -1/2$, and further neighbor exchange interactions being negligibly small (Zinn 1976), see Fig. 1.11. Therefore the phase transition line from the paramagnetic state (PM) to the spin glass state (SG) in Fig. 1.10 reaches zero temperature at an Europium concentration of $x_p^{nnn} \cong 0.13$, the next nearest neighbor percolation threshold of the face-centered cubic crystal (the Eu ions in EuS form a fcc lattice). For $x < x_p^{nnn}$ only finite clusters of magnetically coupled ions exist, instead of an infinite percolating network, and hence no long range correlations of spin glass type can be maintained.[3]

Spin glasses can also be obtained by diluting antiferromagnets instead of ferromagnets, and $Eu_xSr_{1-x}Se$ or $Eu_xSr_{1-x}Te$ (Westerholt and Bach 1981) are examples of such systems, or if ferromagnets and antiferromagnets are randomly mixed, such as the mixture of EuS with EuSe (Westerholt and Bach 1982). Qualitatively, this behavior has been interpreted by Binder *et al.* (1979) who discussed the phase diagram of an Ising system on a square lattice with nearest neighbor interaction J_1 and next nearest neighbor interaction J_2 over a wide range of the variables J_2/J_1 and dilution x, see Fig. 4.18. However we mention that all experimental systems mentioned so far are similar to the Heisenberg model (isotropic exchange $\vec{S}_i \cdot \vec{S}_j$ of vector spins) rather than to the Ising model. In addition, one does not expect any spin glass phases to occur at nonzero temperature in $d = 2$ dimensions (Binder and Young 1986).

While all systems mentioned so far are randomly mixed crystals, one also can have spin glass behavior due to other structural disorder, e.g. insulating glasses containing magnetic ions such as $(CoO)_{0.4}(Al_2O_3)_{0.1}(SiO_2)_{0.5}$ (Morgownik *et al.* 1982, Wenger 1983) behave as a spin glass, and also some metallic glasses are at the same time spin glasses, e.g. $(Fe_xMn_{1-x})_{75}P_{16}B_6Al_3$ (Salamon *et al.* 1981). Of course, in these latter systems neither the detailed structure nor the magnetic interactions are known as well as in the crystalline systems discussed before.

Similar behavior is also found in diluted dielectric crystals, the "spin" is then given by an electric dipole moment rather than a magnetic dipole moment. An example of such an electric "dipolar glass" (Höchli *et al.* 1990) is the mixed crystal $Rb_{1-x}(NH_4)_xH_2PO_4$ (Courtens 1982). Note that pure RbH_2PO_4 orders ferroelectrically, while $NH_4H_2PO_4$ orders antiferroelectrically. It should be noted, however, that the displacement of (charged) ions away from their ideal lattice points (of the undiluted crystal) induced by the dilution creates random electric fields in these dipolar glasses. Therefore one often has to model these systems as spin glasses in the presence of random magnetic fields, if one invokes the analogy to the magnetic systems. Similar problems occur when one considers diluted molecular crystals such

[3]Strictly speaking, this argument is of course only true if one ignores the long range magnetic dipolar interactions. These interactions are always present, but very weak. In fact, they give rise to a spin glass transition in the range $0 < x < x_p^{nnn}$ as well, but this transition occurs in the 10m K range (Eiselt *et al.* 1979), and thus on the scale of Fig. 1.10 this transition line is indistinguishable from the abscissa.

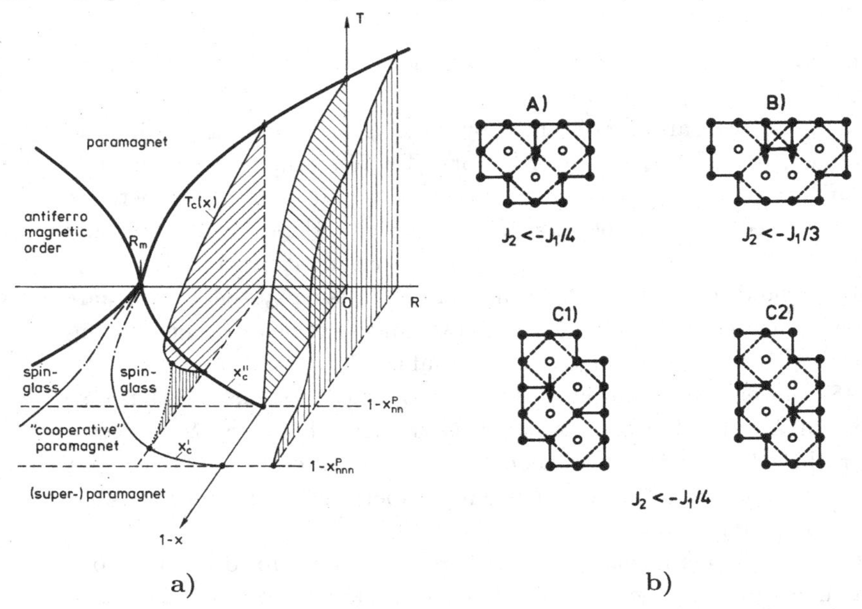

Fig. 4.18 a) Schematic phase diagram of an Ising model with nearest (J_{nn}) and next nearest neighbor exchange (J_{nnn}) as function of temperature, concentration of magnetic ions x, and the ratio J_{nnn}/J_{nn}. b) Some configurations of an Ising square ferromagnet with $J_{nn} > 0$, $J_{nnn} < 0$ (and $-\frac{1}{2} < J_{nnn}/J_{nn} < -\frac{1}{4}$) near dilution sites (open circles). Note that the configuration with 4 dilution sites next nearest neighbors to each other is a kind of magnetic "two-level system", since the ground state is two-fold degenerate. From Binder *et al.* (1979).

as ortho-parahydrogen mixtures (Fig. 1.12) for which the disorder created by dilution has both random bond and random field character. We return briefly to this problem in Sec. 4.6.

Figure 4.19 shows the "hallmark" of spin glass behavior, the sharp cusp in the susceptibility (Mulder *et al.* 1981). While at high temperatures the susceptibility follows the simple Curie law that is found for an ideal paramagnet, for $T < T_f$ the spins can no longer follow the oscillating magnetic field, and hence $\chi'(\omega)$ shows a sharp drop. However, on an expanded scale one sees that the peak of $\chi'(\omega)$ clearly is rounded, and that its position depends weakly on frequency. Figure 4.19 already indicates that the extrapolation to the true static limit ($\omega \to 0$) is subtle. However, Fig. 4.20 demonstrates that also "static" susceptibilities show anomalies (Nagata *et al.* 1979) in that the freezing is seen as a sudden onset of nonergodic behavior: The zero-field cooled susceptibility and the susceptibility measured after cooling the system in a field differ.

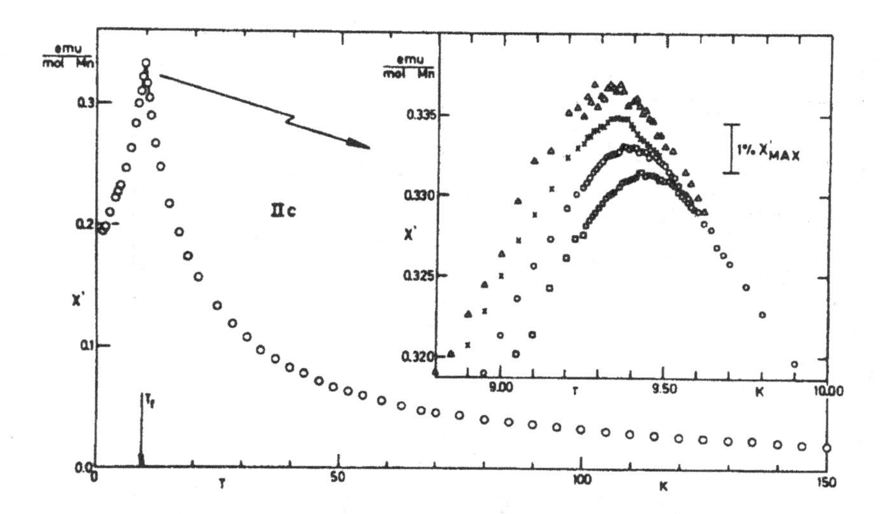

Fig. 4.19 Real part χ' of the complex susceptibility $\chi(\omega)$ as a function of temperature for sample IIc (CuMn with 0.94 at % Mn, powder). The inset shows the frequency dependence and rounding of the cusp by use of strongly expanded coordinate scales. Measuring frequencies: 1.33 kHz (squares), 234 Hz (circles), 104 Hz (crosses), and 2.6 Hz (triangles). From Mulder *et al.* (1981).

Note that for $T < T_f$ all susceptibility data of spin glasses are affected by subtle effects due to aging (Nordblad and Svedlindh 1998) which we will, however, not discuss here, and also the various types of hysteresis loops that have been found in spin glasses, see Binder and Young (1986) for details, will not be reviewed here.

Regarding the specific heat, a quantity that for conventional phase transitions always is a good indicator for the presence of a transition, Fig. 4.21 shows that there is no detectable anomaly at the spin glass transition (Wenger and Keesom 1976). Instead of a real singularity, the specific heat has only a broad peak at temperatures somewhat higher than T_f and an (almost) linear variation with temperature T for $T < T_f$. Remarkably, this also holds for the specific heat of isolating spin glasses such as $Eu_{0.4}Sr_{0.6}S$ (Wosnitza *et al.* 1986).

The only clear experimental proof for the existence of a static phase transition of spin glasses is due to an analysis of the nonlinear magnetic response for $T > T_f$. Writing for the magnetization M in a field H the expansion

$$\frac{M}{H} = \chi_0^{(\text{n.i.})} \left\{ a_1 - \frac{1}{3} a_3 \left(\frac{\mu H}{k_B T} \right)^2 + \frac{2}{15} a_5 \left(\frac{\mu H}{k_B T} \right)^4 + \cdots \right\}, \tag{4.133}$$

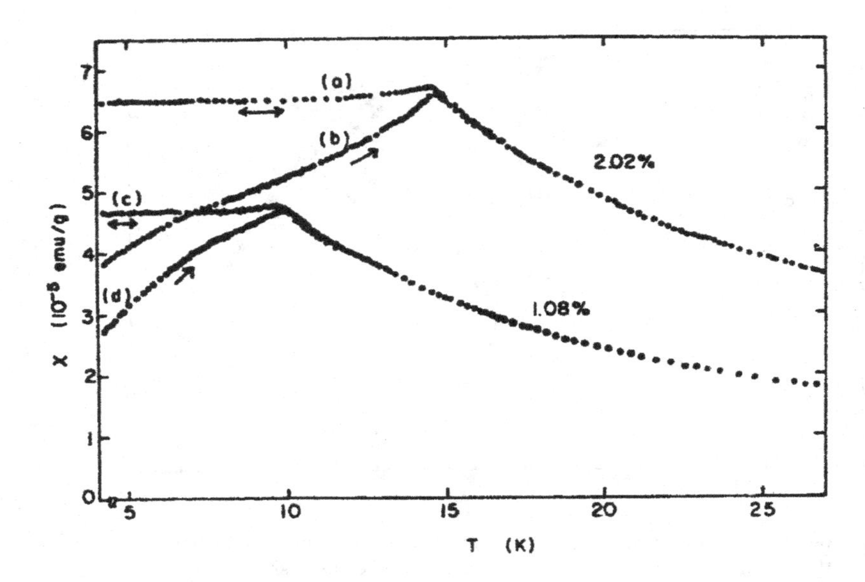

Fig. 4.20 Temperature dependence of static susceptibilities of CuMn for 1.08 and 2.02 at % Mn. After zero-field cooling (H <0.05 Oe), initial susceptibilities (b) and (d) were taken for increasing temperature in a field of H =5.9 Oe. The susceptibilities (a) and (c) were obtained in the field H =5.9 Oe, which was applied above T_f before cooling the samples. From Nagata *et al.* (1979).

where $\chi_0^{(\text{n.i.})}$ is the zero-field susceptibility of a noninteracting paramagnet, and μ is the magnetic moment per spin, it turns out that the coefficients $a_3, a_5, \cdots$ are divergent at T_f, although a_1 is not. (The normalization of the expansion coefficients has been chosen such that $a_1 = a_3 = a_5 = \cdots = 1$ for an ideal paramagnet.) In order to analyze these divergent terms, Omari *et al.* (1983) have studied the quantity $1 - M/(\chi_0 H)$, where $\chi_0 \equiv \chi_0^{(\text{n.i.})} a_1$. At a static phase transition at T_f one then expects a scaling behavior (Chalupa 1977, Suzuki 1977a, Binder 1977)

$$1 - \frac{M}{\chi_0 H} = (1 - T/T_f)^{\beta}\, \widetilde{M} \left\{ \left(\frac{H}{T}\right)^2 (1 - T/T_f)^{-(\gamma+\beta)} \right\}, \qquad (4.134)$$

as we will discuss below. Figure 4.22 presents corresponding data for the CuMn -spin glass in a scaling representation (Omari *et al.* 1983). One sees that one finds rather convincing evidence for a singular behavior consistent with scaling. Similar analyzes are now available for several spin glass systems, although the accuracy with which the static spin glass exponents β, γ

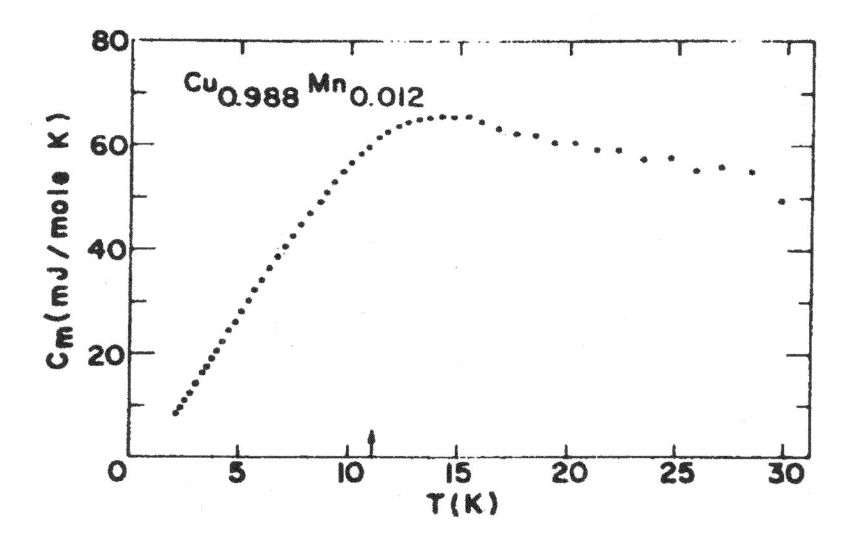

Fig. 4.21 The specific heat of a Cu-Mn alloy plotted vs. temperature. The arrow shows where the susceptibility has its cusp. From Wenger and Keesom (1976).

can be estimated still does not match the accuracy reached for conventional phase transitions.

Of course, the information on the dynamics of spin freezing near the spin glass transition is very rich and many experimental techniques have made valuable contributions: Dynamic susceptibility measurements (Fig. 4.19), Mössbauer effect, nuclear magnetic resonance and electron spin resonance, muon relaxation, and various neutron inelastic scattering techniques (Binder and Young 1986). For instance, the neutron spin echo technique yields the intermediate scattering function $F(k,t)$, see Chap. 2,

$$F(k,t) = S(k,t)/S(k,0)\,, \quad S(k,t) \propto \sum_{ij} \langle \vec{S}_i(t) \cdot \vec{S}_j(0) \rangle_T \exp[i\vec{k} \cdot (\vec{r}_i - \vec{r}_j)] \tag{4.135}$$

over a time domain of roughly 10^{-12} s $< t <$ 10^{-8} s (Mezei 1981). Figure 4.23 presents typical data for CuMn spin glasses (Mezei 1981). It turns out that the data show only a very weak dependence on the wave-vector k, and hence are dominated by the time dependent generalization of the Edwards-Anderson (1975) spin glass order parameter (discussed in

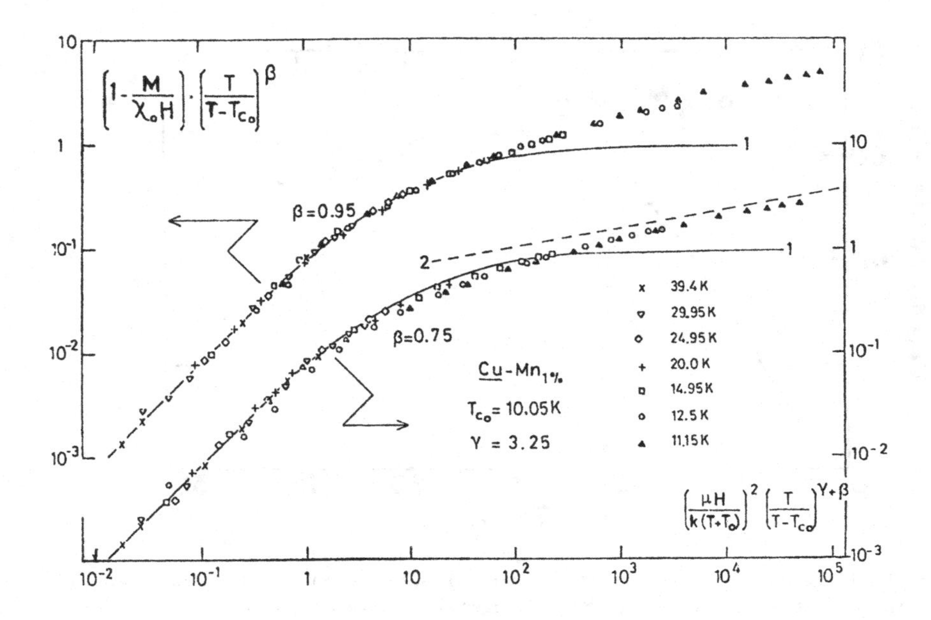

Fig. 4.22 Scaling plot of the nonlinear part of the magnetization for CuMn with 1 at % Mn for two choices of the exponent β (as indicated) and choosing $\gamma = 3.25$, $T_f(= T_c) = 10.05$ K. From Omari *et al.* (1983).

more details later),

$$q(t) = \frac{1}{N}\sum_i \langle \vec{S}_i(t)\cdot\vec{S}_i(0)\rangle_T = [\langle \vec{S}_i(t)\cdot S_i(0)\rangle_T]_{\rm av}\,. \qquad (4.136)$$

Here $[.]_{\rm av}$ denotes again, see Chap. 1, the average over the distribution of the coupling constants. Note that the average over all the spins in the system is equivalent to the averaging over the disorder in a large system. Hence one can deduce from Fig. 4.23 that $q(t)$ develops a very slow decay when the temperature T approaches the freezing temperature T_f, which for this system is around 27.5 K. Of course the accuracy of the data shown in Fig. 4.23 is too limited to reliably deduce the functional form of the decay, and hence cannot be used to distinguish whether this behavior is just the gradual onset of freezing or instead an indication for an underlying phase transition. While such neutron spin echo experiments would be the most direct probe of $q(t)$, more information can in fact be extracted from a joint analysis of, $\chi'(\omega), \chi''(\omega)$, the real and imaginary parts of the complex dynamic magnetic susceptibility, respectively. Using the static

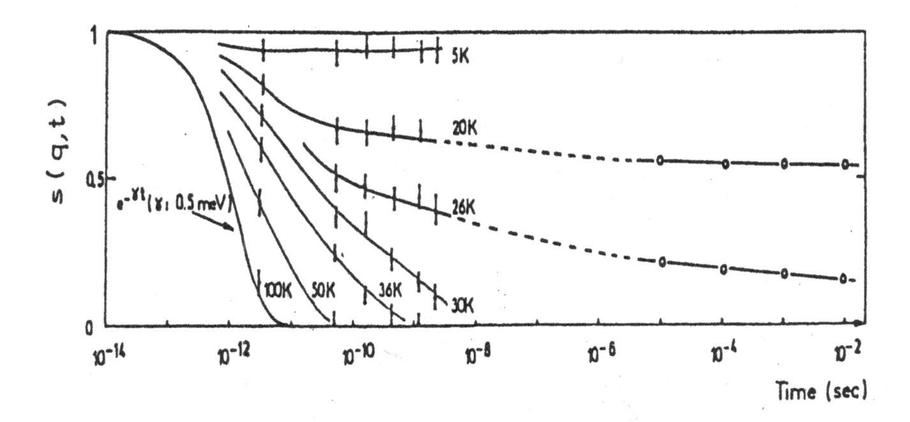

Fig. 4.23 Spin relaxation in CuMn with 5% Mn at various temperatures. Data points at times shorter than 10^{-8} s were directly measured by neutron spin echo techniques at $k = 0.093$ Å^{-1}; those beyond 10^{-6} s were calculated from ac susceptibility results. Note that in this figure the intermediate scattering function $F(k, t)$ of the text is denoted by $s(q, t)$. The lines are guides to the eye. From Mezei (1981).

isothermal and adiabatic susceptibilities χ_T, and χ_s, respectively, one can write (e.g. Wenger 1983)

$$\chi'(\omega) = \chi_s + (\chi_T - \chi_s) \int_{\tau_{\min}}^{\tau_{\max}} g(\tau) \frac{d(\ln \tau)}{1 + \omega^2 \tau^2}, \tag{4.137}$$

$$\chi''(\omega) = (\chi_T - \chi_s) \int_{\tau_{\min}}^{\tau_{\max}} \omega \tau g(\tau) \frac{d(\ln \tau)}{1 + \omega^2 \tau^2}, \tag{4.138}$$

and thus define an effective distribution of relaxation times $g(\tau)$. Here we have assumed that there occurs a minimum relaxation time $\tau_{\min}$ and a maximum relaxation time $\tau_{\max}$ in the spectrum. From experimental data one can derive then estimates for $\tau_{\min}$ and $\tau_{\max}$ as well as the average relaxation time τ_{av}, defined from the frequency at which $\chi''(\omega)$ is maximal, $\tau_{\text{av}} = 1/\omega_{\max}$, or even the whole spectrum $g(\tau)$, at least approximately (e.g. Wenger 1983). Figure 4.24 gives an example for the cobaltaluminosilicate spin glass (Wenger 1983). One sees that for all temperatures $\tau_{\min}$ is a microscopic time, in the range 10^{-12} s $< \tau_{\min} < 10^{-10}$ s, and its $T-$dependence is compatible with simple Arrhenius behavior. In contrast to this, upon approach of T_f the relaxation time $\tau_{\max}$ increases rapidly

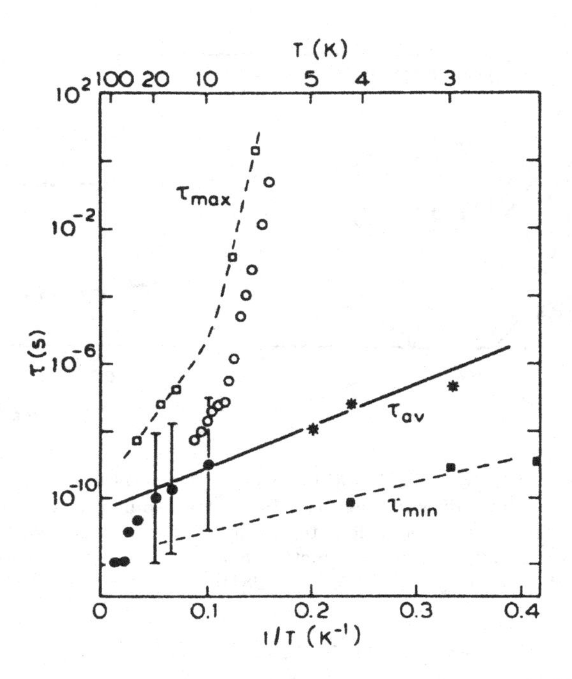

Fig. 4.24 Arrhenius plot of the relaxation times of $(CoO)_{0.4}(Al_2O_3)_{0.1}(SiO_2)_{0.5}$ spin glasses, as obtained from an analysis of dynamic magnetic susceptibilities in the frequency range from 0.64 Hz to 30 MHz. Open squares denote $\tau_{\max}$, full squares $\tau_{\min}$, asterisks denote τ_{av}. Open circles give the position $1/\omega_{\max}$ where $\chi''(\omega)$ has its maximum, while full circles are muon spin resonance measurements. From Wenger (1983).

from 10^{-8} sec to macroscopic times, and the increase is more rapid than an Arrhenius law and can in fact be fitted well with a Vogel-Fulcher-Tammann law, $\tau_{\max} \propto \exp[E_{\text{VF}}/k_B(T - T_0)]$. Note that Fig. 4.24 is indeed qualitatively very similar to experimental findings on relaxation times near the structural glass transition.

Clearly, also the data of Fig. 4.24 are not sufficiently accurate to allow for a quantitative analysis of the dynamics near the spin glass transition. However, for model systems such as $Eu_{0.4}Sr_{0.6}S$ a more accurate data analysis has been made by Bontemps *et al.* (1984) who studied lines $T_f(H, \omega)$ in the (H, T) plane, defining $T_f(H, \omega)$ from the maximum of $\chi''(\omega)$ at a superimposed static magnetic field H. Thus while the traditional view has been that $T_f(H, \omega)/T_f(0, \omega)$ for $\omega \to 0$ can be taken as an estimate for the transition that occurs in mean field theory along the so-called "AT line" (de Almeida and Thouless 1978), Bontemps *et al.* (1984) extract from $T_f(H, \omega)$

a relaxation time $\tau(T, H)$ from the definition $T_f(H, \omega \equiv 1/\tau(T = T_f, H))$, and study the scaling properties of τ assuming that there is a static spin glass transition at nonzero temperature. Dynamic scaling (Hohenberg and Halperin 1977) then would imply {cf. also Eq. (4.134)}

$$\tau(T, H)/\tau_0 = (T/T_f - 1)^{-z\nu}\, \tilde{\tau}\left\{\left(\frac{H}{T}\right)^2 (T/T_f - 1)^{-(\gamma+\beta)}\right\}, \qquad (4.139)$$

where $z\nu$ is the critical exponent of the relaxation time (recall that usually one writes $\tau \propto \xi^z$ and the correlation length $\xi \propto (T/T_f - 1)^{-\nu}$). In Eq. (4.139) $\tilde{\tau}$ is a dimensionless scaling function, while the constant τ_0 sets the time scale. Figure 4.25 shows two variants of a resulting scaling plot: One assuming $T_f = 0$ (a) and the other assuming $T_f = 1.5$ K (b). One sees that both assumptions give fairly reasonable collapse, with large dynamic exponents in both cases, $z\nu \approx 7.2 \pm 0.5$ if $T_f = 1.5$ K while $z\nu = 8.0 \pm 0.5$ if $T_f = 0$, see Bontemps *et al.* (1984). The main reason to prefer $T_f = 1.5$ K over $T_f = 0$ K, with the data used in Fig. 4.25 is that the estimate for τ_0 ($= 2.10^{-7}$ s for $T_f = 1.5$ K, while $\tau_0 \approx 10^{-5}$ s if $T_f = 0$) is physically more reasonable for the case $T_f = 1.5$ K. Thus, one sees that using information on dynamic properties alone it would be very difficult also for the spin glass problem to conclude that there is an underlying finite temperature transition. Fortunately, the analysis of the nonlinear susceptibility, using Eq. (4.134), has allowed a rather convincing statement on this issue. Note however, that for other spin glasses somewhat different values for critical exponents were found (Nordblad and Svedlindh 1998).

4.5.2 *Theoretical Models*

We now turn our attention to the main theoretical model of spin glasses, which is the Edwards-Anderson model (Edwards and Anderson 1975)

$$\mathcal{H} = -\sum_{i \neq j} J_{ij}(\vec{S}_i \cdot \vec{S}_j) - H \sum_i S_i^z \,. \qquad (4.140)$$

Here one assumes a regular lattice of sites labeled by i, each lattice carrying a m-component spin $\vec{S}_i$ of length $|\vec{S}_i| = 1$ (thus $m = 1$: Ising model, $m = 2$: XY model; $m = 3$: Heisenberg model), and the disorder comes in via the choice of a suitable distribution $P(J_{ij})$ of the random exchange interactions $\{J_{ij}\}$. Thus the model does not attempt to describe the interactions of metallic (Fig. 4.17) or nonmetallic (Figs. 1.10 and 1.11) spin glasses in a realistic way. (Dealing with more realistic models having site disorder

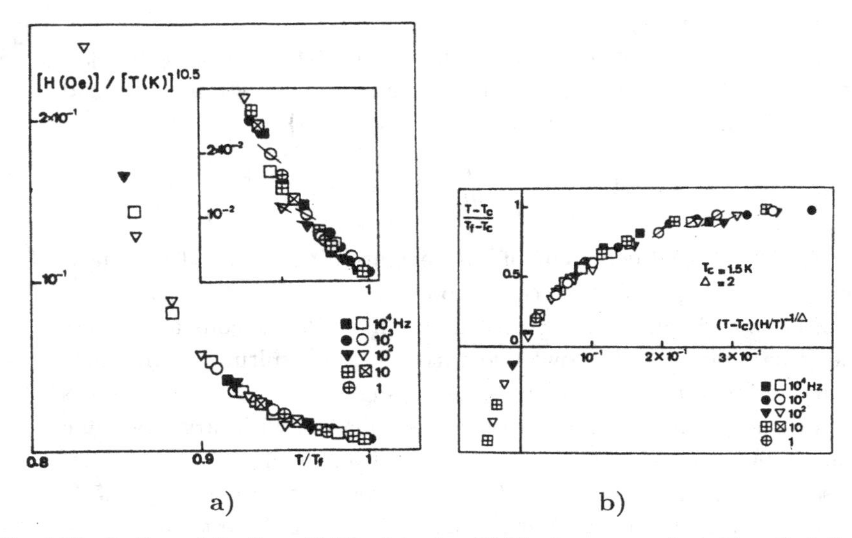

Fig. 4.25 Scaling of the lines $T_f(H, \omega)$ in the (H, T) plane, assuming either that the static transition temperature T_f (denoted in the plot by T_c) is zero (a), or nonzero with $T_c = 1.5$ K (b), using data for $Eu_{0.4}Sr_{0.6}S$. The exponent $\Delta = (\gamma + \beta)/2$ is chosen to be $\Delta = 10.5$ in panel (a) and $\Delta = 2.0$ in panel (b). From Bontemps *et al.* (1984).

rather than bond disorder is considerably more involved, and hence out of consideration here.) The widespread belief is that in view of the great qualitative similarity that one encounters between the physical properties of many different spin glass systems there is no need to aim at a realistic description of one particular system, and hence Eq. (4.140) should capture the essential physics.

Normally, one chooses for the distribution $P(J_{ij})$ either a gaussian, Eq. (1.14) from Chap. 1, or a $\pm J$ distribution, Eq. (1.13). A very important property results for the special case that these distributions are exactly symmetric, $P(J_{ij}) = P(-J_{ij})$. Then it immediately follows that (remember that $\vec{S}_i$ is an m-dimensional unit vector)

$$[\langle \vec{S}_i \cdot \vec{S}_j \rangle_T]_{\text{av}} = \delta_{ij} \tag{4.141}$$

and hence the fluctuation relation for the susceptibility becomes (setting the magnetic moment $\mu \equiv 1$)

$$k_B T \chi = N^{-1} \sum_{i,j} [\langle \vec{S}_i \cdot \vec{S}_j \rangle_T - \langle \vec{S}_i \rangle_T \langle \vec{S}_j \rangle_T]_{\text{av}} = 1 - q\,, \tag{4.142}$$

where we have introduced the so-called Edwards-Anderson order parameter which is defined as

$$q \equiv [\langle \vec{S}_i \rangle_T^2]_{\mathrm{av}} \,. \tag{4.143}$$

Note that Eq. (4.141) does *not* imply that the spin glass does not show any correlations whatsoever. This can be seen by defining the spin correlation function $g_{\mathrm{SG}}(\vec{r})$

$$g_{\mathrm{SG}}(\vec{r}) = [\langle \vec{S}_i \cdot \vec{S}_j \rangle_T^2]_{\mathrm{av}} \quad \text{with } \vec{r} = \vec{r}_i - \vec{r}_j \,. \tag{4.144}$$

Due to the symmetry $P(J_{ij}) = P(-J_{ij})$, it is equally likely to find pairs of spins that are ferromagnetically correlated and other pairs that are antiferromagnetically correlated and Eq. (4.144) takes both possibilities properly into account. The normalized nonlinear part of the magnetic equation of state considered in Eq. (4.133) can now be expressed in terms of the spin glass susceptibility χ_{SG}, which is defined as

$$\chi_{\mathrm{SG}} = \frac{1}{N} \sum_{i,j} [\langle \vec{S}_i \cdot \vec{S}_j \rangle_T^2]_{\mathrm{av}} = \sum_{\vec{r}} g_{\mathrm{SG}}(\vec{r}) \,, \tag{4.145}$$

and hence

$$1 - \frac{M}{\chi_0 H} = \frac{a_3}{3} \left(\frac{H}{k_B T} \right)^2 + O(H/k_B T)^4 \quad \text{with} \quad a_3 = 3\chi_{\mathrm{SG}} - 2 \,. \tag{4.146}$$

Thus, one sees that in this model a singular behavior of a_3 is generally caused by a divergence of the spin glass susceptibility χ_{SG}, and in turn χ_{SG} diverges when the correlation functions $g_{\mathrm{SG}}(\vec{r})$ becomes long ranged. These considerations can be extended to the case where the symmetry $P(J_{ij}) = P(-J_{ij})$ is not present, and thus in addition to the (long range) spin glass-type correlations also some ferro– or antiferromagnetic short range correlations are present (Binder and Young 1986).

Some of the above considerations can in fact be exemplified by a simple caricature of the Edwards-Anderson spin glass, the so-called Mattis (1976) model. In this model the random exchange constants $\{J_{ij}\}$ are defined as

$$J_{ij} = J(\vec{r}_i - \vec{r}_j)\varepsilon_i \varepsilon_j \text{ with } \quad \varepsilon_i = \pm 1 \,(\text{random}) \text{ and } \quad [\varepsilon_i]_{\mathrm{av}} = 0 \,. \tag{4.147}$$

Due to this special choice for the disorder it can be eliminated by a gauge transformation in which one transforms from the original $\vec{S}_i$ in Eq. (4.140) to pseudo spins $\vec{\tau}_i$ defined by

$$\vec{\tau}_i = \varepsilon_i \vec{S}_i \,, \qquad \text{or} \quad \vec{S}_i = \varepsilon_i \vec{\tau}_i \quad (\text{since } \varepsilon_i^2 \equiv 1) \,. \tag{4.148}$$

Thus the problem is mapped onto a ferromagnet without bond disorder in a random magnetic field $H_i = H\varepsilon_i$:

$$\mathcal{H}_F = -\sum_{i \neq j} J(\vec{r}_i - \vec{r}_j)\vec{\tau}_i \cdot \vec{\tau}_j - \sum_i H_i \tau_i^z \,. \tag{4.149}$$

If $H = 0$, we have a standard ferromagnet with no disorder, which hence will exhibit a spontaneous magnetization $M_\tau = \sum \tau_i^z/N \propto (1 - T/T_c)^{\beta_m}$, and also a singularity in the specific heat $C \propto |1 - T/T_c|^{-\alpha_m}$, the "susceptibility" $\chi_\tau = \sum_{i,j} \langle \vec{\tau}_i \cdot \vec{\tau}_j \rangle / N \propto |1 - T/T_c|^{-\gamma_m}$, and the correlation length $\xi_\tau \propto |1 - T/T_c|^{-\nu_m}$. Note that here we have given to all exponents an index "m" to remind the reader that these exponents refer to the critical behavior of the ordinary ferromagnetic m-vector model not to those of the spin glass. However, the latter can be easily obtained, since

$$q = [\langle S_i^z \rangle_T^2]_{\text{av}} = \langle \tau_i^z \rangle^2 [\varepsilon_i^2]_{\text{av}} = \langle \tau_i^z \rangle^2 \propto (1 - T/T_c)^{2\beta_m} \,, \tag{4.150}$$

so we have for the spin glass order parameter exponent $\beta = 2\beta_m$. Similarly, we conclude that Eq. (4.141) still holds and that

$$g_{\text{SG}}(\vec{r}) = \langle \vec{\tau}_i \cdot \vec{\tau}_j \rangle_T^2 [\varepsilon_i^2 \varepsilon_j^2]_{\text{av}} = \langle \vec{\tau}_i \cdot \vec{\tau}_j \rangle_T^2 \,. \tag{4.151}$$

As a consequence, when $\langle \vec{\tau}_i \cdot \vec{\tau}_j \rangle \propto \exp(-r/\xi_\tau)$ for $r \to \infty$, we conclude that $g_{\text{SG}}(\vec{r}) \propto \exp(-2r/\xi_\tau)$, i.e. $\xi_{\text{SG}} = \xi_\tau/2$, and hence $\nu = \nu_m$. Of course, Eq. (4.149) implies that here the "susceptibility" χ_τ is not a response function to a uniform field but rather to a special "randomly staggered" field $H_\tau = H_i = H\varepsilon_i$. Using Eq. (4.151) in Eq. (4.145) one can also find the divergence of the spin glass susceptibility for the Mattis model.

It is clear, however, that the Mattis model is not a good model for a spin glass since it predicts the same singular behavior of the specific heat as for a ferromagnet, whereas we know that for spin glasses it is absent (Fig. 4.21). Monte Carlo simulations of dynamic versions of this model (Stauffer and Binder 1978) do not show the slow relaxation characteristic of spin glasses, in contrast to Monte Carlo simulations of the original Edwards-Anderson model (Binder and Schröder 1976, Ogielski 1985, Binder and Young 1986). It is clear that the Mattis model is just a ferromagnet in disguise, and q in the Mattis model, Eq. (4.150), is just a "secondary order parameter" to M_τ which is the "primary order parameter". Furthermore we also note that the "frustration" variable, Fig. 1.14, (for a frustrated loop $\phi_p \equiv \text{sign}\,\Pi\, J_{ij} = -1$ where the product extends over all bonds of a closed loop in the lattice) is a gauge-invariant disorder, and thus cannot be eliminated by gauge transformations as it has been possible for the Mattis model.

Another mechanism that for a while has been considered as a possible alternative to account for the anomalous behavior of the frequency-dependent susceptibility $\chi(\omega)$ is the idea that these anomalies are due to the "blocking" of super-paramagnetic clusters. Thus, one assumes that metallic spin glasses such as CuMn can be described in terms of an assembly of (ferromagnetic) clusters having s magnetic ions per cluster, with a distribution $P(s)$. The interaction among the clusters is neglected. If the concentration of magnetic ions is x, and $\chi_{\text{n.i}}$ is the susceptibility of an ideal paramagnet, the zero-field susceptibility becomes (cf. Sec. 3.2.2)

$$\chi/\chi_{n.i.} = \frac{1}{x}\sum_{s=1}^{\infty} P(s)s^2 \approx \frac{1}{x}\int_0^{\infty} dsP(s)s^2 \,. \tag{4.152}$$

Now the key assumption is that the clusters exhibit an uniaxial anisotropy energy $E_A = Ks$, where K is a constant. Hence each cluster is a two-level system with an energy barrier E_A between the two states. Assuming that tunneling does not play a role, the time for reorienting a cluster by a thermally activated process is then given by an Arrhenius law, i.e.

$$\tau_s = \nu_a^{-1}\exp(Ks/k_BT)\,, \tag{4.153}$$

where ν_a is an "attempt frequency".

Considering now a dynamic susceptibility $\chi(t)$, one argues that not all clusters {as assumed in Eq. (4.152)} contribute to $\chi(t)$ but only those which can reorient during the time t, i.e. those for which $\tau_s < t$. Hence one defines a cutoff $s_c(t)$ for the cluster size distribution from

$$\nu_a t = \exp[Ks_c(t)/k_BT], \quad \text{or } s_c = (k_BT/K)\ln(\nu_a t)\,, \tag{4.154}$$

and thus

$$\chi(t)/\chi_{n.i.} = (1/x)\int_0^{s_c(t)} dsP(s)s^2 = \frac{1}{x}\int_0^{(k_BT/K)\ln(\nu_a t)} dsP(s)s^2 \,. \tag{4.155}$$

Since in a dilute system the distribution $s^2P(s)$ will have a peak at some finite size $s_{\max}$, it is clear that for sufficiently low temperature $s_c(t)$ will be less than $s_{\max}$. Since $\chi_{n.i.} \propto 1/T$, one expects that $\chi(t)$ will rise with decreasing temperature up to some maximum and then decrease again, and the position of this maximum according to Eq. (4.155) will depend logarithmically on the observation time t (Souletie 1983, Shtrikman and Wohlfarth 1981).

While it is clear that some diluted magnetic systems can be prepared for which the description in terms of Eqs. (4.152)- (4.155) is a reasonable model, it is now generally accepted that most spin glass systems cannot be described as an assembly of independent clusters, and if well-defined clusters are present, also cluster interactions need to be considered (Binder 1977). Therefore, one should not consider the spin glass transition as a simple percolation-type phenomenon, where the effective size of clusters grows as the temperature is lowered (Smith 1974, Cyrot 1981, Mookerjee and Chowdhury 1983).

4.5.3 *The Replica Method and the Mean Field Theory of the Ising Spin Glass*

We now want to investigate the properties of the Ising spin glass, i.e. Eq. (4.140) for the case $m = 1$, by means of statistical mechanics and face the problem how to average over the random quenched disorder in the exchange constants. As discussed already in the Introduction (Chap. 1), the task is to evaluate the average free energy $[F]_{\text{av}} = -k_BT[\ln Z\{J_{ij}\}]_{\text{av}}$, Eq. (1.20).

The quenched averaging of the free energy can be calculated utilizing the so-called "replica trick" (Edwards 1971, Emery 1975, Edwards and Anderson 1975). For this we recall that

$$Z^n = \exp(n \ln Z) \approx 1 + n \ln Z, \quad n \to 0\,, \tag{4.156}$$

and thus we can write

$$[\ln Z\{J_{ij}\}]_{\text{av}} = \lim_{n\to 0} \frac{1}{n}([Z^n\{J_{ij}\}]_{\text{av}} - 1) = \lim_{n\to 0} \frac{\partial}{\partial n}[Z^n\{x\}]_{\text{av}}\,. \tag{4.157}$$

(Note that in the second equality we have used l'Hôpital's rule.) The name "replica trick" stems from the fact that in practice one interprets Z^n as a product of n identical replicas of the system, for positive integer n,

$$Z^n\{J_{ij}\} \equiv \prod_{\alpha=1}^{n} Z_\alpha\{J_{ij}\} = \prod_{\alpha=1}^{n} \operatorname*{Tr}_{S_i^\alpha} \exp\left[-\mathcal{H}\{J_{ij}, S_i^\alpha\}/k_BT\right]$$
$$= \operatorname*{Tr}_{S_i^\alpha} \exp\left[-\sum_{\alpha=1}^{n} \mathcal{H}\{J_{ij}, S_i^\alpha\}/k_BT\right]. \quad (4.158)$$

Since the last expression has the appearance of an ordinary Hamiltonian, namely

$$\mathcal{H}_n\{J_{ij}, S_i^\alpha\} = \sum_{\alpha=1}^{n} \mathcal{H}\{J_{ij}, S_i^\alpha\}, \quad (4.159)$$

for positive integer n the averaging of $[Z^n\{J_{ij}\}]_{\text{av}}$ reduces to the standard annealed average {cf. Eq. (1.21)}. This yields an effective Hamiltonian $\mathcal{H}_n^{\text{eff}}$, which no longer contains any disorder, and hence is translationally invariant:

$$Z_n \equiv [Z^n\{J_{ij}\}]_{\text{av}} \equiv \operatorname*{Tr}_{S_i^\alpha} \exp[-\mathcal{H}_n^{\text{eff}}\{S_i^\alpha\}/k_BT]. \quad (4.160)$$

This effective Hamiltonian is formulated in terms of the variables S_i^α, i.e. each spin at site i now also carries a "replica index" α. While prior to the (annealed) averaging $\mathcal{H}_n\{J_{ij}, S_i^\alpha\}$ is simply a sum of the Hamiltonians of the n different replicas, Eq. (4.159), i.e. the replicas do not interact with each other, the averaging over the disorder will introduce a coupling of the replicas. Thus, one must compute (for $H = 0$)

$$Z_n = \operatorname*{Tr}_{S_i^\alpha} \prod_{\langle i,j\rangle} \int dJ_{ij} P(J_{ij}) \exp\left[\frac{J_{ij}}{k_BT} \sum_{\alpha=1}^{n} S_i^\alpha S_j^\alpha\right]. \quad (4.161)$$

Expanding $\exp(y) = 1 + y + y^2/2! + \ldots$ one obtains the k'th moment $[J_{ij}^k]_{\text{av}}$ for the term y^k, when we carry out the average. The resulting series can again be written as the argument of an exponential function, if the moments are replaced by the corresponding cumulants:

$$\mathcal{H}_n^{\text{eff}}\{S_i^\alpha\}/k_BT = -\sum_{i\neq j} \sum_{k=1}^{\infty} \frac{1}{k!} \frac{J_{ij}^{\text{cum}}(k)}{(k_BT)^k} \left[\sum_{\alpha=1}^{n} S_i^\alpha S_j^\alpha\right]^k. \quad (4.162)$$

Here the sum $i \neq j$ extends over all pairs once, and the k'th cumulant $J_{ij}^{\text{cum}}(k)$ is defined as follows

$$J_{ij}^{\text{cum}}(1) = [J_{ij}]_{\text{av}} \equiv \bar{J}, \quad (4.163)$$

$$J_{ij}^{\text{cum}}(2) = [J_{ij}]_{\text{av}} - [J_{ij}]_{\text{av}}^2 \equiv (\Delta J)^2 \;, \tag{4.164}$$

etc. It is obvious from Eq. (4.162) that the terms with $k > 1$ couple the various replicas to each other, as mentioned above. Equation (4.162) is particularly simple for a gaussian distribution $P(J_{ij})$, since then all cumulants with $k \geq 3$ vanish. Then the explicit form of the effective Hamiltonian becomes

$$\mathcal{H}_n^{\text{eff}}\{S_i^\alpha\}/k_BT = -\frac{\bar{J}}{k_BT}\sum_{i\neq j}\sum_{\alpha=1}^{n} S_i^\alpha S_j^\alpha - \frac{1}{2}\left(\frac{\Delta J}{k_BT}\right)^2 \sum_{i\neq j}\sum_{\alpha,\beta} S_i^\alpha S_j^\alpha S_i^\beta S_j^\beta \;. \tag{4.165}$$

Since $\mathcal{H}_n^{\text{eff}}\{S_i^\alpha\}$ is an effective Hamiltonian of a (fictitious) translationally invariant problem without disorder, all methods of statistical mechanics for ideal pure systems can now be applied. One expects that the simplest step will be a mean field approximation in which one replaces - roughly speaking - the terms $S_i^\alpha S_j^\alpha$ by $S_i^\alpha \langle S_j^\alpha\rangle$, $S_i^\alpha S_i^\beta S_j^\alpha S_j^\beta$ by $S_i^\alpha S_i^\beta \langle S_j^\alpha S_j^\beta\rangle$, and determines the "order parameters"

$$m_\alpha \equiv \langle S_j^\alpha \rangle \text{ and } q_{\alpha\beta} \equiv \langle S_j^\alpha S_j^\beta \rangle_{\alpha\neq\beta} \tag{4.166}$$

self-consistently.

We remind, however, that all these results refer to the case of a positive integer n. But the crucial step needed in Eqs. (4.157) and (4.156) is to continue $\mathcal{H}_n^{\text{eff}}$ analytically to arbitrary positive real n. Unfortunately, such an analytic continuation is not unique. While it is obvious that $\mathcal{H}_n^{\text{eff}}$ is invariant under permutations of the indices of the replicas, as long as n is a positive integer, it is not obvious that this symmetry is preserved if n takes non-integer values and one takes the limit $n \to 0$. This fact has led to the idea of "replica symmetry breaking" (Bray and Moore 1978, de Almeida and Thouless 1978, Blandin 1978, Parisi 1979, 1980).

Before the mean field theory can be worked out in detail, one needs to clarify how one can calculate the magnetization and the Edwards-Anderson order parameter within this replica formalism. The magnetization is given by

$$\begin{aligned} M = [\langle S_i\rangle_T]_{\text{av}} &= \left[\frac{1}{Z\{J_{ij}\}}\text{Tr}S_i\exp(-\mathcal{H}\{J_{ij}\}/k_BT)\right]_{\text{av}} \\ &= \left[\frac{Z^{n-1}\text{Tr}S_i\exp(-\mathcal{H}\{J_{ij}\}/k_BT)}{Z^n}\right]_{\text{av}} = \langle S_i^\alpha\rangle \;, \end{aligned} \tag{4.167}$$

where we denote $\langle(\ldots)\rangle \equiv \lim_{n\to 0} \text{Tr}(\ldots)\exp(-\mathcal{H}_n^{\text{eff}}\{S_i^\alpha\}/k_BT)$ and note that

$Z^n \to 1$ in the limit $n \to 0$. Analogously one obtains for the order parameter

$$q = [\langle S_i \rangle_T^2]_{\rm av} = [\frac{1}{Z^2}{\rm Tr} S_i \exp(-\mathcal{H}\{J_{ij}\}/k_BT)\, {\rm Tr} S_i \exp(-\mathcal{H}(\{J_{ij}\}/k_BT))]_{\rm av}$$

$$= \langle S_i^\alpha S_i^\beta \rangle \qquad (\alpha \neq \beta)\,. \tag{4.168}$$

Here, one of the spins which appear explicitly in a trace in the upper line of Eq. (4.168) has been associated with replica α, the other with replica β. However, no index should be singled out, and hence Eqs. (4.167) and (4.168) need to be amended as (de Dominicis and Young 1983)

$$M = \lim_{n\to 0} \frac{1}{n} \sum_{\alpha=1}^{n} \langle S_i^\alpha \rangle\,, \quad q = \lim_{n \to 0} \frac{1}{n(n-1)} \sum_{\alpha \neq \beta} \langle S_i^\alpha S_i^\beta \rangle\,. \tag{4.169}$$

Since one expects that mean field theories become exact for infinite interaction range, it is of interest to consider a model, the so-called Sherrington-Kirkpatrick (SK) model (Sherrington and Kirkpatrick 1975), in which every spin interacts with every other spin with a distribution $P(J_{ij})$ that is completely independent of the distance between the spins, i.e. one takes the same $P(J_{ij})$ for all pairs of spins S_i, S_j. In order to guarantee the existence of the thermodynamic limit, we have to let the cumulants $J_{ij}^{\rm cum}$ go to zero with appropriate powers of N, namely defining new constants J_0, J,

$$J_{ij}^{\rm cum}(1) = \bar{J} \equiv J_0/N\,, \quad J_{ij}^{\rm cum}(2) = (\Delta J)^2 \equiv J^2/N\,. \tag{4.170}$$

Then Z_n becomes, see Eqs. (4.160) and (4.165),

$$Z_n = [Z^n\{J_{ij}\}]_{\rm av} = \mathop{\rm Tr}_{S_i^\alpha} \exp\left\{ \frac{1}{2N} \sum_{i \neq j} \left[\frac{1}{2} \left(\frac{J}{k_BT} \right)^2 \sum_{\alpha,\beta} S_i^\alpha S_j^\alpha S_i^\beta S_j^\beta \right.\right.$$

$$\left.\left. + \frac{J_0}{k_BT} \sum_\alpha S_i^\alpha S_j^\alpha \right] + \frac{1}{k_BT} \sum_i H_i \sum_\alpha S_i^\alpha \right\}, \tag{4.171}$$

where we have also allowed for a coupling to a local field H_i. Making use of the fact that $(S_i^\alpha)^2 = 1$, and neglecting terms of relative order $1/N$,

Eq. (4.171) can be rewritten as

$$Z_n = \exp\left[\frac{1}{4}\left(\frac{J}{k_BT}\right)^2 nN\right] \sum_{S_i^\alpha} \exp\left\{\frac{(J/k_BT)^2}{2N} \sum_{\alpha<\beta}\left(\sum_i S_i^\alpha S_i^\beta\right)^2\right.$$

$$\left. + \frac{J_0/k_BT}{2N}\sum_\alpha\left(\sum_i S_i^\alpha\right)^2 + \frac{1}{k_BT}\sum_i H_i\left(\sum_\alpha S_i^\alpha\right)\right\}. \quad (4.172)$$

Note that only the presence of infinite range interactions makes it possible to represent terms $S_i^\alpha S_j^\alpha$ as mixed terms resulting from $(\sum_i S_i^\alpha)^2$.

Now the Hubbard-Stratonovich identity

$$\exp(\lambda a^2/2) = \sqrt{\lambda/2\pi}\int_{-\infty}^{+\infty} dx \exp[-\frac{\lambda x^2}{2} + a\lambda x] \quad (4.173)$$

can be used to eliminate the quadratic terms in Eq. (4.172), and using the expression Eq. (4.167) for the magnetization and (4.168) for the order parameter one thus obtains

$$Z_n = \exp\left[\frac{1}{4}\left(\frac{J}{k_BT}\right)^2 nN\right]\left(\prod_{\alpha<\beta}\int_{-\infty}^{+\infty}\sqrt{N/2\pi}\,\frac{J}{k_BT}\,dq_{\alpha\beta}\right)$$

$$\left(\prod_\alpha\int_{-\infty}^{+\infty}\sqrt{\frac{NJ_0/k_BT}{2\pi}}dm_\alpha\right)\exp\left\{-N\frac{(J/k_BT)^2}{2}\sum_{\alpha<\beta}q_{\alpha\beta}^2\right.$$

$$-\frac{N(J_0/k_BT)}{2}\sum_\alpha m_\alpha^2 + N\ln\sum_{S^\alpha}\exp\left[\left(\frac{J}{k_BT}\right)^2\sum_{\alpha<\beta}q_{\alpha\beta}S^\alpha S^\beta\right.$$

$$\left.\left.+\frac{1}{k_BT}\sum_\alpha(J_0m_\alpha + H)S^\alpha\right]\right\}. \quad (4.174)$$

Here m_α is the magnetization in replica α and for the sake of simplicity we have used here a uniform magnetic field $H_i = H$.

If we assume now that one may interchange the limit $N \to \infty$ with the limit $n \to 0$ so that $\lim_{n\to 0}\lim_{N\to\infty}$ is calculated, the integrals in Eq. (4.174) can be evaluated by the method of steepest descent, since the argument of the exponential function is proportional to N. Therefore the neighborhood of the saddle point dominates the integral, and the only problem is to locate the saddle point. One thus finds for the free energy per spin

$$-f/k_BT = \lim_{n\to 0}\left\{\frac{1}{4}\left(\frac{J}{k_BT}\right)^2\left(1-\frac{1}{n}\sum_{\alpha\neq\beta} q_{\alpha\beta}^2\right)+\frac{J_0}{2k_BT}\frac{1}{n}\sum_\alpha m_\alpha^2\right.$$

$$+\frac{1}{n}\ln\sum_{S^\alpha}\exp\left[\left(\frac{J}{k_BT}\right)^2\sum_{\alpha<\beta} q_{\alpha\beta}S^\alpha S^\beta\right.$$

$$\left.\left.+\frac{1}{k_BT}\sum_\alpha\left(J_0 m_\alpha + H\right)S^\alpha\right]\right\}, \tag{4.175}$$

where $q_{\alpha\beta}, m_\alpha$ are given by the saddle-point conditions

$$\partial f/\partial q_{\alpha\beta} = 0 \quad \text{and} \quad \partial f/\partial m_\alpha = 0\,. \tag{4.176}$$

For the sake of self-consistency we also must have

$$m_\alpha = \langle S^\alpha\rangle = \lim_{n\to 0}\frac{\mathop{\mathrm{Tr}}_{S^\alpha} S^\alpha \exp[(\frac{J}{k_BT})^2\sum_{\alpha<\beta} q_{\alpha\beta}S^\alpha S^\beta + \frac{1}{k_BT}\sum_\alpha(J_0 m_\alpha + H)S^\alpha]}{\mathop{\mathrm{Tr}}_{S^\alpha}\exp[(\frac{J}{k_BT})^2\sum_{\alpha<\beta} q_{\alpha\beta}S^\alpha S^\beta + \frac{1}{k_BT}\sum_\alpha(J_0 m_\alpha + H)S^\alpha]} \tag{4.177}$$

and

$$q_{\alpha\beta} = \langle S^\alpha S^\beta\rangle = \lim_{n\to 0}\frac{\mathop{\mathrm{Tr}}_{S^\alpha} S^\alpha S^\beta \exp[\ldots]}{\mathop{\mathrm{Tr}}_{S^\alpha}\exp[\ldots]}\,, \tag{4.178}$$

where the arguments of the exponentials are the same as in Eq. (4.177).

Finally, it is of interest to take a further derivative with respect to H to find also the susceptibility and which is thus given by

$$\chi = \lim_{n\to 0}\frac{1}{k_BT}\left(1+\frac{1}{n}\sum_{\alpha\neq\beta} q_{\alpha\beta}\right). \tag{4.179}$$

While the expressions given by Eqs. (4.177) and (4.178) are a generalization of the self-consistency condition of a ferromagnet,

$$M = \langle S\rangle = \frac{\mathop{\mathrm{Tr}}_{S^\alpha} S\exp[\frac{1}{k_BT}(J_0M+H)S]}{\mathop{\mathrm{Tr}}_{S^\alpha}\exp[\frac{1}{k_BT}(J_0M+H)S]} = \tanh\left[\frac{1}{k_BT}(J_0M+H)\right], \tag{4.180}$$

the situation is now much more complicated, since $q_{\alpha\beta}$ is an infinite dimensional matrix, when one takes the analytic continuation from integer n

to real n. In order to make progress Sherrington and Kirkpatrick (1975) assumed that one can postulate that the solution is "replica-symmetric", i.e.

$$m_\alpha = M \quad \text{and} \quad \text{and } q_{\alpha\beta} = q \quad (\alpha \neq \beta)\,. \tag{4.181}$$

Noting that $\frac{1}{n}n(n-1)q$ tends to $-q$ for $n \to 0$, Eq. (4.179) then yields immediately

$$\chi = (1-q)/k_BT \tag{4.182}$$

from which we see that the susceptibility has a cusp if q starts to be nonzero, a result that agrees with the one we have obtained previously for spin glasses with symmetric coupling constants, see Eq. (4.142). However, note that Eq. (4.182) requires only that no spontaneous magnetization exists, but not that $J_0 = 0$. From Eq. (4.181) it also follows that

$$\sum_{\alpha<\beta} q_{\alpha\beta} S_\alpha S_\beta = \frac{1}{2} q \sum_{\alpha\neq\beta} S_\alpha S_\beta = \frac{1}{2} q \left[\Big(\sum_\alpha S_\alpha \Big)^2 - n \right] \tag{4.183}$$

and thus one can evaluate the traces in Eqs. (4.177) and (4.178), using once more the Hubbard-Stratonovich identity, Eq. (4.173), to get

$$M = \frac{1}{\sqrt{2\pi}} \int_{-\infty}^{+\infty} dz e^{-z^2/2} \tanh \frac{1}{k_BT} \left[J\sqrt{q}z + J_0 M + H \right] \tag{4.184}$$

$$q = \frac{1}{\sqrt{2\pi}} \int_{-\infty}^{+\infty} dz e^{-z^2/2} \Big(\tanh \frac{1}{k_BT} \left[J\sqrt{q}z + J_0 M + H \right] \Big)^2, \tag{4.185}$$

and the free energy is obtained to be given by

$$\begin{aligned} -\frac{f}{k_BT} &= \Big(\frac{J}{2k_BT}\Big)^2 \Big(1-q\Big)^2 - \frac{J_0}{2k_BT} M^2 \\ &+ \frac{1}{\sqrt{2\pi}} \int_{-\infty}^{+\infty} dz e^{-z^2/2} \ln \left[2\cosh \frac{1}{k_BT} \Big(J\sqrt{q}z + J_0 M + H \Big) \right]. \end{aligned} \tag{4.186}$$

It is straightforward to solve these replica-symmetric equation numerically. For the special case $J_0 = 0$, $H = 0$ we also have $M = 0$, and for small q we may expand Eq. (4.185) in a power series in q to find the solution analytically. One finds that a solution with $q > 0$ exists only for

$T < T_f = J/k_B$. The free energy then becomes, introducing the quantity $Q = q(J/k_BT)^2$,

$$f/k_BT = -\ln 2 - \frac{1}{4}\Big(J/k_BT\Big)^2 + \frac{1}{4}\Big[1 - \Big(k_BT/J\Big)^2\Big]Q^2 - \frac{Q^3}{3} + \frac{17}{24}Q^4 + - \cdots, \tag{4.187}$$

and the spin glass order parameter for $T < T_f$ simply becomes

$$q = (1 - T/T_f) + \frac{1}{3}(1 - T/T_f)^2 + \cdots . \tag{4.188}$$

The specific heat also gets a cusp at T_f, which is as pronounced as the cusp in the susceptibility (remember that this is not in accord with the experimental findings). It is also interesting to study the magnetic equation of state for small nonzero fields H. One finds

$$M = \frac{H}{k_BT}\left\{1 - \frac{1}{3}\left(\frac{H}{k_BT}\right)^2 \frac{T^2 + 2T_f^2}{T^2 - T_f^2} + O(H^4)\right\}, \quad T > T_f \tag{4.189}$$

$$M = \frac{H}{k_BT}\left\{1 - \frac{|H|}{\sqrt{2}k_BT_f} + \cdots\right\}, \quad T = T_f . \tag{4.190}$$

Defining now a nonlinear susceptibility χ_{nl}, that is essentially equivalent to χ_{SG} from Eqs. (4.145), as

$$\chi_{\text{nl}} \equiv \left(\frac{\partial^2 \chi(H)}{\partial H^2}\right)_T\Bigg|_{H=0}, \tag{4.191}$$

one finds from Eq. (4.189) that χ_{nl} diverges as $T \to T_f$,

$$\chi_{\text{nl}} \propto (T - T_f)^{-1} . \tag{4.192}$$

If we now define a critical behavior at the spin glass transition in terms of the power-laws

$$C \propto (T - T_f)^{-\alpha}, \quad q \propto (1 - T/T_f)^{\beta}, \quad \chi_{\text{nl}} \propto (T - T_f)^{-\gamma}, \tag{4.193}$$

we see from the above explicit results that these exponents are

$$\alpha = -1,\ \beta = 1,\ \gamma = 1 \quad \text{(SK-model)} . \tag{4.194}$$

Although the replica-symmetric solution fails for $T < T_f$ as we will see below, the result Eq. (4.194) remains valid, in the framework of the mean field theory. Figure 4.26 shows then the full numerical result for the replica-symmetric solution for C and χ in the case $H = 0, J_0 = 0$, while Fig. 4.27 shows the phase diagram when $J_0 \neq 0$ (Sherrington and Kirkpatrick 1975).

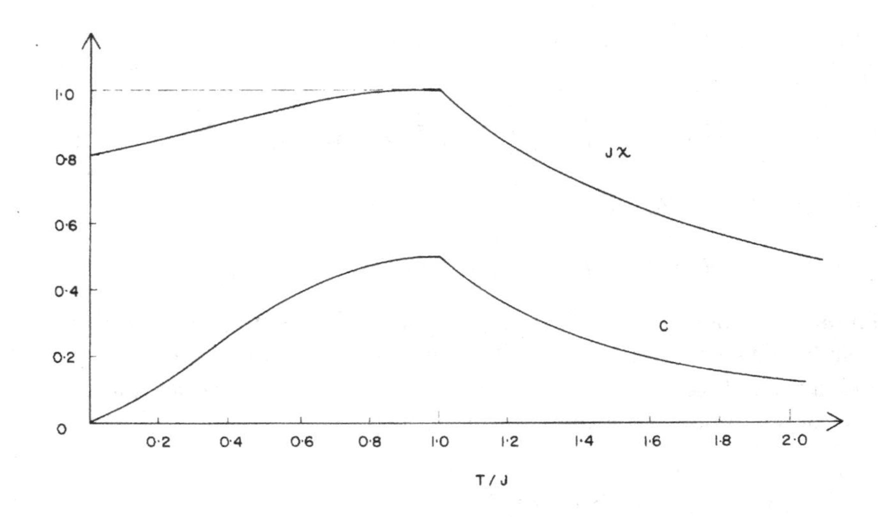

Fig. 4.26 Temperature dependence of the susceptibility χ and the specific heat C for the replica-symmetric solution of the Sherrington-Kirkpatrick (1975) model with $J_0 = H_0 = 0$. Note that $T_f = J$. The dashed line is the prediction of the Parisi theory for χ.

4.5.4 *Replica Symmetry Breaking*

Soon after Sherrington and Kirkpatrick (1975) had proposed the replica-symmetric solution for the infinite range Ising spin glass, it turned out that for $T < T_f$ this solution is not acceptable for various physical reasons. First of all, one finds that at low temperatures the entropy becomes negative although for any Ising system at $T > 0$ the entropy surely has to be non-negative. Secondly, one finds that the nonlinear susceptibility χ_{SG} becomes negative for $T < T_f$ below the AT line (de Almeida and Thouless 1978), see Fig. 4.28. Since χ_{SG} can be expressed as a sum of squares of fluctuations, see Eq. (4.145), it must be nonnegative, and the Sherrington-Kirkpatrick (SK) solution cannot be correct. In fact, the problem is that for $T < T_f$ the symmetry is broken: While in the paramagnetic phase there is one "valley" in phase space that dominates the thermodynamics, for $T < T_f$ there are many "valleys" in the "energy landscape", see Fig. 1.16. In the framework of the replica formalism developed in the previous section, the way in which a broken symmetry can be introduced is a nontrivial issue and therefore we will discuss it in more detail in this subsection.

While for ordinary phase transitions the character of the ordered phase is normally rather transparent and hence leads to obvious choices for the appropriate way in which the symmetry has to be broken, for spin glasses

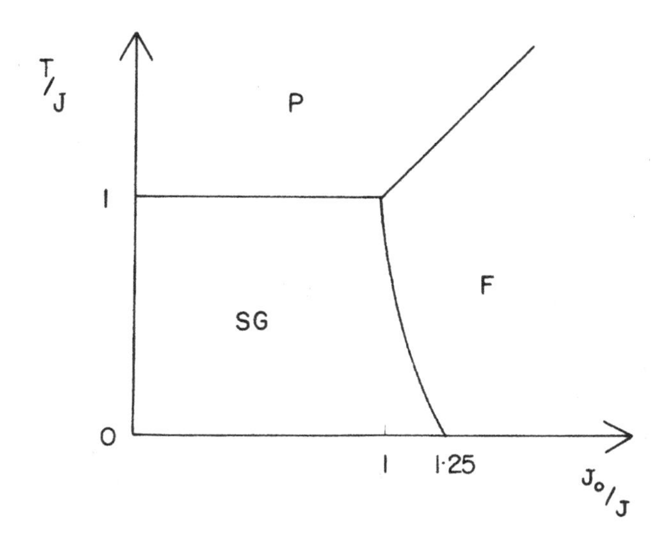

Fig. 4.27 Phase diagram for the Sherrington-Kirkpatrick (1975) solution of the infinite range Ising spin glass. The phases are paramagnetic (P), ferromagnetic (F), and spin glass (SG). Note that for $1 < J_0/J < 1.25$ the system goes from paramagnet to ferromagnet to a "reentrant" spin glass phase as the temperature is lowered. This reentrant behavior does not occur in the Parisi solution of the Sherrington-Kirkpatrick model, however. From Sherrington and Kirkpatrick (1975).

such a guidance from simple symmetry considerations to choose the symmetry breaking is missing. One therefore needs to make an educated guess! Of course, this guessed solution has subsequently to be tested for its stability (e.g., χ_{SG} must be nonnegative, etc.), and only if all stability tests are passed can the solution be maintained as a reasonable guess for the true solution (see Talagrand 2006, 2007 for a proof that with the Parisi replica symmetry breaking scheme the relevant solution for the statistical mechanics averages has indeed been found).

A guess for the order parameter has been made by Parisi (1979, 1980) who replaced the single order parameter q, that resulted from the assumption that all $q_{\alpha\beta}$ are equivalent, see Eq. (4.181), by an order parameter function $\tilde{q}(x)$. Remember that $q_{\alpha\beta}$ is an infinite dimensional matrix, with $0 \leq \alpha \leq n$, $0 \leq \beta \leq n$ being continuous variables. If one takes the limit $n \to 0$, one can still have a nontrivial variable, say, $\alpha/(\alpha+\beta)$, to parametrize the entries of this matrix. The scheme proposed by Parisi (1979, 1980) involves a hierarchical tree structure of order parameters that implies an "ultrametric symmetry" between them. We here skip all details

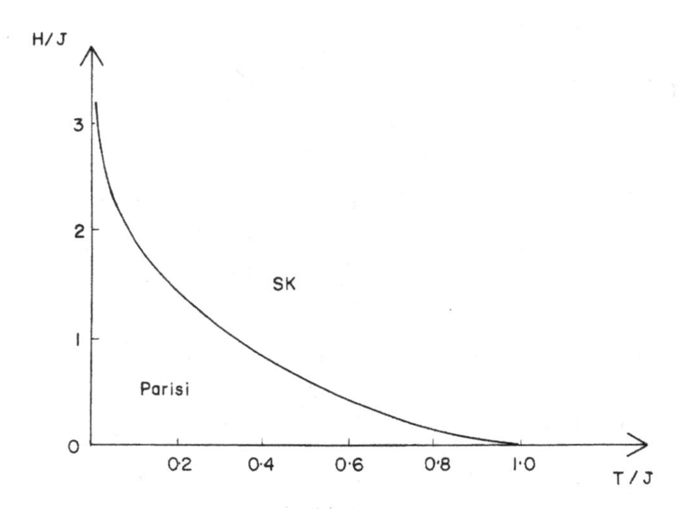

Fig. 4.28 Plot of the AT line for the Sherrington-Kirkpatrick (1975) infinite range Ising spin glass model with $J_0 = 0$. To the right of this line the replica-symmetric solution of Sherrington and Kirkpatrick [SK] (1975) with a single-order parameter is correct, while to the left of this line the replica symmetry must be broken, and one beliefs that the solution proposed by Parisi applies. This solution represents the many-valley-structure of phase space and nonergodic behavior, while the SK solution means there is a single valley. Therefore the AT line signals the onset of irreversibility. From de Almeida and Thouless (1978).

of his procedure and only quote the final result, for $H = 0$ and T near T_f,

$$\tilde{q}(x) = x/2,\ 0 < x \leq x_1 \ \text{ and } \ \tilde{q}(x) = x_1/2\ , x_1 \leq x \leq 1\ . \qquad (4.195)$$

One can show that in general

$$\chi = [1 - \int_0^1 \tilde{q}(x)dx]/(k_B T) \qquad (4.196)$$

and that the order parameter representing full equilibrium according to statistical mechanics (i.e. one averages over all "ergodic components", see Eqs. (1.37)-(1.40), Chap. 1) is

$$q = \lim_{H\to 0} \lim_{N\to\infty} [\langle S_i \rangle_T^2]_{\text{av}} = \int_0^1 \tilde{q}(x)dx\ . \qquad (4.197)$$

From the last two equations we thus see immediately that the relation

$\chi = [1-q]/(k_B T)$ remains valid. Since $\langle S_i \rangle_T = \sum_\ell P_\ell \langle S_i \rangle_T^{(\ell)}$, where we have decomposed $\langle S_i \rangle_T$ into the contributions from different "ergodic components" or "valleys" according to their probabilities $\{P_\ell\}$, q may pick up "interference terms" of the form $P_\ell P_{\ell'} \langle S_i \rangle_T^{(\ell)} \langle S_i \rangle_T^{(\ell')}$, involving different valleys. However, recalling that the notion of "ergodic components" implies that different ergodic components are separated by infinitely high barriers in phase space, it is clear that an experiment (or a computer simulation) measures only the property of a single valley. We therefore need an order parameter which is calculated from a single valley and which is subsequently averaged over all the valleys since they are in principle all equivalent. Such an order parameter can be defined as

$$q_{\text{EA}} = \left[\sum_\ell P_\ell (\langle S_i \rangle_T^{(\ell)})^2 \right]_{\text{av}} . \tag{4.198}$$

In order to calculate q_{EA} using statistical mechanics, one can introduce a second copy (also called "real replica") of the system, i.e. which has exactly the same interactions. Denoting by $S_{i,1}$ and $S_{i,2}$ the spin variables of the original and the replica, respectively, one can introduce a coupling term in the Hamiltonian

$$\delta \mathcal{H} = -\widetilde{H} \sum_i S_{i,1} S_{i,2} \tag{4.199}$$

that couples the two copies together, i.e. for $\widetilde{H} > 0$ the two systems prefer to be in the same state. In other words, whatever valley system 1 is in, it causes a very special "random field" to act on system 2, so that the latter prefers to be in the same valley as 1. If we take the limit $N \to \infty$ before $\widetilde{H} \to 0$, we expect that the only contribution to the average will be where the two systems are in the same valley. Consequently q_{EA} can also be evaluated as

$$q_{\text{EA}} = \lim_{\widetilde{H} \to 0} \lim_{N \to \infty} [\langle S_{i,1} S_{i,2} \rangle_T]_{\text{av}} . \tag{4.200}$$

An alternative definition for the order parameter can be given in terms of dynamics: If we define an effective time-dependent order parameter $q(t)$ as

$$q(t) \equiv [\langle S_i(0) S_i(t) \rangle_{t'}]_{\text{av}} , \tag{4.201}$$

where $\langle . \rangle_{t'}$ is the time average, we have

$$q_{\mathrm{EA}} = \lim_{t\to\infty} \lim_{N\to\infty} q(t) \,. \tag{4.202}$$

Note that the barriers between the valleys can only be infinitely high if $N \to \infty$, so in order to stay an infinite time in a single valley, we must first let $N \to \infty$ and then $t \to \infty$. Conversely we have from Eq. (4.197)

$$q = \lim_{\widetilde{H}\to 0} \lim_{N\to\infty} \lim_{t\to\infty} q(t) \;. \tag{4.203}$$

One now can show that within the replica symmetry breaking scheme one obtains

$$q_{\mathrm{EA}} = \tilde{q}(1) \;. \tag{4.204}$$

One should note that in the SK solution (or in a Mattis spin glass), there would of course be no difference between q and q_{EA}: Apart from the trivial spin reversal symmetry, there is a single valley in the Ising Mattis spin glass. A nonzero difference

$$\Delta = q_{\mathrm{EA}} - q \tag{4.205}$$

hence signifies a multi-valley structure in phase space and a breakdown of ergodicity since the time average, q_{EA}, is not equal to the thermal average q. In this context, it is also interesting to ask how Eq. (4.205) manifests itself in the context of dynamic properties. From standard linear response theory one obtains

$$\frac{dq(t)}{dt} = -k_B T \chi(t), \quad H \to 0, \quad \bar{J} \equiv 0 \tag{4.206}$$

where $\chi(t)$ is the usual linear response function and it was used that for a symmetric bond distribution we have $[\langle S_i(t) S_j(t') \rangle_T]_{\mathrm{av}} = 0$ for $i \neq j$. Integrating Eq. (4.206) to obtain

$$k_B T \int_0^t \chi(t')dt' = 1 - q(t) \quad (H \to 0, \bar{J} = 0) \,, \tag{4.207}$$

we can now consider the limits $t \to \infty$, $N \to \infty$ and thus see that

$$\lim_{\widetilde{H}\to 0} \lim_{N\to\infty} \lim_{t\to\infty} \int_0^t \chi(t')dt' = \chi \,, \tag{4.208}$$

since for the considered limit a time average must coincide with the statistical mechanics ensemble average χ. On the other hand one has

$$\lim_{t\to\infty}\lim_{N\to\infty}\int_0^t \chi(t') = \lim_{\omega\to 0}\chi(\omega)\,, \tag{4.209}$$

where $\chi(\omega)$ is the dynamic susceptibility. From Eqs. (4.202), (4.207), and (4.209) we then can conclude that

$$k_BT\lim_{\omega\to 0}\chi(\omega) = 1 - q_{\text{EA}} \quad (H=0, \bar{J}=0)\,. \tag{4.210}$$

and together with Eqs. (4.203), (4.207), and (4.208) we obtain

$$\Delta = k_BT[\chi - \lim_{\omega\to 0}\chi(\omega)] \quad (H\to 0, \bar{J}=0)\,. \tag{4.211}$$

Hence we see that there is a difference between the equilibrium (isothermal) susceptibility χ and the zero-frequency limit of the dynamic susceptibility. This is because $\chi(\omega)$ only probes the response in a single valley, whereas χ includes an extra contribution, Δ/k_BT, from changes in the statistical weight of the valley due to the applied perturbation (de Dominicis and Young 1983).

In computer simulations order parameters are defined as time averages over large but finite observation time intervals t_{obs} (Binder and Schröder 1976, Binder 1977):

$$\bar{q}(t_{\text{obs}}) = \frac{1}{N}\left[\sum_i\left(\frac{1}{t_{\text{obs}}}\int_0^{t_{\text{obs}}} S_i(t)dt\right)^2\right]_{\text{av}} = \frac{2}{t_{\text{obs}}}\int_0^{t_{\text{obs}}}\left(1-\frac{t}{t_{\text{obs}}}\right)q(t)dt\,, \tag{4.212}$$

$$\begin{aligned}\bar{\chi}(t_{\text{obs}}) &= \frac{1}{Nk_BT}\left[\frac{1}{t_{\text{obs}}}\int_0^{t_{\text{obs}}}\Big(\sum_i S_i(t)\Big)^2 dt - \left(\frac{1}{t_{\text{obs}}}\int_0^{t_{\text{obs}}}\sum_i S_i(t)dt\right)^2\right]_{\text{av}} \\ &= \frac{1}{k_BT}[1-\bar{q}(t_{\text{obs}})]\end{aligned} \tag{4.213}$$

where we have again used that $[\langle S_j(t_1)S_k(t_2)\rangle_{t'}]_{\text{av}}$ vanishes for a symmetric bond distribution, if $j\neq k$. In Eqs. (4.212), the overbar stands for the Monte Carlo time average, as defined here. In principle, again all quantities of interest can be defined in terms of limits analogous to those defined above,

$$q_{\text{EA}} = \lim_{t_{\text{obs}}\to\infty}\lim_{N\to\infty}\bar{q}(t_{\text{obs}}), \quad q = \lim_{\widetilde{H}\to 0}\lim_{N\to\infty}\lim_{t_{\text{obs}}\to\infty}\bar{q}(t_{\text{obs}})\,, \tag{4.214}$$

$$\chi = \lim_{\widetilde{H}\to 0} \lim_{N\to\infty} \lim_{t_{\rm obs}\to\infty} \bar{\chi}(t_{\rm obs}), \quad \lim_{\omega\to 0} \chi(\omega) = \lim_{t_{\rm obs}\to\infty} \lim_{N\to\infty} \bar{\chi}(t_{\rm obs}). \tag{4.215}$$

Note that the existence of a $\Delta \neq 0$ {Eqs. (4.205) and (4.211)} is sometimes also referred to as a violation of the fluctuation-dissipation relation. However, from our treatment it should be clear that here linear response theory does apply but that one must distinguish whether the limits are taken such that one picks up properties from a single valley or properly averages over all valleys.

An important aspect of this multi-valley description is that the states in different valleys are *not* "orthogonal" to each other in the sense that their projection onto each other does not necessarily vanish. Instead these states may have a nonzero overlap, which we will denote by $q^{\ell\ell'}$. Thus if we have valleys (states) ℓ and ℓ', then their overlap $q^{\ell\ell'}$ is defined as

$$q^{\ell\ell'} = \frac{1}{N}\sum_i \langle S_i\rangle_T^{(\ell)} \langle S_i\rangle_T^{(\ell')} . \tag{4.216}$$

Let us introduce the probability $P(q)$ that in equilibrium (where the states ℓ and ℓ' have weights P_ℓ, $P_{\ell'}$, respectively) an overlap q occurs,

$$P(q) = \left[\sum_{\ell,\ell'} P_\ell P_{\ell'}\, \delta(q - q^{\ell\ell'})\right]_{\rm av} . \tag{4.217}$$

We now consider a quantity $q^{(k)}$ defined as

$$q^{(k)} = \left[\sum_{\ell,\ell'} P_\ell P_{\ell'} \langle S_1 S_2 \cdots S_k\rangle_T^{(\ell)} \langle S_1 S_2 \cdots S_k\rangle_T^{(\ell')}\right]_{\rm av} , \tag{4.218}$$

where all sites $1, 2, \cdots, k$ of the spins are different. Since for an infinite-range system in the thermodynamic limit a "clustering property" holds, i.e. $\lim_{N\to\infty} \langle S_i S_j\rangle_T^{(\ell)} = \langle S_i\rangle_T^{(\ell)} \langle S_j\rangle_T^{(\ell)}$, if $i \neq j$, which follows from the fact that no single pair of spins is singled out, Eq. (4.218) can be reduced to

$$q^{(k)} = \left[\sum_{\ell,\ell'} P_\ell P_{\ell'} (q^{\ell\ell'})^k\right]_{\rm av} = \int q^k P(q) dq , \tag{4.219}$$

where we have made use of Eq. (4.217).

On the other hand, $q^{(k)}$ can be evaluated within the replica formalism as follows

$$q^{(k)} = \left[\langle S_1 S_2 \cdots S_k \rangle_T^2\right]_{\text{av}} = \lim_{n\to 0} \frac{1}{n(n-1)} \sum_{\alpha\neq\beta} \langle S_1^\alpha S_1^\beta S_2^\alpha S_2^\beta \cdots S_k^\alpha S_k^\beta \rangle \,. \tag{4.220}$$

Since in mean field theory averages on different sites decouple, Eq. (4.220) becomes

$$q^{(k)} = \lim_{n\to 0} \frac{1}{n(n-1)} \sum_{\alpha\neq\beta} q_{\alpha\beta}^k = \int_0^1 (\tilde{q}(x))^k dx \;, \tag{4.221}$$

where in the last equation the step from $q_{\alpha\beta}$ to the distribution function $\tilde{q}(x)$ can be taken as a definition of $\tilde{q}(x)$. Note that Eqs. (4.218)-(4.221) hold for arbitrary k. Since we can rewrite Eq. (4.221) to give

$$q^{(k)} = \int_0^1 (\tilde{q}(x))^k dx = \int q^k \frac{dx}{dq} dq \;, \tag{4.222}$$

a comparison between Eqs. (4.219) and (4.222) shows that we must have in general (Parisi 1983, Houghton *et al.* 1983)

$$\frac{dx}{dq} = P(q) \,. \tag{4.223}$$

So it follows that the inverse of the Parisi function $q = \tilde{q}(x)$ is the probability that an overlap q between different states occurs. If there is a single state, $P(q)$ is simply a delta function. For the case given by Eq. (4.195), we expect that $P(q)$ has a delta function at $q_1 = x_1/2$ and a horizontal part extending from $0 \leq q \leq q_1$. This behavior of $P(q)$ has been checked by Monte Carlo simulations for the SK model with various values of N ($32 \leq N \leq 192$) by Young (1983) and later by Billoire and Marinari (2002) who considered systems with up to 4096 spins. While the strong finite size effects that are present for such small values of N prevent too strong conclusions, a behavior of $P(q)$ consistent with the general theoretical expectations could in fact be obtained.

A very interesting aspect of $P(q)$ is that it is not self-averaging, i.e. different bond configurations give different results for $P(q)$ even for $N \to \infty$. While one can show that quantities like energy and magnetization are self-averaging, susceptibility (for $T < T_f$) and spin glass susceptibility exhibit also lack of self-averaging (see Binder and Young (1986) for more details).

One can prove that the free energy density f of mean field spin glasses is self-averaging, however, i.e. $[\langle f^2\rangle_T]_{\rm av} - [\langle f\rangle_T]^2_{\rm av} \to 0$ as $N \to \infty$. But the precise behavior how the thermodynamic limit is approached is still under discussion (Aspelmeier *et al.* 2008): While one can prove that for $f < \bar{f} \equiv [\langle f\rangle_T]^2_{\rm av}$ the probability $P_N(f)$ decays exponentionally for large deviations (Parisi and Rizzo 2008), $P_N(f) \propto \exp\{-{\rm const}N(\bar{f}-f)^{6/5}\}$, the full probability distribution $P_N(f)$ is not yet known. Finite size scaling of the form $P_N(f) = N^{-5/6}\tilde{p}\{N^{5/6}(f-\bar{f})\}$ where $\tilde{p}$ is some scaling function has been conjectured (Kondor 1983, Parisi 2009).

Finally, we discuss the extension of the SK-model to dynamics. Within the context of Monte Carlo simulations, one interprets the simulations in terms of a master equation for the kinetic Ising model, which describes the time evolution of the probability $P(\vec{X},t)$ that a state $\vec{X} = \{S_1, S_2, \cdots, S_N\}$ is realized at time t,

$$\frac{dP(\vec{X},t)}{dt} = -\sum_{\vec{X}'} W(\vec{X}\to\vec{X}')P(\vec{X},t) + \sum_{\vec{X}'} W(\vec{X}'\to\vec{X})P(\vec{X}',t)\,. \tag{4.224}$$

Here $W(\vec{X}\to\vec{X}')$ is the transition probability to go from state $\vec{X}$ to state $\vec{X}'$. The first term on the right hand side of Eq. (4.224) describes the loss in probability through all processes that lead away from the considered state, while the second term includes the gain via all reverse processes. For $W(\vec{X}\to\vec{X}')$ one can, e.g., follow Glauber (1963) and set

$$W(\vec{X}\to\vec{X}') = \frac{1}{2\tau_0}\left\{1-\tanh\left(\frac{\delta\mathcal{H}}{2k_BT}\right)\right\}, \tag{4.225}$$

where the parameter τ_0 sets the time scale (so W can be interpreted as transition probability per unit time; note that Ising models do not have any intrinsic dynamics of their own) and $\delta\mathcal{H}$ is the energy change due to the move $\vec{X}\to\vec{X}'$ (which usually is just a flip of a randomly selected spin, $S_i \to -S_i$).

In the context of analytical derivations, it is more convenient to work with a Langevin equation rather than a master equation. Then the constraint $S_i^2 = 1$ is relaxed into a "soft spin" version, in which each spin can take any value between $-\infty$ and $+\infty$, but values $S_i = \pm 1$ are still preferred due to a suitable choice of parameters r, u in the following Hamiltonian (de Dominicis 1978, Sompolinsky and Zippelius 1982):

$$-\frac{\mathcal{H}}{k_BT} = \frac{r}{2}\sum_i S_i^2 + \frac{u}{4}\sum_i S_i^4 + \frac{1}{k_BT}\sum_{\langle i,j\rangle} J_{ij}S_iS_j + \frac{H}{k_BT}\sum_i S_i\,. \tag{4.226}$$

Thus if one makes the choice that $r < 0$ and $u > 0$, the first two terms on the right hand side form a double well potential for S_i, hence making that at low temperatures $|S_i|$ fluctuates very little. The corresponding equation of motion is then postulated as (Hohenberg and Halperin 1977)

$$\frac{\partial S_i(t)}{\partial t} = -\frac{1}{\tau_0}\frac{\partial(\mathcal{H}/k_BT)}{\partial S_i} + \eta_i(t) , \tag{4.227}$$

where τ_0 sets again a time scale, and $\eta_i(t)$ is a gaussian random noise with a variance given by

$$\langle \eta_i(t)\eta_j(t')\rangle = 2\tau_0^{-1}\delta_{ij}\delta(t-t') , \tag{4.228}$$

thus obeying the fluctuation-dissipation theorem. The case of discrete spins can be recovered by taking the limit $r \to -\infty$, $u \to \infty$ with $|r/u| = 1$. For a finite system the discrete as well as the soft-spin model give at long times a Boltzmann distribution which is independent of the initial state of the spins.

Here we focus on the behavior of the autocorrelation function $q(t)$, see Eq. (4.201), near the spin glass transition. Dynamic scaling (Hohenberg and Halperin 1977) suggests that

$$q(t) \propto t^{-\lambda}\tilde{q}_{\pm}(t/\tau) , \tag{4.229}$$

where λ is a characteristic exponent, $\tilde{q}_{\pm}$ are universal scaling functions for $T > T_f$ and $T < T_f$, and τ is the relaxation time that diverges at T_f. Usually one writes

$$\tau \propto \xi_{\text{SG}}^z , \tag{4.230}$$

where z is a dynamic exponent, and ξ_{SG} is the spin glass correlation length (for a system with short range interactions $\xi_{\text{SG}} \propto (T - T_f)^{-\nu}$). Strictly speaking, it is not possible to define a correlation length for an infinite range model, but formally Eq. (4.230) remains valid even for the infinite range case if $\nu = 1/2$ is taken. (This choice will be justified later.) From the expected behavior below T_f one obtains λ, since

$$\lim_{t\to\infty} q(t) = q_{\text{EA}} \propto (T_f - T)^{\beta} , \tag{4.231}$$

where we have made use of Eqs. (4.193) and (4.194), so one requires that $\tilde{q}_-(\zeta \to \infty) \propto \zeta^{\lambda}$ to cancel the time dependence. This yields

$$\lambda = \beta/(z\nu) . \tag{4.232}$$

It is also useful to define an "average relaxation time" $\tau_{\rm av}$ (Binder 1977)

$$\tau_{\rm av} = \int_0^\infty q(t)dt \tag{4.233}$$

for $T > T_f$ where $q_{\rm EA} = 0$. Using Eq. (4.229) one finds

$$\tau_{\rm av} \propto (T - T_f)^{-z_{\rm av}\nu} \quad \text{with} \quad z_{\rm av} = z - \beta/\nu \,. \tag{4.234}$$

Explicit calculations (Sompolinsky and Zippelius 1982, Sommers and Fischer 1985) yield

$$q(t) \propto t^{-1/2}\tilde{q}_+\left\{\frac{t}{\tau_0}(T/T_f - 1)^2\right\}, \tag{4.235}$$

which shows that $z = 4$, $z_{\rm av} = 2$, and since $\beta = 1$ (see Eq. (4.194)), $\nu = 1/2$, Eq. (4.234) is fulfilled.

For $T < T_f$ the formulation of a valid theory is much more difficult. Sompolinsky (1981) introduced an approach which in the static limit becomes equivalent to the Parisi (1979, 1980) replica symmetry breaking solution, without introducing any replica method. In addition, this dynamical theory shows that for $T < T_f$ the approach to equilibrium is not exponential $\{\propto \exp(-t/\tau)$ as it would be true for $T > T_f\}$ but instead $q(t) - q_{\rm EA}$ vanishes with a (non-universal) power of t. In addition, for finite N one expects time scales that diverge with N as $\ln \tau \propto N^x$, but it is still not fully resolved which value the exponent x takes. So there are still features of the infinite range Ising spin glass that are not understood well. A particular active field of research on spin glasses in recent years has been the discussion of the consequences of violations of the fluctuation-dissipation theorem, see Crisanti and Ritort (2003) and Cugliandolo (2003) for a recent review in a more general context, Beletti *et al.* (2008) for recent simulations on aging spin glasses, and Ocio and Hérisson (2003) for a recent example of experimental studies of such issues in spin glasses.

It is clearly evident from the introductory discussion in this subsection that the mean field theory for spin glasses, which correctly describes the Sherrington-Kirkpatrick infinite range model, has rather unconventional properties. Should one expect that features predicted by the replica symmetry breaking approach such as the nontrivial spin glass order parameter distribution $P(q)$ and the resulting "multivalley" structure of the phase space for $T < T_f$ with "ultrametric" topology all carry over to finite-dimensional spin glasses with short-range forces? Due to the absence of exactly solvable models for short-range spin glasses, one has to rely largely on Monte Carlo simulations to answer this question. At the time of writing, one still can find somewhat controversial statements on the proper conclusions to be drawn from the results of these simulations in the literature (Parisi 2008, Young 2008). Thus we shall give only a rather brief introduction to short-range spin glasses in the following section.

4.5.5 *Spin Glasses Beyond Mean Field Theory*

For the statistical mechanics of standard phase transitions (see, e.g., Fisher 1974), the usual approach is to start with mean field theory and then to extend it by adding fluctuations (e.g. proceeding from the infinite range limit to models with a finite range of interactions). The consideration of fluctuations can then be used to check the self-consistency in terms of the so-called Ginzburg (1960) criterion. Normally one finds that mean field theory remains qualitatively correct, i.e., the critical exponents remain those of the mean field theory, if the dimensionality d exceeds the so-called "upper critical dimensionality" d_u. E.g., one finds $d_u = 4$ for the ordinary paramagnetic-ferromagnetic transition of the m-vector model, while $d_u = 6$ for the percolation transition. The renormalization group (Fisher 1974, Domb and Green 1977) provides a systematic way to calculate critical exponents for $d < d_u$ in an expansion in terms of $\varepsilon = d_u - d$.

Of course, it seems natural to apply these ideas to the spin glass problem as well. Ignoring the problem of replica symmetry breaking, which should be fine for temperatures $T > T_f$, the partition function is written as a functional integral (Harris *et al.* 1976, Chen and Lubensky 1977)

$$Z = \int \mathcal{D}q \; \exp\{-\mathcal{F}[q]/k_BT\}\,, \tag{4.236}$$

where the free energy functional $\mathcal{F}$ is constructed from the replica method. For this one defines a local analog of the order parameter $q_{\alpha\beta}$ or its replica-symmetric version q (but short wavelength fluctuations of the order parameter field $q(\vec{r})$ are assumed to be averaged over)

$$\mathcal{F}[q]/k_BT = F_0/k_BT - \int d^d r \Big\{ \frac{1}{4} v[q(\vec{r})]^2 + w[q(\vec{r})]^3 + \cdots [\nabla q(\vec{r})]^2 \Big\}\,, \tag{4.237}$$

where F_0 is the free energy of the disordered phase, v, w are phenomenological coefficients, and the spatial coordinates have been rescaled such that the coefficient of the $(\nabla q)^2$ term becomes unity. If one ignores the r-dependence of $\vec{q}(\vec{r})$, Eq. (4.237) is indeed of the form obtained from the replica-symmetric theory, Eq. (4.187), so Eq. (4.237) can be considered as a phenomenological extension of that theory. In the inhomogeneous case, one can consider a correlation function

$$G_{\text{SG}}(\vec{r}) = [\langle q(\vec{r}')q(\vec{r}' + \vec{r})\rangle_T]_{\text{av}} \propto \exp(-r/\xi_{\text{SG}}), \tag{4.238}$$

which is the analog of the correlation function defined in Eq. (4.144). From Eq. (4.237) one immediately predicts (note that for $T > T_f$ the term wq^3 can be ignored, and $v \propto T/T_f - 1$, as the comparison with Eq. (4.187) shows) that, as usual in mean field theory, $\xi_{\text{SG}} \propto (T/T_f - 1)^{-1/2}$, and hence $\nu = 1/2$, as noted in the context of Eq. (4.230).

Now the Ginzburg criterion for the self-consistency of the mean field theory, in terms of critical exponents, leads to the result that at d_u the hyperscaling relation (Fisher 1974)

$$d\nu = 2\beta + \gamma \tag{4.239}$$

should be fulfilled. (See Eqs. (4.193) and (4.194) for a definition of the exponents.) Using the results $\beta = 1$, $\gamma = 1$, $\nu = 1/2$ one finds from Eq. (4.239) that

$$d_u = 6 \tag{4.240}$$

for spin glasses. Harris *et al.* (1976) were the first to compute exponents in a renormalization group approach in first order in $\varepsilon = d_u - d$. We quote here their result for the m-vector spin glass:

$$\nu = \frac{1}{2} + \frac{5m\varepsilon}{12(2m-1)}, \quad \eta = \frac{-m\varepsilon}{3(2m-1)}, \tag{4.241}$$

where η is a critical exponent describing the decay of G_{SG} at $T = T_f$, $G_{\text{SG}}(\vec{r}) \propto r^{-(d-2+\eta)}$. Although in the meantime this expansion has been extended to order ε^3 (Green 1985), its applicability to physically relevant dimensionalities ($d = 3$ and $d = 2$) is very questionable. The problem is that usually the ε-expansion is accurate only for $d \gg d_\ell$, the lower critical dimensionality for the considered transition, and this is not the case here. At the lower critical dimensionality d_ℓ one expects that $T_f = 0$ and that the spin glass correlation length shows an exponentially strong divergence in T (McMillan 1984),

$$\ln \xi_{\text{SG}} \propto T^{-\sigma}, \quad \sigma = 2, \quad d = d_\ell, \quad T \to 0, \tag{4.242}$$

while for $d < d_\ell$ the spin glass correlation length also diverges as $T \to 0$ but only with a power-law,

$$\xi_{\text{SG}} \propto T^{-\nu'}, \quad d < d_\ell, \quad T \to 0, \tag{4.243}$$

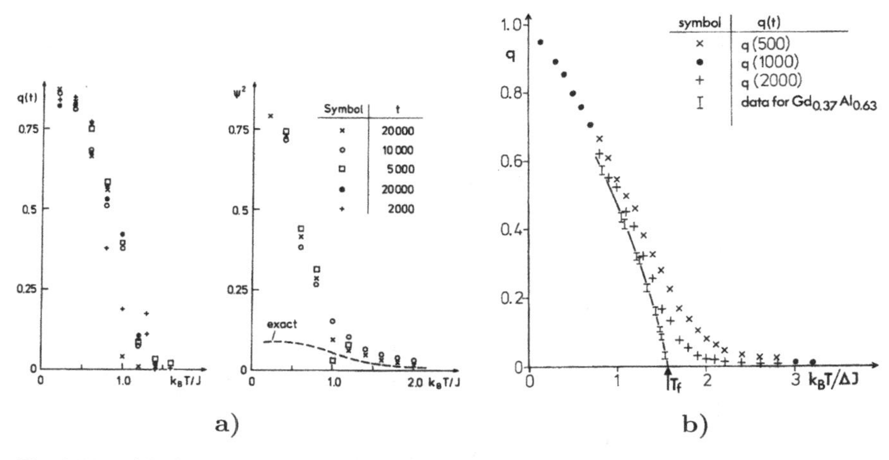

Fig. 4.29 a) Left part: Time-dependent Edwards-Anderson order parameter $q(t)$ plotted vs. temperature in 16×16 square $\pm J$ Ising lattices [data for $t = 2000$ are from an 80×80 lattice]. The right part shows the square of a Mattis-like order parameter ψ^2, where the projection of a spin on its corresponding ground state orientation is sampled. From Morgenstern and Binder (1980). b) Time-dependent Edwards-Anderson order parameter $q(t)$ vs. temperature for a $16 \times 16 \times 16$ gaussian Ising spin glass at various observation times t, measured always in attempted Monte Carlo steps per spin. Experimental data for $Gd_{0.37}$ $Al_{0.63}$ (Mizoguchi *et al.* 1977) are included, choosing $k_BT_f/\Delta J$ arbitrarily. From Binder (1977).

where ν' is a phenomenological exponent.

Now the identification of d_ℓ for spin glasses has been a controversial matter until today. So far analytical methods provided little guidance on this issue, and so the evidence comes mainly from Monte Carlo simulations, which are, however, hampered by the large relaxation times of these systems. Some time ago it has been suggested that even the nearest neighbor Edwards-Anderson spin glass on the square lattice has a nonzero freezing temperature (Binder and Schröder 1976), but it is now clear that this finding was due to the use of observation times $t_{\rm obs}$, see Eq. (4.212), that were by far too short. In fact, if the relaxation time τ of $q(t)$, Eq. (4.229), becomes comparable to $t_{\rm obs}$, $\bar{q}(t_{\rm tobs})$ starts to be nonzero, see Eq. (4.212). However, extrapolation of the data to $t_{\rm obs} \to \infty$ turned out to be rather difficult.

As an example we show in Fig. 4.29 Monte Carlo results for $q(t)$ as a function of temperature for the $\pm J$ model in $d = 2$ (Morgenstern and Binder 1980) and the gaussian Ising spin glass in $d = 3$ (Binder 1977, including experimental data for the $Gd_{0.37}Al_{0.63}$ spin glass from Mizoguchi

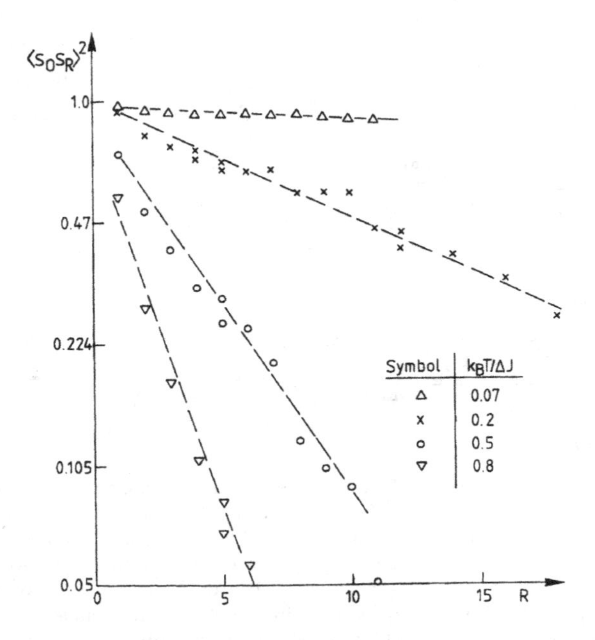

Fig. 4.30 Averaged spin glass correlation function $g_{\text{SG}}(R)$ as a function of R, for the two-dimensional gaussian Ising spin glass as obtained from a transfer matrix calculation. Note the logarithmic scale of the ordinate. Broken straight lines indicate a simple exponential decay, $g_{\text{SG}}(R) \propto \exp(-R/\xi_{\text{SG}})$. From Morgenstern and Binder (1980).

et al. 1977). On the basis of this figure, one might think that in the $\pm J$ model in $d = 2$ one has $k_BT_f/J \approx 1$ and in the gaussian model in $d = 3$ one has $k_BT_f/\Delta J \approx 1.5$. However, one now knows that $T_f = 0$ in $d = 2$, and also in the case of the gaussian spin glass in $d = 3$ the critical temperature T_f was severely overestimated, since one now believes that in that model $k_BT_f/\Delta J \approx 0.96$ (Marinari *et al.* 2000). The first clear-cut evidence that $T_f = 0$ in $d = 2$ actually came from exact recursive transfer matrix type calculations of the spin glass correlation $g_{\text{SG}}(\vec{r})$, see Eq. (4.144) (Morgenstern and Binder 1980, Fig. 4.30), since this method yields equilibrium results by construction, but unfortunately is limited to rather small system sizes (and is hardly applicable in $d = 3$).

Although the nearest neighbor Ising spin glass in $d = 2$ does not have a phase transition to a glass-like phase at nonzero temperature T, it is still a useful "laboratory" for numerical study where glassy behavior gradually emerges as $T \to 0$, and very nontrivial ground states and low-lying excitations occur. Using exact ground state algorithms from optimization techniques (Hartmann and Rieger 2002), one can obtain exact values for the ground state energy of rather large lattices (such as $L \times L$ lattices with $L = 256$) as well as their low-lying excited states (Hartmann and Moore 2003, Hartmann 2008). Such low-lying excited states differ from the ground state by the presence of domain walls running across the system, or due to droplets of linear extent ℓ. Of course, both the contour enclosing the droplet and the contour representing a domain wall are not smooth lines but rather fractals. The energy cost of such contours scales as $E \propto \ell^{\theta}$ and $E \propto L^{\theta'}$, with L the linear size of the system, and scaling implies that $\theta = \theta' = -1/\nu'$ {cf. Eq. (4.243)}. The best estimate for θ seems to be $\theta \approx -0.29$ (Hartmann and Moore 2003), but there is still considerable uncertainty on the precise value of θ, since corrections to scaling seems to be large, and earlier work (e.g. Kawashima 2000) seemed to be at variance with the above scaling relation $\theta = -1/\nu'$. Note that in this description the lower critical dimension d_ℓ corresponds to $\theta = -1/\nu' = 0$; for $d > d_\ell$ one has $\theta > 0$, and hence a spin glass order is stable at low temperature.

Amoruso *et al.* (2006) use conformal field theory methods to argue that domain walls in $d = 2$ can be understood as stochastic Loewner evolution process to derive a relation between the fractal dimension of these walls ($d_f \approx 1.27$) and the exponent θ, namely $d_f = 1 - 3/[4(3 + \theta)]$.

For $d > d_\ell$ the exponent θ is positive and the variation of θ with d has also found much interest (Boettcher 2004, 2005, Jörg and Katzgraber 2008). By considering dilute spin glasses near the percolation threshold, rather large values of L could be studied, yielding the estimates $\theta(d = 3) = 0.24(1)$, $\theta(d = 4) = 0.61(1)$, $\theta(d = 5) = 0.88(5)$, and $\theta(d = 6) = 1.1(1)$. Boettcher (2005) suggested a fitting formula $\theta(d) = -1.988 + 1.125d - 0.1533d^2 + 0.0086d^3$, which reproduces the above numbers correctly and furthermore predicts the exactly known result, $\theta(d = 1) = -1$. The latter value can be understood by realizing that since for a $d = 1$ chain the barrier is due to the absolutely weakest bond (which is of order $1/L$, for L bonds in a continuous bond distribution). Since this analysis implies $\theta(d = 2.4986) = 0$, it strengthens the hypothesis that actually $d_\ell = 5/2$. Note, however, that a study of $\theta(d)$ at high dimensionality by replica methods (Aspelmeier *et al.* 2003) does not agree with the above result for $\theta(6)$.

Jörg and Katzgraber (2008) have presented a finite size scaling analysis of the 4th order cumulant of the spin glass order parameter distribution functions, $g(L, T) = (3 - [\langle q^4 \rangle]_{\rm av}/[\langle q^2 \rangle]^2_{\rm av})/2$. As is well known (Binder and Young 1986), the critical temperature can be found from the intersection point $g^* = g(L, T_c)$ at which curves $g(L, T)$ intersect.

Plotting $\ln[g(2L,T)/g(L,T)]$ vs. $g(L,T)$ one constructs scaling functions, from which estimates $\theta(d \approx 3) \approx 0.2$ and $\theta(d \approx 4) \approx 0.75$ could be extracted.

Another very interesting aspect of two-dimensional $\pm J$ models is the onset of ferromagnetic order if the fraction $1-p$ of antiferromagnetic bonds is sufficiently small. One finds that the ferro-para transition line $T_c(p)$ decreases with increasing $1-p$ until the Nishimori (1981) line is reached, which is given by the equation $\tanh(J/k_BT) = 2p-1$. It is believed that $T_c(p)$ ends at $p = p* \approx 0.89$ (with $T^*/J \approx 0.95$) at a multicritical point (see e.g. Hasenbusch *et al.* 2008a,b, and references therein). No ferromagnetic order is possible for $p < p*$ and the critical behavior of the ferro-para transition for $T < T^*$ is controlled by a "strong disorder fixed point", with exponents $\nu \approx 1.5$, $\eta \approx 0.128$ (Parisen Toldin *et al.* 2009). For $T > T^*$, the transition is believed to belong to the same universality class as the site-diluted Ising ferromagnet, i.e. the critical behavior of the pure Ising ferromagnet modified by logarithmic corrections (Shankar 1987, Ludwig 1988). In contrast to this, at the multicritical point the exponents are very different, e.g. $\nu \approx 4.0$ (Hasenbusch *et al.* 2008a,b). This multicritical point hence separates two ferro-para transitions of different character.

Last but not least we mention that in the $\pm$ J model one has also obtained evidence (Matsuda *et al.* 2008) for the existence of Griffiths singularities for $T < T_c(p=0)$ (Griffiths 1969). These singularities fall in the class of "essential singularities" but are nevertheless very weak, due to the very rarely occurring large clusters of unfrustrated interactions.

The existence of a spin glass transition of the $d = 3$ Ising spin glass has remained controversial until in 1985 Ogielski and Morgenstern presented more precise simulation results, generated on a special purpose computer, that suggested that there was a transition at $T_f \approx 1.22$ where ξ_{SG} diverges according to a power-law (Fig. 4.31). Ogielski (1985) also analyzed the dynamics (Fig. 4.32) and obtained the estimate $z\nu \approx 6 \pm 1$ for the dynamic exponent z, cf. Eq. (4.230). All these exponent estimates quoted here are still preliminary, however: More recent even more extensive Monte Carlo simulations of the Ising spin glass (Kawashima and Young 1996) applying the finite size scaling methods have shown that Ogielski and Morgenstern (1985) still overestimated the freezing temperature somewhat, and put forward the value $T_f \approx 1.1$. Similarly, the exponent ν quoted in Fig. 4.31 no longer is believed to be the most accurate estimate, but rather one has $\nu \approx 2.5$ (Hasenbusch *et al.* 2008a, 2008b, Katzgraber *et al.* 2006) or even $\nu \approx 2.7$ (Hukushima *et al.* 2009). Similar methods showed that also XY

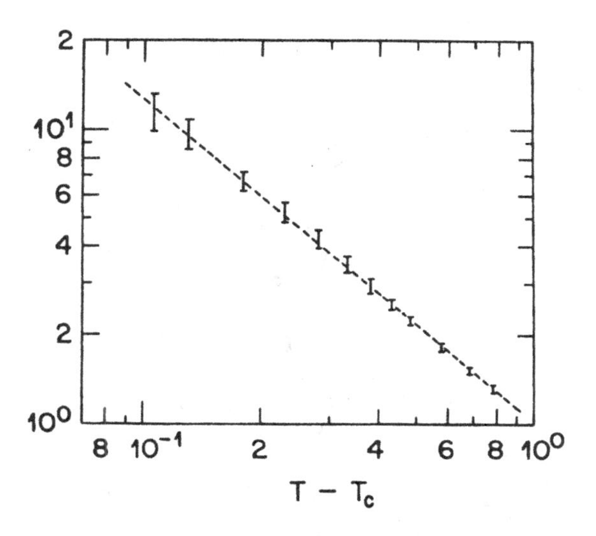

Fig. 4.31 Temperature dependence of the spin glass correlation length $\xi_{SG}(R)$ of the three-dimensional $\pm J$ model, assuming $T_f = 1.22$ on the log-log plot. The dashed straight line implies an exponent $\nu \approx 1.12 \pm 0.12$. From Ogielski and Morgenstern (1985).

and Heisenberg spin glasses have nonzero (though unusually low) transition temperatures, contrary to longstanding belief (see Binder and Young 1986 for a discussion) that in these models $T_f = 0$. So the original assertion that $d_\ell = 4$ for vector spin glasses is incorrect, and probably one has $d_\ell \approx 2.5$ for both Ising and vector spin glasses! The finding (Lee and Young 2003) that vector spin glasses have a freezing transition with a nonzero freezing transition is of course gratifying, since experimental spin glass systems like CuMn-spin glasses or $Eu_xSr_{1-x}S$ are much closer in character to Heisenberg spin glasses, and attempts to identify an anisotropy that would turn these systems Ising-like have not been convincing.

An important complication arises for short-range spin glasses with more component spins (XY spin glasses and isotropic Heisenberg spin glasses), because apart from the spin glass order described by the Edwards-Anderson-type order parameter a completely different ordering comes into play, involving the "chirality" degree of freedom. This degree of freedom is most easily understood if we consider a single frustrated plaquette (i.e., a square with 4 spins at the corners, and 3 bonds are $+J$ while one bond is $-J$, to create "frustration", see Chap. 1). If we consider the Ising case, the ground states are found such that 3 out of the four bonds lead

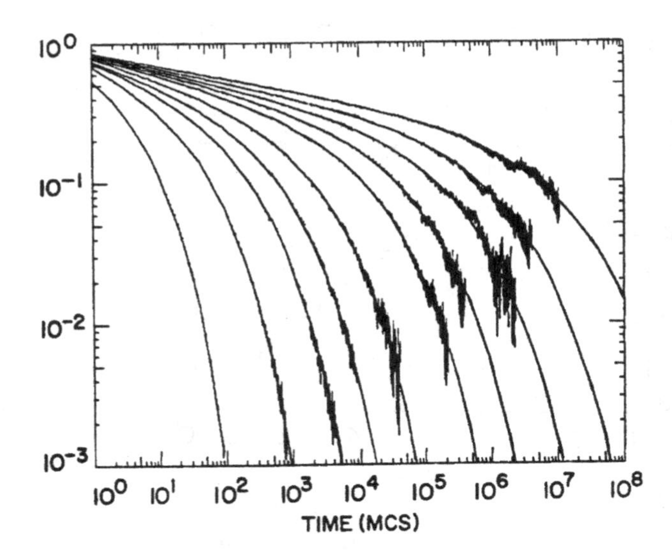

Fig. 4.32 Self-correlation function $q(t)$ of the three-dimensional $\pm J$ model plotted vs. time for lattice size 64^3 and temperatures (from left to right) $T/J = 2.5$, 2.0, 1.8, 1.7, 1.6, 1.5, 1.45, 1.40, 1.35 and 1.3. Solid curves represent fit of the data with the formula $Ct^{-x}\exp[-(t/\tau)^y]$. From Ogielski (1985).

to an energy $-J$ while the remaining one contributes $\pm J$, so that the (8-fold degenerate) ground state of such an isolated plaquette has energy $-2J$ (unlike the only 2-fold degenerate unfrustrated plaquettes, which have energy $-4J$). However, if we deal with classical XY-spins, for instance, the ground state of such a frustrated plaquette is no longer collinear: if we describe the spin orientations by the components $(\cos\theta_i, \sin\theta_i)$, where θ_i is the angle with some quantization direction, the spin arrangement is canted, $\theta_i - \theta_{i-1} = \pm\pi/4$, for a frustrated plaquette while for an unfrustrated plaquette it would be collinear. There are two different senses of rotation when one goes around the plaquette, and already Villain (1977) introduced a variable "chirality" τ to distinguish them. If one puts a single frustrated plaquette in a two-dimensional, nonfrustrated "sea" of plaquettes at zero temperature, one finds that the spin direction rotates quasicontinuously along a large loop around the frustrated plaquette. The total rotation angle is again $\pm\pi$, and hence the chirality can again be introduced. The picture of the spin configuration is reminiscent of the vortices introduced by Kosterlitz and Thouless (1973) to describe excitations of two-dimensional

XY magnets (apart from the fact that around a vortex the spin rotation is $\pm 2\pi$, however.) Although the positions of frustrated plaquettes are frozen, the chiralities are dynamical variables, since they can "jump" between the two values $+\pi$ and $-\pi$.

It turns out that this concept can be carried over to the Heisenberg spin glasses in $d = 3$ dimensions as well, and Kawamura (1992,1998), Campbell and Kawamura 2007, Viet and Kawamura 2010 have suggested a "chiral glass order scenario" for Heisenberg spin glasses. For Heisenberg spins on a simple cubic lattice, the local chirality at the i'th site and along the μ'th axis is called $\chi_{i\mu}$ and defined from three neighboring Heisenberg spins in terms of the product $\chi_{i\mu} = \vec{S}_{i+\vec{r}} \cdot (\vec{S}_i \times \vec{S}_{i-\vec{r}})$ where $\vec{r}$ denotes the unit vector along the μ'th axis. Thus for N spins we have then $3N$ local chirality variables to describe the "handedness" of the noncollinear structures induced by frustration. It has been suggested (although this suggestion has been debated) that chiral glass order in fully isotropic magnets sets in at higher temperatures then the standard spin-glass order. Particularly interesting is the presence of a transition in a nonzero magnetic field (analogous to the Gabay-Toulouse (1981) line obtained in the mean field theory for isotropic spin glasses). Presumably many experimental spin glass systems should be modelled as Heisenberg spin glasses with weak anisotropy, and the chiral ordering scenario is a promising model for such systems (Campbell *et al.* 2010), although there are still some unsolved problems (Shirakura *et al.* 2009).

These comments show that many aspects of spin glasses have remained unclear until the time of writing. This remark is even more true with respect to the properties of spin glasses at temperatures $T < T_f$. The question is whether replica symmetry breaking, which has yielded the multi-valley structure of the low temperature phase of spin glasses in the mean field limit, remains also relevant for spin glasses with short range interactions at high enough dimensionalities, and if so, does "high enough dimensionalities" include all dimensionalities $d > d_\ell$?

A particularly interesting issue in this context is whether "ultrametricity" (Rammal *et al.* 1986) also holds for short range spin glasses. An ultrametric space is a special kind of metric space in which the triangular inequality $d_{\alpha\gamma} \leq d_{\alpha\beta} + d_{\beta\gamma}$ ($d_{\alpha\beta}$ represents the distance between two points α and β) is replaced by a stronger condition where $d_{\alpha\gamma} \leq \max \{d_{\alpha\beta}, d_{\beta\gamma}\}$. Ultrametricity is an intrinsic property of Parisi's replica symmetry breaking scheme. Katzgraber and Hartmann (2009) attempted to test this property for a one-dimensional spin glass with random power-law interactions (changing the power law exponent is similar like changing the dimensionlalities) and found that for this system ultrametricity can indeed be observed.

Fisher and Huse (1988) have suggested an alternative description of the low temperature phase of an Ising spin glass, based on a variety of "droplet model"-type arguments, which implies that the low-temperature phase is only two-fold degenerate in the thermodynamic limit, and hence the Parisi replica symmetry breaking scheme would be irrelevant for short range spin glasses in low dimensionalities. Some mathematical arguments that support this view have been proposed by Newman and Stein (1996). However, Marinari *et al.* (2000) refute this criticism and present various evidence in favor of the replica symmetry breaking scenario in short range Ising spin glasses. In our view, the issue is not yet definitely settled, and also the possibility that neither the droplet theory of Fisher and Huse (1988) nor the replica symmetry breaking mean field type description of Parisi (1979, 1980, 1983) and coworkers applies, also should be considered seriously (Krzakala and Martin 2000, Palassini and Young 2000). Whether or not this so-called "trivial-nontrivial (TNT)" scenario holds is still controversial, see Katzgraber and Hartmann (2009). Therefore, we do not enter here into a more detailed discussion of these still controversial matters. We mention however that a first-principle approach which distinguishes between these two scenarios can again be based on the method of coupling two copies (1 and 2) of the system with an identical bond configuration ("real replicas") together, so that the total Hamiltonian (including also an external field) becomes (Parisi and Virasoro 1989)

$$\mathcal{H} = -\sum_{\langle i,j \rangle} J_{ij} S_{i,1} S_{j,1} - \sum_{\langle i,j \rangle} J_{ij} S_{i,2} S_{j,2} - H \sum_i (S_{i,1} + S_{i,2}) - \varepsilon \sum_i S_{i,1} S_{i,2} \,. \tag{4.244}$$

Considering now the overlap q between the copies 1 and 2, which we denote as $q(\varepsilon)$, one finds that in the symmetry breaking scenario a discontinuity occurs as $\varepsilon \to 0$, namely

$$\lim_{\varepsilon \to 0^+} q(\varepsilon) = q_{\mathrm{EA}} \,, \quad \lim_{\varepsilon \to 0^-} q(\varepsilon) = q_m \,, \tag{4.245}$$

where q_m (for $H \neq 0$) is the minimum value that occurs in the overlap function $q(x)$ (for $H \to 0$ we also have $q_m \to 0$). In contrast to this a theory without replica symmetry breaking yields also $\lim\limits_{\varepsilon \to 0^-} q(\varepsilon) = q_{\mathrm{EA}}$, so there is no singular response. In fact, one expects the ε-dependence for

$\varepsilon > 0$ to be singular (Marinari *et al.* 2000)

$$q(\varepsilon) = q_{\text{EA}} + a\varepsilon^b \,, \tag{4.246}$$

where in the infinite range case $b = 1/2$, while in the short range case b is a nontrivial exponent. For the $d = 4$ gaussian Ising spin glass Marinari *et al.* (2000) present rather convincing evidence in favor of Eq. (4.246), with $b \approx 1/4$ using lattices up to $L = 24$, and the resulting extrapolation for q_{EA} could be confirmed independently. Obtaining conclusive results for $d = 3$ is more difficult, since the system is already rather close to its lower critical dimension.

We also note that the idea to couple two real replicas of the same system together, as done in Eq. (4.244), and investigate an order parameter q that measures the overlap of their configurations, is useful as a general concept to discuss glassy systems, even if there is no quenched disorder: see Mézard and Parisi (2009) for a comparative discussion of this concept for both spin glasses and the structural glass transition of undercooled fluids.

One distinctive feature of the replica symmetry breaking mean field type description of Parisi (1979, 1980, 1983) is the existence of the Almeida-Thouless line $H_{AT}(T)$ in the H-T phase diagram of Ising spin glasses (Fig. 4.28). In contrast, according to the Fisher-Huse (1988) droplet model no AT line should exist for spin glasses with short range interactions. Recent Monte Carlo simulations (Young and Katzgraber 2004, Jörg *et al.* 2008, Katzgraber *et al.* 2009) have provided evidence that an AT line exists for $d > d_u = 6$ only, but not for the physically relevant dimension $d = 3$. Of course, there is always the possibility to argue that an AT line exists but occurs for smaller fields than have been accessible hitherto to the simulations studies.

Another approach to test for specific predictions of the replica symmetry breaking scenario is based on the study of dynamic quantities. Consider a time-dependent perturbation of the Hamiltonian

$$\mathcal{H}' = \mathcal{H} + \varepsilon(t)A(t)\,, \tag{4.247}$$

where e.g. $A(t) = (1/N)\sum_i S_i(t)$. In addition to the autocorrelation function $C(t,t') \equiv [\langle A(t)A(t')\rangle]_{\text{av}}$ one can also define a response function $R(t,t')$

$$R(t,t') \equiv \left.\frac{\partial\langle A(t)\rangle}{\partial\varepsilon(t')}\right|_{\varepsilon=0} . \tag{4.248}$$

If time translational invariance holds, in the unperturbed state, one can derive the so-called fluctuation-dissipation theorem which we write here as follows (Chandler 1987)

$$k_B T R(t,t') = \Theta(t-t')\frac{\partial C(t,t')}{\partial t'}, \tag{4.249}$$

where $\Theta(t-t')$ is the Heaviside step function. Now it is of interest to generalize Eq. (4.249) to out of equilibrium situations by defining a function $X(t,t')$ that describes the violation of the fluctuation-dissipation theorem (Cugliandolo and Kurchan 1993, Cugliandolo 2003, Crisanti and Ritort 2003)

$$k_B T R(t,t') = X(t,t')\Theta(t-t')\frac{\partial C(t,t')}{\partial t'}. \tag{4.250}$$

It was conjectured that at long times $X(t,t') \equiv X(C(t,t'))$ and that the function $X(Q)$ is nothing but the $x(q)$ of the replica symmetry breaking theory (i.e., the inverse function of $\tilde{q}(x)$. While numerical data (Marinari *et al.* 1998) seem to be compatible with these conjectures, we are not aware of a definite proof that actually $\tilde{q}(x)$ can be computed in this way.

At the end of this section, we emphasize that we only intended to present a tutorial introduction to the very rich field of spin glasses, and mention that many interesting aspects have not been touched upon. Those include problem of finding true ground states of spin glass models (Hartmann and Rieger 2002, 2004, Schneider and Kirkpatrick 2006, Roma *et al.* 2009, and references therein), the chaotic dependence of properties on the magnetic field and possibly other parameters (temperature, system size), see Katzgraber and Krzakala (2007) for a study of this problem and for references, the more precise physical interpretation of overlaps q and their distribution $P(q)$ (one should not simply interpret a spin glass state as an ordinary mixture of pure phases occupying the available volume, with well-defined interfaces separating these pure phases since such a picture is only reasonable if the interfaces cover a finite fraction of the volume), the physical consequences of ultrametric topology in the space of replicas, and physically very interesting out of equilibrium phenomena such as aging (Bouchaud *et al.* 1998, Cugliandolo 2003).

4.6 Variants and Extensions of Spin Glasses

In this section we consider the statistical mechanics of various other systems that contain quenched disorder. Some of these models are just variations on the theme of spin glasses, such as spin glasses with p-spin interactions, the random energy models, or the p-state Potts glasses. Others are motivated by the desire to explain the physical properties of real systems, such as models for quadrupolar glasses (also called "orientational glasses"), or the random field Ising model (RFIM) which is a generic model for the description of systems such as randomly diluted antiferromagnets in a homogeneous external magnetic field.

4.6.1 *p-Spin Interaction Spin Glasses and the Random Energy Model*

In the absence of external magnetic fields the Edwards-Anderson model (Edwards and Anderson 1975) of spin glasses has a special spin-reversal symmetry: Reversing the sign of all the spins $\{S_i \to -S_i\}$ keeps the Hamiltonian invariant. Of course, there is no counterpart of such a symmetry whatsoever in the problem of the structural glass transition. Thus, it is also desirable to consider generalizations of the spin glass problem that lack this particular symmetry.

One way to get rid of this symmetry is to consider a generalization to p-spin couplings (Derrida 1980, 1981)

$$\mathcal{H} = \sum_{i_1 \ldots i_p} J_{i_1 \ldots i_p} S_{i_1} \ldots S_{i_p} \ , \quad S_i = \pm 1 \ , \tag{4.251}$$

where the sum is over all groups of p spins in the system. The interactions have a gaussian distribution suitably scaled with N and p to obtain a sensible limit as N tends to infinity.

A special case occurs when one considers the limit $p \to \infty$ (Derrida 1980, 1981): Then the energy levels become independent random variables, and therefore this model is called the "random energy model" (Derrida 1980, 1981). As a result the free energy can be obtained in a straightforward manner and one finds that there is a transition at

$$T_f/(\Delta J) = (4 \ln 2)^{-1/2} \ , \tag{4.252}$$

where ΔJ is the width of the gaussian distribution for the coupling constants. Furthermore one can show that for $T > T_f$ the free energy is given by the same expression as for the Sherrington-Kirkpatrick (1975) model in the paramagnetic phase. However, for $T \leq T_f$ the free energy is independent of temperature, showing that the entropy vanishes at T_f, and the system is in a frozen ground state for all $T < T_f$.

Gross and Mézard (1984) have shown that the order parameter function $\tilde{q}(x)$ of the model given by Eq. (4.251) can be obtained by the first step of the Parisi replica symmetry breaking scheme. For $p \to \infty$, $\tilde{q}(x)$ is particularly simple, namely $\tilde{q}(x) = 0$ for $0 \leq x < x_1$, while $\tilde{q}(x) = 1$ for $x_1 < x \leq 1$, with $x_1 = T/T_f$. So the spin glass order parameter

$$q = \int_0^1 \tilde{q}(x)dx = 1 - T/T_f \tag{4.253}$$

starts to increase continuously at $T = T_f$, and the order parameter distribution function for $T < T_f$ is a sum of two delta functions,

$$P(q) = (1 - w(q_0))\delta(q) + w(q_0)\delta(q - q_0) \ , \tag{4.254}$$

with $q_0 = 1$, $w(q_0) = 1 - T/T_f$. Although the overlap at $x = x_1$ jumps discontinuously from zero to unity, the transition of the random energy model is not a first order transition in the usual sense since the order parameter q increases continuously at T_f, see Eq. (4.253), and there is also no latent heat. In any case, the random energy model is the rare case of a model with a complicated phase space with many valleys, for which the statistical mechanics can be worked out exactly, both with and without the replica method (Gross and Mézard 1984).

As it stands, the random energy model does not relate to the physical real space of a system, and hence is in a sense of mean field character. But since its properties are so well understood, it is desirable to search for generalizations of the random energy model, which are formulated in real space, e.g. on a lattice. Such a model on a one-dimensional lattice was proposed by Brunet and Derrida (1997): There are L groups of m spins S_k each such that only neighboring groups of spins interact (so the interaction range is $2m$). For each group $i = 1, \cdots, L$ one defines a state variable σ_i taking values $1, \cdots, 2^m$. In these variables $\{\sigma_i\}$ the interactions are restricted to nearest neighbors so that the Hamiltonian becomes $\mathcal{H}(\sigma) = \sum\limits_{i=0}^{L} E_i(\sigma_i, \sigma_{i+1})$. For each bond the 2^{2m} interaction energies E_i are quenched random variables taken from a Gaussian distribution of zero mean and variance $E^2 = m/2$. In the limit $m \to \infty$ the standard random energy model results, with distribution $P(E) \propto \exp(-E^2/mL)$, and the transition temperature is given by Eq. (4.252), with $\Delta J = 1$. For large but finite m it is possible to define correlation functions that detect a growing length scale as T approaches T_f (Franz *et al.* 2008). A useful correlation is the projection of a state of a subsystem ("window" of length ℓ) on the ground state of this window. One expects that in the limit $m \to \infty$ this correlation is unity for $\ell < \ell_c(T)$ and zero for $\ell > \ell_c(T)$, $\ell_c(T \to T_f) \to \infty$. In this limit this model has thus a spin glass transition that resembles the "mosaic picture" (Kirkpatrick and Wolynes 1987) of the structural glass transition: The system is frozen in regions of size $\ell_c(T)$, while at larger length scales it still behaves like an ordinary disordered system (or fluid, respectively). Unfortunately, for finite m the crossover between both limits is extremely gradual and thus it is not clear whether one should expect clear evidence for this scenario in physically more realistic models.

4.6.2 *Potts Glasses*

Another interesting generalization of the Ising spin glass is the Potts spin glass (Potts 1952, Elderfield and Sherrington 1983a-c, Erzan and Lage 1983, Gross *et al.* 1985). The Hamiltonian of the p-state mean field Potts glass of N interacting Potts spins σ_i $(i = 1, 2, \ldots, N)$, that can take p discrete values $\sigma_i \in \{1, 2, \ldots, p\}$, is defined as

$$\mathcal{H} = -\sum_{i<j} J_{ij}(p\delta_{\sigma_i \sigma_j} - 1)\,. \tag{4.255}$$

Here the term $\sum_{i<j} J_{ij}$ is added for the sake of convenience since it makes the mean energy to vanish in the limit $T \to \infty$ for each realization of the disorder. The "exchange constants" (bonds) J_{ij} are quenched random variables which are assumed to be distributed according to a gaussian distribution $P(J_{ij})$, as it is the case in standard spin glasses,

$$P(J_{ij}) = \left[\sqrt{2\pi}(\Delta J)\right]^{-1} \exp\{-(J_{ij} - J_0)^2/[2(\Delta J)^2]\}\,. \tag{4.256}$$

The first two moments J_0 and ΔJ are chosen such as to ensure a sensible thermodynamic limit,

$$J_0 \equiv [J_{ij}]_{\rm av} = \tilde{J}_0/(N-1)\,, \tag{4.257}$$

$$(\Delta J)^2 \equiv [J_{ij}^2]_{\rm av} - [J_{ij}]_{\rm av}^2 = 1/(N-1)\,. \tag{4.258}$$

Note that for $p = 2$ this model reduces to the mean field Ising spin glass model of Sherrington and Kirkpatrick (1975).

For the choice Eq. (4.258) for the parameter ΔJ, the replica-symmetric mean field theory predicts for $p \leq 6$ that either (for sufficiently negative

$\tilde{J}_0$) a transition to a spin glass phase occurs at $T_s = 1$, or a transition to a ferromagnetic phase occurs at T_f (if $T_f > 1$), with

$$T_f^{-1} = \frac{\tilde{J}_0}{(p-2)}\left[-1+\sqrt{1+2(p-2)/\tilde{J}_0^2}\right]. \tag{4.259}$$

In addition, a second transition to a different type of spin glass phase, sometimes also called "randomly canted ferromagnetic phase", is predicted at a transition temperature T_2 which is given by

$$T_2 = (p/2-1)/(1-\tilde{J}_0). \tag{4.260}$$

However, if one chooses $\tilde{J}_0$ sufficiently negative, such as $\tilde{J}_0 = 3-p$, one finds that $T_2 = \frac{1}{2}T_s$ $(= 1/2)$, i.e. is independent of p, and hence for temperatures close to T_s this second transition can be disregarded.

In the disordered phase, the internal energy per spin e and entropy per spin s are given by

$$e = -(p-1)(T_s/2T) \quad \text{and} \quad s = \ln p - (p-1)(T_s/2T)^2. \tag{4.261}$$

Note that Eq. (4.261) predicts a Kauzmann temperature T_K at which s vanishes, namely $T_K/T_s = \sqrt{(p-1)/\ln p}/2$, but it turns out that in fact for all $p < \infty$ there occurs a transition at a temperature higher than T_K, and therefore this Kauzmann temperature has no physical significance. Actually, also for the p-spin interaction spin glass with p finite, the entropy is nonzero (and positive) for $0 < T < T_f$.

One finds from a one-step-replica symmetry breaking approach (Gross *et al.* 1985, Kirkpatrick and Wolynes 1987, Kirkpatrick and Thirumalai 1988a, 1988b, Thirumalai and Kirkpatrick 1988, Cwilich and Kirkpatrick 1989, Cwilich 1990) that actually a second order spin glass transition to a spin glass phase does occur only for $p \leq 4$, while for $p > 4$ a new type of first order transition occurs, similar to that of the random energy model, at a temperature T_0 which exceeds both T_s and T_K. Thus Eq. (4.254) is again valid, with $w(q_0) \propto 1 - T/T_0$, but now q_0 has a nontrivial value less than unity. Making an expansion in p around $p = 4$ one can calculate both q_0 and T_0 analytically (Cwilich and Kirkpatrick 1989)

$$q_0 = \frac{2}{7}(p-4) + O(p-4)^2, \quad T_0 - T_s \propto (p-4)^2 + O(p-4)^3\,, \tag{4.262}$$

while for integer values $p > 4$ the correct values for q_0 and T_0 can only be obtained numerically (de Santis *et al.* 1995). For instance, for $p = 10$ one

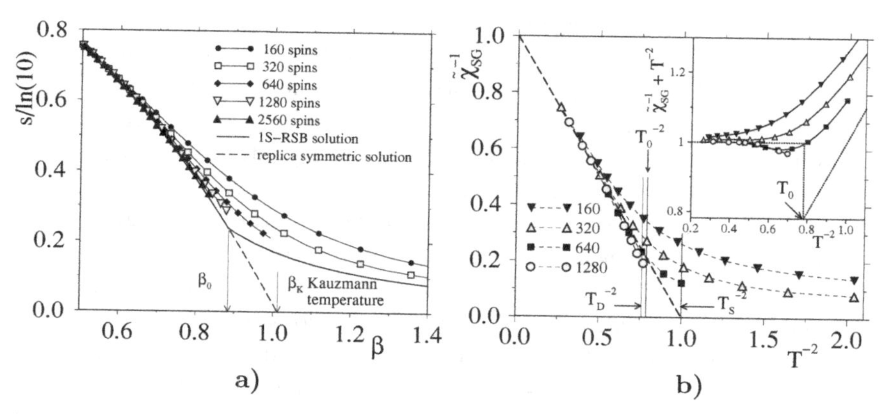

Fig. 4.33 Thermodynamic quantities for a 10-state Potts glass. a) Monte Carlo results for the entropy per spin, normalized by its high-temperature value, plotted versus inverse temperature for different system sizes (curves with symbols). The bold dashed curve and the bold solid curve are the replica-symmetric solution, Eq. (4.261), and the one-step replica-symmetry breaking solution of de Santis *et al.* (1995), respectively. Vertical arrows indicate the static inverse transition temperature $\beta_0 = 1/T_0$ and the inverse of the Kauzmann temperature T_K, $\beta_K = 1/T_K$, where the entropy of the replica-symmetric solution vanishes. From Brangian *et al.* (2002a). b) Inverse of the reduced spin glass susceptibility χ_{SG}^{-1} versus the square of the inverse temperature, for different system sizes obtained from Monte Carlo simulations (curves with symbols). The broken solid line shows the result from the replica-symmetric theory, Eq. (4.264). Vertical straight lines indicate the "spinodal temperature" $T_s = 1$ as well as the static (T_0) and dynamical (T_D) transition temperatures, respectively. The inset shows a plot of $\chi_{SG}^{-1} + (T_s/T)^2$, to illustrate the nonmonotonic convergence toward Eq. (4.264). From Brangian *et al.* (2002a).

finds

$$T_0 = 1.1312\ , \quad q_0(T_0) = 0.452\ . \tag{4.263}$$

The spin glass susceptibility $\chi_{SG} = N[\langle q^2 \rangle]_{av}/(p-1) = N \int q^2 P(q)dq/(p-1)$ then remains finite as $T \to T_0^+$, since the replica-symmetric result for χ_{SG}, which is correct for $T > T_0$, is simply

$$\chi_{SG} = [1 - (T_s/T)^2]^{-1}, \quad T > T_0\ , \tag{4.264}$$

while $\chi_{SG}^{-1} \equiv 0$ for $T \leq T_0$ due to the occurrence of the spin glass order parameter.

To illustrate this we show in Fig. 4.33 the entropy per spin as well as χ_{SG}^{-1} for the case $p = 10$, and compare the exact numerical results of de Santis *et al.* (1995) with Monte Carlo results due to Brangian *et al.* (2002a) for finite N. A particularly interesting feature of these Monte Carlo results are the very pronounced finite size effects associated with such an unconventional type of glass transition. Brangian *et al.* (2002a) also obtained qualitative evidence for the double-peak structure of the spin glass order parameter distribution, Eq. (4.254), although for finite N the delta functions are rounded into rather broad peaks for finite height, and are also shifted away to somewhat larger values of q.

The most interesting point about the mean field p-state Potts glass with $p > 4$ as well as the p-spin interaction Ising mean field spin glass is the discovery due to Kirkpatrick and Wolynes (1987) and Kirkpatrick and Thirumalai (1988a, 1988b) that the dynamic versions of these models exhibit an ergodic-to-nonergodic transition at a dynamical transition temperature $T_D > T_0$. This can be seen by considering relaxation functions such as the autocorrelation function $C(t)$ defined as

$$C(t) = \frac{1}{N(p-1)} \sum_{i=1}^{N} [\langle \vec{S}_i(t) \cdot \vec{S}_i(0) \rangle]_{\text{av}} \,. \qquad (4.265)$$

Here the Potts spins σ_i are given in the so-called simplex representation $\vec{S}_i(t)$ which takes into account the symmetry between the p states (see Zia and Wallace 1975 and Wu 1982 for details). One finds that $C(t)$ develops a decay in two steps (cf. Fig. 1.15 of Chap. 1), i.e. $C(t)$ first decays to a plateau which has a lifetime $\tau(T)$ and decays to zero only in a second step. When one approaches T_D, this lifetime diverges according to a power-law,

$$\tau(T) \propto (T - T_D)^{-\Delta} \,, \qquad (4.266)$$

with an exponent Δ which is non-universal (typically the value of Δ is around two). For $T \leq T_D$, the lifetime of the plateau is infinite, i.e. the autocorrelation function no longer decays to zero but stays at a finite height which is given by the Edwards-Anderson parameter q_{EA}, see Eq. (4.198). Since this nonzero value reflects the nonergodic behavior of the system it is often called the "nonergodicity parameter". A particularly intriguing fact is that the equations that one can derive to describe the time dependence of $C(t)$ are formally identical to those of a schematic model proposed in the framework of the mode coupling theory of the structural glass transition (Götze 1984, 1985, 1990, 2009, Leutheusser 1984) and which will be discussed in more detail in Chap. 5. Thus this points out a potentially deep link between the dynamical transition occurring in spin glasses, i.e. systems with quenched disorder, and the transition found in structural glasses which have no frozen-in disorder.

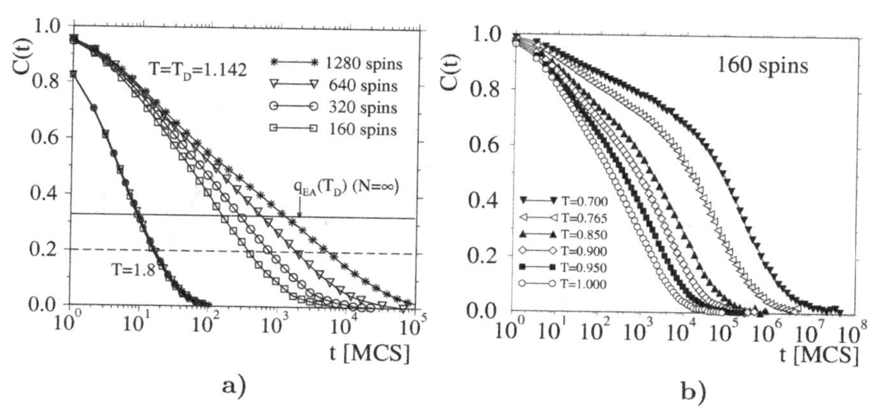

Fig. 4.34 a) Time-dependence of the autocorrelation function $C(t)$ of the Potts spins in the 10-state mean field Potts glass for $T = 1.8$ and for $T = T_D = 1.142$ for several values of N, as indicated. The horizontal solid straight line shows the theoretical value (de Santis *et al.* 1995) of the Edwards-Anderson order parameter $q_{\mathrm{EA}}(T_D)$ for $N \to \infty$, while the dashed horizontal line shows the value $C(t = \tau) = 0.2$ that was used to define the "plateau lifetime" τ which is used in Fig. 4.35. Time is measured in units of Monte Carlo steps per spin (MCS). From Brangian *et al.* (2001). b) Plot of $C(t)$ versus t for $N = 160$ at temperatures below the predicted critical temperature T_D. From Brangian (2002).

It is again an interesting question, that so far only has been addressed by numerical work (see, e.g., Brangian *et al.* 2001, 2002a), to ask how this nonergodic behavior emerges if one considers systems with a finite number N of spins and lets N tend to infinity. As an example, Fig. 4.34 shows Monte Carlo data for the autocorrelation function $C(t)$ of the Potts spins (Brangian 2002, Brangian *et al.* 2001). One sees that even for system sizes N of the order of 10^2 to 10^3, the predicted ergodic-to-nonergodic transition at $T = T_D$ is not observed, in that one cannot yet clearly recognize a presence of a long-lived plateau in $C(t)$. Instead one observes only a slow, non-exponential decay. Also the characteristic signatures of glassy relaxation, such as the presence of a two-step relaxation, are only found for T distinctly below T_D (Fig. 4.34b).

The reasons for the strong finite size effects that have been seen in the simulations of this model for the static as well as dynamic quantities, are not well understood yet. Figure 4.35 shows attempts to extrapolate the results for finite system sizes to $N \to \infty$, and to verify that the behavior of the model is in fact compatible with the presence of the predicted ergodic-to-nonergodic transition. To this aim Brangian *et al.* (2001) suggested the validity of a dynamical finite size scaling hypothesis (Hohenberg and

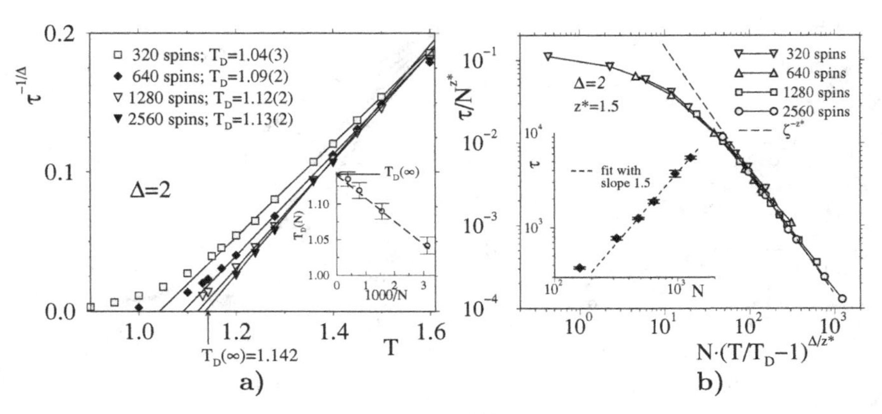

Fig. 4.35 a) Temperature dependence of $\tau^{-1/\Delta}$ for the 10-state mean field Potts glass, using $\Delta = 2$ as a trial value, for different system sizes as indicated in the figure. The straight lines are fits to Eq. (4.266) to a proper subset of points. The resulting extrapolated values for $T_D(N)$ are given in the figure, while $T_D(\infty) = 1.142$ is highlighted by an arrow on the abscissa axis. From Brangian *et al.* (2002a). b) A log-log plot of the scaled relaxation time τ/N^{z^*} versus the scaled distance $N(T/T_D - 1)^{\Delta/z^*}$ from the dynamical transition temperature T_D, choosing $z^* = 1.5$ and $\Delta/z^* = 1.3$. The inset is a log-log plot of $\tau(T = T_D)$ versus N showing that $\tau(N, T_D) \propto N^{z^*}$. From Brangian *et al.* (2001).

Halperin 1977, Suzuki 1977b, Binder 1992)

$$\tau = N^{z^*} \tilde{\tau}\{N(T/T_D - 1)^{\Delta/z^*}\} \, , N \to \infty \, , \; T/T_D - 1 \to 0 \, , \qquad (4.267)$$

where z^* is an unknown exponent and $\tilde{\tau}(\zeta)$ is a scaling function which must obey $\tilde{\tau}(\zeta \to \infty) \propto \zeta^{-z^*}$ in order that the proper thermodynamic limit results, i.e. Eq. (4.266). Figure 4.35b shows that the simulations are reasonably consistent with such a scaling, with a value of $z^* \approx 1.5$. In this context, we note that related finite size scaling ideas have been very successful (Berthier 2003) for the analysis of the glass transition (occurring at $T = 0$) in the one-dimensional spin-facilitated kinetic Ising model (Fredrickson and Andersen 1984) and in the kinetically constrained lattice gas (Kob and Andersen 1993).

Brangian (2004) has also studied the dynamical susceptibility $\chi(t)$ of the mean field Potts glass, which is defined as $\chi(t) = N[\langle q(t)^2 \rangle - \langle q(t) \rangle^2]$, where $q(t)$ denotes the overlap between two equilibrium configurations separated by a time t (Franz and Parisi 2000). One finds that $\chi(t)$ exhibits a maximum of (height χ^*) at time t^*, and these quantities satisfy power laws, $\chi^* \propto (T - T_D)^{-1}$ and $t^* \propto (T - T_D)^{-\Delta}$, with $\Delta = 2.3 \pm 0.3$. The dynamic exponent Δ estimated in this way from $\chi(t)$ is thus compatible with the estimate extracted from the autocorrelation function (Fig. 4.35). Plotting the data in scaled form, $\chi(t,T)/\chi^*$ versus $t/t^*(T)$, reveals a nice "data collapse" (Brangian 2004). As will become apparent in the following chapters, a similarly behaving dynamical susceptibility can also be defined for the structural glass transition (see Chap. 6).

An important fact about the dynamical transition at T_D is that it does not show up at all in static quantities such as spin glass susceptibility, specific heat, entropy, etc. However, there is an entropy-like quantity that is sensitive to this ergodic-to-nonergodic transition, namely the complexity which measures the entropy associated with the number of distinct valleys. As pointed out already in Chap. 1, the partition function Z can be decomposed into contributions from different valleys ℓ as follows,

$$Z = \sum_{\ell} Z_\ell \,, \quad Z_\ell = \sum_{\lambda \in \ell} \exp(-E_\lambda/k_BT) = P_\ell Z \,, \tag{4.268}$$

where P_ℓ is the statistical weight associated with the ℓ'th valley. Of course, in order that such a grouping of microstates λ into distinct valleys ℓ is physically meaningful, it is required that these valleys are separated in phase space by infinitely high energy barriers (in the thermodynamic limit, $N \to \infty$) from each other. Any average in the canonical ensemble of some observable A_λ can then be constructed as an average $\langle A \rangle_T^{(\ell)}$ taken in a particular valley ℓ, averaged afterwards over all valleys with their appropriate weights,

$$\langle A \rangle_T = \sum_{\ell} P_\ell \langle A \rangle_T^{(\ell)}, \quad \langle A \rangle_T^{(\ell)} = \frac{1}{Z_\ell} \sum_{\lambda \in \ell} A_\lambda \exp(-E_\lambda/k_BT) \,. \tag{4.269}$$

Note that Eqs. (4.268) and (4.269) yield of course the standard statement $\langle A \rangle_T = (1/Z) \sum A_\lambda \exp(-E_\lambda/k_BT)$, where now the unrestricted sum runs over all microstates λ.

Although the canonical ensemble average of an extensive observable thus is simply an average of this observable over the various valleys (also referred to as "thermodynamic states", or "ergodic components"), this is no longer true for the root mean square width Δ_A characterizing the width of fluctuations (Palmer 1982), because in addition to the "intravalley variance" there is also an *intervalley* contribution $\Delta_A^{(\mathrm{inter})}$:

$$\langle A^2\rangle_T - \langle A\rangle_T^2 = \sum_\ell P_\ell \langle A^2\rangle_T^{(\ell)} - \Big(\sum_\ell P_\ell \langle A\rangle_T^{(\ell)}\Big)^2 \tag{4.270}$$

$$= \sum_\ell P_\ell \Big\{\langle A^2\rangle_T^{(\ell)} - \Big[\langle A\rangle_T^{(\ell)}\Big]^2\Big\} + \Big[\Delta_A^{(\mathrm{inter})}\Big]^2 \tag{4.271}$$

$$\text{with}\quad \Big[\Delta_A^{(\mathrm{inter})}\Big]^2 = -\sum_\ell \sum_{\ell'} P_\ell P_{\ell'} \Big\{\langle A\rangle_T^{(\ell)}\langle A\rangle_T^{(\ell')} - [\langle A\rangle_T^{(\ell)}]^2\Big\}. \tag{4.272}$$

This result has interesting consequences for various fluctuation relations. Consider, e.g., the specific heat at constant field H

$$C_H = (\partial E/\partial T)_H = \sum_\ell P_\ell (\partial \langle \mathcal{H}\rangle_T^{(\ell)}/\partial T)_H + \Delta C\,, \tag{4.273}$$

with $\Delta C = k_B[\Delta_E^{(\mathrm{inter})}/T]^2$. From Eq. (4.272) we see that this intervalley contribution to the specific heat vanishes only if P_ℓ is independent of temperature. Furthermore it needs of course to be clarified whether or not such an intervalley contribution actually is extensive.

Most interesting is the behavior of the free energy F. We emphasize that, unlike Eq. (4.269), F is not the average $\bar{F}$ over all the individual valley free energies $F_\ell = -k_BT\ln Z_\ell$. To see this we note that for each valley ℓ we can write $F = -k_BT\ln Z = -k_BT\ln Z_\ell + k_BT\ln P_\ell = F_\ell + k_BT\ln P_\ell$. Thus averaging this expression over the valleys we obtain

$$\bar{F} = \sum_\ell P_\ell F_\ell = F + TI \quad \text{with} \quad I = -k_B \sum_\ell P_\ell \ln P_\ell\,, \tag{4.274}$$

$$\text{or}\quad F = \bar{F} - TI\,. \tag{4.275}$$

Note that $K^* = \exp(I/k_B)$ can be interpreted as the effective number of valleys that contribute to the thermodynamic average. For models such as the p-spin interaction spin glass or the mean field p-state Potts glass with $p > 4$, the quantity I starts to become nonzero at the ergodic-to-nonergodic transition, i.e. at T_D, in a discontinuous manner, see Fig. 4.36, and then decreases again as the temperature is lowered and ultimately vanishes at the transition temperature T_0 of the static transition. From Eq. (4.275) it follows that we have $I = S - \bar{S}$, i.e. I is the difference between the total entropy and the average entropy of a valley, and therefore it is often also called a "configurational entropy" or "complexity" of a glassy system, since it is related to the number of different configurations that are identified as distinct thermodynamic states of the glass. For this reason, the complexity is

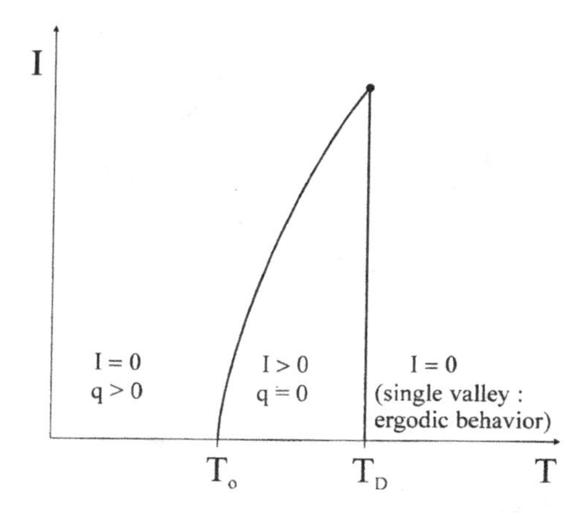

Fig. 4.36 Temperature dependence of the complexity I for systems undergoing an ergodic-to-nonergodic transition (schematic).

often compared to the "configurational entropy" defined by the experimentalists as the difference $\Delta S(T)$ between the total entropy of a (supercooled) fluid and the corresponding crystal (assuming that the vibrational entropy of the crystal and the glass are equal). Extrapolating this difference down to the temperature at which it vanishes defines the Kauzmann temperature T_K, $\Delta S(T = T_K) = 0$ (Figs. 1.17, 1.18). If one could indeed identify $\Delta S(T)$ with $I(T)$, T_K would be the analog of the static transition temperature T_0 of the mean field models considered in this section. We emphasize, however, that we feel that this analogy is misleading, since the complexity is nonzero only below the ergodic-to-nonergodic transition, while $\Delta S(T)$ is found to be nonzero in the full temperature range up to the melting temperature. Furthermore experiments show that $\Delta S(T)$ is completely smooth near the critical temperature $T_c \equiv T_D$ of the mode coupling theory (which in this context is interpreted as a "rounded" dynamical transition). Thus, we think that it makes more sense to compare the experimental $\Delta S(T)$ with the total entropy of the Potts glass (Fig. 4.33), which is a configurational entropy of the spin configurations, since phonon-like excitations are absent in discrete spin models by construction, rather than to associate it with the complexity. In any case, explicitly the behavior shown in Fig. 4.36 has only been calculated for the p-spin interaction model in the spherical limit (Crisanti *et al.* 1993).

In the framework of the description of the theory of the (spin)-glass transition in Fig. 4.36, where the glass order parameter q describes the

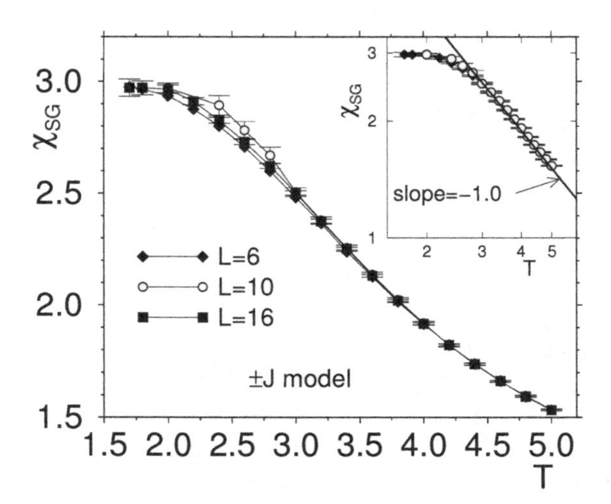

Fig. 4.37 T−dependence of the spin glass susceptibility $\chi_{\rm SG} = N[\langle q^2\rangle]_{\rm av}/(p-1)$ for the $p = 10$-state nearest neighbor $\pm J$ Potts glass with $J_0 = [J_{ij}]_{\rm av} = -1$ and $(\Delta J)^2 = [J_{ij}^2]_{\rm av} - [J_{ij}]^2_{\rm av} = +1$, leading to $J = \sqrt{2}$ and a fraction $x = (1 - 1/\sqrt{2})/2 \approx 0.146$ of ferromagnetic bonds. Three choices of L are included in the figure, as indicated. The inset shows that at high T the data is compatible with the $1/T$ expected for a system below its critical dimension. From Brangian *et al.* (2003).

overlap between two real replicas of the system, one can obtain a free energy $f(q)$, which has two minima for $T_0 < T < T_D$ (Mézard and Parisi 2009, Franz and Parisi 1998): the minimum at $q = 0$ represents the stable equilibrium state, while the minimum at $q \neq 0$ (with $f(q) > f(0)$ in this regime of temperatures) represents the (infinitely many) metastable states. In this mean-field description, T_D is a "spinodal" (limit of metastability)), similar to standard first-order transitions. However, for short-range systems spinodals are still ill-defined (Binder 1987), and thus one expects also that the definition of a "complexity" counting the number of metastable states must become "fuzzy" (Mézard and Parisi 2009) for systems with short-range interactions.

A very interesting generalization of the p-spin interaction model in the spherical limit to the case of long but finite range R of interaction was proposed by Franz and Toninelli, (2004). Specifically, one considers the dynamics of finite cubic lattices with linear dimensions L satisfying the inequality $\ln(L) \ll R \ll L$, and takes the limits $L \to \infty$, $R \to \infty$ jointly. Considering the dynamics in the regime $T_0 < T < T_D$, Franz (2007) showed that relaxation occurs on two time-scales: the shorter time scale remains finite in the limit $R \to \infty$, and can thus be interpreted as relaxation inside a single "valley". The longer time-scale diverges in the considered limits, and tentatively can be associated with processes where the system moves from one valley to another.

Now we turn to the question what is left from this interesting behavior if one considers models with short range interactions instead of these infinite range mean field models. Unfortunately, relatively little is known about this question. For $q = 10$ a Monte Carlo study (Brangian *et al.* 2002b, 2003) gives rather clear evidence, that this model has no transition of any kind, not even at $T = 0$: The spin glass susceptibility at $T = 0$ clearly stays finite (Fig. 4.37), and there are almost no finite size effects, for lattice linear dimensions of the order $L = 10$, and the specific heat also shows no sign of any anomaly (Fig. 4.38) and becomes essentially zero for the temperatures where the spin glass susceptibility χ_{SG} starts to saturate at its finite maximum value.

Actually, one can show that the specific heat is dominated by a Schottky-like peak due to strongly ferromagnetically coupled pairs of Potts spins (Brangian *et al.* 2003). These clusters (and triplets, quadruplets, etc.) of strongly coupled ferromagnetic Potts spins dominate also the dynamics of the model, see Fig. 4.39. One observes a decay of the self-correlation function $C(t)$ of the Potts spins, see Eq. (4.265), in several steps, and at low temperatures a first long-lived plateau develops at $C(t) \approx 0.6$. While at a first glance this is qualitatively reminiscent to the behavior predicted by the mean field theory, a quantitative analysis of the temperature dependence of the relaxation time shows that τ does not at all follow a power-law behavior, but instead shows an Arrhenius law, $\tau \propto \exp(E_{\text{act}}/k_B T)$ with $E_{\text{act}} = 14.6$ (see Fig. 4.39b). (Here τ was defined by the condition that $C(\tau) = 0.4$.)

In fact, the numerical value of E_{act} is easily understood since dimers of Potts spins coupled by ferromagnetic bonds involve a "binding energy" $p|J| = 10\sqrt{2} \approx 14.13$ which needs to be overcome when a ferromagnetic pair is broken up, and using the concentration of ferromagnetic bonds one can also explain the height of this plateau (Brangian *et al.* 2003). The second plateau in $C(t)$, Fig. 4.39a, is correspondingly attributed to spins coupled by two ferromagnetic bonds to their neighbors, involving hence an energy $2p|J| = 20\sqrt{2} \approx 28.3$, in excellent agreement with the data from the simulation (Fig. 4.39b).

The absence of a transition for the $p = 10$ short range Potts glass has been confirmed by Lee *et al.* (2006) who calculated also the spin glass

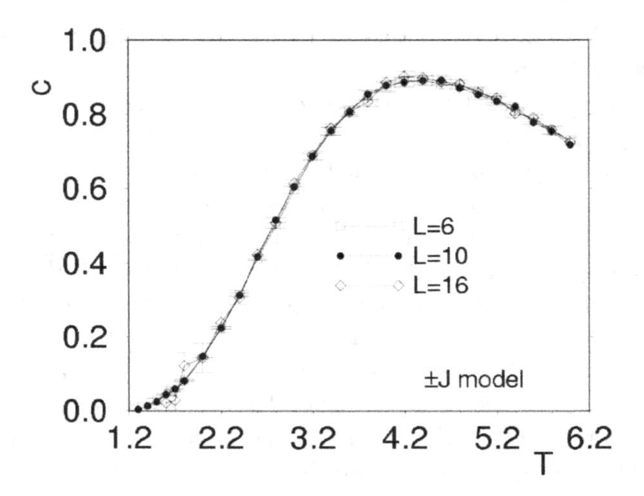

Fig. 4.38 Temperature dependence of the specific heat of the nearest neighbor 10-state $\pm J$ Potts glass, for the same parameters as used in Fig. 4.37. From Brangian *et al.* (2002b).

correlation length and showed that it is of the order of a single lattice spacing at low temperature. However, it is an interesting issue what happens for other values of p: while Lee *et al.* (2006) found a transition with $\nu \approx 1.18$ for $p = 3$, Cruz *et al.* (2009) tentatively suggest also a spin glass critical point for $p = 4$, at $k_B T_c/J \approx 0.25$ for the $\pm J$ model. However, they note that a related model (the 4-state permutation Potts glass, see Marinari *et al.* 1999) is found to have a Kosterlitz-Touless-type transition, indicative that the model is at its lower critical dimension.

Hence, the conclusion about the nearest neighbor 10-state Potts glass is that both the dynamical ergodic-to-nonergodic transition and the static glass transition that occur in the mean field version of the model are completely wiped out, no trace whatsoever being left of these phenomena. However, one should not overemphasize these findings too much: It has been argued (Eastwood and Wolynes 2002) that different short range models may differ in the quantitative extent to which the singularities of the mean field limit are rounded off. Thus, other models could have a behavior where one almost observes two transitions (a dynamic one at T_D and a static one at T_0), and only in the close vicinity of the apparent transition temperatures the singular behavior is rounded off. Such a behavior is likely, for instance, for a Potts glass model with a large but finite range of the interactions. Clearly, more work is required to clarify all these issues.

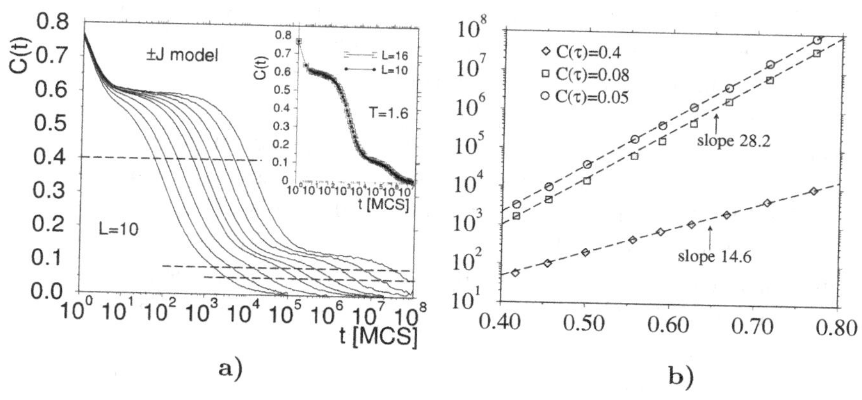

Fig. 4.39 a) Time dependence of the spin autocorrelation function $C(t)$ of the $p = 10$-state nearest neighbor $\pm J$ Potts glass, for temperatures (left to right) $T = 2.4$, 2.2, 2.0, 1.8., 1.7, 1.6, 1.5, 1.4, and 1.3 ($L = 10$). The horizontal dashed lines are used to define the relaxation times (τ_i) (see text for details). In the inset $C(t)$ for $L = 10$ and $L = 16$ is compared at $T = 1.6$ to show that finite size effects are absent. b) Arrhenius plot of the relaxation times τ_i extracted from part a). Straight lines are fits to the data with an Arrhenius law (the numbers give the activation energies). From Brangian *et al.* (2002b).

4.6.3 *Quadrupolar Glasses as Models for Diluted Molecular Crystals*

Already in Chap. 1 we have mentioned that in molecular crystals the long range order of the interacting quadrupole moments may be disrupted by dilution with ions that do not carry a quadrupole moment, similar to the case of spin glasses where the long range order of ferromagnets with competing interactions is destroyed by dilution with nonmagnetic ions. A simple example is ortho-hydrogen diluted with para-hydrogen (Sullivan 1983, see also Figs. 1.12 and 1.13); other related systems are N_2 diluted with Ar, KCN diluted with KBr, etc. (Knorr 1987, Loidl 1989, Höchli *et al.* 1990). Since the theoretical aspects of such systems have been reviewed in detail elsewhere (Harris and Meyer 1985, Binder and Reger 1992, Binder 1998), we will here only briefly review some salient features which are of interest in the context of the other glassy systems discussed in this book.

The generalization from spin glasses to quadrupolar glasses is nontrivial for several reasons.

- In the absence of external magnetic fields, the Hamiltonian of standard spin glasses (Sec. 4.5) is invariant under reversal of all spins. As a consequence, the free energy is an even function of the field.

However, quadrupolar glasses do not have such a symmetry, and so the free energy, when it is expanded as a power series in the field, must contain both even and odd terms (Vollmayr *et al.* 1994, Haas *et al.* 1996).

- While in magnetic systems random magnetic fields can occur only in exceptional cases, such as in diluted antiferromagnets in uniform external fields (Fishman and Aharony 1979), the lower symmetry of orientational glasses allows in the effective Hamiltonian the presence of linear terms in the orientational order parameter (Harris and Meyer 1985, Michel 1987a, 1987b). Thus orientational glasses are not analogous to standard spin glasses in zero random field but instead correspond to spin glasses in nonzero random field for which the sharp onset of Edwards-Anderson type ordering is smeared out (Pirc *et al.* 1987, 1994, Tadic *et al.* 1994).
- While in magnetic spin glasses the lattice can be treated as perfect and rigid, in orientational glasses the rotation-translation coupling is important (Michel and Naudts 1977, 1978, de Raedt *et al.* 1981, Lynden-Bell and Michel 1994). In fact, in crystals such as KCN the coupling between the orientations of the $(CN)^-$ dumbbells to acoustic phonons gives rise to an effective long range elastic interaction. Therefore, the first order transition of KCN from the cubic "plastic crystal" phase (in which the $(CN)^-$ can rotate and are orientationally disordered) to the orientationally ordered orthorhombic phase is a prototype of an "elastic phase transition" (Cowley 1976, Folk *et al.* 1976), associated with the softening of an elastic constant.
- Often the orientational glass phase and the orientationally long range ordered phases have different crystallographic symmetries (e.g. the ordered phase of ortho-hydrogen is cubic while the quadrupolar glass and plastic crystal phases are hexagonal close packed). Sometimes it is even claimed that real orientational glasses should be viewed as a random arrangement of nano-domains of the ordered phase (Knorr 1987, Loidl 1989, Höchli *et al.* 1990). Despite this simplified mean-field treatments of quadrupolar order in ortho-hydrogen/para-hydrogen mixtures have been found to be helpful for the interpretation of NMR experiments (Kokshenev 1996).

In view of the above problems, there is presently no consensus on the appropriate theoretical modeling of orientational glasses (Binder 1998).

Therefore we will here discuss only the description of isotropic quadrupolar glasses in terms of Edwards-Anderson-like models (Goldbart and Sherrington 1985, Carmesin and Binder 1987, 1988, Hammes *et al.* 1989, Mancini and Sherrington 2006). Note that the Potts glass can be considered as a model of anisotropic orientational glasses, e.g. one can associate the three states of the 3-state Potts model with the orientations of a linear molecule along the x, y, z axes of a cubic crystal. A more general description associates with each lattice site i an (electrical) quadrupole moment tensor

$$f_i^{\mu\nu} = \int \rho_i(\vec{x})(x_\mu x_\nu - \frac{\delta_{\mu\nu}}{m} \sum_{\lambda=1}^{m} x_\lambda^2) d\vec{x} \,. \tag{4.276}$$

Here $\rho_i(\vec{x})$ is the charge distribution of the i-th molecule, $\mu, \nu \in \{1, 2, \ldots, m\}$, and $\vec{x}$ is an m-dimensional vector. We are mainly interested in the case $m = 3$, since planar isotropic orientational glasses ($m = 2$) are equivalent to XY spin glasses (Carmesin 1987). Assuming a bilinear coupling between quadrupole moments the Hamiltonian becomes

$$\mathcal{H}_{\rm QG} = -\sum_{\langle i,j\rangle} \sum_{\mu\nu} \sum_{\mu'\nu'} J_{ij}^{\mu\nu\mu'\nu'} f_i^{\mu\nu} f_j^{\mu'\nu'} \,. \tag{4.277}$$

For axially symmetric charge distributions $f_i^{\mu\nu}$ can be written in terms of the components $\{S_i^\mu\}$ of a unit vector along this preferred axis,

$$f_i^{\mu\nu} = (S_i^\mu S_i^\nu - \delta_{\mu\nu}/m) f \,, \tag{4.278}$$

where f describes the strength of the quadrupole moments (we will choose units such that $f = 1$). For an isotropic interaction we also have $J_{ij}^{\mu\nu\mu'\nu'} = J_{ij}\, \delta_{\mu\mu'} \delta_{\nu\nu'}$ and hence Eq. (4.277) reduces to

$$\mathcal{H}_{\rm IQG} = -\sum_{\langle i,j\rangle} J_{ij} [(\vec{S}_i \cdot \vec{S}_j)^2 - 1/m] \,. \tag{4.279}$$

For a discussion of the more general Hamiltonian, Eq. (4.277) we refer the reader to more detailed reviews (e.g. Binder and Reger 1992). Here we only note the special case that Eq. (4.277) exhibits cubic anisotropy,

$$\mathcal{H}_{\rm CQG} = -\sum_{\langle i,j\rangle} J_{ij} \sum_{\mu=1}^{3} [(S_i^\mu S_j^\mu)^2 - 1/3] \,, \tag{4.280}$$

and which is believed to be in the same universality class as the 3-state Potts glass (Carmesin 1989).

While the replica-symmetric mean field theory of the quadrupolar glasses was already presented a long time ago (Goldbart and Sherrington 1985) and although it is clear that it suffers from similar instabilities as the replica-symmetric mean field theories of spin glasses (Sec. 4.5) and Potts glasses (Sec. 4.6.2.), an extension to formulate a 1-step replica symmetry breaking theory has only been given about 20 years later (Mancini and Sherrington 2006). Treating the vector dimensionality m as a continuous variable, it is found that for $m > m^* \approx 3.37$ a scenario with two successive transitions occurs as in the mean-field Potts glass (with $p > 4$ states). Again there occurs a static transition at $T = T_0$ (without latent heat) and a dynamical transition at a somewhat higher temperature T_D. For $m < m^*$, however, there occurs a single continuous transition. Of course, as in the case of spin glasses and Potts glasses, it is uncertain which one of these transitions (if any) survives for more realistic models with short range interactions at physical dimensionalities.

Although systems with the fully isotropic interaction, Eq. (4.279), hardly occur in nature, their behavior is of interest as a limiting case of the anisotropic systems, in analogy to isotropic spin glasses (Binder and Young 1986). While in these latter systems the spin pair correlation function is trivial for a symmetric distribution of bonds, $P(J_{ij}) = P(-J_{ij})$, since $g_{\rm F}(\vec{r}) = [\langle \vec{S}_i \cdot \vec{S}_j \rangle_T]_{\rm av} = \delta_{ij}$, where $\vec{r} = \vec{r}_i - \vec{r}_j$, this is not true for quadrupolar glasses due to the lower symmetry of their phase space: While there is a single orientation for which a uniaxial molecule is parallel to a given other molecule, there are infinitely many perpendicular orientations. Therefore, both ferromagnetic as well as glass-like correlations need to be considered (Hammes *et al.* 1989):

$$g_{\rm F}(\vec{r}) = \Big[\langle \sum_{\mu\nu} f_i^{\mu\nu} f_j^{\nu\mu} \rangle_T\Big]_{\rm av} = \Big[\langle (\vec{S}_i \cdot \vec{S}_j)^2 - 1/m \rangle_T\Big]_{\rm av} \propto \exp\Big[-r/\xi_{\rm F}(T)\Big], \tag{4.281}$$

$$g_{\rm G}(\vec{r}) = \Big[\langle \sum_{\mu\nu} f_i^{\mu\nu} f_j^{\nu\mu} \rangle_T^2\Big]_{\rm av} = \Big[\langle (\vec{S}_i \cdot \vec{S}_j)^2 - 1/m \rangle_T^2\Big]_{\rm av} \propto \exp\Big[-r/\xi_{\rm G}(T)\Big]. \tag{4.282}$$

Monte Carlo simulations (Hammes *et al.* 1989) suggest that both correlation lengths $\xi_{\rm F}(T)$ and $\xi_{\rm G}(T)$ diverge as $T \to 0$ with (presumably) the same power-law $\xi_{\rm F}(T) \propto T^{-\nu_0}$, $\xi_{\rm G}(T) \propto T^{-\nu_0}$, and that only the prefactor in the power-law for $\xi_{\rm G}(T)$ is somewhat larger. The estimates for the exponent ν_0 are $\nu_0 \approx 0.63$ ($d = 2$) and $\nu_0 \approx 1.02$ ($d = 3$), and hence it is implied that the model Eq. (4.279) exhibits a glass transition at zero

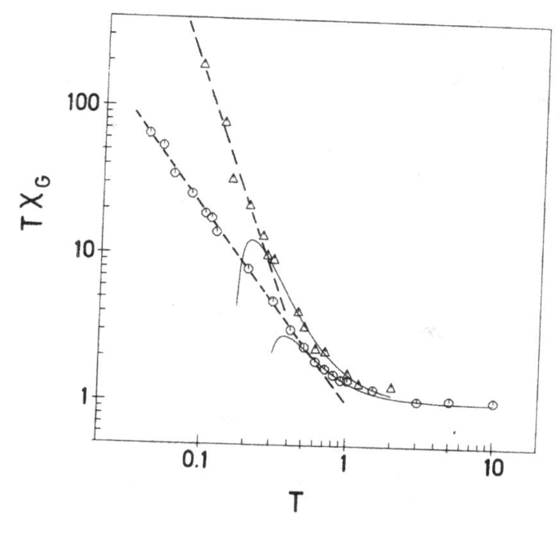

Fig. 4.40 Log-log plot of the glass susceptibility χ_G of the isotropic orientational glass with nearest neighbor random gaussian symmetric interaction versus temperature. Broken straight lines show the power-laws $\chi_G \propto T^{-\gamma_0}$ with the exponents as quoted in the text, while full curves show the first 4 terms of a high temperature expansion (Carmesin 1987). Triangles refer to $d = 3$ and circles to $d = 2$. From Hammes *et al.* (1989).

temperature only. The fact that also $\xi_F(T)$ seems to diverge at $T = 0$ may be related to the fact that in the mean field version of this model (Goldbart and Sherrington 1985) one finds at low temperatures a "canted ferromagnetic phase", i.e. a mixed phase of ferromagnetic order and transverse glass order.

In analogy to the case of spin glasses, see Eq. (4.145), one can define also for these systems a "glass susceptibility", again defined as a sum over all glass correlations

$$\chi_G = 1 + \sum_{\vec{r} \neq 0} g_G(\vec{r}), \tag{4.283}$$

and, consistent with the behavior of the correlation length, a power-law (Fig. 4.40) behavior has been detected, $\chi_G \propto T^{-\gamma_0}$, with $\gamma_0 \approx 1.35$ ($d = 2$) and $\gamma_0 \approx 2.7$ ($d = 3$). Within the (considerable) errors of all theses estimates the data are compatible with the scaling law appropriate for zero temperature transitions, i.e. $\gamma_0 = d\nu_0$. However, one should remark that all the quoted exponents for this model are at best "effective exponents", since they are extracted from a temperature range where ξ_G is only of the order

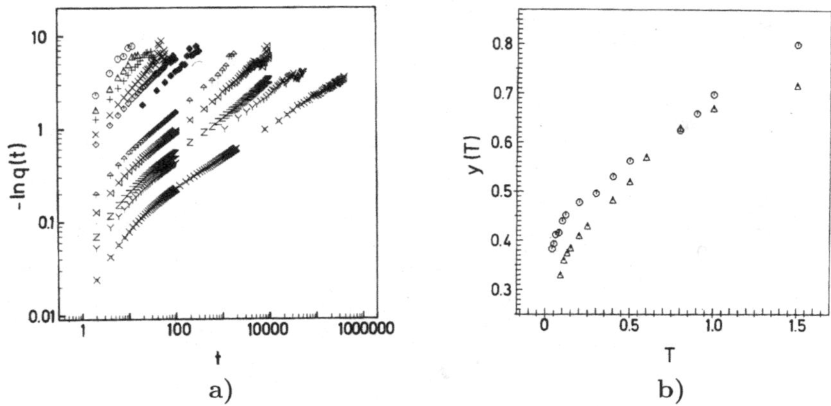

Fig. 4.41 a) Plot of the time-dependent Edwards-Anderson order parameter $q(t)$ for isotropic orientational glasses in a log-log form for $d = 3$, $m = 3$ at various temperatures, T =1.5 (circles), 1.0 (triangles), 0.8, 0.6, 0.5, 0.4, 0.3, 0.2, 0.15, 0.12, 0.09 (crosses). b) Temperature dependence of the KWW exponent $y(T)$ for $d = 2$ (circles) and $d = 3$ (triangles). From Hammes *et al.* (1989).

of 1-2 lattice spacings! (Lower temperatures were inaccessible because of the dramatic slowing down of the dynamics.) As a further warning, we mention that also the early work on isotropic spin glasses (Jain and Young 1986, Olive *et al.* 1986, Morris *et al.* 1986) claimed a similar transition at $T = 0$, while recent work (Lee and Young 2003) established that the transition temperature is small but finite.

Generalizing the concept of a time-dependent Edwards-Anderson order parameter $q(t)$ in terms of a autocorrelation function of the relevant degrees of freedom from spins (in the case of spin glasses) to quadrupole moments (in the case of orientational glasses), we have

$$q(t) = \left[\frac{1}{N}\frac{1}{1-1/m}\sum_i \langle\{\vec{S}_i(0)\cdot\vec{S}_i(t)\}^2 - 1/m\rangle_T\right]_{\mathrm{av}}. \tag{4.284}$$

Figure 4.41 shows data for $q(t)$ for the case $d = 3$, $m = 3$. Since one finds that the decay of $q(t)$ is well described by the Kohlrausch-Williams-Watts-function (KWW) $q(t) \propto \exp\{-[t/\tau(T)]^{y(T)}\}$, a log-log plot of $\{-\ln q(t)\}$ vs. t is shown, since then the data fall on straight lines, the slope of which is the stretching exponent $y(T)$. Unlike the case of supercooled fluids, for which one also finds that the KWW-law gives a good description of the various relaxation functions with a $y(T)$ that depends only moderately on T, one finds here that $y(T)$ exhibits a pronounced dependence on temperature (Fig. 4.41b). The relaxation time $\tau(T)$ is found to diverge as $\tau \propto T^{-z\nu}$, with $z\nu \approx 4.3$ (for $d = 2$) and $z\nu \approx 6.8$ (for $d = 3$), see Hammes *et al.* (1989). For $d = 3$ it is also possible to fit the data with an Arrhenius law, $\tau(T) \propto \exp(\Delta E/T)$, with an activation energy in the range $5 \leq \Delta E \leq 8$. However, again one has to be aware that it is presently unclear whether or not sufficiently low enough temperatures have been reached to allow well founded statements on the asymptotic behavior. Since it is still a matter of debate whether or not models of the type of Eq. (4.277) capture the essential features of real orientational glasses, relatively little attention has been devoted to clarify the behavior of these models.

4.6.4 *Atomistically Realistic Models of Diluted Molecular Crystals*

Since there is no consensus on the basic ingredients of coarse-grained models for orientational glasses (e.g. it is unclear whether "random bonds" or "random fields" are more important, etc.), it is of interest to consider chemically realistic models of real materials in an attempt to clarify some of these basic questions about the proper theoretical modeling. The first studies along this line were done by Lewis and Klein (1986, 1989) who studied $(KBr)_{1-x}(KCN)_x$ and Klee *et al.* (1988) who investigated $Ar_{1-x}(N_2)_x$ by means of molecular dynamics (MD) simulations. While such methods are in principle very valuable tools to obtain information on the details of the structure as well as the dynamics of the investigated systems, one must bear in mind that there are also important limitations: (i) The desired direct comparison with experiment is often hampered due to the use of inaccurate potentials. Sometimes, these potentials are not known with sufficient accuracy, sometimes one uses a simplified potential to reduce the needed computer time to do the simulations. E.g., for the N_2-Ar system simple Lennard-Jones potentials have been used, which for pure N_2 give a phase transition from the ordered Pa3 phase to the disordered plastic crystal phase at $T_c = 17.5$ K, while the experimental value is $T_c = 35$ K. Also, in principle one should take into account electrostatic interactions between "partial charges" residing on the nitrogen atoms, but the Ewald summations needed to treat such Coulomb interactions correctly would have slowed down the program by more than an order of magnitude, so this was not done either. (ii) MD means solving Newton's equations of motion, i.e. it is a purely classical approach that leaves out all the quantum effects needed for an accurate description of solids at the low temperatures of interest. (iii) The available time scales are extremely short: Lewis and

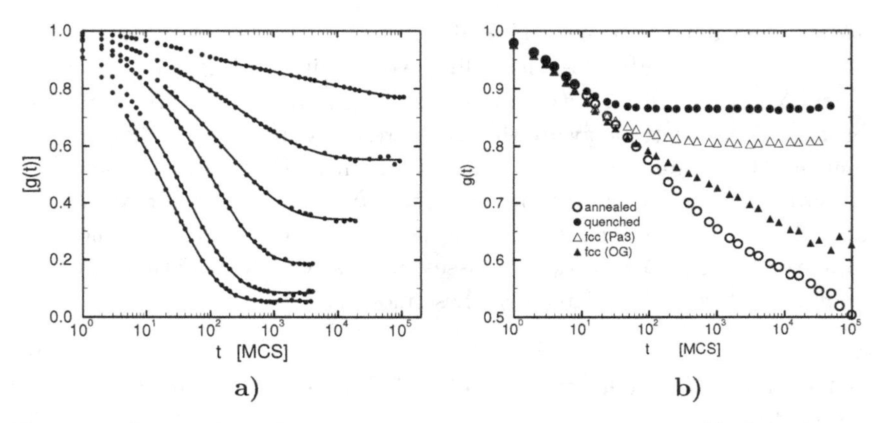

Fig. 4.42 a) Time-dependent Edwards-Anderson order parameter $q(t)$, here denoted by $g(t)$, for $Ar_{1-x}(N_2)_x$ for $x = 0.33$ and the temperatures T =5, 10, 15, 20, 30, and 40 K, from top to bottom. Dots are the MC data, lines are fits with $q(t) = q_{\mathrm{EA}} + (1 - q_{\mathrm{EA}})\exp[-(t/\tau)^y]$. From Müser and Nielaba (1995). b) Same as a), but for $x = 0.7$ at $T = 12$ K. Open circles show the case when translational degrees of freedom are "annealed" (i.e., are fully mobile), and full circles are for a system in which the translational degrees of freedom were frozen. The open triangle show the case in which the translational degrees of freedom are constrained to the ideal fcc lattice, with the orientations having initially the Pa3 symmetry, while full triangles refer to a fcc lattice for positions and orientations from an "annealed" state as initial configuration. From Müser (1996).

Klein (1986) used an MD integration time step of $\Delta t = 2 \cdot 10^{-15}$s, and the total observation time was only 7.10^{-12}s!

Here we will not review this MD work but instead focus on more recent Monte Carlo (MC) studies on the same system (Müser and Nielaba 1995; Müser 1996) which also suffer from the limitations (i) and (ii) but can proceed to somewhat longer time scales. Of course, MC does not give a realistic description of the physical dynamics since it captures the thermally activated transitions of the rotator molecules from one crystal field potential minimum to another one, but not the librational motions and other phonons. Regarding the reorientational dynamics, a comparison with an MD study (Müser and Ciccotti 1995) of an isolated N_2 molecule in an Ar matrix showed that one MC step corresponds approximately to 10^{-13}s of real time in the dilute limit. Typically MC runs were carried out over 10^6 Monte Carlo steps (MCS), allowing accurate estimations of $q(t)$ up to $t \approx 10^5$ MCS which thus corresponds to 10 nanoseconds.

Using the same potentials as Klee *et al.* (1988), data for $q(t)$ was obtained for this quadrupolar glass (Müser 1996), see Fig. 4.42. One can

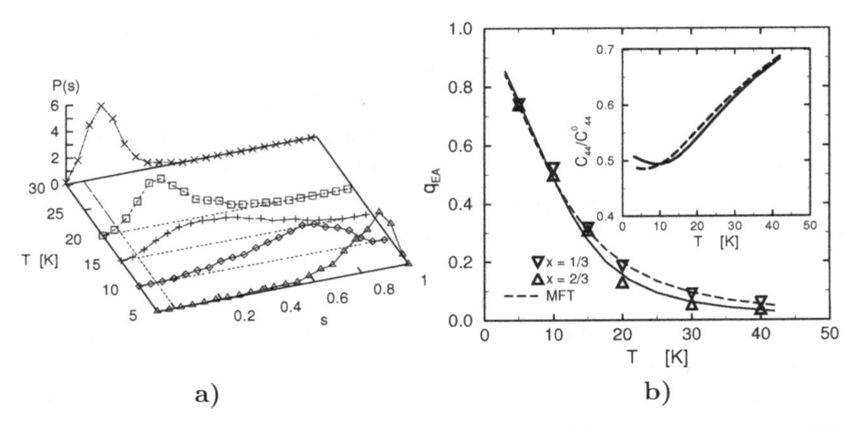

Fig. 4.43 a) Evolution of the distribution $P(s)$ of the order parameter $s = q_{\mathrm{EA}}$ with temperature, as deduced from a MC simulation of $\mathrm{Ar}_{1-x}(\mathrm{N}_2)_x$ for $x = 0.33$ (Müser and Nielaba 1995). b) Average order parameter q_{EA} at concentrations $x = 1/3$ and $x = 2/3$ for $\mathrm{Ar}_{1-x}(\mathrm{N}_2)_x$ plotted vs. temperature. Curves are fits according to a replica-symmetric mean field theory for a four state Potts glass in a random field h, fitting the width of the exchange distribution J as well as h ($x = 1/3 : J = 9.8$, $h = 0.9$; $x = 2/3 : J = 13.2$, $h = 0.28$). The inset shows the elastic constant $C_{44}(T)$ relative to the bare elastic constant C_{44}^0, which according to mean field theory is given by $C_{44}(T)/C_{44}^0 = [1+\bar{\gamma}(1-q_{\mathrm{EA}})/k_BT]^{-1}$, $\bar{\gamma}$ being a constant describing the rotation-translation coupling. From Müser 1996.

extract the temperature dependence of the static Edwards-Anderson order parameter, $q_{\mathrm{EA}} = \lim_{t\to\infty} q(t)$, and its distribution (Fig. 4.43), as well as the relaxation time and the KWW exponent $y(T)$. The latter decreases from about 0.6 at high temperatures to about 0.2 at low temperatures, similar to the behavior of the isotropic quadrupolar glass model (Fig. 4.41). The relaxation time is found to increase by several orders of magnitude when the temperature is lowered, but the statistical uncertainties do not allow to make any firm statements on the functional form of this temperature dependence.

Note that in these simulations only $N = 250 - 500$ fcc lattice sites were studied, and that the sample average consisted of a few independent runs only, whereas for the case of the isotropic quadrupolar glass, averages were done over 80 to 2000 samples of random bond configurations. This difference clearly demonstrates the computational advantages one has when simulating coarse grained lattice models instead of more realistic atomistic models.

Of course, the advantage of these simulations is not just the ability to "fit" experimental data, but to gain qualitative insight into the nature of the phenomena. E.g., one can check to what extent the rotation-translation coupling is important for the behavior of orientational glasses. In the framework of the present model, this question was studied by "quenching" the positions of the atoms or molecules either at their local equilibrium positions or at positions of an ideal rigid lattice. While the initial decay of $q(t)$ is not much affected, there is a pronounced effect at large times in that one finds that the quenching of the translational degrees of freedom stabilizes the static component of the orientational glass order parameter, Fig. 4.42b. Given the fact that the rotation-translation coupling acts like a local random field, one thus can conclude that annealed random fields have a much weaker effect on the ordering than quenched random fields. Therefore the problem of carrying out the correct statistical averages has to be carefully considered by the analytical theories.

Figure 4.43b shows in addition that the temperature dependence of the order parameter can indeed be fitted very well by an infinite range random bond-random field Potts model. This model would have a sharp glass transition at a nonzero temperature in zero random field (Elderfeld and Sherrington 1983a-c). Similar fits of real experimental data in terms of the random bond-random field Ising spin glass were done by Pirc *et al.* (1994) and Tadic *et al.* (1994). However, from this good agreement one should not conclude too quickly that mean field theory is a correct description of real quadrupolar glasses since also data for nearest-neighbor Potts glasses in a random field (Haas *et al.* 1996) show qualitatively the same behavior. Thus the smooth decay of $q_{\text{EA}}(T)$ with increasing temperature is a too unspecific information to discriminate between different theories.

The simulations of Müser and Nielaba (1995) also yielded much more detailed information, such as the local distribution of the order parameter, see Fig. 4.43. Again, these data are in very good qualitative agreement with corresponding information extracted from NMR experiments (Esteve *et al.* 1982). Also the temperature dependence of elastic constants (which, following Michel (1987a, 1987b), can be predicted from $q_{\text{EA}}(T)$ with a simple mean field argument) is in good qualitative accord with the corresponding experiments (Knorr 1987). From the study of these chemically realistic models of orientational glasses, we thus have evidence that a modeling in terms of Eq. (4.277) makes sense if one includes linear terms $h_i^{\mu\nu} f_i^{\mu\nu}$, that describe the effects of local random fields $\{h_i^{\mu\nu}\}$.

Furthermore we mention that also pure systems can show glassy dynamics. E.g. molecular crystals which exhibit a phase where molecules can freely rotate, the so-called "plastic crystals", may exhibit a slow relaxation similar to spin glasses and supercooled fluids, if their transition to a phase with long-range orientational order is avoided (Brand *et al.* 2002). Although there is no quenched disorder in these systems, and one has an underlying lattice structure, these systems show nevertheless a glass-like dynamics. Thus we see that plastic crystals are intesting systems to study how phonons couple to the (slowly relaxing) rotational degrees of freedom. In the case of *molecular* glass-forming liquids the situation is much more complex since i) each molecule has a different local environment and ii) this environment relaxes in a non-trivial way. Depending on the ratio between the timescales of the orientational and translational degrees of freedom, one can have various types of glass transitions (Letz *et al.* 2000, Chong and Götze 2002a,2002b, Chong *et al.* 2005, Chong 2008, Chong and Kob 2009, Zhang and Schweizer 2009). However, dwelling more on this subject would lead us too far away from the main focus of the book and therefore we refer the interested reader to the quoted references.

4.6.5 *Spin Models with Quenched Random Fields*

In this subsection we discuss the classical problem of the random field Ising model (RFIM, Imry and Ma 1975) and related systems which are generic models to study the effects of quenched disorder on phase transitions in solids, in which impurities exhibit a linear coupling to the order parameter. The simplest model to consider is a m-component ferromagnet with nearest neighbor exchange coupling J and exposed to a random field $\vec{H}_i$ which acts on every lattice site i and is completely uncorrelated with zero mean and variance h^2:

$$\mathcal{H} = -J \sum_{\langle i,j \rangle} \vec{S}_i \cdot \vec{S}_j - \sum_i \vec{H}_i \cdot \vec{S}, \quad \left[\vec{H}_i \cdot \vec{H}_j\right]_{\text{av}} = h^2 \delta_{ij} \,. \qquad (4.285)$$

Before discussing the statistical mechanics of this model, we briefly illustrate how this type of disorder can be physically realized by diluted antiferromagnets in a uniform magnetic field. The Hamiltonian of a non-diluted antiferromagnet in a uniform field is similar to Eq. (4.285), but now $\vec{H}_i = \vec{H}$ (i.e. independent of i) and $J < 0$. We then assume that we have a square (or simple cubic) lattice, which can be straightforwardly decomposed into two sublattices 1 and 2 that connect second nearest neighbor sites. One then transforms to new variables $\vec{S}_j^+$, $\vec{S}_j^-$ using the spins $\vec{S}_{j1}$,

$\vec{S}_{j2}$ from the two sublattices:

$$\vec{S}_j^+ = (\vec{S}_{j1} + \vec{S}_{j2})/\sqrt{2}, \quad \vec{S}_j^- = (\vec{S}_{j1} - \vec{S}_{j2})/\sqrt{2}\,. \tag{4.286}$$

From this definition we see that $\vec{S}_j^-$ is the local antiferromagnetic order parameter (the "staggered magnetization") while $\vec{S}_j^+$ yields the local magnetization.

Now one asks the question whether or not there can be a term in the Hamiltonian that provides a bilinear coupling between these two local order parameters:

$$\mathcal{H}_{\text{F-AF}} = -\sum_{\langle i,j\rangle} J_{ij}^*(\vec{S}_i^+)_z(\vec{S}_j^-)_z\,. \tag{4.287}$$

It is evident that, for symmetry reasons, in an ideal perfect lattice Eq. (4.287) must be identically zero since the Hamiltonian of an antiferromagnet must be invariant against an interchange of sublattices 1 and 2, because the labelling of the sublattices is arbitrary. Consequently we must have $J_{ij}^* \equiv 0$, since $\vec{S}_j^-$ changes sign when we interchange the sublattices while $\vec{S}_i^+$ does not, and hence $\mathcal{H}_{\text{F-AF}}$ would change sign if J_{ij}^* is non-zero.

In contrast to this in a diluted system the symmetry against the interchange of sublattices is locally broken, because sometimes a site on sublattice 1 is taken by a nonmagnetic impurity, and sometimes a site on sublattice 2. Thus we can have $J_{ij}^* \neq 0$, although this symmetry must of course still be present for the configurational average, i.e. $[J_{ij}^*]_{\text{av}} = 0$, because the diluted antiferromagnet still is invariant against a change of sign of the total staggered magnetization.

Now if a uniform magnetic field H_z is present it will create a static magnetization, $\langle(\vec{S}_i^+)_z\rangle_T$, and hence $\mathcal{H}_{\text{F-AF}}$ will act like a random staggered field on the antiferromagnetic order parameter

$$\mathcal{H}_{\text{F-AF}} = -\sum_{\langle i,j\rangle} J_{ij}^*\langle(\vec{S}_i^+)_z\rangle_T(\vec{S}_j^-)_z - \ldots = -\sum_j h_j^s(\vec{S}_j^-)_z - \ldots\,, \tag{4.288}$$

where the random staggered field $h_j^s = \sum_{i(\neq j)} J_{ij}^*\langle(\vec{S}_i^+)_z\rangle_T$ is a quenched random variable, since due to the presence of quenched nonergodic impurities some sites i do not contribute in this sum, and also the number of nonmagnetic neighbors of a site j fluctuates.[4]

[4]Note that in Eq. (4.288) the "..." stands for the thermally fluctuating part $(\vec{S}_i^+)_z - \langle(\vec{S}_i^+)_z\rangle$, since this term can be ignored here because it describes "harmless" annealed disorder.

Depending on the anisotropy of the original antiferromagnetic system, both the case of the RFIM (one spin component, $m = 1$), the cases of XY model ($m = 2$), and the Heisenberg model ($m = 3$) in random fields can be realized. Hence, there have been extensive experimental studies of suitable magnetic model systems (Belanger 1998), including also crystals where magnetic interactions are restricted to two-dimensional planes (Ferreira *et al.* 1983). So both random field systems in $d = 3$ and $d = 2$ have been studied. A further experimental realization of the case $d = 2$ is provided by adsorbed mono-layers, that form "registered superstructures" (i.e. orderings that are commensurate with the corrugation of the substrate crystalline lattice). Using the well known mapping between lattice gas models and Ising ferromagnets, one recognizes that also such systems correspond to Ising antiferromagnets in a field (that relates to the chemical potential of the adsorbate) as well. If some of the preferred sites on which particles could be adsorbed are blocked by irreversibly chemisorbed impurities, one again effectively creates a random field. For example, carbon monoxide (CO) adsorbed on graphite exhibits an Ising-type transition (head-tail ordering of these uniaxial molecules) at $T_c = 5.18\mathrm{K}$ (Wiechert and Krömker 2002). Dilution with Ar or N_2 causes a rounding of the transition, which can be quantitatively accounted for with the theory of the RFIM in $d = 2$ dimensions (Wiechert and Krömker 2002). Of course, both in this example and in Eq. (4.288), it is assumed that the real situation can be idealized by letting an independent random field of suitable strength act on every site of the lattice, Eq. (4.285), i.e. one invokes again a kind of "universality hypothesis", as is standard in all phase transition problems (Fisher 1974), which means that such details of a model are irrelevant as far as qualitative statements (existence of phase transitions, classification of critical behavior, etc.) are concerned.

We first focus on the case of the RFIM. Figure 4.44 shows the phase diagram obtained by Aharony (1978) from a molecular field calculation for a $\pm h$ distribution. The figure refers to $d = 2$, although the molecular field theory yields qualitatively the same phase diagram in all dimensions. (Note that in the general case the phase boundary between the ferromagnetic and the paramagnetic state ends at the ordinate at $h_c = dJ$.) One should also note that the RFIM exhibits a nonzero Edwards-Anderson order parameter $q_{\mathrm{EA}} = [\langle S_i \rangle_T^2]_{\mathrm{av}}$ at all temperatures $T < \infty$, since the random fields h_i have some alignment effect on the spins at the site i in the direction of the fields. This observation is of course no surprise, since from the theory of spin glasses (Binder and Young 1986) it is well known that $[h_i^2]_{\mathrm{av}}$ can be

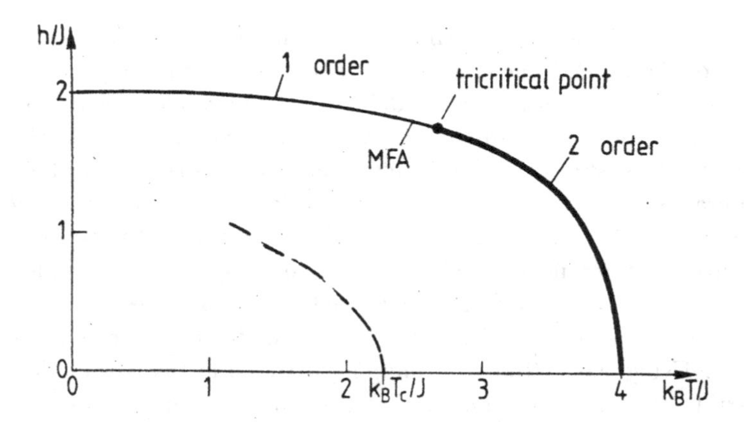

Fig. 4.44 Phase diagram of the nearest neighbor Ising model on the square lattice exposed to a random field with distribution $P(h) = [\delta(h_i - h) + \delta(h_i + h)]/2$. The full curve is the molecular field calculation due to Aharony (1978), which yields a second order line up to a tricritical point and then continues up to the point $T = 0$, $h_c = 2J$ as a first order transition. The broken curve shows the (rounded) transition located from specific heat maxima obtained from transfer matrix calculations. From Morgenstern *et al.* (1981).

considered as the "conjugate field" to the order parameter q.

Now it turns out that at best the molecular field theory describes the behavior of the RFIM qualitatively correctly only for $d > 2$, but not at $d = 2$, since this is the lower critical dimensionality for the RFIM. This implies, that arbitrarily small random fields are sufficient to disrupt the long range order and the system is broken up in a random arrangement of (large) domains (Fig. 4.45). The larger the strength of the random field becomes, the smaller the typical linear dimension of these domains will be, and the more the singularity of the specific heat C of the pure system ibecomes smeared out {remember that $C = A \ln |1 - T/T_c| + B$, where A,B are constants and T_c is the critical temperature of the pure system (Onsager 1944)} (Fig. 4.46). Very similar experimental data have been found for the quasi-two-dimensional diluted antiferromagnet $Rb_2Co_{0.85}Mg_{0.15}F_4$ in a uniform magnetic field (Fig. 4.47), which, as discussed above, should be a physical realization of a $d = 2$ RFIM (Ferreira *et al.* 1983).

The instability of long range ferromagnetic order against weak random fields was first discussed in a seminal paper by Imry and Ma (1975). They discussed the excess of one sign of the random field $\pm h$ in a compact domain of volume L^d having all its linear dimensions of order L. They suggested

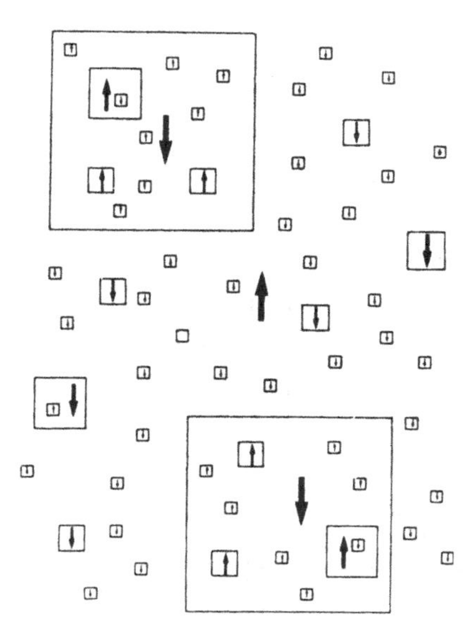

Fig. 4.45 Schematic description of the state of a two-dimensional Ising model in a weak random field at low temperatures. Arrows show the direction of the magnetization in the respective domains. The random field induced roughness of the domain walls is ignored in this sketch. From Morgenstern *et al.* (1981).

that this excess should be distributed according to a gaussian and should be of order of $L^{d/2}$, the square root of the considered volume. Depending on the sign of the excess, the system eventually could lower its (free) energy by the Zeeman energy $hL^{d/2}$ if it overturns the magnetization in the volume L^d, but it would have to pay a (free) energy due to the creation of domain walls of the order of JL^{d-1}. Imry and Ma (1975) argued that formation of domains in parts of the system where the random excess has a suitable sign will always be favorable if $d/2 > d - 1$, since then for large enough L the Zeeman energy gain will always outweigh the domain wall energy cost, irrespective of how small h is. This argument hence yields an instability of the long range ferromagnetic order of the RFIM for all $d < 2$, a result which is remarkable because it is known that for pure Ising ferromagnets without a random field an analogous instability against spontaneous formation of domains occurs only in $d = 1$.

Actually this simple argument does not yet clarify what happens exactly in the RFIM for $d = 2$. The situation can be clarified if one considers the domain walls between coexisting regions of positive and negative

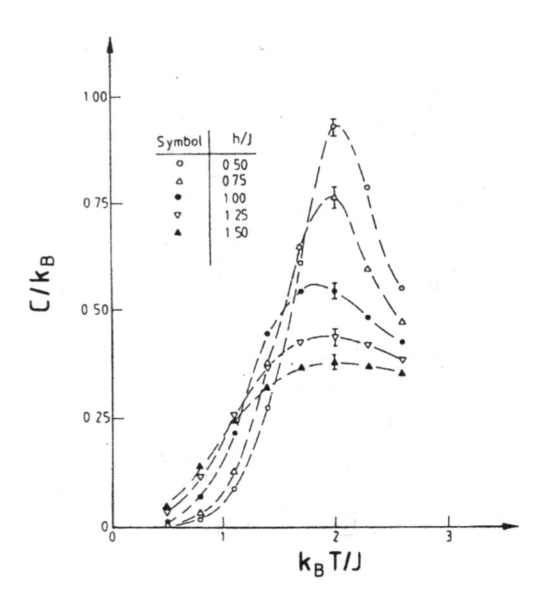

Fig. 4.46 Specific heat of the two-dimensional nearest Ising model plotted versus temperature for various strengths h of the random field, as indicated. Data are from exact transfer matrix calculations for finite lattices (large enough so that finite size rounding is still negligible against the random-field induced rounding). The statistical errors shown are due to the sampling of a finite number (at most a few hundred were used) of random field configurations. From Morgenstern *et al.* (1981).

magnetization and asks how the structure of the walls and their free energy cost is affected by the random field (Grinstein and Ma 1982, 1983, Binder 1983, Grinstein 1984, Nattermann 1984, 1985a, 1985b, Villain 1985, Nattermann and Villain 1988, Moore *et al.* 1996). For this one can use a generalization of the Imry-Ma argument and see whether or not it is favorable to displace the interface on a length scale a forward or backward by a distance $b(a)$, in view of the possible gain of random field Zeeman energy (Binder 1983). It turns out that one must consider the "decoration" of the interface by such random excursions on all scales starting from some minimum scale $a_{\min}$ up to the maximum scale L which is given either by the size of the system or the condition that on this scale the total energy gain compensates the domain wall (free) energy, as shown qualitatively in Fig. 4.48. This procedure can be done both on the lattice (Binder 1983) and in the continuum (Nattermann 1984, 1985a, 1985b, Moore *et al.* 1996).

Adapting the notation of Moore *et al.* (1996), we introduce four exponents κ, ζ, Θ and Γ to describe the interfacial roughening and the associated

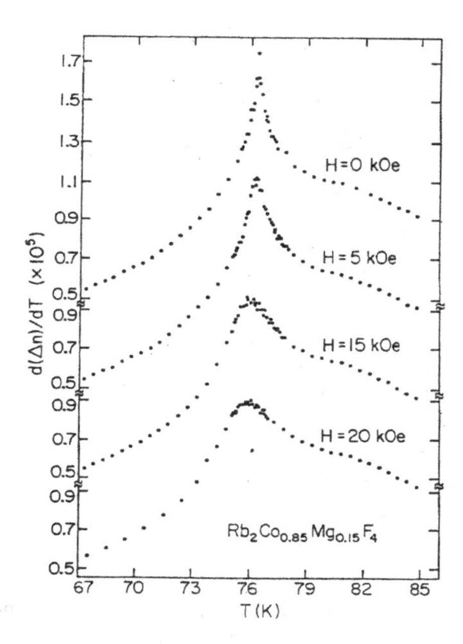

Fig. 4.47 Specific heat of the diluted quasi-two-dimensional Ising-like antiferromagnet $Rb_2Co_{0.85}Mg_{0.15}F_4$ plotted vs. temperature for several choices of the uniform magnetic field. From Ferreira *et al.* (1983).

free energy gain $\delta\mathcal{F}$ due to the random field h:

$$b \propto h^{\kappa} a^{\zeta}, \quad \delta\mathcal{F} \propto h^{\Gamma} a^{\Theta}. \tag{4.289}$$

The exponent ζ is often referred to as the "wandering exponent". One finds for these exponents on the square lattice (Binder 1983, Moore *et al.* 1996)

$$\zeta = 3 - d, \quad \kappa = 2, \quad \Gamma = 2, \quad \Theta = 3 - d, \quad d \leq 3. \tag{4.290}$$

Hence, we see that arbitrary small random fields lead at low temperatures to a roughening of interfaces for $d \leq 3$, while for $h = 0$ in $d = 3$ interfaces are non-rough in the Ising model at low temperatures. In $d = 2$, $\zeta = 1$, which means b/a is independent of a, irrespective of how large a is, which shows again that $d = 2$ is the lower critical dimension. In the continuum, the corresponding results are (Moore *et al.* 1996)

$$\zeta = (5 - d)/3, \quad \kappa = 2/3, \quad \Gamma = \frac{4}{3}, \quad \Theta = (d + 1)/3, \quad d \geq 2. \tag{4.291}$$

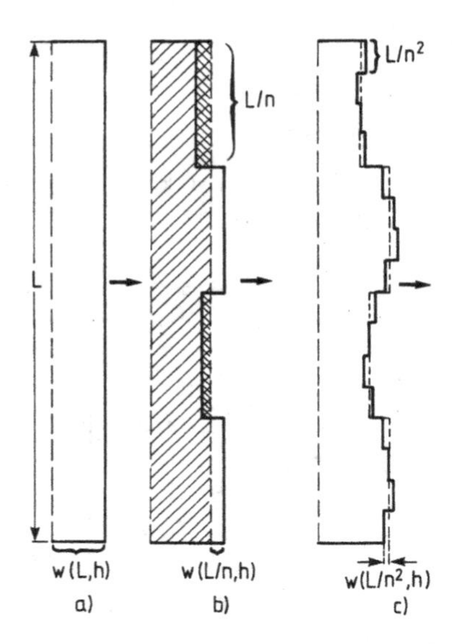

Fig. 4.48 Schematic illustration of how an interface is "decorated" by excursions of the interface backward and forward, occurring on all lateral length scales and performed such that the system gains Zeeman energy from the random field due to these interfacial excursions. From Binder (1983).

At this point it is worth to note that a field-theoretic approach based on a replica-symmetric approach gave the incorrect result $\zeta = (5-d)/2$ (Parisi and Sourlas 1979), while the inclusion of replica symmetry breaking (Mézard and Parisi 1990) gives the same value of ζ as quoted in Eq. (4.291). To obtain their result Parisi and Sourlas (1979) applied the simple rule that a d-dimensional Ising system in a random field has the same critical behavior as a pure system in $d-2$ dimensions, which would have implied that $d = 3$ is the lower critical dimension of the RFIM. This conclusion was proved to be wrong by rigorous arguments (Imbrie 1984) showing the existence of a spontaneous magnetization in $d = 3$ in weak random fields at sufficiently low temperatures.

Summing up the free energy changes $\delta\mathcal{F}$ on all scales up to a maximum scale L one finds that in $d = 2$ there exists a scale $\xi = L$ at which the free energy gains compensate the "bare" interfacial (free) energy, which is of order LJ. This correlation length ξ is (Binder 1983, Nattermann 1984, 1985a, 1985b, Moore *et al.* 1996)

$$\xi \propto \exp[\text{const} \cdot (h/J)^{-\Gamma}] \,, \tag{4.292}$$

where, see Eqs. (4.290) and (4.291), we have $\Gamma = 2$ for the lattice while $\Gamma = 4/3$ in the continuum. Note however, that at $T > 0$ and $h \to 0$ one always has a crossover from the lattice result to the continuum result due to thermal fluctuations (Binder 1983, Nattermann 1984, 1985a, 1985b, Moore *et al.* 1996). The scale ξ hence is the minimum length scale on which domains as sketched in Fig. 4.45 must appear spontaneously.

In this section we have put much emphasis on the description of the behavior at the lower critical dimension, because the RFIM is the rare case of a model where randomness can indeed destroy standard long range order, stabilizing states which are locally ordered (Fig. 4.45). We have seen that domains with rough walls are a key concept to understand the behavior. Therefore, the consideration of domain-like excitations has also found a lot of interest in other systems (e.g. the "droplet model" of Fisher and Huse (1988) which attempts to describe the low temperature phase of spin glasses with short range interactions, or the approach of Kirkpatrick and Wolynes (1987) who emphasized the importance of the interfaces between local minimas in the free energy for describing structural glass-formers). However, neither in spin glasses nor in structural glasses has one succeeded to derive a conclusive argument of this kind to clarify what the lower critical dimension for a phase transition to the glassy phase at nonzero temperature should be.

On the other hand, the Imry-Ma (1975) domain argument can easily be carried over to isotropic magnets. One argues (Imry and Ma 1975) that the energy associated with a wall on a scale L is not JL^{d-1} but rather JL^{d-2} in d dimensions: This happens because in an isotropic system it is energetically favorable to spread out the variation from $+M$ to $-M$ over the whole scale L, so only magnetization gradients of order $1/L$ arise. Hence if one uses a Ginzburg-Landau-Wilson Hamiltonian, the term $\int d\vec{x}(\nabla \vec{S})^2$ contributes with an energy of order $L^d(1/L)^2 = L^{d-2}$. Requiring now that $hL^{d/2} > JL^{d-2}$ for large enough L, irrespective of how small h is, yields the condition $d < 4$ and thus one concludes that for isotropic ferromagnets in random fields the lower critical dimension is $d_\ell = 4$. Thus, isotropic spin systems in $d = 3$ exposed to random fields are systems below their lower critical dimension, and thus one expects phase transitions to occur only at $T = 0$, while the RFIM in $d = 3$ is expected to show ferromagnetic long range order up to some nonzero critical temperature $T_c > 0$. Note that here we have assumed that the RFIM exhibits a second order transition from the paramagnetic to the ferromagnetic state, which is an assumption that also has been questioned (alternatives would be that the RFIM always exhibits a first order transition, or that a spin-glass-like phase intervenes in between the ferro- and paramagnetic phases). A renormalization group treatment of the Ginzburg-Landau-Wilson Hamiltonian subject to a weak gaussian random field shows that the marginal dimension d^*, above which mean field critical exponents apply, gets shifted from the value $d^* = 4$ for the pure system to the value $d^* = 6$ for the random field system (Nattermann 1998). This behavior is fundamentally different from the critical behavior of systems with randomly quenched disorder that couples to the *square* of the order parameter only, rather than *linearly*: In that case $d^* = 4$ remains true also for the "impure" system, although for $d < d^*$ the critical exponents get altered. This weaker effect of the disorder occurs, e.g., for a randomly diluted ferromagnet (above the percolation threshold, of course) or for diluted antiferromagnets in the absence of magnetic fields. In the Ginzburg-Landau-Wilson Hamiltonian, $\mathcal{H}_{\text{GLW}}(\vec{S}) = \int d\vec{x}[(\nabla\vec{S})^2 + r\vec{S}^2 + u(\vec{S}^2)^2]$, this disorder is described by giving the coefficient r a random component, $r = \bar{r} + \delta r(\vec{x})$ with $\delta r(\vec{x})$ being gaussian distributed and uncorrelated, $[\delta r(\vec{x})\,\delta r(\vec{x}')]_{\text{av}} = \Delta^2\delta(\vec{x} - \vec{x}')$. In contrast, the random field type disorder $h(\vec{x})S^z(\vec{x})$ with $[h(\vec{x})h(\vec{x}')]_{\text{av}} = h^2\delta(\vec{x} - \vec{x}')$ which is the subject of the present subsection has a very drastic effect on the phase transition in $d = 3$, and since this physical dimensionality now is much closer to the lower critical dimensionality $d_\ell = 2$ rather than the marginal dimensionality $d^* = 6$, one learns little from the application of the renormalization group expansions concerning the actual critical behavior of the RFIM. The most detailed information on the phase transition of the RFIM comes from numerical work such as Monte Carlo calculations (e.g. Rieger 1995) and extrapolations of high temperature series expansions (e.g. Gofman *et al.* 1996).

While rigorous results on the order of the phase transition of the RFIM in weak random fields in $d = 3$ are still lacking, there has been growing consensus from numerical studies (e.g. Newman and Barkema 1996, Vink *et al.* 2006, 2008) that the transition is of second order. If it is of second order, it must have very unconventional critical behavior (Nattermann 1998): The key point is that the standard hyperscaling relation between critical exponents (cf. Eqs. (3.79) and (4.239), for instance)

$$2 - \alpha = \gamma + 2\beta = \nu d \tag{4.293}$$

does not hold, but is replaced by (Villain 1985, Fisher 1986)

$$2 - \alpha = \gamma + 2\beta = \nu(d - \theta) \tag{4.294}$$

where the "violation of hyperscaling exponent" θ is believed to be given by (Schwartz 1985)

$$\theta = \frac{\gamma}{\nu} = 2 - \eta \,. \tag{4.295}$$

Here $\alpha, \beta, \gamma, \nu$ have the standard meaning as critical exponents of the specific heat, magnetization, susceptibility and correlation length ξ. In addition, critical slowing down no longer is described by the usual power law for the relaxation time τ, $\tau \propto \xi^z$ {Hohenberg and Halperin (1977); cf. also Eq. (4.230)}, but rather there occurs "thermally activated critical slowing down", namely (Villain 1985; Fisher 1986)

$$\ln \tau \propto \xi^{\theta} \propto (T/T_c - 1)^{-\gamma} \,. \tag{4.296}$$

Apart from the presence of an exponent γ ($\gamma > 1$ beyond mean field theory), Eq. (4.296) is reminiscent of the well-known Vogel-Fulcher-Tammann law (which corresponds to $\gamma = 1$ in Eq. (4.296)) of the structural glass transition, see Chaps. 5 and 6.

Since relations of the type of Eq. (4.296) with exponents γ different from unity are discussed in various contexts, e.g. the so-called "mosaic picture" of the structural glass transition (Kirkpatrick and Wolynes 1987), we briefly describe in the following the physical interpretation of Eqs. (4.294)-(4.296) in more detail. First, we recall that the relation $2 - \alpha = \gamma + 2\beta$ simply follows from the fact that near criticality the singular part of the free energy F_{sing} depends on the two variables H and $t = T/T_c - 1$ in the scaled form

$$F_{\text{sing}}(t, H) = |t|^{2-\alpha} \tilde{F}(H/|t|^{\beta+\gamma}) \tag{4.297}$$

where the scaling function $\tilde{F}(\varsigma)$ for $t > 0$ behaves like $\tilde{F}(0) + \text{const} \cdot \zeta^2 + \cdots$. Thus $\chi = -(\partial^2 F_{\text{sing}}/\partial H^2)|_t \propto |t|^{2-\alpha-2\beta-2\gamma}$ and since $\chi \propto |t|^{-\gamma}$, the thermodynamic scaling relation $2 - \alpha = \gamma + 2\beta$ follows. While the hyperscaling relation equating $2 - \alpha$ to νd can be rather generally justified by the renormalization group theory of critical phenomena (Domb and Green 1977), we quote here a "handwaving" plausibility argument only: Near T_c at $H = 0$ the singular part of the free energy $F_{\text{sing}}(t, 0)$ can be attributed to correlated regions of spins, of linear dimension ξ. Each such region has essentially one Ising degree of freedom (magnetization direction up or down, and can orient independently from its neighbors). Thus, while for $T \to \infty$ the free energy of the system is due to entropy of N non-interacting spins, $F = -k_B T N \ln 2$, near T_c we can attribute F_{sing} to the entropy of the N/ξ^d

independent clusters of spins which have the typical size ξ,

$$F_{\text{sing}}(t,0)/N \approx -\xi^{-d}k_B T \ln 2 \propto |t|^{2-\alpha}\,. \tag{4.298}$$

where in the last step we have used Eq. (4.297). Since $\xi \propto |t|^{-\nu}$, hyperscaling $d\nu = 2-\alpha$ follows.

However, for an Ising system in a random field $\pm h$ near T_c the situation is different: While we still can split the system into regions of linear dimension ξ, such that each region can be considered as essentially independent of its neighbors, the main contribution to F_{sing} now is not the entropy but the Zeeman energy due to the coupling to the random field. More precisely, since in each region of volume ξ^d there is an excess of one sign of the random field, on average a random field of size $H_r = \pm h\xi^{-d/2}$ acts on each spin of such a region (ignoring prefactors of order unity throughout). Thus a magnetization $m = \chi H_r = \pm\chi h \xi^{-d/2}$ per spin results, and this leads to the Zeeman energy

$$F_{\text{sing}}(t,0)/N = -mH_r = -h^2\xi^{-d}\chi \propto \xi^{\gamma/\nu-d}\,, \tag{4.299}$$

where in the last step we have used $\chi \propto t^{\gamma} \propto \xi^{\gamma/\nu}$. Thus a scaling relation $2-\alpha = d\nu - \gamma$ follows, in agreement with Eqs. (4.294) and (4.295).

Next we consider the interfacial excess free energy, f_{int}, which is known to vanish as T approaches T_c from below as

$$f_{\text{int}} \propto (-t)^{\mu} = (-t)^{(d-1)\nu} = (-t)^{2-\alpha-\nu}\,. \tag{4.300}$$

To understand the first equality one can argue that $F_{\text{sing}}(t,0)$ is a free energy per volume but f_{int} is a free energy per area, and since near T_c there is only one relevant length scale in the problem, we must have $F_{\text{sing}}(t,0) \propto \xi^{-d}$, and thus $f_{\text{int}} \propto F_{\text{sing}}\xi \propto \xi^{-(d-1)}$. Again, we stress that Eq. (4.300) can be derived by renormalization group methods in a much more rigorous way, but here we wanted to give only a plausibility argument. The important point of the present discussion is, however, that the relation $\mu = (d-1)\nu$ also is violated in the RFIM, and Eqs. (4.293), (4.294), and (4.299) imply rather

$$\mu = (d-1)\nu - \gamma = (d-3+\eta)\nu\,, \tag{4.301}$$

since the relation $f_{\text{int}} \propto F_{\text{sing}}\xi$ should still hold.

Clearly, the violation of the hyperscaling relation expressed in Eqs. (4.294) and (4.301), is surprising — other disordered magnets, such as ferromagnets which are randomly diluted with nonmagnetic atoms (Stinchcombe 1983, Berche *et al.* 2004) or spin glasses (Sec.4.5) do satisfy standard hyperscaling, i.e. Eqs. (4.293) and (4.300), like the pure systems. This fact holds even though system with randomly quenched disorder at criticality exhibit lack of self-averaging (Wiseman and Domany 1995, 1998). Hence one wonders what feature is special about random field criticality?

The basic insight that explains why random-field systems are different comes from the existence of a second kind of "susceptibility", the so-called "disconnected susceptibility" χ_{dis}. If we define (for a particular realization of the random field configuration) the thermal average of the order parameter $\langle m \rangle_T = L^{-d}\langle \sum S_i \rangle$, for a system of volume L^d, it will in general be of order $\pm L^{-d/2}$, i.e. nonzero even in the paramagnetic region, due to the excess H_r of the random field in this realization in the considered volume. Thus, while $M = [\langle m \rangle_T]_{\text{av}} = 0$, we have a non-trivial disconnected susceptibility

$$\chi_{\text{dis}} \equiv L^d [\langle m \rangle_T^2]_{\text{av}} > 0 \tag{4.302}$$

in addition to the standard ("connected") susceptibility, $\chi = L^d[\langle m^2 \rangle_T - \langle m \rangle_T^2]_{\text{av}}$. In fact, estimating as above that $\langle m \rangle_T \approx \pm \chi H_r \propto \pm \chi L^{-d/2}$ we find that $\chi_{\text{dis}} \propto \chi^2$. Hence for temperatures T near the critical temperature T_c the disconnected susceptibility diverges more strongly:

$$\chi_{\text{dis}} \propto (T/T_c - 1)^{-\bar{\gamma}}, \quad \bar{\gamma} = 2\gamma\,. \tag{4.303}$$

While for pure systems near criticality the combination $X = M^2 \xi^d / \chi$ is a universal constant (i.e., a universal combination of critical amplitudes, since the temperature dependence cancels out, $2\beta - d\nu + \gamma = 0$) for the RFIM a corresponding combination is $X_r = M^2 \xi^d / \chi_{\text{dis}}$, since it is $\bar{\gamma}$ rather than γ which satisfies the standard hyperscaling relation,

$$\bar{\gamma} + 2\beta = d\nu\,. \tag{4.304}$$

This result is in fact plausible if one considers a finite size scaling context, as is appropriate for the analysis of computer simulation studies (Eichhorn and Binder 1996a, 1996b, Vink *et al.* 2006, 2008). At the critical point, we expect that in a finite system χ saturates at a value depending on system size as $\chi \propto L^{\gamma/\nu}$, and hence $\langle m \rangle_T \propto \pm H_r \chi \propto \pm L^{-d/2} L^{\gamma/\nu} \propto L^{-\beta/\nu}$, implying that

$$\gamma + \beta = d\nu/2 \tag{4.305}$$

which yields Eq. (4.304), if the above result $\bar{\gamma} = 2\gamma$ is used. A consequence of this result is that the two peaks of the order parameter distribution $P_L(m)$ at criticality occur at positions near $\pm[\langle|m|\rangle_{T_c}]_{\text{av}} \propto \pm L^{-\beta/\nu}$. The width of these peaks, however, scales with a different exponent, namely

$$\Delta m \propto L^{-(d-\gamma/\nu)/2} . \tag{4.306}$$

For a pure system we have $d - \gamma/\nu = 2\beta/\nu$, and therefore at criticality there is a universal, nontrivial shape of the order parameter distribution. As a consequence, moment ratios such as $[\langle m^4\rangle_T]_{\text{av}}/[\langle m^2\rangle_T]^2_{\text{av}}$ for pure systems take nontrivial, size independent values, useful for the unbiased estimation of T_c (Binder 1992), while for the RFIM such moment ratios tend to the trivial value unity (Eichhorn and Binder 1996a, 1996b, Vink *et al.* 2006, 2008).

Since for a large RFIM the distribution $P_L(m)$ resembles a distribution of two narrow Gaussians located at $\pm[\langle|m|\rangle_{T_c}]_{\text{av}} \propto \pm L^{-\beta/\nu}$, it is of interest to consider the free energy barrier that one has to cross if the system changes the sign of its magnetization. Similarly as for pure systems for $T < T_c$, the states at the barrier, near $m \approx 0$, can be described as a slab-like state, where a domain with state $+[\langle|m|\rangle_{T_c}]_{\text{av}}$ is separated by two (rough) domain walls from a domain with state $-[\langle|m|\rangle_{T_c}]_{\text{av}}$ (both domains are connected into themselves via the periodic boundary conditions, so the domain walls are planar, on average). Thus we can write

$$\ln P_L(m \approx 0) \approx 2L^{d-1} f_{\text{int}}(T_c) \propto 2L^{d-1}L^{-(d-3+\eta)} \propto L^{2-\eta} = L^{\theta} , \tag{4.307}$$

where the exponent θ is defined by the last equality. Here we have used the fact that $f_{\text{int}}(T) \propto \xi^{-(d-3+\eta)}$ in the RFIM, see Eqs. (4.300) and (4.301), and the finite size scaling principle that "L scales with ξ" means that the critical vanishing of $f_{\text{int}}(T)$ at $T = T_c$ is rounded off in a behavior $f_{\text{int}}(T_c) \propto L^{-(d-3+\eta)}$. While in a pure system, where $f_{\text{int}}(T) \propto \xi^{-(d-1)}$ cf. Eq. (4.300), we find that $f_{\text{int}}(T_c) \propto L^{-(d-1)}$ and hence $\ln P_L(m \approx 0)$ is of order unity, i.e. the system changes the sign of the magnetization without crossing any free barriers, for the RFIM at criticality magnetization reversals in a finite system require crossing of free energy energy barriers of order $\Delta F \propto L^{\theta}$.

Since we expect that $P_L(m)$ satisfies finite size scaling, i.e. temperature dependencies enter via an argument L/ξ, it is clear that for $T > T_c$ we still expect free energy barriers $\Delta F \propto \xi^{\theta}$ rather than L^{θ}, and thus we understand qualitatively, where Eq. (4.296) comes from: due to the action of the random field, even for $T > T_c$ we still have two rather well separated states, domains of size ξ have a typical magnetization $\pm h\chi\xi^{-d/2} \propto \pm h\xi^{\gamma/\nu-d/2}$ and to reverse their magnetization a domain wall of area ξ^{d-1} must be formed, involving a free energy cost $\Delta F \propto \xi^{d-1} f_{\text{int}}(T) = \xi^{d-1}\xi^{-d+3-\eta} = \xi^{2-\eta} = \xi^{\theta}$. Since we expect that such free energy barriers give rise to an Arrhenius law for the relaxation time, $\tau \propto \exp(\Delta F/k_BT)$, Eq. (4.296) results.

Evidence for the fact that the distribution $P_L(m)$ of the RFIM at criticality has two widely separated peaks, with a deep minimum in between, has recently been obtained for a related model, the Asakura-Oosawa model for a colloid-polymer mixture in a random medium (Vink *et al.* 2006, 2008). Already de Gennes (1984) pointed out that a binary (AB) mixture, which exhibits a unmixing transition into A-rich and B-rich phases in the bulk, when it is brought into a gel or another porous medium, the gel (or the pore walls, respectively) will show a preferential attraction for one of the components. Due to the quenched random structure of the gel (or porous medium), the problem can be "translated" to an off-lattice version of the RFIM. The same argument applies to vapor-liquid transitions in porous media. While early experiments studying such transitions in silica aerogels (Wong and Chan 1990, Wong *et al.* 1993) and simulations (Alvarez *et al.* 1999, Sarkisov and Monson 2000, De Grandis *et al.* 2004) failed to verify the RFIM character of these systems, recent simulations of the Asakura-Oosawa model succeeded to verify it in detail (Vink *et al.* 2006, 2008). In this work, the random medium was modelled simply by freezing the small fraction of the colloid particles in random positions.

Such phase separation phenomena in disordered systems exhibit also other unconventional properties, that are typical for systems with complex free energy landscapes (Kierlik *et al.* 2002). Particularly intruiging is the out-of-equilibrium dynamics of the RFIM at zero temperature (Liu and Dahmen (2009) and references therein). Using the kinetic Ising model with single spin-flip dynamics, the variation of the magnetization upon changing the magnetic field has been studied. One finds "avalanches" of widely varying sizes. Here the term "avalanche" stands for a flip of a group of neighboring spins during the magnetization process, leading to a jump singularity in the curve $M(H)$ vs. H. These "avalanches" are enclosed by contours with fractal dimensionalities (Liu and Dahmen 2007, 2009; Perez-Reche and Vives 2004), and their size and surface area distributions exhibit interesting scaling properties. The physical picture behind these singularities is that the domain walls (which are rough for $d < 5$, see also Fig. 4.48) are pinned in energetically favorable configurations; one such favorable configuration is separated from the next one by an energy barrier (the qualitative picture corresponds to Fig. 1.16, upper part). Only when the change in Zeeman energy exceeds the barrier can the domain wall jump to the next position, involving the "avalanche" of overturned spins mentioned above.

Very intriguing questions arise also if we consider models that undergo a first order transition (in the absence of any quenched disorder), and add to such a model a weak random field. E.g., such a case can be realized by the three-state Potts ferromagnet in a random field h_i (Eichhorn and Binder 1996a, 1996b):

$$\mathcal{H} = -J \sum_{\langle i,j \rangle} (\vec{S}_i \cdot \vec{S}_j) - \sum h_i (\vec{u}^{(1)} \cdot \vec{S}_i) , \qquad (4.308)$$

where (according to the simplex representation of the Potts model, Zia and Wallace 1975 and Wu 1982) the Potts spins $\vec{S}_i$ can take one of the three discrete vectors $\vec{u}^{(1)} = (1, 0)$, $\vec{u}^{(2)} = (-1/2, \sqrt{3}/2)$ and $\vec{u}^{(3)} = (-\frac{1}{2}, -\sqrt{3}/2)$. Recall that the 3-state Potts model can qualitatively represent a cubic molecular crystal where parallel orientation of the molecules along one of the cubic axes is preferred, see Sec. 4.6.3. Equation (4.308) can hence be useful as a coarse-grained model for an orientational glass, if one assumes that random field disorder rather than random bond disorder dominates (Binder 1998). Now it is well known that the model given by Eq. (4.308) for zero random fields has a first order transition from the "ferromagnetic" to the disordered phase. Eichhorn and Binder (1996a, 1996b) presented various theoretical arguments and Monte Carlo evidence that with increasing strength of the random field the first order character of the transition of this model becomes weaker, until one reaches a tricritical point and there is a region of field strengths where the random field Potts model actually undergoes a second order transition. Even more exciting is the case of Potts models with $p > 4$ states in $d = 2$ dimensions: While in the pure case it is known exactly (Wu 1982) that these models undergo first order transitions, arbitrarily weak disorder suffices to turn these first order transitions in $d = 2$ into second order transitions (Aizenman and Wehr 1989).

An interesting case of a physical system undergoing a first order transition are fluids of very elongated molecules that show an isotropic-nematic transition. In porous media such fluids are exposed to a random field coupling linearly to the tensor order parameter describing the nematic order (Maritan *et al.* 1994, Khasanov 2005) and a glass-like phase seems to occur (Wu *et al.* 1992).

As a last variation on this theme, we mention the random axis model (Harris *et al.* 1973)

$$\mathcal{H} = -\sum_{\langle i,j \rangle} J_{ij}(\vec{S}_i \cdot \vec{S}_j) - \sum_i (\vec{a}_i \cdot \vec{S}_i)^2 . \tag{4.309}$$

Here the $\vec{a}_i$ are vectors that describe a random uniaxial anisotropy, such that $[\vec{a}_i]_{\text{av}} = 0$, $[\vec{a}_i \cdot \vec{a}_j]_{\text{av}} = \Delta^2 \delta_{ij}$, and the components of $\vec{a}_i$ are gaussian distributed. It has been suggested that for low temperatures and large Δ the system exhibits a spin glass phase, in addition to the ferromagnetic phase at smaller Δ.

Thus we have seen that the statistical mechanics of systems with random fields is a very rich and interesting topic, and we refer the reader to the articles of Belanger (1998) and Nattermann (1998) for more details.

4.6.6 *From Spin Glasses to Computer Science and Information: An Outlook to the Random Satisfiability Problem*

In recent years there has been a very fruitful interaction between the statistical mechanics of random systems on the one hand, and problems in theoretical computer science (information theory, combinatorial optimization, etc.) on the other hand. Thus by now this subject is vast and fills whole books (e.g. Hartmann and Rieger 2002, 2004, Schneider and Kirkpatrick 2006, Mézard and Montanari 2009); thus we shall mention only briefly one example, which has surprisingly close analogies to the mean field Potts spin glass (Sec. 4.6.2) and related p-spin interaction models (Sec. 4.6.1): The random satisfiability problem (also denoted SAT).

Suppose we have N Boolean variables x_i, $i = 1, \ldots, N \in (0,1)$; 1 means "true" and 0 means "false"; the negation of x_i hence is $\bar{x}_i \in (0,1)$, $i = 1, \ldots, N$. In addition, we have M constraints in the form of a clause. A clause is a logical OR of some variables or their negations. A clause $a \in (1, \ldots, M)$ involving K_a variables forbids exactly one among the 2^{K_a} assignments to these K_a variables. The logical OR commonly is denoted by the $\wedge$ symbol, e.g. for the 3-SAT problem, $C_1 = x_1 \wedge \bar{x}_2 \wedge x_3$ would be an example for a clause. A satisfiablility problem then is summarized by a conjunctive logical formula, $F = C_1 \wedge C_2 \wedge \cdots \wedge C_M$.

While 2-SAT problems are solvable by algorithms in polynomial time and less interesting, for $K \geq 3$ the satisfiability problems are "NP complete", i.e. a complete algorithm to solve them will need a computer time growing exponentially in N (in the worst case). We here consider only random $K-$SAT problems, containing only clauses of length K, which are generated uniformly at random from the set of $\binom{N}{K}2^K$ possible such clauses. A crucial parameter then is the clause density $\alpha = M/N$, and one considers as in statistical physics the "thermodynamic limit", $N \to \infty$, $M \to \infty$, with $\alpha =$ const. One also introduces a cost function $E(\vec{X})$, where $\vec{X}$ is the set of all 2^N configurations of the $\{x_i\}$: if all clauses are satisfied in a state, $E(\vec{X}) = 0$, analogous to the ground state energy of a physical system. The more clauses are violated, the larger $E(\vec{X})$.

One now is interested in understanding the "clustering property" of all solutions of such a random satisfiability problem for $E = 0$ (or a given low-lying nonzero value of E, analogous to a "metastable state" rather than a "ground state"). In terms of the projection of different solutions onto each other one can define a "Hamming distance" between these states (i.e, the number of coordinates $\{x_i\}$ in which two states $\vec{X}$, $\vec{X}'$ differ). With the help of this definition, one can analyze the configuration space of these solutions and distinguish whether it can be decomposed into distinct "pure states" (analogous to distinct "valleys" in the spin glass problem) or whether no such nontrivial decomposition exists. It turns out that the approach is rather technical but one can analyze the problem with one-step replica symmetry breaking methods, just as the models of Secs. 4.6.1 and 4.6.2! Thus we skip all details, but only sketch the main results qualitatively (Mézard and Montanari 2009): In a sense, the parameter α plays a role similar to temperature in these spin-glass-type problems: for $K \geq 4$ and $\alpha < \alpha_d(K)$, one has a replica-symmetric phase, i.e. there does not exist any nontrivial decomposition of the solution phase space and hence the "number of valleys" is $\mathcal{N}_N = 1$. For $\alpha_d(K) < \alpha < \alpha_c(K)$, one is in a state with 1-step replica symmetry breaking, but one has an exponentially high number $\mathcal{N}_N \propto \exp(NI)$ of disjunct "pure states". Here I is an entropy-type quantity which in the spin-glass context was called "complexity" (see Fig. 4.36). At $\alpha_c(K)$ one finds that I vanishes; in the regime $\alpha_c(K) < \alpha < \alpha_s(K)$ only a small number of pure states with $E = 0$ can be found, while for $\alpha > \alpha_s(K)$ one is in the UNSAT phase, i.e. no solution to the satisfiability problem does exist (this transition has no analogy in the spin glass models discussed in Secs. 4.6.1 and 4.6.2, of course). Since one expects that this phase space structure has important consequences on the efficiency of algorithms to find solutions of the satisfiablity problem we can conclude that the techniques developed for studying spin glasses have unexpected applications in the domain of optimisation problems.

References

Abrahams, E., Anderson, P.W., Licciardello, D.C., and Ramakrishnan, T.V. (1979) *Phys. Rev. Lett.* **42**, 673.

Aharony, A. (1978) *Phys. Rev.* **B18**, 3318.

Aizenman, M., and Wehr, J. (1989) *Phys. Rev. Lett.* **62**, 2503.

Alexander, S. (1989) *Phys. Rev.* **B40**, 7953.

Alexander, S., and Orbach, R. (1982) *J. Phys. Lett. (Paris)* **43**, L625.

Alvarez, M., Levesque, D., and Weis, J.J. (1999) *Phys. Rev.* **E60**, 5495.

Amoruso, C., Hartmann, A.K., Hastings, M.B., and Moore, M.A. (2006) *Phys. Rev. Lett.* **97**, 267202.

Anderson, P.W. (1958) *Phys. Rev.* **109**, 1942.

Anderson, P.W., Halperin, B.I., and Varma, C.M. (1972) *Phil. Mag.* **25**, 1.

Angell, C.A. (1991) *J. Non-Cryst. Solids* **131-133**, 13.

Arai, M., Inamura, Y., and Hannon, A.C. (1999) *Physica* **B263-264**, 268.

Aspelmeier, T., Moore, M., and Young, A.P. (2003) *Phys. Rev. Lett.* **90**, 127202.

Aspelmeier, T., Billoire, A., Marinari, E., and Moore, M.A. (2008) *J. Phys. A: Math. Theor.* **41**, 3240008.

Bach, H., and Krause, D. (eds.) (1999) *Analysis of the Composition and Structure of Glass and Glass Ceramics* (Springer, Berlin).

Bach, H., and Neuroth, N. (eds.) (1998) *The Properties of Optical Glass* (Springer, Berlin).

Baschnagel, J., Bennemann, C., Paul, W., and Binder, K. (2000) *J. Phys.: Condens. Matter* **12**, 6365.

Baumgärtner, A., and Binder, K. (1981) *J. Chem. Phys.* **75**, 2994.

Beck, H., and Güntherodt, H.J. (eds.) (1983) *Glassy Metals II: Atomic Structure and Dynamics, Electronic Structure, Magnetic Properties* (Springer, Berlin).

Beck, H., and Güntherodt, H.J. (eds.) (1994) *Glassy Metals III: Amorphization Techniques, Catalysis, Electronic and Ionic Structure* (Springer, Berlin).

Belanger, D.P. (1998) in *Spin Glasses and Random Fields* (A.P. Young, ed.) p. 251 (World Scientific, Singapore).

Bell, R.J., and Dean, P. (1972) *Phil. Mag.* **25**, 1381.

Belletti, F., Cotallo, M., Cruz, A., Fernandez, L.A., Gordillo-Guerrero, A., Guidetti, M., Maiorano, A., Mantovani, F., Marinari, E., Martin-Mayor, V., Sudupe, A.M., Navarro, D., Parisi, G., Perez-Gaviro, S., Ruiz-Lorenzo, J.J., Schifano, S.F., Sciretti, D., Tarancon, A., Tripiccione, R., Velasco, J.L., and Yllanes, D. (2008) *Phys. Rev. Lett.* **101**, 157201.

Benguigui, L. (1984) *Phys. Rev. Lett.* **53**, 2028.

Berche, P.E., Chatelain, C., Berche, B., and Janke, W. (2004) *Eur. Phys. J.* **B38**, 463.

Bergenholtz, J., and Fuchs, M. (1999) *Phys. Rev.* **E59**, 5706.

Bergman, D.J., and Kantor, Y. (1984) *Phys. Rev. Lett.* **53**, 511.

Berman, R. (1976) *Thermal Conductivity of Solids* (Clarendon Press, Oxford).

Berthier, L. (2003) *Phys. Rev. Lett.* **91**, 055701.

Billoire, A., and Marinari, E. (2002) *Europhys. Lett.* **60**, 775.

Binder, K. (1977) Festkörperprobleme **17**, 55.

Binder, K. (1983) *Z. Phys.* **B50**, 343.

Binder, K. (1987) *Rep. Progr. Phys.* **50**, 783.

Binder, K. (1992) in: *Computational Methods in Field Theory* (H. Gausterer and C.B. Lang, eds.) p. 59 (Springer, Berlin).

Binder, K. (1998) in: *Spin Glasses and Random Fields* (A.P. Young, ed.) p. 99 (World Scientific, Singapore).

Binder, K., and Reger, J.D. (1992) *Adv. Phys.* **41**, 547.

Binder, K., and Schröder, K. (1976) *Phys. Rev.* **14**, 2142.

Binder, K., and Landau, D.P. (1984) *Phys. Rev.* **B30**, 1477.

Binder, K., and Young, A.P. (1986) *Rev. Mod. Phys.* **58**, 801.

Binder, K., Kinzel, W., and Stauffer, D. (1979) *Z. Phys.* **B36**, 161.

Black, J.L. (1978) *Phys. Rev.* **B17**, 2740.

Black, J.L., and Halperin, B.I. (1977) *Phys. Rev.* **16**, 2879.

Blandin, A. (1978) *J. Phys.* (Paris) Colloq. **C6-39**, 1499.

Blavatska, V., and Janke, W. (2009) *J. Phys. A: Math. Theor.* **22**, 015001.

Boettcher, S. (2004) *Europhys. Lett.* **67**, 453.

Boettcher, S. (2005) *Phys. Rev. Lett.* **95**, 197205.

Bontemps, N., Rajchenbach, J., Chamberlin, R.V., and Orbach, R. (1984) *Phys. Rev.* **B30**, 6514.

Boon, J.P., and Yip, S. (1980) *Molecular Hydrodynamics* (Dover Publ., New York).

Bouchaud, J.-P., Cugliandolo, L.F., Kurchan, J., and Mézard, M. (1998) in *Spin Glasses and random Fields* (Young, A.P., ed.) p. 161 (World Scientific, Singapore).

Boutet de Monvel, A., and Bovier A. (eds.) (2009) *Spin Glasses: Statics and Dynamics* (Springer, Berlin).

Brand, R., Lunkenheimer, P., and Loidl, A. (2002) *J. Chem. Phys.* **116**, 10386.

Brangian, C. (2002) *Ph.D. Thesis* (Johannes Gutenberg Universität Mainz).

Brangian, C. (2004) *Physica* **A338**, 471.

Brangian, C., Kob, W., and Binder, K. (2001) *Europhys. Lett.* **53**, 756.

Brangian, C., Kob, W., and Binder, K. (2002a) *J. Phys. A: Math. Gen.* **35**, 191.

Brangian, C., Kob, W., and Binder, K. (2002b) *Europhys. Lett.* **59**, 546.

Brangian, C., Kob, W., and Binder, K. (2003) *J. Phys. A: Math. Gen.* **36**, 10847.

Bray, A.J., and Moore, M.A. (1978) *Phys. Rev. Lett.* **41**, 1068.

Brinker, C.J., and Scherer, G.W. (1990) *Sol-Gel Science: The Physics and Chemistry of Sol-Gel Processing* (Academic Press, New York).

Broderix, K., Goldbart, P.M., and Zippelius, A. (1997) *Phys. Rev. Lett.* **79**, 3688.

Broderix, K., Löwe, H., Müller, P., and Zippelius, A. (2000) *Phys. Rev.* **E63**, 011510.

Brodsky, M.H. (1979) (ed.) *Amorphous Semiconductors* (Springer, Berlin).

Brunet, E., and Derrida, B. (1997) *Phys. Rev.* **E56**, 2597.

Buchenau, U. (2001) *J. Phys.: Condens. Matter* **13**, 7827.

Buchenau, U., Nücker, N., and Dianoux, A.J. (1984) *Phys. Rev. Lett.* **53**, 2316.

Buchenau, U., Galperin, Yu.M., Gurevich, V.L., and Schober, H.R. (1991) *Phys. Rev.* **B43**, 5039.

Buchenau, U., Galperin, Yu.M., Gurevich, D.A., Parshin, A.A., Ramos, M.A., and Schober, H.R. (1992) *Phys. Rev.* **B46**, 2798.

Bunde, A., Kantelhardt, J.W., and Russ, S. (2002) *J. Non-Cryst. Solids* **307-310**, 96.

Campbell, I.A., and Kawamura, H. (2007) *Phys. Rev. Lett.* **99**, 019701.

Campbell, I.A., and Petit, D.M.C. (2010) *J. Phys. Soc. Jap.* **79**, 011006.

Cannella, V., and Mydosh, J.A. (1972) *Phys. Rev.* **B6**, 4220.

Carmesin, H.-O. (1987) *Phys. Lett.* **A125**, 294.

Carmesin, H.-O. (1989) *J. Phys.* **A22**, 297.

Carmesin, H.-O., and Binder, K. (1987) *Europhys. Lett.* **4**, 269.

Carmesin, H.-O., and Binder, K. (1988) *Z. Phys.* **B68**, 375.

Carmesin, I., and Kremer, K. (1988) *Macromolecules* **21**, 2819.

Cates, M.E., Fuchs, M., Kroy, K., Poon, W.C.K., and Puertas, A.M. (2004) *J. Phys.: Condens. Matter* **16**, S4861.

Chalupa, J. (1977) *Solid State Commun.* **24**, 429.

Chandler, D. (1987) *Introduction to Modern Statistical Mechanics* (Oxford University Press, Oxford).

Chen, J.H., and Lubensky, T.C. (1977) *Phys. Rev.* **B16**, 2106.

Chong, S.-H. (2008) *Phys. Rev.* **E78**.

Chong, S.-H., and Götze, W. (2002a) *Phys. Rev.* **E65**, 041503.

Chong, S.-H., and Götze, W. (2002b) *Phys. Rev.* **E65**, 051201.

Chong, S.-H., and Kob, W. (2009) *Phys. Rev. Lett.* **102**, 025702.

Chong, S.-H., Moreno, A.J., Sciortino, F., and Kob, W. (2005) *Phys. Rev. Lett.* **94**, 215701.

Courtens, E. (1982) *J. Phys. Lett. (Paris)* **43**, L199.

Courtens, E., Pelous, J., Phalippou, J., Vacher, R., and Woignier, T. (1987) *Phys. Rev. Lett.* **58**, 128.

Courtens, E., Foret, M., Hehlen, B., and Vacher, R. (2001) *Solid State Commun.* **117**, 181.

Courtens, E., Foret, M., Hehlen, B., Rufflé, B., and Vacher, R. (2003) *J. Phys.: Condens. Matter* **15**, S1279.

Cowley, R.A. (1976) *Phys. Rev.* **B13**, 4877.

Crisanti, A., Horner, H., and Sommers, H.-J. (1993) *Z. Phys.* **B92**, 257.

Crisanti, A., and Ritort, F. (2003) *J. Phys. A: Math. Gen.* **36**, R181.

Cruz, A., Fernandez, L.A., Gordillo-Guerrero, A., Guidetti, M., Maiorano, A., Mantovani, F., Marinari, E., Martin-Mayor, V., Sudupe, A.M., Navarro, D., Parisi, G., Perez-Gaviro, S., Ruiz-Lorenzo, J.J., Schifano, S.F., Sciretti, D., Tarancon, A., Tripiccione, R., Velasco, J.L., Yllanes, D., and Young, A.P. (2009) *Phys. Rev.* **B79**, 184408.

Cugliandolo, L.F. (2003) in *Lecture Notes for "Slow relaxations and nonequilibrium dynamics in condensed matter", Les Houches July, 1–25, 2002; Les Houches Session LXXVII* (J.-L. Barrat, M. Feigelman, J. Kurchan, and J. Dalibard, eds.) p. 367 (Springer, Berlin).

Cugliandolo, L.F., and Kurchan, J. (1993) *Phys. Rev. Lett.* **71**, 173.

Cwilich, G. (1990) *J. Phys. A: Math. Gen.* **A23**, 5029.

Cwilich, G., and Kirkpartick, T.R. (1989) *J. Phys. A: Math. Gen.* **22**, 4971.

Cyrot, M. (1981) *Solid State Commun.* **39**, 1009.

Daoud, M. (2000) *Macromolecules* **33**, 3019.

Daoud, M., and Coniglio, A. (1981) *J. Phys.* **A14**, L301.

De Almeida, J.R.L., and Thouless, D.J. (1978) *J. Phys.* **A11**, 983. 5B

De Arcangelis, L., del Gado, E., and Coniglio, A. (2002) *Eur. Phys. J.* **E9**, 277.

De Dominicis, C. (1978) *Phys. Rev.* **B18**, 4913.

De Dominicis, C., and Giardina, I. (2006) *Random Fields and Spin Glasses* (Cambridge University Press, Cambridge).

De Dominicis, C., and Young, A.P. (1983) *J. Phys.* **A16**, 2063.

de Gennes, P.G. (1967) *Physics* **3**, 37.

de Gennes, P.G. (1976a) *J. Phys. (Paris)* **37**, L1.

de Gennes, P.G. (1976b) *La Recherche* **7**, 919.

de Gennes, P.G. (1978) *C.R. Acad. Sci. (Paris)* **286B**, 131.

de Gennes, P.G. (1979a) *Scaling Concepts in Polymer Physics* (Cornell University Press, Ithaca).

de Gennes, P.G. (1979b) *J. Phys. Lett. (Paris)* **40**, L197.

de Gennes, P.G. (1984) *J. Phys. Chem.* **88**, 6469.

De Grandis, V., Gallo, P., and Rovere, M. (2004) *Phys. Rev.* **E70**, 061505.

del Gado, E., de Arcangelis, L., and Coniglio, A. (2002) *Phys. Rev.* **E65**, 041803.

de Raedt, B., Binder, K., and Michel, K.H. (1981) *J. Chem. Phys.* **75**, 2977.

Derrida, B. (1980) *Phys. Rev. Lett.* **45**, 79.

Derrida, B. (1981) *Phys. Rev.* **B24**, 2613.

de Santis, E., Parisi, G., and Ritort, F. (1995) *J. Phys. A: Math. Gen.* **28**, 3025.

des Cloizeaux, J., and Jannink, G. (1990) *Polymers in Solutions: Their Modeling and Structure* (Oxford University Press, Oxford).

Deutsch, H.-P., and Binder, K. (1991) *J. Chem. Phys.* **94**, 2294.

Doi, M., and Edwards, S.F. (1986) *The Theory of Polymer Dynamics* (Clarendon Press, Oxford).

Domb, C., and Green, M.S. (1977) (eds.) *Phase transitions and critical phenomena,* Vol. 6 (Academic Press, London).

Donth, E. (2001) *The Glass Transition Relaxation Dynamics in Liquids and Disordered Materials* (Springer, Berlin).

Doussineau, P., Legnes, P., Levelut, A., and Robin, A. (1978) *J. Phys. Lett.* **39**, L265.

Eastwood, M.P., and Wolynes, P.G. (2002) *Europhys. Lett.* **60**, 587.

Edwards, S.F. (1971) in *Polymer Networks* (A.J. Chompff and S. Newman, eds.) p. 83 (Plenum, New York).

Edwards, S.F., and Anderson, P.W. (1975) *J. Phys.* **F5**, 965.

Eichhorn, K., and Binder, K. (1996a) *Z. Phys.* **B99**, 413.

Eichhorn, K., and Binder, K. (1996b) *J. Phys.: Condens. Matter* **8**, 5209.

Eiselt, G., Kötzler, J., Maletta, H., Stauffer, D., and Binder, K. (1979) *Phys. Rev.* **B19**, 2664.

Elderfield, D., and Sherrington, D. (1983a) *J. Phys.* **C16**, L491.

Elderfield, D., and Sherrington, D. (1983b) *J. Phys.* **C16**, L971.

Elderfield, D., and Sherrington, D. (1983c) *J. Phys.* **C16**, L1169.

Emery, V.J. (1975) *Phys. Rev.* **B11**, 239.

Erman, B., and Mark, J.E. (1997) *Structures and Properties of Rubberlike Networks* (Oxford University Press, Oxford).

Erzan, A., and Lage, E.J.S. (1983) *J. Phys.* **C16**, L55.

Esteve, D., Sullivan, N.S., and Devoret, M. (1982) *J. Phys. Lett.* **43**, 793.

Farago, O., and Kantor, Y. (2000a) *Phys. Rev.* **E62**, 6094.

Farago, O., and Kantor, Y. (2000b) *Phys. Rev.* **E80**, 2533.

Farago, O., and Kantor, Y. (2002) *Europhys. Lett.* **57**, 458.

Feldman, J.L., Allen, P.B., and Bickham, S.R. (1999) *Phys. Rev.* **B59**, 3551.

Feng, S., and Sen, P. (1984) *Phys. Rev. Lett.* **52**, 216.

Ferreira, I.B., King, A.R., Jaccarino, V., Candy, J.L., and Guggenheim, H.J. (1983) *Phys. Rev.* **B28**, 5192.

Fischer, K.H., and Hertz, J.A. (1991) *Spin Glasses* (Cambridge Univ. Press, Cambridge).

Fisher, D.S. (1986) *Phys. Rev. Lett.* **56**, 416.

Fisher, D.S., and Huse, D.A. (1988) *Phys. Rev.* **B38**, 386.

Fisher, M.E. (1974) *Rev. Mod. Phys.* **46**, 597.

Fishman, S., and Aharony, A. (1979) *J. Phys.* **C12**, L729.

Fleurov, V.N., and Trakhtenberg, L.I. (1982) *Solid State Commun.* **44**, 187.

Fleurov, V.N., and Trakhtenberg, L.I. (1983) *Solid State Commun.* **46**, 755.

Fleurov, V.N., and Trakhtenberg, L.I. (1986) *J. Phys. C: Solid State* **19**, 5529.

Flory, P.J. (1941a) *J. Am. Chem. Soc.* **63**, 3083.

Flory, P.J. (1941b) *J. Am. Chem. Soc.* **63**, 3091.

Flory, P.J. (1941c) *J. Am. Chem. Soc.* **63**, 3906.

Flory, P.J. (1953) *Principles of Polymer Chemistry* (Cornell Univ. Press, Ithaca).

Folk, R., Iro, H., and Schwabl, F. (1976) *Z. Phys.* **B25**, 69.

Franz, S. (2007) *J. Stat. Phys.* **126**, 765.

Franz, S., and Parisi, G. (1998) *Physica* **A61**, 317.

Franz, S., and Parisi, G. (2000) *J. Phys.: Condens. Matter* **12**, 6335.

Franz, S., and Toninelli, P.I. (2004) *J. Phys. A: Math. Gen.* **37**, 7433.

Franz, S., Parisi, G., and Ricci-Tersenghi, F. (2008) *J. Phys. A: Math. Theor.* **41**, 324011.

Fredrickson, G.H., and Andersen, H.C. (1984) *Phys. Rev. Lett.* **53**, 1244.

Gabay, M., and Toulouse, G. (1981) *Phys. Rev. Lett.* **47**, 201.

Gilroy, K.S., and Phillips, W.A. (1981) *Phil. Mag.* **B43**, 735.

Ginzburg, V.L. (1960) *Soviet Phys. Solid State* **2**, 1824.

Glauber, R.J. (1963) *J. Math. Phys.* **4**, 294.

Gofman, M., Adler, J., Aharony, A., Harris, A.B., and Schwartz, M. (1996) *Phys. Rev.* **B53**, 6362.

Goldbart, P.M., and Goldenfeld, N.D. (1987) *Phys. Rev. Lett.* **58**, 2676.

Goldbart, P.M., and Goldenfeld, N.D. (1989) *Macromolecules* **22**, 948.

Goldbart, P.M., and Sherrington, D. (1985) *J. Phys.* **C18**, 1923.

Goldenfeld, N.D., and Goldbart, P.M. (1992) *Phys. Rev.* **A45**, R5343.

Golding, B., Graebner, J.E., Halperin, B.I., and Schutz, R.J. (1973) *Phys. Rev. Lett.* **30**, 223.

Golding, B., Graebner, J.E., Kane, A.B., and Black, J.L. (1978) *Phys. Rev. Lett.* **41**, 1487.

Götze, W. (1984) *Z. Phys.* **B56**, 139.

Götze, W. (1985) *Z. Phys.* **B60**, 195.

Götze, W. (1990) in *Liquids, Freezing and the Glass Transition* (J.-P. Hansen, D. Levesque, and J. Zinn-Justin, eds.) p. 287 (Amsterdam, North-Holland).

Götze, W. (2009) *Complex Dynamics of Glass-Forming Liquids. A Mode Coupling Theory* (Oxford University Press, Oxford).

Götze, W., and Mayr, M.R. (2000) *Phys. Rev.* **E55**, 3183.

Graessley, W.W. (2008) *Polymeric Liquids and Networks: Dynamics and Rheology* (Garland Science, London).

Green, J.E. (1985) *J. Phys.* **A18**, L43.

Griffiths, R.B. (1969) *Phys. Rev. Lett.* **23**, 17.

Grinstein, G. (1984) *J. Appl. Phys.* **55**, 2371.

Grinstein, G., and Ma, S.-K. (1982) *Phys. Rev. Lett.* **49**, 685.

Grinstein, G., and Ma, S.-K. (1983) *Phys. Rev.* **B28**, 2588.

Grosberg, A.Yu., and Khokhlov, A.R. (1994) *Statistical Physics of Macromolecules* (American Inst. Physics, Woodbury).

Gross, D.J., and Mézard, M. (1984) *Nuc. Phys.* **B240**, 431.

Gross, D.J., Kanter, I., and Sompolinsky, H. (1985) *Phys. Rev.* **55**, 304.

Guillot, B., and Guissani, Y. (1997) *Phys. Rev. Lett.* **78**, 2401.

Güntherodt, H.J., and Beck, H. (eds.) (1981) *Glassy Metals I: Ionic Structure, Electronic Transport, and Crystallization* (Springer, Berlin).

Gurevich, V.L., Parshin, D.A., and Schober, H.R. (2003) *Phys. Rev.* **B67**, 094203.

Gutzow, I., and Schmelzer, J. (1995) *The Vitreous State: Thermodynamics, Structure, Rheology, and Crystallization* (Springer, Berlin).

Haas, F.F., Vollmayr, K., and Binder, K. (1996) *Z. Phys.* **B99**, 393.

Hamakawa, Y. (ed.) (1982) *Amorphous Semiconductor Technologies and Devices* (Ohmsha Ltd., Tokyo).

Hammes, D., Carmesin, H.-O., and Binder, K. (1989) *Z. Phys.* **B76**, 115.

Harris, A.B., and Lubensky, T.C. (1987) *Phys. Rev.* **B35**, 6964.

Harris, A.B., and Meyer, H. (1985) *Can. J. Phys.* **63**, 3.

Harris, A.B., Plischke, M., and Zuckermann, J. (1973) *Phys. Rev. Lett.* **31**, 160.

Harris, A.B., Lubensky, T.C., and Chen, J.H. (1976) *Phys. Rev. Lett.* **36**, 415.

Harris, A.B., Meir, Y., and Aharony, A. (1987) *Phys. Rev.* **B36**, 8752.

Hartmann, A.K. (2008) *Phys. Rev.* **B77**, 144418.

Hartmann, A.K., and Moore, M.A. (2003) *Phys. Rev. Lett.* **90**, 127207.

Hartmann, A.K., and Rieger, H. (2002) *Optimization Algorithm in Physics* (Wiley-VCH, Berlin).

Hartmann, A.K., and Rieger, H. (2004) *New Optimization Algorithms in Physics* (Wiley-VCH, Berlin).

Hasenbusch, M., Parisen Toldin, F., Pelissetto, A., and Vicari, E. (2008a) *Phys. Rev.* **E77**, 051115.

Hasenbusch, M., Pelissetto, A., and Vicari, E. (2008b) *Phys. Rev.* **B78**, 214205.

Hehlen, B., Courtens, E., Vacher, R., Yamanaka, A., Karaoka, M., and Inoue, E. (2000) *Phys. Rev. Lett.* **84**, 5355.

Höchli, U.T., Knorr, K., and Loidl, A. (1990) *Adv. Phys.* **39**, 405.

Höfling, F., and Franosch, T. (2007) *Phys. Rev. Lett.* **98**, 140601.

Hohenberg, P.C., and Halperin, B.I. (1977) *Rev. Mod. Phys.* **49**, 435.

Horbach, J., and Kob, W. (1999) *Phys. Rev.* **B60**, 3169.

Horbach, J., Kob, W., and Binder, K. (1998) *J. Non-Cryst. Solids* **235-238**, 320.

Horbach, J., Kob, W., and Binder, K. (1999), *J. Phys. Chem.* **B103**, 4104.

Horbach, J., Kob, W., and Binder, K. (2001) *Eur. Phys. J.* **B19**, 531.

Houghton, A., Jain, S., and Young, A.P. (1983) *J. Phys.* **C16**, L375.

Hukushima, K., Campbell, I.A., and Takayama, H. (2009) *Int. J. Mod. Phys.* **C20**, 1313.

Hunklinger, S., and Arnold, W. (1976) in: *Physical Acoustics XII* (W. P. Mason and R. N. Thurstan, eds.) p. 155 (Academic Press, New York).

Hunklinger, S., and Piché, L. (1975) *Solid State Commun.* **17**, 1189.

Ilyin, V., Procaccia, I., Regev, I., and Shokef, Y. (2009) *Phys. Rev.* **B80**, 174201.

Imbrie, J. (1984) *Phys. Rev. Lett.* **53**, 1747.

Imry, Y., and Ma, S.K. (1975) *Phys. Rev. Lett.* **35**, 1399.

Ioffe, A.F., and Regel, A.R. (1960) *Progr. Semicond.* **4**, 237.

Jain, S., and Young, A.P. (1986) *J. Phys.* **C19**, 3913.

Jörg, T., and Katzgraber, H.G. (2008) *Phys. Rev. Lett.* **101**, 197205.

Jörg, T., Katzgraber, H.G, and Krzakala, F. (2008) *Phys. Rev. Lett.* **100**, 197202.

Kantelhardt, J.W., Russ, S., and Bunde, A. (2001) *Phys. Rev.* **B63**, 064301.

Kantor, Y. (1985) in *Scaling Phenomena in Disordered Systems* (R. Pynn and A. Skjeltorp, eds.) p. 391 (Plenum Press, New York).

Kantor, Y., and Webman, I. (1984) *Phys. Rev. Lett.* **52**, 1891.

Karpov, V. G., Klinger, M. I., and Ignatiev, F. N. (1982) *Solid State Commun.* **44**, 333.

Karpov, V.G., Klinger, M.I., and Ignatiev, F.N. (1983) *Zh. Eksp. Teor. Fiz.* **84**, 760.

Kasuya, T. (1956) *Progr. Theor. Phys.* **16**, 45.

Katzgraber, H.G., and Hartmann, A.K. (2009) *Phys. Rev. Lett.* **102**, 037207.

Katzgraber, H.G., and Krzakala, F. (2007) *Phys. Rev. Lett.* **98**, 017201.

Katzgraber, H.G., Larson, D., and Young, A.P. (2009) *Phys. Rev. Lett.* **102**, 177205.

Katzgraber, H.G., Körner, M., and Young, A.P. (2006) *Phys. Rev.* **B73**, 224432.

Kawamura, H. (1992) *Phys. Rev. Lett.* **68**, 3785.

Kawamura, H. (1998) *Phys. Rev. Lett.* **80**, 5421.

Kawamura, H. (2010) *J. Phys. Soc. Jap.* **79**, 011007.

Kawashima, N., and Young, A.P. (1996) *Phys. Rev.* **B53**, R484.

Kawashima, N. (2000) *J. Phys. Soc. Jap.* **69**, 987.

Khasanov, B.M. (2005) *JETP Lett.* **81**, 24.

Kierlik, E., Monson, P.A., Rosinberg, M.I., and Tarjus, G. (2002) *J. Phys.: Condens. Matter* **14**, 9295.

Kirkpatrick, T.R., and Thirumalai, D. (1988a) *Phys. Rev.* **B37**, 5342.

Kirkpatrick, T.R., and Thirumalai, D. (1988b) *Phys. Rev.* **B37**, 4439.

Kirkpatrick, T.R., and Wolynes, P.G. (1987) *Phys. Rev.* **B36**, 8552.

Klee, H., Carmesin, H.-O., and Knorr, K. (1988) *Phys. Rev. Lett.* **61**, 1855.

Klemens, P.G. (1955) *Proc. Phys. Soc. London* **68A**, 1113.

Klinger, M.I., and Karpov, V.G. (1983) *Zh. Eksp. Teor. Fiz.* **84**, 425.

Knorr, K. (1987) *Physica Scripta T* **19**, 531.

Kob, W., and Andersen, H.C. (1993) *Phys. Rev.* **E48**, 4364.

Kokshenev, V.B. (1996) *Phys. Rev.* **B53**, 2191.

Kondor, I. (1983) *J. Phys.* **A16**, L127.

Kosterlitz, J.M., and Thouless, D.J. (1973) *J. Phys.* **C6**, 1181.

Kovalenko, N.P., Krasny, Y.P., and Krey, V. (2001) *Physics of Amorphous Metals* (Wiley-VCH, Berlin).

Kremer, K., and Binder, K. (1984) *J. Chem. Phys.* **81**, 6381.

Kroy, K., Cates, M.E., and Poon, W.C.K. (2004) *Phys. Rev. Lett.* **92**, 148302.

Krzakala, F., and Martin, O.C. (2000) *Phys. Rev. Lett.* **85**, 3013.

Last, B.J., and Thouless, D.J. (1971) *Phys. Rev. Lett.* **27**, 1719.

Lee, L.W., Katzgraber, H.G., and Young, A.P. (2006) *Phys. Rev.* **B74**, 104416

Lee, L.W., and Young A.P. (2003) *Phys. Rev. Lett.* **90**, 227203.

Letz, M., Schilling, R., and Latz, A. (2000) *Phys. Rev.* **E62**, 5173.

Leutheusser, E. (1984) *Phys. Rev.* **A29**, 2765.

Lewis, T.J., and Klein, M.L. (1986) *Phys. Rev. Lett.* **57**, 2698.

Lewis, T.J., and Klein, M.L. (1989) *Phys. Rev.* **B40**, 7080.

Liu, Y., and Dahmen, K.A. (2007) *Phys. Rev.* **E76**, 031106.

Liu, Y., and Dahmen, K.A. (2009) *Europhys. Lett.* **86**, 56003.

Loidl, A. (1989) *Annu. Rev. Phys. Chem.* **40**, 19.

Loponen, M.T., Dynes, R.C., Narayanmurti, V., and Garno, J.P. (1980) *Phys. Rev. Lett.* **45**, 457.

Ludwig, A.W.W. (1988) *Phys. Rev. Lett.* **61**, 2388.

Ludwig, S., Enss, C., Stechlow, P., and Hunklinger, S. (2002) *Phys. Rev. Lett.* **88**, 75501.

Lynden-Bell, R.M., and Michel, K.H. (1994) *Rev. Mod. Phys.* **66**, 721.

Mancini, F.P., and Sherrington, D. (2006) *J. Phys. A: Math. Gen.* **39**, 13393.

Marinari, E., Mossa, S., and Parisi, G. (1999) *Phys. Rev.* **B59**, 8401.

Marinari, E., Parisi, G., Ricci-Tersenghi, F., and Ruiz-Lorenzo, J.J. (1998) *J. Phys. A.: Math. Gen.* **31**, 2611.

Marinari, E., Parisi, G., Ricci-Tersenghi, F., Ruiz-Lorenzo, J.J., and Zuliani, F. (2000) *J. Stat. Phys.* **98**, 973.

Maritan, A., Cieplak, M., Bellini, T., and Banavar, J.R. (1994) *Phys. Rev. Lett.* **72**, 4113.

Martin, E., and Adolf, D. (1991) *Annu. Rev. Phys. Chem.* **42**, 311.

Matsuda, Y., Nishimori, H., and Hukushima, K. (2008) *J. Phys. A: Math. Theor.* **41**, 324012.

Mattis, D.C. (1976) *Phys. Lett.* **A56**, 421.

McMillan, W.J. (1984) *J. Phys.* **C17**, 3179.

Meißner, M., and Spitzmann, K. (1981) *Phys. Rev. Lett.* **46**, 265.

Mézard, M., and Montanari, A. (2009) *Information, Physics and Computation* (Oxford University Press, Oxford).

Mézard, M., and Parisi, G. (1990) *J. Phys. A.: Math. Gen.* **23**, L1229.

Mézard, M., and Parisi, G. (2009) arXiv: 0910.2838v1.

Mezei, F. (1981), in *Recent Developments in Condensed Matter Physics, Vol. 1* (J.R. Devreese, ed.) p. 679 (Plenum, New York).

Michel, K.H. (1987a) *Phys. Rev.* **B35**, 1405.

Michel, K.H. (1987b) *Phys. Rev.* **B35**, 1414.

Michel, K.H., and Naudts, J. (1977) *J. Chem. Phys.* **67**, 547.

Michel, K.H., and Naudts, J. (1978) *J. Chem. Phys.* **68**, 216.

Milchev, A., and Binder, K. (1996) *J. Phys. (Paris) II* **6**, 21.

Miller, M., and Liaw, P. (2008) *Bulk Metallic Glasses* (Springer, Berlin).

Mizoguchi, T., McGuire, T.R., Kirkpatrick, S., and Gambino, J.R. (1977) *Phys. Rev. Lett.* **38**, 89.

Mookerjee, A., and Chowdhury, D. (1983) *J. Phys.* **F13**, 431.

Moore, E.D., Stinchcombe, R.B., and de Queiroz, S.L.A. (1996) *J. Phys. A: Math. Gen.* **29**, 7409.

Morgenstern, I., and Binder, K. (1980) *Phys. Rev.* **B22**, 288.

Morgenstern, I., Binder, K., and Hornreich, R.M. (1981) *Phys. Rev.* **B23**, 287.

Morgownik, A.F.J., Mydosh, J.A., and Wenger, L.E. (1982) *J. Appl. Phys.* **53**, 2211.

Morris, B.W., Colborne, S.G., Moore, M.A., Bray, A.J., and Canisius, J. (1986) *J. Phys.* **C19**, 1157.

Mott, N.F. (1966) *Phil. Mag.* **13**, 93.

Mott, N.F., and Davis, E.A. (1971) *Electronic Processes in Noncrystalline Materials* (Clarendon Press, Oxford).

Mulder, C.A.M., van Duyneveldt, A.J., and Mydosh, J.A. (1981) *Phys. Rev.* **B23**, 1384.

Müser, M.H. (1996) *J. Phys.: Condens. Matter* **8**, 913.

Müser, M.H., and Ciccotti, G. (1995) *J. Chem. Phys.* **103**, 4273.

Müser, M.H., and Nielaba, P. (1995) *Phys. Rev.* **B52**, 7201.

Nagata, S., Keesom, P.H., and Harrison, H.R. (1979) *Phys. Rev.* **B19**, 1633.

Nakayama, T. (2002) *Rep. Progr. Phys.* **65** 1195.

Nattermann, T. (1984) *Z. Phys.* **B54**, 247.

Nattermann, T. (1985a) *Phys. Status Solidi (b)* **131**, 563.

Nattermann, T. (1985b) *Phys. Status Solidi (b)* **132**, 125.

Nattermann, T. (1998) in *Spin Glasses and Random Fields* (A.P. Young, ed.) p. 277 (World Scientific, Singapore).

Nattermann, T., and Villain, J. (1988) *Phase Transitions* **11**, 5.

Newman, M.E., and Barkema, G.I. (1996) *Phys. Rev.* **E53**, 393.

Newman, C.M., and Stein, D.L. (1996) *Phys. Rev. Lett.* **76**, 515, 4821.

Nishimori, T. (1981) *Progr. Theor. Phys.* **66**, 1169.

Nordblad, P., and Svedlindh, P. (1998) in *Spin Glasses and Random Fields* (A.P. Young, ed.) p.1 (World Scientific, Singapore)

Normand, J.M., Herrmann, H.J., and Hajjar, M. (1988) *J. Stat. Phys.* **52**, 441.

Ocio, M., and Hérisson, D. (2003) in *Slow relaxations and nonequilibrium dynamics in condensed matter* (J.L. Barrat, M. Feigelman, J. Kurchan, and J. Dalibard, eds.) p. 605 (Springer, Berlin).

Ogielski, A.T. (1985) *Phys. Rev.* **B32**, 7384.

Ogielski, A.T., and Morgenstern, I. (1985) *Phys. Rev. Lett.* **54**, 928.

Okun, K., Wolfgardt, M., Baschnagel, J., and Binder, K. (1997) *Macromolecules* **30**, 3075.

Olive, J.A., Sherrington, D., and Young, A.P. (1986) *Phys. Rev.* **B34**, 6341.

Omari, R., Prejean, J.J., and Souletie, J. (1983) *J. Phys. (Paris)* **44**, 1069.

Onsager, L. (1944) *Phys. Rev.* **65**, 117.

Palassini, M., and Young, A.P. (2000) *Phys. Rev. Lett.* **85**, 3017.

Palmer, R.G. (1982) *Adv. Phys.* **31**, 669.

Parisen Toldin, F., Pelissetto, A., and Vicari, E. (2009) *J. Stat. Phys.* **135**, 1039.

Parisi, G. (1979) *Phys. Rev. Lett.* **43**, 1754.

Parisi, G. (1980) *J. Phys.* **A13**, 1101, 1887.

Parisi, G. (1983) *Phys. Rev. Lett.* **50**, 1946.

Parisi, G. (2008) *J. Phys. A: Math. Theor.* **41**, 324002.

Parisi, G. (2009) *Lett. Math. Phys.* **88**, 255.

Parisi, G., and Rizzo, T. (2008) *Phys. Rev. Lett.* **101**, 117205.

Parisi, G., and Sourlas, N. (1979) *Phys. Rev. Lett.* **43**, 744.

Parisi, G., and Virasoro, M. (1989) *J. Phys. (Paris)* **50**, 3317.

Parshin, D.A. (1994) *Phys. Solid State* **36**, 991.

Paul, W., Heermann, D.W., Kremer, K., and Binder, K. (1991) *J. Phys. (Paris) II* **1**, 37.

Perez-Reche, F.J., and Vives, E. (2004) *Phys. Rev.* **B70**, 214422.

Phillips, W.A. (1972) *J. Low Temp. Phys.* **7**, 351.

Phillips, W.A. (1981) (ed.) *Amorphous Solids: Low Temperature Properties* (Springer, Berlin).

Piché, L., Maynard, R., Hunklinger, S., and Jäckle, J. (1974) *Phys. Rev. Lett.* **32**, 1426.

Pirc, R., Tadic, B., and Blinc, R. (1987) *Phys. Rev.* **B36**, 18607.

Pirc, R., Tadic, B., and Blinc, R. (1994) *Physica* **B193**, 109.

Polian, A., Vo-Tanh D., and Richet, P. (2002) *Europhys. Lett.* **57**, 375.

Potts, R. B. (1952) *Proc. Camb. Phil. Soc.* **48** 106.

Rammal, R., and Toulouse, G. (1983) *J. Phys. Lett. (Paris)* **44**, L13.

Rammal, R., Toulouse, G., and Virasoro, M. (1986) *Rev. Mod. Phys.* **58**, 765.

Raychaudhuri, A.K. (1989) *Phys. Rev.* **B39**, 1927.

Richet, P., Bottinga, Y., Denielou, D., Petitet, J.P., and Tegui, C. (1982) *Geochim. et Cosmochim. Acta* **46**, 2639.

Rieger, H. (1995) *Phys. Rev.* **B52**, 6659.

Rogge, S., Natelson, D., and Osheroff, D.D. (1996) *Phys. Rev. Lett.* **76**, 3136.

Romá, F., Risau, S., Ramircz-Pastor, A.J., Nieto, F., and Vogel, E.E. (2009) *Physica* **A388**, 2821.

Rouse, P. (1953) *J. Chem. Phys.* **21**, 127.

Rubinstein, M., and Colby, R. (2003) *Polymer Physics* (Oxford University Press, Oxford).

Ruderman, M.A., and Kittel, C. (1954) *Phys. Rev.* **96**, 99.

Ruocco, G., Sette, F., di Leonardo, R., Monaco, G., Sampoli, M., Scopigno, T., and Viliani, G. (2000) *Phys. Rev. Lett.* **84**, 5788.

Rufflé, B., Foret, M., Courtens, E., Vacher, R., and Monaco G. (2003) *Phys. Rev. Lett.* **90**, 095502.

Rufflé, B., Parshin, D.A., Courtens, E., Vacher, R. (2008) *Phys. Rev. Lett.* **100**, 015501.

Salamon, M.B., Rao, K.V., and Yeshurun, Y. (1981) *J. Appl. Phys.* **52**, 1687.

Sarkisov, L., and Monson, P.A. (2000) *Phys. Rev.* **E61**, 7231.

Schirmacher, W. (2006) *Europhys. Lett.* **73**, 892.

Schirmacher, W., Ruocco, G., and Scopigno, T. (2007) *Phys. Rev. Lett.* **98**, 025501.

Schirmacher, W., Diezemann, G., and Ganter, C. (1998) *Phys. Rev. Lett.* **81**, 136.

Schmid, B., and Schirmacher, W. (2008) *Phys. Rev. Lett.* **100**, 137402.

Schneider, J.J., and Kirkpatrik, S. (2006) *Stochastic Optimization* (Springer, Berlin).

Schober, H.R., and Oligschleger, C. (1996) *Phys. Rev.* **B53**, 11469.

Scholze, H. (1990) *Glass-Nature, Structure, and Properties* (Springer, Berlin).

Schwartz, M. (1985) *J. Phys. C: Solid. State Phys.* **18**, 135.

Sciortino, F., and Sastry, S. (1994) *J. Chem. Phys.* **100**, 3881.

Shankar, R. (1987) *Phys. Rev. Lett.* **58**, 2466.

Sheng, P., Zhou, M., and Zhang, Z.-Q. (1994) *Phys. Rev. Lett.* **72**, 234.

Sherrington, D., and Kirkpatrick, S. (1975) *Phys. Rev. Lett.* **35**, 1972.

Shintani, H., and Tanaka, H. (2008) *Nature Mat.* **7**, 870.

Shirakura, T., Ninomiya, D., Iyama, Y., and Matsubara, F. (2009) *J. Phys. A: Math. Theor.* **41**, 115602.

Shtrikman, S., and Wohlfarth, E.P. (1981) *Phys. Lett.* **A85**, 467.

Singh, K., and Shimakawa, K. (2003) *Advances in Amorphous Semiconductors* (CRC Press, New York).

Smith, D.A. (1974) *J. Phys.* **F4**, L266.

Sokolov, A.P. (1999) *J. Phys.: Condens. Matter* **11**, A213.

Sokolov, A.P., Calenczuk, R., Salce, B., Kisliuk, A., Quitmann, D., and Duval, E. (1997) *Phys. Rev. Lett.* **78**, 2405.

Sokolov, A.P., Rössler, E., Kisliuk, A., and Quitmann, D. (1993) *Phys. Rev. Lett.* **71**, 2062.

Sommers, H.J., and Fischer, K.H. (1985) *Z. Phys.* **B58**, 125.

Sompolinsky, H. (1981) *Phys. Rev. Lett.* **47**, 935.

Sompolinsky, H., and Zippelius, A. (1982) *Phys. Rev.* **B25**, 6860.

Sosman, R.B. (1927) *The Properties of Silica* (Chemical Catalog Co., New York).

Souletie, J. (1983) *J. Phys. (Paris)* **44**, 1095.

Stauffer, D. (1976) *J. Chem. Soc. Faraday Trans. 2* **72**, 1354.

Stauffer, D., and Aharony, A. (1992) *Introduction to Percolation Theory* (Taylor and Francis, London).

Stauffer, D., and Binder, K. (1978) *Z. Phys.* **B30**, 313.

Stauffer, D., Coniglio, A., and Adam, M. (1982) *Adv. Polym. Sci.* **44**, 103.

Stechlow, P., Enns, C., and Hunklinger, S. (1998) *Phys. Rev. Lett.* **80**, 5361.

Stinchombe, R.B (1983) *Phase Transitions and Critical Phenomena, Vol. 7*, (C. Domb and J.L. Lebowitz, eds.) p. 151 (London: Academic).

Stinchcombe, R.B. (1985) in *Scaling Phenomena in Disordered Systems* (R. Pynn and A. Skjeltorp, eds.) p. 465 (Plenum Press, New York).

Stockmayer, W.H. (1943) *J. Chem. Phys.* **11**, 45.

Strobl, G. (1996) *The Physics of Polymers. Concepts for Understanding their Structure and Behavior* (Springer, Berlin).

Sullivan, N.S. (1983) *AIP Conf. Proc.* **103**, 121.

Suryanarayana, C., and Inoue, A. (2010) *Bulk Metallic Glasses* (CRC Press, New York).

Sussmann, J.A. (1964) *Phys. Kond. Mat.* **2**, 146.

Suzuki, M. (1977a) *Progr. Theor. Phys.* **58**, 1151.

Suzuki, M. (1977b) *Progr. Theor. Phys.* **58**, 1142.

Tadic, B., Pirc, R., Peterson, J., and Wiotte, W. (1994) *Phys. Rev.* **B50**, 9824.

Talagrand, M. (2006) *Ann. Math.* **163**, 221.

Talagrand, M. (2007) *J. Stat. Phys.* **126**, 837.

Taraskin, S.N., and Elliott, S.R. (1997) *Phys. Rev.* **B56**, 8605.

Taraskin, S.N., and Elliott, S.R. (1999a) *J. Phys.: Condens. Matter* **11**, A219.

Taraskin, S.N., and Elliott, S.R. (1999b) *Phys. Rev.* **B59**, 8572.

Taraskin, S.N., and Elliott, S.R. (2000a) *Phys. Rev.* **B61**, 12017.

Taraskin, S.N., and Elliott, S.R. (2000b) *Phys. Rev.* **B61**, 12031.

Taraskin, S.N., Allen, E., and Elliot, S.R. (2002) *J. Non-Cryst. Solid* **307-310**, 92.

Taraskin, S.N., Loh, Y.L., Natarajan, G., and Elliott, S.R. (2001) *Phys. Rev. Lett.* **86**, 1255.

Theenhaus, T., Schilling, R., Latz, A., and Letz, M. (2001) *Phys. Rev.* **E64**, 051505.

Thirumalai, D., and Kirkpartick, T.R. (1988) *Phys. Rev.* **B38**, 4881.

Toulouse, G. (1977) *Commun. Phys.* **2**, 115.

Treloar, L.R.G (1975) *The Physics of Rubber Elasticity* (Clarendon Press, Oxford).

van Beest, B.H.W., Kramer, G.J., and van Santen, R.A. (1990) *Phys. Rev. Lett.* **64**, 1955.

Viehman, D.C., and Schweizer, K.S. (2008) *J. Chem. Phys.* **128**, 084509.

Viet, D.X., and Kawamura, H. (2009) *Phys. Rev. Lett.* **102**, 027202.

Villain, J. (1977) *J. Phys.* **C10**, 4793.

Villain, J. (1985) *J. Phys.* **46**, 1843.

Vink, R.L.C., Binder, K., and Löwen, H (2006) *Phys. Rev. Lett.* **97**, 230603.

Vink, R.L.C., Binder, K., and Löwen, H. (2008) *J. Phys.: Condens. Matter* **20**, 404222.

Vollmayr, K., Schreider, G., Reger, J.D., and Binder, K. (1994) *J. Non-Cryst. Solids* **172-174**, 488.

von Löhneysen, H. (1981) *Phys. Rep.* **79**, 161.

Wenger, L.E. (1983), in *Proceedings of the Heidelberg Colloquium on Spin Glasses, Lecture Notes in Physics Vol. 192* (J.L. van Hemmen and I. Morgenstern, eds.) p. 60 (Springer, Berlin).

Wenger, L.E., and Keesom, P.H. (1976) *Phys. Rev.* **B13**, 4053.

Westerholt, K., and Bach, H. (1981) *J. Magn. Magn. Mater.* **24**, 191.

Westerholt, K., and Bach, H. (1982) *J. Phys.* **F12**, 1227.

Wiechert, H., and Krömker, B. (2002) *J. Non-Cryst. Solids* **307-310**, 538.

Wiseman, S., and Domany, E. (1995) *Phys. Rev.* **E52**, 3469.

Wiseman, S., and Domany, E. (1998) *Phys. Rev.* **E58**, 2938.

Wong, A.P.Y., and Chan, M.H.W (1990) *Phys. Rev. Lett.* **65**, 2567.

Wong, A.P.Y, Kim, S.B., Goldburg, W.I., and Chan, M.H.Q. (1993) *Phys. Rev. Lett.* **70**, 954.

Wosnitza, J., von Löhneysen, H., Zinn, W., and Krey, U. (1986) *Phys. Rev.* **B33**, 3436.

Wu, F.Y. (1982) *Rev. Mod. Phys.* **54**, 235.

Wu, X.-I., Goldburg, W.I., Liu, M.X., and Xue, J.Z. (1992) *Phys. Rev. Lett.* **69**, 470.

Wyart, M., Nagel, S.R., and Witten, T.A. (2005) *Europhys. Lett.* **72**, 486.

Yosida, K. (1957) *Phys. Rev.* **106**, 893.

Young, A.P. (1983) *Phys. Rev. Lett.* **51**, 1206.

Young, A.P. (1998) (ed.) *Spin Glasses and Random Fields* (World Scientific, Singapore).

Young, A.P. (2008) *J. Phys. A: Math. Theor.* **41**, 324016.

Young, A.P., and Katzgraber, H.G. (2004) *Phys. Rev. Lett.* **93**, 207203.

Zallen, R. (1983) *The Physics of Amorphous Materials* (Academic, New York).

Zeller, R.C., and Pohl, R.O. (1971) *Phys. Rev.* **B4**, 2029.

Zhang, R., and Schweizer, K.S. (2009) *Phys. Rev.* **E80**, 011502.

Zhu, T.C., Maris, H.J., and Tauc, J. (1991) *Phys. Rev.* **B44**, 4281.

Zia, R.K.P., and Wallace, D.J. (1975) *J. Phys. A: Math Gen.* **8**, 1495.

Zimm, B. (1956) *J. Chem. Phys.* **24**, 269.

Zimmermann, J., and Weber, G. (1981) *Phys. Rev. Lett.* **46**, 661.

Zinn, W. (1976) *J. Magn. Magn. Mater.* **3**, 23.

Chapter 5

Supercooled Liquids and the Glass Transition

In most elementary textbooks on crystalline solid state physics the making of the structure of interest, i.e. the crystal, is rarely discussed since it is considered to be a trivial matter: Just take a liquid, cool it down and at the melting temperature the system will crystallize since the free energy of the crystalline state becomes lower than the one of the liquid state. Although for some materials the growth of *good*, i.e. almost perfect, crystals is an art, most of the crystalline properties do not depend crucially on the way the crystal has been produced and thus on the density of defects. Therefore it is justified to start a theoretical treatment with the discussion of the perfect crystal, its properties, and consider only subsequently the effect of defects etc. (Ashcroft and Mermin 1976). For glasses the situation is very different: Although also here the production of a glass is fairly simple, in most cases it is sufficient to cool down the sample sufficiently quickly, there are several issues that have so far made it impossible to obtain a full theoretical understanding of glasses. Firstly there is the important question regarding the mechanism (or mechanisms!) that leads to the glass transition: In contrast to the case of the liquid to crystal transition which is related to the crossing of two branches of the free energy, it has so far not been possible to identify the real reason for the occurrence of the glass transition. Secondly it has been found that the properties of the glass depend quite strongly on the way they have been produced, also this in strong contrast to the case of a crystal. Last but not least these properties also depend on time, i.e. glasses show the phenomenon of aging, a behavior that can be very important in technical applications. All these issues are currently far from being understood and hence constitute a very active field of research. Consequently it is for the moment not possible to give a conclusive discussion on this subject. The goal of this chapter is therefore to give an introduction to the salient features of glass-forming systems and to discuss the so-called "mode coupling theory of the glass transition" (MCT), a theory that has been found to be useful for understanding many aspects of glass-forming systems (Götze 2009). Further theoretical models will be discussed in Chap. 6. Note that since it is neither appropriate nor possible to give here a complete overview of glass-forming liquids and glasses, we refer the reader who wants to learn more on this subject to some other textbooks and review articles (Zarzycki 1991, Feltz 1993, Varshneya 1993, Debenedetti 1997, Donth 2002, Dyre 2006, Greaves and Sen 2007).

5.1 Phenomenology of Glass-Forming Systems

The goal of the present section is to discuss some of the salient features of glass-forming systems. Note that only once we have presented these features it will be possible to explain in a more precise way what we mean by "glass-forming" systems since *a priori* it is rather difficult to come up with an accurate definition of this term.

One of the most important class of glass-forming systems are liquids and therefore we will start our discussion with this paradigm. Other examples will be given below. At sufficiently high temperatures the typical relaxation time of liquids, τ, is on the order of a picosecond, i.e. microscopically small. This short time scale reflects the fact that in a liquid at high temperatures the particles move quickly, bump into each other and make a motion that resembles, on the time scale of picoseconds, to the trajectory of a Brownian particle. It is of course no surprise that if the temperature is lowered the relaxation time increases, since the mean thermal velocity of the particles decreases like $T^{1/2}$. However, if one studies the temperature dependence of τ in a more quantitative way one finds that this dependence is much stronger than this trivial one in that a change of T by a factor of two leads to an increase of τ by many orders of magnitude. Typical examples for this increase are shown in Fig. 5.1 where we present the $T-$dependence of the viscosity of several liquids in an Arrhenius plot, i.e. $\log(\eta)$ vs. $1/T$. (Note that for most systems the viscosity shows a very similar temperature dependence as τ and since experimentally it is simpler to measure η than τ one often discusses $\eta(T)$ instead of $\tau(T)$.) This figure shows that the viscosity increases by about 15 decades if one changes temperature by a factor of 3-4. Although at first sight such an increase might seem to be reminiscent to the one found in standard second order phase transitions,

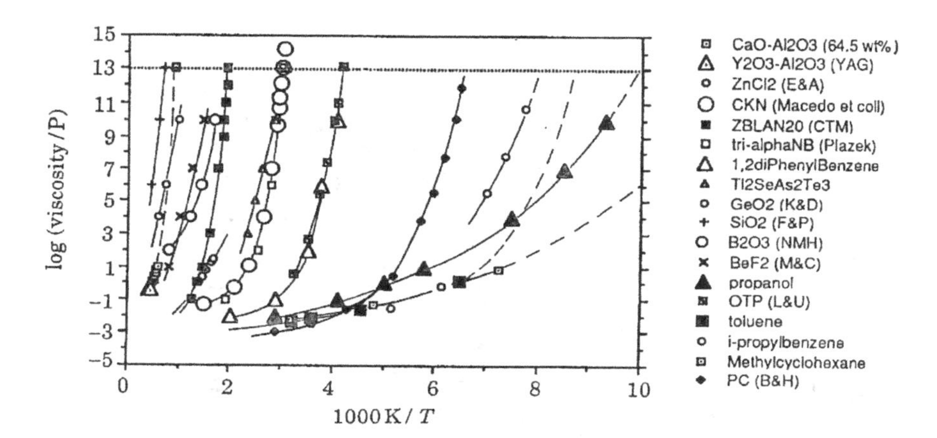

Fig. 5.1 Arrhenius plot of the temperature dependence of the viscosity of various glass-forming liquids. After Angell *et al.* (1994).

where the phenomenon is known as "critical slowing down" (Hohenberg and Halperin 1977), there is the important difference that the temperature range over which the critical slowing down is observed is only a few percent of the critical temperature, whereas in glass-forming systems it is about 100 times larger. Identifying the microscopic mechanism that gives rise to the dramatic slowing down of the relaxation dynamics is the most important challenge in the field of glasses. Although some aspects of this slowing down can be understood quite well within the framework of MCT, see Sec. 5.2, there is so far no theoretical approach that is able to describe *all* aspects of the relaxation dynamics in the whole temperature range.

Due to their very different chemical compositions, and hence interaction energies, of the liquids shown in Fig. 5.1, the strong increase in $\eta(T)$ occurs at very different temperature scales, which makes the comparison of the various curves difficult. In order to overcome this problem it is customary to introduce a reduced temperature scale. This is usually done by defining a so-called "glass transition temperature" T_g by means of $\eta(T_g) = 10^{12}$ Pa·s, which is equivalent to 10^{13}P, and to replot the data as a function of T_g/T.[1] Figure 5.2 shows that this type of rescaling of the temperature axis does not lead to a collapse of the individual data sets onto a master curve (Uhlmann 1972, Laughlin and Uhlmann 1972). Instead

[1]Note that the choice of 10^{12}Pa·s is quite arbitrary. It reflects the fact that a system with this viscosity has a relaxation time of about one minute, i.e. a times scale that is convenient for humans.

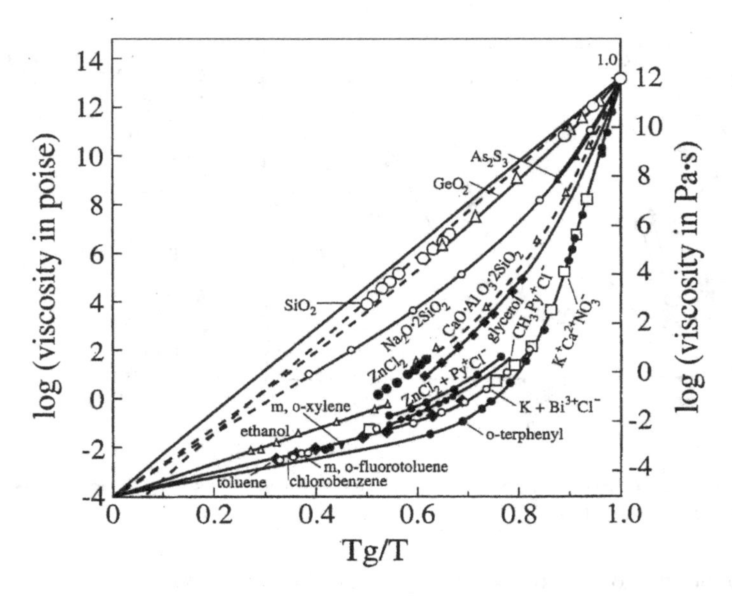

Fig. 5.2 Viscosity of various glass-forming liquids as a function of rescaled temperature T_g/T. After Angell (1985).

one finds that certain liquids seem to show in the whole experimentally accessible temperature range an Arrhenius law, such as SiO_2, whereas other glass-formers show a pronounced curvature. Following Angell we call these two limiting cases "strong" and "fragile" glass-formers (Angell 1985) and mention that graphs like the one shown in Fig 5.2 are usually called "Angell plot".[2] One popular possibility to characterize the so-called "fragility" of a glass-former in a more quantitative way is to define it as follows:

$$m = \left. \frac{d \log_{10} \eta}{d T_g/T} \right|_{T=T_g} , \tag{5.1}$$

i.e. m is just the slope of the curve $\eta(T)$ in the Angell-plot at T_g. With this

[2]We emphasize that the terms "strong" and "fragile" have nothing to do with the mechanical properties of the glass. Instead the terms attempt to express the fact that the activation energy for the relaxation process, which is nothing else than the local slope of the $\eta(T)$−curves in an Arrhenius plot, is independent/depends on temperature. Since this activation energy must be related in some (so far unknown!) way to the structural properties of the liquid, the corresponding structure of a strong glass-former has a structure that is independent of T, i.e. it is "strong", whereas the one of a fragile glass-former depends on T, i.e. it is "fragile". See Coslovich and Pastore (2007) and Berthier and Witten (2009) for results of computer simulations that address this point.

definition m ranges between 15 for strong glass formers up to 200 for fragile glass formers. Although it has been found that many dynamical properties of glass-forming liquids, such as the stretching parameter discussed below, correlate with the fragility index m (Böhmer *et al.* 1993), it is so far not clear whether this correlation is of fundamental importance or not. In particular we mention that the presentation of the data given in Fig. 5.2 has the flaw that at hight temperatures all the curves seem to converge to one common point. This can, however, not really be the case but is instead only an artifact of the presentation of the data which compresses all the high temperatures in a very small $1/T$−range. One possibility to avoid this problem will be discussed in Sec. 5.2 in the context of Fig. 5.16.

Although we have discussed so far only the temperature dependence of the viscosity and the relaxation time, most other dynamical quantities, such as the diffusion constant, the conductivity, the dielectric relaxation, etc., show a qualitatively similar strong T−dependence (Elliott 1987, Stickel *et al.* 1995, Lunkenheimer *et al.* 1996a). This is in contrast to the one of thermodynamic quantities (specific heat, enthalpy, compressibility, etc.) or structural quantities (density, static structure factor, etc.) whose temperature dependence is quite weak in that these observables change, in the same temperature range, typically only by a few percent or a factor of 2-5. (Note that also this weak T−dependence is in strong contrast to the situation encountered in a second order phase transition for which a correlation length exists that diverges at the critical temperature.) Although it can of course not be excluded that in glass-forming liquids complex higher order correlation functions (multi-particle correlation functions, etc.) do show a strong dependence on T, all the functions investigated so far have shown only a weak temperature dependence (Dasgupta *et al.* 1991, Ernst *et al.* 1991, Baschnagel and Binder 1995, Scheidler *et al.* 2002, Berthier and Garrahan 2003a). This result for structural glasses differs thus from the behavior found in spin glasses for which it is indeed possible to identify a divergent length scale and for which the description of the slowing down related to a qualitative change of the form of the free energy is feasible (see Sec. 4.5). Thus from this point of view the slow relaxation of spin glasses is better understood than the one of structural glasses. Note, however, that the observed absence of an increasing length scale is mainly based on investigations that considered *two-point* correlation functions (radial distribution functions, static structure factor, etc.). Some recent theoretical approaches do, however, predict that certain multipoint correlation functions, which are called "point to set correlations" and which give the correlation between one particle with a certain number of fixed particles (Montanari and Semerjian 2006, Franz Montanari 2007), should show upon cooling an increasing lengthscale that is related to the size of the mosaic tiles introduced in the context of the random first order theory developped by Kirkpatrick, Thirumalai, and Wolynes (Kirkpatrick and Thirumalai 1987, Kirkpatrick and Wolynes 1987a, 1987b, Kirkpatrick *et al.* 1989, Biroli and Bouchaud 2009) and which will be discussed in more detail in Chap. 6. Evidence that this static length scale does indeed increase upon cooling has recently been found by Biroli *et al.* (2008).

Last but not least we mention that in recent years one has found strong evidence that in glassy systems *dynamical* length scales do indeed increase significantly with decreasing temperature due to the presence of dynamical heterogeneities (for a review see Berthier *et al.* 2010). This aspect of the dynamics will be discussed in more detail in Chap. 6.

Since for structural glasses the scenario of a second order phase transition does not find much support in the experimental data or in computer simulations, one needs other mechanisms that are responsible for the slowing down of these systems. A first step is therefore to understand the precise form for the increase of the relaxation time with decreasing T. One very popular function which seems to describe the data rather well is the so-called "Vogel-Fulcher-(Tammann)-law" (Vogel 1921, Fulcher 1925, Tammann and Hesse 1926):

$$\eta(T) = \eta_0 \exp\left(\frac{B}{T-T_0}\right) . \tag{5.2}$$

This functional form thus predicts a divergence of the viscosity at $T = T_0$, the so-called "Vogel-temperature" and a super-Arrhenius increase of η close to T_0. Hence, the parameter B/T_0 determines whether $\eta(T)$ shows the Arrhenius dependence of strong glass-formers, which corresponds to the case $T_0 = 0$ and hence a large B/T_0, or the strong curvature found in the fragile glass-formers , which corresponds to a small B/T_0. We emphasize that the Vogel-Fulcher law does not have a solid theoretical foundation but gives only a good but certainly not perfect representation of the data (see, e.g., Stickel *et al.* 1995). The same flaw is also inherent to the so-called Bässler-law which is given by the functional form $\eta(T) = \eta_0 \exp(B/T^2)$ (Bässler 1987, Richert and Bässler 1990). In fact Angell *et al.* (1994) list and review about 10 (!) different temperature dependencies that have been proposed in the literature, most of them have a certain merit, i.e. there exist substances or $T-$intervals for which they give a good representation of the data. However, only very few of these functional forms can be justified theoretically and hence most of these expressions must be considered as purely phenomenological laws. One notable exception is the functional form predicted by mode coupling theory, see Sec. 5.2, which is given by

$$\eta(T) = \eta_0 (T - T_c)^{-\gamma}, \tag{5.3}$$

where T_c is the critical temperature of the theory and γ is an exponent whose value depends on the glass-former considered. Although this power-law usually describes the experimental data only in a quite limited range, typically 2-4 decades in $\eta(T)$, it has the remarkable merit that, at least in principle, the critical temperature as well as the exponent can be calculated *a priori* and hence are not fit parameters. More details will be given in Sec. 5.2.

A further important conclusion that can be drawn from Fig 5.2 is that all the curves are very smooth and that in particular they do not show any sign of a singularity at the melting temperatures of the liquids. In fact it is found that the viscosity of the liquid at its melting temperature varies by many orders of magnitude in that it is, e.g. $O(10^6 \text{Pa·s})$ for SiO_2 and $O(10^{-1})$Pa·s for ortho-terphenyl. Thus this is a strong indication that the reason for the slowing down of the dynamics is *not at all* related to the fact that the liquid is supercooled. Hence the presence of the crystalline state might be important for practical reasons, since the liquid might be prone to crystallization, but it is certainly not relevant for a theoretical understanding of the slowing down of the dynamics.

In order to obtain a better understanding of the relaxation dynamics of glass-forming systems it is necessary to investigate it on the microscopic level. Experimentally this can be done by means of light or neutron scattering in which one has direct access to the intermediate scattering functions introduced in Sec. 2.1 (Berne and Pecora 2000, Lovesey 1994). In Fig. 5.3 we show schematically the time dependence of a typical correlation function, such as the coherent intermediate scattering function $F(q,t)$, where q is the wave-vector, see Eq. (2.83). Two curves are shown: The one to the left corresponds to the relaxation dynamics of a liquid at high temperatures, whereas the one to the right to the case of a low temperature. If one makes a Taylor expansion in time of the trajectories of the particles, $\mathbf{r}_i(t) = \mathbf{r}_i(0) + \mathbf{v}_i t + \cdots$, and plugs this into the definition of $F(q,t)$, one finds immediately that this correlation function must show at small time a t^2 dependence. (Note that here we have assumed that the dynamics is

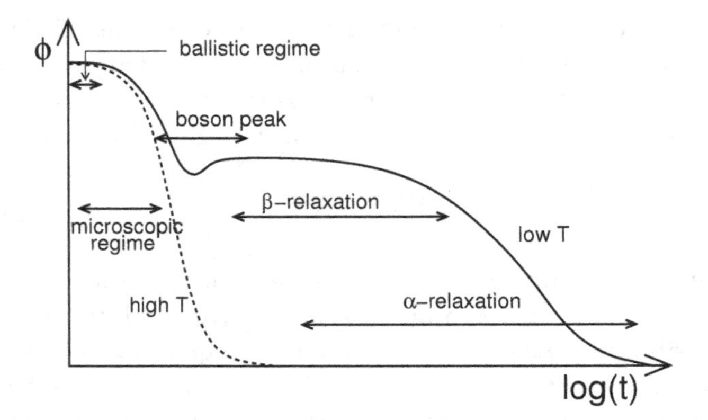

Fig. 5.3 Typical time dependence of a correlation function. The curve to the left corresponds to the case of a normal liquid and the one to the right to a strongly supercooled liquid.

Newtonian.) Since this time dependence is the same as the one found in a system of non-interacting particles, the time window in which it is observed is usually called "ballistic regime". Because in a dense liquid the particles will very quickly feel the presence of the other particles, the ballistic regime is only short and typically lasts for $O(30\text{fs})$. (The exact value depends on the strength of the interactions.) At longer times the particles undergo collisions with other particles and hence their trajectory is rather erratic, similar to the one found for a Brownian particle, and therefore the correlation function decays exponentially.

In the case of low temperatures the time correlation function is more complex. Although we find also here at very short times a ballistic regime, at its end the correlation function does not show an exponential decay to zero. Instead the correlator shows a crossover to a plateau, a behavior that is reminiscent to the one found in a crystal for which the long time motion of the particles is of vibrational nature, i.e. the particles oscillate around their equilibrium position and there is no relaxation. The reason why this vibrational motion is also present in glass-forming liquids is that, for intermediate time scales, each particle finds itself trapped in a (disordered) neighborhood of particles that prevents it to make a large displacement and hence force it to rattle around in this cage, a phenomenon that is sometimes also called "cage effect". (Note that the neighboring particles that form this cage are themselves trapped and thus also they can perform only a vibrational motion.) Thus in this time regime the relaxation dynamics of the liquid is very similar to the one of the glass of the same material and hence shows all the unusual vibrational features of these solids, such as, e.g., the so-called "boson peak", a vibrational motion with a frequency somewhat below the usual microscopic excitations, i.e. is around 1THz, and whose microscopic origin has been strongly debated (Winterling 1975, Buchenau *et al.* 1986, Benassi *et al.* 1996, Schirmacher *et al.* 1998, Rat *et al.* 1999, Sokolov 1999, Götze and Mayr 2000, Hehlen *et al.* 2000, Courtens *et al.* 2001, 2003, Grigera *et al.* 2001, Horbach *et al.* 2001, Ruffl é *et al.* 2008, Schmid and Schirmacher 2008), see also the discussion in Section 4.4.2. This slow vibrational motion is the reason for the small dip in the time correlation function just at the beginning of the plateau. Only for much larger times, note the logarithmic time axis in Fig. 5.3!, the particles can finally escape from their cage and hence the correlation function starts to decay again. This final decay is, however, usually not given by an exponential function but is instead described well by the Kohlrausch-Williams-Watts (KWW) function given by

$$\Phi(t) = A \exp\left(-(t/\tau)^{\beta}\right) , \tag{5.4}$$

where the so-called "stretching exponent", or "Kohlrausch exponent", β is smaller than 1.0 (Kohlrausch 1847, Williams and Watts 1980). We remark that although this functional form gives usually a good description of the final decay of the time correlation functions, there is no microscopic justification for it. (Note that for the case of spin glasses the final decay is often better described by a KWW-function multiplied by a power-law, see Sec. 4.5, Ogielski and Stein (1985), and Binder and Young (1986).)

The existence of the stretching in the time correlation function has intrigued researchers for quite a bit of time and has in recent years given rise to a large body of literature on the so-called "dynamical heterogeneities". By this term one means the following: Imagine a macroscopic sample of a glass forming liquid for which a (mean) relaxation function $\Phi(t)$ that probes the relaxation dynamics on a local scale (intermediate scattering function, dielectric relaxation, ...) is found to be stretched. There are two extreme scenarios to explain the stretching (Richert 1994): The sample is "heterogeneous", i.e. due to the disorder in the system each particle has a slightly different environment and hence the time scale for the (exponential) relaxation differs from particle to particle. The *mean* relaxation function is thus given by an average over exponential functions that have different time scales and thus can be approximated well by a stretched exponential (under the assumption that the distribution of relaxation times is not too exotic). In the other extreme case, the "homogeneous scenario", all the particles relax in the same way, i.e. show a stretched exponential with the same KWW-exponent. Thus the average correlation function will be a KWW-law with the same exponent. Which one of these two scenarios is realized in a real system is presently not yet clear but most likely the truth lies between these two extreme cases, at least for the majority of glass-forming systems. Experiments and computer simulations have indeed given evidence that the relaxation dynamics is not completely homogeneous although the exact nature of the heterogeneities as well as the reason for their existence are not well understood yet (Butler and Harrowell 1991, Schmidt-Rohr and Spiess 1991, Richert 1994, Cicerone *et al.* 1995, Perera and Harrowell 1996, Kob *et al.* 1997, Doliwa and Heuer 1998, Yamamoto and Onuki 1998, Sillescu 1999, Büchner and Heuer 2000, Ediger 2000, Kegel and van Blaaderen 2000, Weeks *et al.* 2000, Berthier 2004, Jung *et al.* 2004). The presence of these heterogeneities has important implications for the relation between the diffusion constant and the relaxation time. Imagine that we decompose our system in regions of mesoscopic size. If the system is heterogeneous the relaxation time in each of these regions, τ_i, will depend on the region i. Assume that within each of these regions the diffusion constant is inversely proportional to τ_i, as it is predicted to be the case from the Stokes-Einstein relation, $D_i \propto T/\eta_i \propto \tau_i^{-1}$, or by mode coupling theory (see Sec. 5.2.3). Although this relation thus holds in each part of the sample we have that $D := \langle D_i \rangle_i$ is *not* inversely proportional to $\tau := \langle \tau_i \rangle_i$, and hence the product of $D \cdot \tau$ is *not* a constant. From a physical point of view the explanation of this simple mathematical fact is that the mean squared displacement, and hence the diffusion constant, is dominated by the *fast* moving particles, whereas the intermediate scattering function, and hence τ, is dominated by the *slow* particles. Thus a few percent of fast particles can lead to a strong increase of D whereas they will hardly affect the (mean) $\alpha-$relaxation. As a consequence the product $D \cdot \tau$ increases with decreasing T. More details on the dynamical heterogeneities can be found in the review articles by Sillescu (1999), Glotzer (2000), and Richert (2002) and in Sec.6.2 we will discuss the dynamical heterogeneities in the context of growing dynamical length scales.

We now return to the average correlation function $\Phi(t)$. For historical reasons its final decay, i.e. the relaxation process that allows the particles to leave their cage, is called the "$\alpha-$process", whereas the time window in which the correlators are close to the plateau is called "$\beta-$process".[3] Note that the end of the $\beta-$process coincides with the beginning of the $\alpha-$process. Last but not least we point out the fact that at low temperatures the time scale for the microscopic dynamics is separated by many decades in time from the $\alpha-$ and $\beta-$process. This observation makes it possible to use a theoretical approach, the so-called Zwanzig-Mori projection operator formalism discussed in Sec. 5.2, to eliminate the fast degrees of freedom and to obtain effective equations of motion for the slow degrees of freedom.

In an experiment a time correlation function $\Phi(t)$ like $F(q, t)$ can be measured, e.g., by photon-correlation. However, in most experimental setups one does not measure $\Phi(t)$ but the imaginary part of its time Fourier transform, which is proportional to the associated dynamic susceptibility (Hansen and McDonald 1986, Barrat and Hansen 2003):

$$\chi''(\omega) = \frac{\omega}{k_B T} \Phi''(\omega) . \tag{5.5}$$

In Fig. 5.4 we show a schematic plot of the frequency dependence of the dynamic susceptibility of the correlators shown in Fig. 5.3. We recognize that the microscopic regime is now seen as peaks that are located at the typical microscopic vibrational frequencies of the system, i.e. in the range of 1-50 THz. At low T the boson peak is seen as a peak at frequencies below the microscopic ones and the $\alpha-$relaxation is seen as a broad peak whose position moves rapidly to lower frequencies if the temperature is lowered. Finally the $\beta-$relaxation corresponds to the frequency range in which $\chi''(\omega)$ show the minimum between the $\alpha-$peak and the microscopic vibrations.

We have mentioned above that thermodynamic quantities do not show a strong dependence on temperature. This does not imply, however, that they are of no interest to understand the physics of glass-forming materials.

[3] We emphasize that this $\beta-$process, which is sometimes also called "fast $\beta-$process", should not be confused with the so-called "Johari-Goldstein $\beta-$process", or "slow $\beta-$process" that is found in many, but not all, glass-formers (Johari and Goldstein 1971, Ngai and Paluch 2004). The latter is seen as a broad peak in the dynamic susceptibility, i.e. ω times the imaginary part of the time-Fourier transform of $\Phi(t)$, see Eq. (5.5). Since the amplitude of this peak is usually quite small, one sees it only at relatively low temperatures, i.e. when it is no longer masked by the $\alpha-$peak (which is the frequency equivalent of the $\alpha-$process). It is found that the location of the Johari-Goldstein peak shows an Arrhenius dependence on temperature, and therefore it is believed that it is related to some sort of local activated process, such as the flipping of a side-group in a polymer, without giving rise to real relaxation.

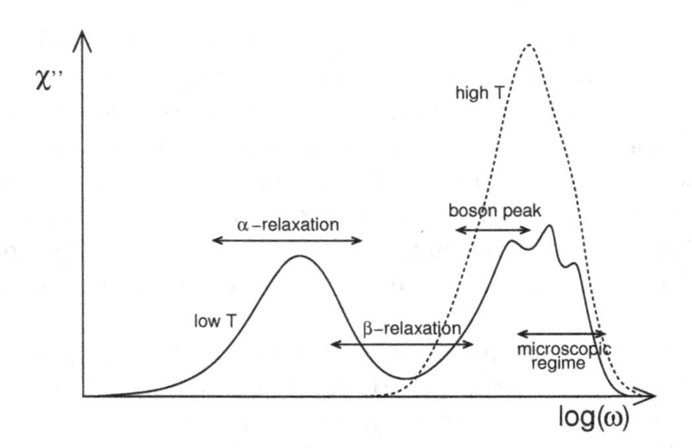

Fig. 5.4 Schematic plot of the frequency dependence of the imaginary part of the dynamic susceptibility corresponding to the time correlation function shown in Fig. 5.3.

Take for example the specific heat C. (Here it is irrelevant whether one considers the situation of constant pressure or constant volume. The same is true for quantum effects.) The $T-$dependence of C of a glass-forming liquid is shown schematically in Fig. 5.5. Also included in the graph is the specific heat of the material in its crystalline state and we see that due to anharmonic effects this latter curve increases smoothly from its value at $T = 0$ to a higher value at $T = T_m$, the melting temperature of the system (Ashcroft and Mermin 1976). From the graph we recognize that with decreasing temperature the specific heat of the liquid decreases rapidly, a dependence that has important implications for the relaxation dynamics as can be seen as follows. As we have seen in the context of Fig. 5.3, at sufficiently low temperatures the motion of the particles at short and intermediate times is of vibrational nature, due to the trapping inside the cage, and the relaxation dynamics takes place only on a much longer time scale. For such a system it is hence possible to split the specific heat into two parts: The vibrational part C_{vib}, which includes harmonic and anharmonic vibrations inside the cage, and C_{conf}, the "configurational part" of the specific heat which is due to the relaxational degrees of freedom, i.e. the type of motion which allows the particles to leave their cage and hence the liquid to flow.[4] It can be expected that C_{vib} is quite similar to

[4]We emphasize that this distinction between vibrational and configurational degrees of freedom can be done not only in hand-wavy way, but that it is indeed possible to make this separation also within statistical mechanics. For this one considers a "local temperature" $T(\mathbf{r},t) = (3k_Bm)^{-1}\sum_i \mathbf{p}_i^2\delta(\mathbf{r}-\mathbf{r}_i(t))$ and uses the Zwanzig-Mori projection

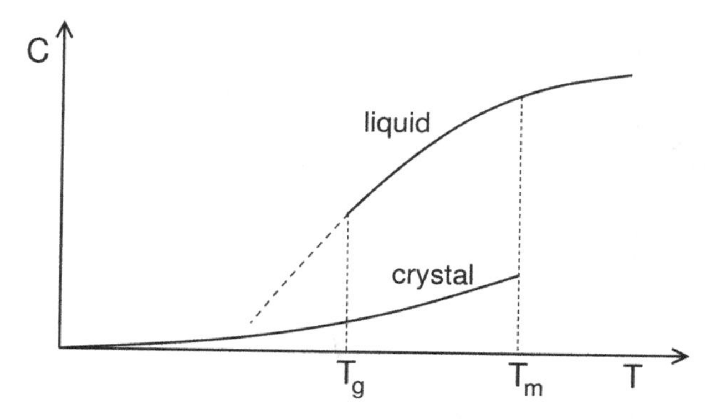

Fig. 5.5 Schematic plot of the temperature dependence of the specific heat of a glass-forming liquid (upper curve). The dashed line is an extrapolation of this curve to temperatures below the glass transition temperature T_g. The lower curve is the specific heat for the crystalline state of the same material that ends at the melting temperatures T_m.

the specific heat of the crystal, which is of course also purely vibrational, and hence the difference between the specific heat of the crystal and the one of the liquid can be used as a good estimate for $C_{\rm conf}$. Thus the fact that this difference seems to decrease rapidly with decreasing T, see Fig. 5.5, is an indication that $C_{\rm conf}$ becomes quickly smaller. Since $C_{\rm conf}$ is related to the number of states accessible to the system and hence to the number of possibilities for a structural rearrangement, a decreasing $C_{\rm conf}$ hints to a slowing down of the dynamics. This line of reasoning can be formalized also more quantitatively (Simon 1931, Kauzmann 1948). For this we use the T−dependence of the specific heat to calculate the entropy $S(T)$ in the crystalline and liquid state:

$$S_\alpha(T) = S_\alpha(T_m) - \int_T^{T_m} \frac{C_\alpha}{T}\, dT \qquad \alpha \in \{\text{liquid}, \text{crystal}\}\,. \tag{5.6}$$

Using this relation it is thus possible to calculate $\Delta S(T)$, the difference between the entropy of the liquid and the one of the crystal. The T−dependence of ΔS, normalized by its value at the melting temperature T_m is shown schematically in Fig. 5.6 (for real experimental data see, e.g., Kauzmann 1948). It is found that, at least for fragile glass-formers, this

operator formalism, see Sec. 5.2, to derive an equation of motion for the slowly varying component of $T(\mathbf{r}, t)$ (Scheidler *et al.* 2001). This approach allows also the interpretation of the results from frequency dependent specific heat measurements in which the sample is heated and cooled with a given frequency and one monitors how its energy contents behaves as a function of frequency (Birge and Nagel 1985, Menon 1996).

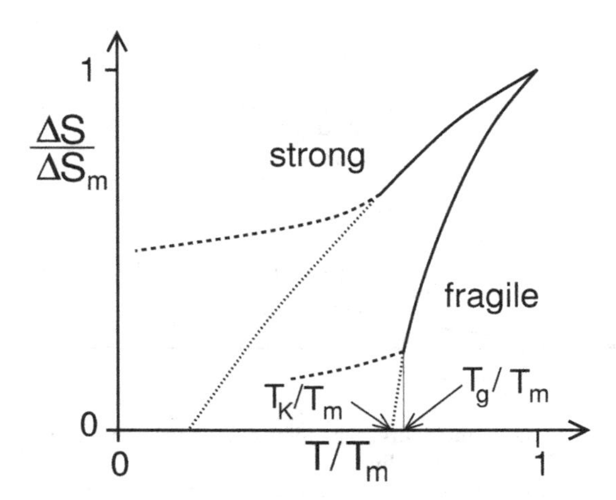

Fig. 5.6 Schematic plot of the temperature dependence of the difference between the entropy of a glass-forming liquid and the one of the corresponding crystal, normalized by its value at the melting point. The two solid lines correspond to the typical behavior of a strong and fragile glass-former. The dashed lines are the non-equilibrium data in the glass, whereas the dotted lines are an extrapolation of the equilibrium data to lower temperatures. These latter lines cross the abscissa at the Kauzmann temperature T_K.

normalized difference decreases rapidly with decreasing T and that it seems to become zero at a *finite* temperature, the so-called "Kauzmann temperature" T_K (Kauzmann 1948). For strong glass-forming liquids, the decrease of ΔS is much slower and T_K seems to be very close to zero. The trend that T_K/T_m increases with increasing fragility can be understood as follows: From Eq. (5.6) it follows that

$$\Delta S(T_m) = \int_{T_K}^{T_m} \frac{C_{\text{liquid}} - C_{\text{crystal}}}{T} \, dT \,. \tag{5.7}$$

Thus at a given value for T_m and a given entropy of fusion $\Delta S(T_m)$ we see that T_K increases with increasing $C_{\text{liquid}} - C_{\text{crystal}}$. Experimentally it is found that the jump in the specific heat between the glass and the liquid state, and hence also the jump in C between the crystal and the liquid, increases rapidly if the fragility is increased (Angell *et al.* 1994). Hence Eq. (5.7) allows indeed to rationalize the observation that fragile glass-formers have a T_K which is close to T_g, whereas strong glass-formers have a T_K that is relatively close to zero.

The observation that ΔS seems to become zero at a finite temperature is somewhat unusual since naively one might have expected that it is not possible to have an entropy smaller than the one of a crystal.[5] There are several possibilities to avoid this "problem". First of all one has to recall that below T_m the liquid is no longer the stable state and since thermodynamics does not make any statements about the entropy of metastable states there is in principle no problem. However, this argument helps only for finite temperatures. If one considers the limit $T \to 0$, the entropy of the (perfect) crystal goes to zero whereas the one of the ideal glass seems to stay at a negative value (if the specific heat of the liquid does not fall significantly below the one of the crystal). A negative entropy is, however, not compatible with Boltzmann's relation $S = k_B \ln \Omega$, where Ω is the number of available states. This apparent contradiction is what is known as "Kauzmann's paradox". A second possibility, proposed already by Kauzmann, is that with decreasing T the liquid becomes less and less stable, i.e. it becomes more and more prone to crystallization. Thus the point T_K marks the stability limit of the liquid and it is not possible to have the system *in quasi-equilibrium* at temperatures below T_K. Although this scenario is in principle a valid possibility, it is presently thought that it does not reflect the situation in real materials since the observed nucleation rates at low temperatures do not increase quickly enough (see Debenedetti 1997 and Cavagna 2009 for a more extensive discussion on this point). The third possibility is that at T_K the liquid just stops to flow, since there are no more states into which it can move, and hence it undergoes a complete structural arrest (apart of course of the vibrational motion), i.e. it becomes an ideal glass (Angell and Tucker 1974). This scenario is also in agreement with the analytical results of mean-field models of spin glasses (see Sec. 4.5) and the theory of Adam and Gibbs (Adam and Gibbs 1965), discussed in Sec. 6.1.1, which makes a quantitative connection between the vanishing of ΔS and the slowing down of the system. A further possibility is that the extrapolation of $\Delta S(T)$ below T_g, i.e. below the temperature close to which the system falls out of equilibrium due to the exceedingly large relaxation times, is wrong. Instead of becoming zero at T_K, $\Delta S(T)$ shows slightly above T_K a sharp bend and crosses over to a much weaker $T-$dependence without ever crossing the abscissa.

It is presently not clear whether for typical structural glasses the third,

[5] This expectation is wrong since, e.g., in a system of hard spheres at constant volume the Helmholtz energy A is given by $-TS$, since there is no potential energy. Thus minimizing A implies the *maximization* of S. Since it is believed that at high densities the crystalline state corresponds to the stable ground state, its entropy must be *larger* than the one of the liquid and hence ΔS will be negative.

i.e. ideal glass state, or the forth scenario, i.e. crossover of ΔS close to T_K, is correct (if any!). Although the concept of an ideal glass is certainly appealing and is also corroborated by the mentioned mean-field calculations as well as experimental data, Stillinger (1988) has put forward a compelling argument against this scenario, and in the following we will reproduce this reasoning. To derive this result, Stillinger used the concept of "inherent structures" which is the set of local minima of the potential energy of a given system (Goldstein 1969, Stillinger and Weber 1982, Debenedetti *et al.* 2001).[6] Computer simulations of systems like Lennard-Jones, analytical calculations for simple models, or very plausible arguments show that the total number of such local minima with a given energy increases exponentially quickly with the number of particles in the system, *even* if the permutation of the particles or simple symmetry operations of the system are factored out, a result which is in contrast to the situation found in a crystal (Stillinger 1999). To see this one considers a system of N particles which have only a short range interaction. Take an inherent structure that corresponds to an energy that is very low, i.e. which is populated significantly for temperatures close to T_K. Now cut the system into M equal pieces, e.g. cubes, with $N \gg M \gg 1$, reassemble these pieces into a new configuration, and relax the structure into its nearest local minima. Since $M \gg 1$ the local arrangement of the particles within each pieces will not be changed very much by this relaxation procedure and also the energy will be very close to the one of the original sample. Thus the only difference in energy between the original inherent structure and the new one stems from

[6] Note that each configuration of particles has exactly *one* associated inherent structure. This structure can be found, e.g., by using the configuration as the starting point of a steepest descent procedure in the potential energy and the local minimum found is then the inherent structure. Thus the configuration space can be partitioned in a unique manner into the sum of basins of attraction of the inherent structures. The relevance of these inherent structures stems from the fact that at low temperatures the system can be thought of making vibrational motion around a given local minimum and a transition to a nearby minimum will only occur rarely. By making this hypothesis of separation of time scales it is therefore possible to express thermodynamic quantities, like the energy or the free energy, as a function of the properties of the local minima (depth, vibrational frequencies) and the number of minima of a given type. Recent computer simulations have shown that this is indeed possible and that hence the concept of inherent structures is very useful (Sastry *et al.* 1998, Sciortino *et al.* 1999, Schrøder *et al.* 2000, Debenedetti *et al.* 2001, La Nave *et al.* 2002, Berthier and Garrahan 2003b, Doliwa and Heuer 2003, Shell *et al.* 2003, Wales 2003, Ruocco *et al.* 2004, Heuer 2008). Note that the number of inherent structures at a given energy can also be used to estimate the complexity I discussed in Chap. 4, see Eq. (4.274), although subtle differences might exist (Biroli and Monasson 2000).

the particles at the interfaces between the cubes. However, since $M \gg 1$ this interface energy (per particle) is small and therefore the energy difference (per particle) between the old and new inherent structure is small also. Thus this procedure shows that if one has *one* inherent structure, it is possible to generate immediately exponentially many (in M) and that therefore it is not possible to have only *one* (or *algebraically* many) ideal glass state(s).[7] As a consequence the temperature T_K below which one has by definition only one glass state, must be zero, a result that finds support from computer simulations (Sastry 2004).

In view of this argument, which seems to be very plausible, one has to see how its consequences, the absence of an ideal glass state, can be reconciled with the experimental data which, after extrapolation (!), seems to indicate the existence of such a state. One possibility is that the energy density of the inherent structures increases only very slowly for energies that are below $k_B T_K$, whereas it starts to increase quickly for energies above this threshold. Thus although strictly speaking there is no ideal glass or Kauzmann temperature, it would in practice nevertheless be possible to identify this temperature since the thermodynamic and dynamic quantities show at T_K a very strong temperature dependence. Evidence for such a strong $T-$dependence of the entropy has, e.g., been found in computer simulations of polymers in which sophisticated Monte Carlo moves were used to equilibrate the system even close to T_K (Wolfgardt *et al.* 1996). The reason why there is a strong change in the behavior of the energy density of the inherent structures at T_K are presently not well understood. However, the calculations for mean-field models of spin glasses show that there are models for which such a thermodynamic transition does indeed exist and it is thus quite plausible that this mean-field mechanism is responsible for the rapid change of the thermodynamic properties of real structural glasses as well, even if at the end the non-mean-field character of the latter systems washes out the real singularity.

Note that in most of the discussions made so far we have assumed that the liquid is in (quasi-)equilibrium, i.e. that it has sampled a representative part of configuration space without having found the "reaction path" to form a crystal. Since we have seen that with decreasing temperature the

[7] Note that this argument no longer holds in the case of long range interactions since then the energy of the interface can grow linearly with the system size. Therefore the above reasoning does not apply for mean-field like systems for which it is indeed found that there is an ideal glass state (Sherrington and Kirkpatrick 1975, Edwards and Anderson 1975). See Sec. 4.5.3 for details.

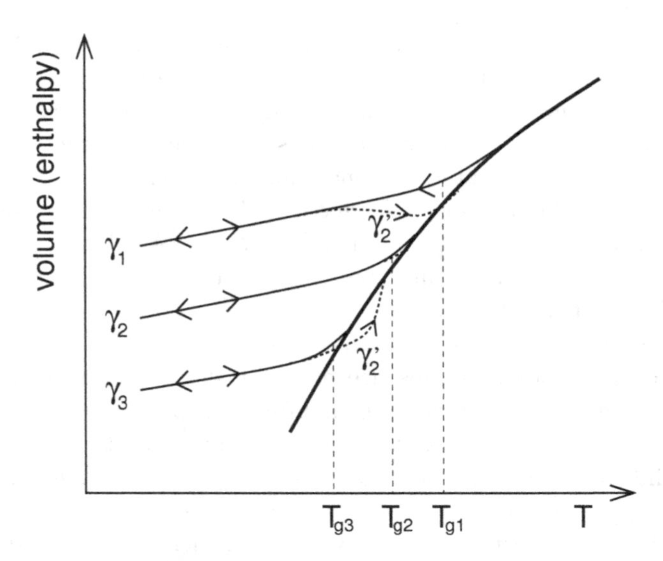

Fig. 5.7 Schematic plot of the temperature dependence of the volume, or the enthalpy, in a cooling and heating experiment on a glass-forming liquid. The bold solid curve is the equilibrium curve. The thin solid lines correspond to the cooling experiment with different cooling rate γ with $\gamma_1 > \gamma_2 > \gamma_3$. The dotted line are the heating curves with a rate $\gamma_2' = \gamma_2$.

typical relaxation times of the liquid increase dramatically, see the discussion made in context of Fig. 5.1, it is in practice not possible to equilibrate it down to (the hypothetical) T_K. The reason for this is that in order to equilibrate a system one usually has to wait for a time which is on the order of the typical relaxation time. Thus the lowest temperature at which a liquid can be studied *in equilibrium* will depend on the patience of the experimenter, which is certainly finite, and hence the mentioned temperature will be above T_K.

Another way to see this is to perform a cooling experiment in which the sample is cooled with a cooling rate γ_1 from a high temperature to a low temperature. This is shown schematically in Fig. 5.7 where we plot the volume, or any other quantity of interest, as a function of the temperature of the bath. For sufficiently high temperatures the curve for the cooling will follow the one of the liquid in equilibrium. However, at a certain temperature T_{g1} the relaxation times of the liquid exceed the time scale of the cooling, which is proportional to γ_1^{-1}, and hence the liquid falls out of equilibrium and forms a glass. This transition to the glass state can be noticed by the relative abrupt change of slope of the curve. If the same experiment is repeated with a cooling rate $\gamma_2 < \gamma_1$ one finds qualitatively the same behavior. However, the temperature at which the liquid falls out of equilibrium is now $T_{g2} < T_{g1}$, since due to the smaller cooling rate the system is able to equilibrate at a somewhat lower temperature. Hence we see that the glass transition temperature depends on the details of the experiment one is doing and hence is *not* an intrinsic temperature of the system, in contrast to, e.g., the melting temperature.[8]

Consider now an experiment in which a glass, which has been prepared with a cooling rate γ_1, is heated up using a heating rate $\gamma_2' = \gamma_2 < \gamma_1$. At low temperature the heating curve will follow the cooling curve, see Fig. 5.7. Slightly below the glass transition temperature T_{g1} the heating curve will, however, start to leave the cooling curve, because the system will start to relax toward the equilibrium curve since due to the small heating rate it has time to do so ($\gamma_2 < \gamma_1$!). Note that in principle this trend to reach equilibrium, a phenomenon that is called "aging", is present also at much lower temperatures. However, at low T the relaxation dynamics is so slow that in practice aging effects are very hard to detect. Since aging is an out of equilibrium phenomenon, it is not possible to apply the well known standard methods of equilibrium statistical mechanics to describe it. Due to this difficulty it is only in recent years that theoretical approaches have been developed that are able to treat the problem of the out of equilibrium relaxation dynamics of disordered systems by using mean-field theories or phenomenological descriptions like the so-called "trap model". More on these theories can be found in various review articles (Cugliandolo and Kurchan 1993, Bouchaud *et al.* 1996, Bouchaud *et al.* 1998, Latz 2000, 2001, Cugliandolo 2003).

If the heating of the glass is done with the same rate with which is has been produced, the mentioned aging effect is only very weak, although it is still present. Therefore the heating curve tracks the one of the cooling quite closely. If the heating is done with a rate that is higher than the cooling rate for the production of the glass, the system does not have the time to start to equilibrate even if T_{g3} is approached. Hence it will stay on the glass branch, which is below the liquid branch, until it has reached a

[8]We emphasize that the glass transition temperatures discussed in this paragraph have nothing to do with the one mentioned in the context of Fig. 5.2 and whose definition was $\eta(T_g) = 10^{12}$Pa·s. The latter is mostly used to make a comparison of the relevant temperature scale for different glass-forming materials, choosing an arbitrary but reasonable value for the viscosity. In contrast to this, the T_g's obtained from the mentioned cooling experiments are much more relevant since they allow to distinguish whether a system can be considered as a liquid or a glass.

temperature around T_{g2} at which it will start to approach the equilibrium curve.

Note that all this dependence on the cooling rate can be rationalized, at least qualitatively, by remembering the strong temperature dependence of the relaxation time. Since the derivative of the volume with respect to temperature is just the thermal expansion coefficient $\alpha(T)$, it is evident that the mentioned dependence of the volume is reflected also in a cooling/heating rate dependence of $\alpha(T)$, or if one considers the enthalpy, in the dependence of the specific heat $C(T)$. Thus by differentiating the curves for the various values of γ one sees immediately that, e.g., a fast heating will produce an overshoot in $C(T)$ just above the glass transition temperature T_{g3}, whereas a slow heating will produce a dip just below T_{g1}.

From Fig. 5.7 it is clear that the properties of the glass at low temperatures will depend on the cooling rate. To a first approximation these properties are strongly related to the ones of the liquid at the temperature at which the system fell out of equilibrium. Therefore this temperature is sometimes also called "fictive temperature". However, this name is somewhat misleading since the concept of temperature implies that it is the same for all observables of the system. For the case of glasses this is, however, not the case since the temperature at which the system falls out of equilibrium will depend on the observable considered. E.g. it is found that, for the same cooling rate!, the volume and the enthalpy do not fall out of equilibrium at the same temperature and hence in principle it is necessary to say that a glass has a given fictive temperature *for a certain observable* (Brüning and Samwer (1992), Vollmayr *et al.* 1996). Due to the kinetic nature of the glass transition it can be expected, however, that observables that have the same (equilibrium) relaxation time also have the same fictive temperature (Cugliandolo *et al.* 1997).

Although these results have been known by real glass makers since many years (Tool and Eichlin 1931), and who have used them to improve the properties of glasses by "annealing" them just below T_g, it is only in recent times that the concept of a fictive temperature has been based on a solid statistical mechanics foundation (Cugliandolo *et al.* 1997, Nieuwenhuizen 1998).[9] In particular it has been proposed to use the fluctuation-dissipation theorem, i.e. the relation between the time derivative of a correlation function and its

[9]Although there is no clear difference between the terms "aging" and "annealing", the former refers often to the situation in which the properties of the material deteriorate whereas in the latter case they improve. Notable counterexamples are, however, the aging of wine and cheese.

associated response, as a basis to define the fictive temperature. Consider an observable A and its autocorrelation function $C(t,t') = \langle A(t)A(t')\rangle$. If H is the conjugate field to A, the response of A to a perturbation H defines the response $R(t,t') = \delta A(t)/\delta H(t')$. (One example is that the application of pressure leads to a change in volume and this can be used to define the compressibility.) In equilibrium the properties of the system are invariant under time translation and hence the response as well as the correlation functions C depend only on the difference $t-t'$. The fluctuation-dissipation theorem states that (Chandler 1987)

$$R(t-t') = \frac{1}{k_B T}\frac{\partial C(t-t')}{\partial t'}\,. \tag{5.8}$$

In the out of equilibrium case this relation no longer holds and the functions C and R will depend not only on the time difference. For example one might take t as the time that has passed since the system has been driven out of equilibrium and $t-t'$ as the time lag for time correlation functions. To take into account this dependence on the two times, one generalizes Eq. (5.8) to

$$R(t,t') = \frac{1}{k_B T}X(t,t')\frac{\partial C(t,t')}{\partial t'}\,, \tag{5.9}$$

where the equation should be viewed as a definition of the unknown function $X(t,t')$, since the functions $R(t,t')$ and $C(t,t')$ can be measured directly in a real experiment. A comparison of Eq. (5.9) with Eq. (5.8) shows that $k_B T/X(t,t')$ can be used to define an "effective temperature" of the (out of equilibrium) system. Since *a priori* not much can be said on how X depends on t,t' or on the observable A, it might seem somewhat arbitrary to make such a definition. However, it can be shown than for certain mean-field models of spin glasses, as well as for coarsening systems, $X(t,t')$ becomes only a function of $C(t,t')$, i.e. $X(t,t') = x(C(t,t'))$ and independent of A (Cugliandolo and Kurchan 1993, Cugliandolo *et al.* 1996, Bouchaud *et al.* 1998). Although for structural glasses it has so far not been possible to calculate the function x, computer simulations have given evidence that it has a very simple form: $x(c) = 1$ for $c > c_{\mathrm{EA}}$ and $x(c) = m < 1$ for $c < c_{\mathrm{EA}}$ (Parisi 1997, Barrat and Kob 1999, Di Leonardo *et al.* 2000, Sciortino and Tartaglia 2001, Berthier and Barrat 2002a, 2002b). Here c_{EA} is the Edwards-Anderson order parameter introduced already in the context of spin glasses, see Sec. 4.5, i.e. it is the height of the plateau in the correlation function C (see also Fig. 5.3). These simulations have indicated also that the constant m is independent of the observable (Berthier and Barrat 2002a,

2002b), in agreement with the prediction of the calculations for the mean-field models. These results thus show that using the generalization of the fluctuation-dissipation theorem, i.e. Eq. (5.9), can indeed be used to define a fictive temperature of glasses and that this definition is reasonable. More details on these issues can be found in the review articles by Bouchaud *et al.* (1996), Bouchaud *et al.* (1998), and Cugliandolo (2003).

Before we conclude this section we point out that so far we have always assumed that the slowing down of the relaxation dynamics of the liquid is due to a decrease in temperature. It has to be emphasized, however, that this is by no means the only way to produce a glass. Other important possibilities include the compression of a fluid or crystalline samples at constant temperature, chemical reactions or evaporation in sol-gel systems, deposition of vapor on cold surfaces, etc. Therefore it is possible to generate glasses without ever going through the liquid state. E.g. strong mechanical grinding of various metals can result in an amorphous solid, and also the damage caused by the neutron irradiation of a crystal will transform it into a glass. Thus it is clear that a glass is certainly not a unique thermodynamic state: Even if certain macroscopic properties, such as density or the compressibility, are the same, it is very likely that the microscopic properties are different. Thus for a complete understanding of the properties of the glass it is in principle necessary to know its complete history, a feature which makes it a very useful technological material (since a change in the production process can be used to improve certain properties), but of course hampers tremendously its theoretical description.

5.2 The Mode Coupling Theory of the Glass Transition

Having discussed some of the main features of glass-forming systems, we now turn our attention to the theoretical description of the phenomenon of glassy dynamics. Due to the scientific importance and the interlectual challenge of the subject, a multitude of theoretical approaches have been proposed to rationalize the slowing down of glass-forming liquids and in Chap. 6 we will discuss some of them. As we will see at that place, in most cases these approaches translate a rather simple physical idea (importance of cooperativity, presence of defects, number of accessible states, etc.) into formulas that subsequently can be used to fit experimental data and, by studying the determined fit parameters as a function of temperature and the system considered, to make the proposed physical picture somewhat more quantitative. Although this approach is certainly useful to rationalize in a relatively simple way some of the experimental findings, it has the disadvantage that it remains qualitative and also has only a very limited ability to make predictions. It is, however, very desirable to have a theory at hand that allows to make a quantitative comparison with experimental data and to make nontrivial predictions. Unfortunately it has not been possible so far to find a reliable theory that is able to give a reliable description of *all* the phenomena encountered in glass-forming liquids. Nevertheless there exists at least one theoretical approach, presently the only one, that allows to make quantitative calculations for the dynamics of glass-forming systems and that furthermore has made many predictions on this dynamics. This theory, the so-called mode coupling theory of the glass transition (MCT), is making use of one important experimental fact: The structural properties of glass-forming liquids are only a very weak function of temperature, despite the fact that the relaxation dynamics changes by many orders of magnitude. In addition the theory takes advantage of the observation that at sufficiently low temperatures the time scale of the $\alpha-$relaxation is much larger than the microscopic one (see Fig. 5.3). Therefore it should in principle be possible to concentrate on the slowly varying variables, i.e. the structural relaxation. One systematic way to derive equations of motion for slow variables is the so-called Zwanzig-Mori projection operator formalism (Zwanzig 1960, 1961, Mori 1965). In the following we will therefore introduce this formalism, describe subsequently how mode coupling theory makes use of it, and finally discuss the predictions of the theory and some of the tests that have been done to check the validity and accuracy of this theoretical approach. More detailed reviews on the mathematical aspects of the mode coupling theory can be found in Götze (1991), Das (2004), and Götze (2009).

5.2.1 *The Zwanzig-Mori Projection Operator Formalism*

The Zwanzig-Mori formalism is a theoretical method that allows to derive an *exact* equation of motion for an arbitrary phase space function of a statistical mechanics system. Using physical insight into the problem of interest, this equation can subsequently be approximated thus allowing to obtain information on the dynamics of the system. Typical applications include the description of the slow dynamics in critical phenomena, of long time tails in the velocity auto-correlation function in liquids, or the dynamics of magnetic systems. More information on this approach can be found in Hansen and McDonald (1986), Balucani and Zoppi (1994), and Balucani *et al.* (2003).

Consider a classical $N-$particle system that is described by a Hamiltonian H. The equation of motion for an arbitrary phase space function g is given by

$$\dot{g} = i\mathcal{L}g = \{H, g\}, \tag{5.10}$$

where $\mathcal{L}$ is the Liouville operator and $\{.,.\}$ are the Poisson brackets. The set of all possible phase space functions form a vector space in which one can define a scalar product between two vectors g and h as

$$(g|h) = \langle \delta g^* \delta h \rangle, \tag{5.11}$$

where $\langle . \rangle$ is the usual canonical average, and $\delta g = g - \langle g \rangle$, i.e. we are considering only the fluctuating part of g.

Let us assume that, thanks to some physical insight, we know that the time dependence of the phase space variables A_n, $n = 1, \ldots, k$ is slow. In the following we will collect these functions into a column vector $\mathbf{A}$ and derive an equation of motion for it. To this aim we define a projection operator $\mathcal{P}$ that projects an arbitrary function g on the space spanned by the subset A_n:

$$\mathcal{P}g = (\mathbf{A}|g)(\mathbf{A}|\mathbf{A})^{-1}\mathbf{A} = \sum_{n,m} |A_n)[(A_i|A_j)]^{-1}_{nm}(A_m|g), \tag{5.12}$$

where $[(A_i|A_j)]^{-1}_{nm}$ is the n, m element of the inverse of the matrix $(A_i|A_j)$. It is easy to verify that $\mathcal{P}$ and $1 - \mathcal{P}$ are projectors, i.e. that they satisfy $\mathcal{P}^2 = \mathcal{P}$, $(1 - \mathcal{P})^2 = 1 - \mathcal{P}$, and that $\mathcal{P}(1 - \mathcal{P}) = (1 - \mathcal{P})\mathcal{P} = 0$.

From Eq. (5.10) it follows that the time dependence of $\mathbf{A}$ is given by $\mathbf{A}(t) = \exp(i\mathcal{L}t)\mathbf{A}$. If we insert in this expression, after the propagator $\exp(i\mathcal{L}t)$, the identity operator $\mathcal{P} + (1 - \mathcal{P})$ and differentiate the result with respect to t we obtain

$$d\mathbf{A}/dt = \exp(i\mathcal{L}t)[\mathcal{P} + (1 - \mathcal{P})]i\mathcal{L}\mathbf{A} \tag{5.13}$$

$$= i\mathbf{\Omega} \cdot \mathbf{A}(t) + \exp(i\mathcal{L}t)(1 - \mathcal{P})i\mathcal{L}\mathbf{A}, \tag{5.14}$$

where the "frequency matrix" $i\mathbf{\Omega}$ is defined by $i\mathbf{\Omega} = (\mathbf{A}|i\mathcal{L}\mathbf{A}) \cdot (\mathbf{A}|\mathbf{A})^{-1}$. We now rewrite the second term on the right hand side of Eq. (5.14). For this we make use of the identity

$$\exp(i\mathcal{L}t) = \exp[i(1 - \mathcal{P})\mathcal{L}t] + \exp(i\mathcal{L}t)S(t), \tag{5.15}$$

where the function $S(t)$ is defined as

$$S(t) = 1 - \exp(-i\mathcal{L}t)\exp[i(1-\mathcal{P})\mathcal{L}t]\,. \tag{5.16}$$

Differentiating this equation with respect to t gives the relation $dS(t)/dt = \exp(-i\mathcal{L}t)i\mathcal{P}\mathcal{L}\exp[i(1-\mathcal{P})\mathcal{L}t]$. Integrating this equation and making use of $S(0) = 0$, which follows from Eq. (5.16), thus gives the representation

$$S(t) = \int_0^t d\tau \exp(-i\mathcal{L}\tau)i\mathcal{P}\mathcal{L}\exp[i(1-\mathcal{P})\mathcal{L}\tau]\,. \tag{5.17}$$

This relation and Eq. (5.15) can now be used to rewrite the last term in Eq. (5.14):

$$\exp(i\mathcal{L}t)(1-\mathcal{P})i\mathcal{L}\mathbf{A} = \int_0^t \exp[i\mathcal{L}(t-\tau)]i\mathcal{P}\mathcal{L}\mathbf{f}(\tau)d\tau + \mathbf{f}(t)\,, \tag{5.18}$$

where the function $\mathbf{f}(t)$ is called "fluctuating force" (see Eq. (5.21) for the motivation for this name) and is given by

$$\mathbf{f}(t) = \exp[i(1-\mathcal{P})\mathcal{L}t]i(1-\mathcal{P})\mathcal{L}\mathbf{A}\,. \tag{5.19}$$

This last expression shows that the initial value of $\mathbf{f}$ is given by $i(1-\mathcal{P})\mathcal{L}\mathbf{A}$ and that its time dependence is given by the propagator $\exp[i(1-\mathcal{P})\mathcal{L}t]$ instead of the usual $\exp[i\mathcal{L}t]$. Due to the factor $(1-\mathcal{P})$ we also see immediately that

$$(\mathbf{A}|\mathbf{f}(t)) = 0\,, \tag{5.20}$$

i.e. $\mathbf{f}(t)$ is always orthogonal to $\mathbf{A}$.

Using in Eq. (5.18) the relations $\mathcal{P}\mathcal{L}\mathbf{f}(\tau) = (\mathbf{A}|\mathcal{L}\mathbf{f}(\tau)) \cdot (\mathbf{A}|\mathbf{A})^{-1}\mathbf{A}$ and $i(\mathbf{A}, \mathcal{L}\mathbf{f}(\tau)) = i((1-\mathcal{P})\mathcal{L}\mathbf{A}|\mathbf{f}(\tau))$, which follows from the fact that $\mathcal{L}$ is a Hermitian operator, i.e. that $(A|\mathcal{L}B) = (\mathcal{L}A|B)$, and realizing that this last term is equal to $-(\mathbf{f}(0), \mathbf{f}(\tau))$, we obtain for the equation of motion (5.14) the expression

$$\dot{\mathbf{A}} = i\mathbf{\Omega} \cdot \mathbf{A}(t) - \int_0^t d\tau \mathbf{M}(\tau) \cdot \mathbf{A}(t-\tau) + \mathbf{f}(t)\,, \tag{5.21}$$

where we have introduced the *memory function*

$$\mathbf{M}(t) = (\mathbf{f}|\mathbf{f}(t)) \cdot (\mathbf{A}|\mathbf{A})^{-1}\,. \tag{5.22}$$

Thus the equation of motion for $\mathbf{A}$ has been transformed into a generalized Langevin equation with $\mathbf{f}(\mathbf{t})$ playing the role of the fluctuating forces and relation (5.22) is just the expression for the fluctuation-dissipation theorem.

To obtain an equation of motion for the time correlation functions $\mathbf{C}(t) = \langle \mathbf{A}^*(0)\mathbf{A}(t)\rangle$, note that this is a $k \times k$ matrix, we take the scalar product of Eq. (5.21) with $\mathbf{A}$ and use the fact, see Eq. (5.20), that $(\mathbf{A}|\mathbf{f}(t)) = 0$:

$$\dot{\mathbf{C}}(t) = i\boldsymbol{\Omega} \cdot \mathbf{C}(t) - \int_0^t d\tau \mathbf{M}(\tau) \cdot \mathbf{C}(t-\tau) \,. \tag{5.23}$$

Note that this, so-called "memory equation", is exact, since no approximations have been made. Its formal solution can be easily obtained by making a Laplace transformation. If we introduce the Laplace transforms of $\mathbf{C}$ and $\mathbf{M}$,

$$\widehat{\mathbf{C}}(z) = i\int_0^\infty dt \exp(izt)\mathbf{C}(t) \quad \text{and} \quad \widehat{\mathbf{M}}(z) = i\int_0^\infty dt \exp(izt)\mathbf{M}(t) \,, \tag{5.24}$$

we obtain immediately the solution

$$\widehat{\mathbf{C}}(z) = -\left[z\mathbf{I} + \boldsymbol{\Omega} - i\widehat{\mathbf{M}}(z)\right]^{-1} \cdot \mathbf{C}(0) \,. \tag{5.25}$$

Here $\mathbf{I}$ is the unit matrix.[10] Since the memory function $\mathbf{M}(t)$ has a very complicated time dependence, because it is given by the modified propagator $\exp[i(1-\mathcal{P})\mathcal{L}t]$, see Eqs. (5.19) and (5.22), this formal solution is not of much use since $\widehat{\mathbf{M}}(z)$ is not known. It is, however, possible to approximate $\mathbf{M}(t)$ and thus to obtain a reasonable approximate solution for $\mathbf{C}(t)$. An alternative is to derive an equation of motion for $\mathbf{M}(t)$ and use its solution as input in Eq. (5.25). Surprisingly it is in fact possible to do this, and in fact quite simple, as we will demonstrate now.

From Eq. (5.19) we see that the time derivative of $\mathbf{f}(t)$ is given by

$$\frac{d\mathbf{f}(t)}{dt} = i(1-\mathcal{P})\mathcal{L}\mathbf{f}(t) \,, \tag{5.26}$$

which from a formal point of view is identical to the equation of motion for a generic space phase variable g in Eq. (5.10). The only difference is that now the Liouville operator $\mathcal{L}$ is replaced by the operator $\mathcal{L}_1 = (1-\mathcal{P})\mathcal{L}$. Hence we can immediately repeat all the steps we have done to derive the equation of motion of the set of variables $\mathbf{A}$ and hence to obtain an equation of motion for $\mathbf{f}(t)$ which reads:

$$\dot{\mathbf{f}} = i\boldsymbol{\Omega}_1 \cdot \mathbf{f}(t) - \int_0^t \mathbf{M}_1(\tau) \cdot \mathbf{f}(t-\tau) d\tau + \mathbf{f}_1(t) \,. \tag{5.27}$$

[10] Here we have made use of the convolution theorem of Laplace transforms (LT) which states that $LT[\int_0^t M(\tau)C(t-\tau)d\tau](z) = -iLT[M(t)](z)\, LT[C(t)](z)$.

Here we have introduced a new frequency matrix $\boldsymbol{\Omega}_1$, a new fluctuating force $\mathbf{f}_1(t)$, and a new memory function $\mathbf{M}_1(t)$ which are given by

$$i\boldsymbol{\Omega}_1 = (\mathbf{f}|i\mathcal{L}_1\mathbf{f}) \cdot (\mathbf{f}|\mathbf{f})^{-1}, \tag{5.28}$$

$$\mathbf{f}_1(t) = \exp[i(1-\mathcal{P}_1)\mathcal{L}_1 t]i(1-\mathcal{P}_1)\mathcal{L}_1\mathbf{f}, \tag{5.29}$$

$$\mathbf{M}_1(t) = (\mathbf{f}_1|\mathbf{f}_1(t)) \cdot (\mathbf{f}|\mathbf{f})^{-1} \tag{5.30}$$

with the new projector $\mathcal{P}_1$

$$\mathcal{P}_1 g = (\mathbf{f}|g)(\mathbf{f}|\mathbf{f})^{-1}\mathbf{f}. \tag{5.31}$$

Taking the scalar product of Eq. (5.27) with $\mathbf{f}$ thus gives us an exact equation of motion for the memory function $\mathbf{M}(t)$,

$$\dot{\mathbf{M}}(t) = i\boldsymbol{\Omega}_1 \cdot \mathbf{M}(t) - \int_0^t d\tau \mathbf{M}_1(\tau) \cdot \mathbf{M}(t-\tau), \tag{5.32}$$

which solution is obtained again by a Laplace transform:

$$\widehat{\mathbf{M}}(z) = -\left[z\mathbf{I} + \boldsymbol{\Omega}_1 - i\widehat{\mathbf{M}}_1(z)\right]^{-1} \cdot \mathbf{M}(0). \tag{5.33}$$

If we plug this expression into Eq. (5.25) for $\widehat{\mathbf{C}}(z)$, we see that we have a double fraction expression for this correlator:

$$\widehat{\mathbf{C}}(z) = \frac{-1}{z\mathbf{I} + \boldsymbol{\Omega} + \dfrac{1}{z\mathbf{I} + \boldsymbol{\Omega}_1 - i\widehat{\mathbf{M}}_1(z)} \cdot i\mathbf{M}(0)} \cdot \mathbf{C}(0). \tag{5.34}$$

It is evident that the described procedure can be iterated indefinitely and hence we will arrive ultimately at a presentation of $\widehat{\mathbf{C}}(z)$ in form of a continued fraction. Note that by construction all the occurring quantities, $\boldsymbol{\Omega}$, $\widehat{\mathbf{C}}(0)$, $\boldsymbol{\Omega}_1$, etc. are *constants*, i.e. independent of time, except for the last memory function considered which will depend on time. If we thus make a reasonable approximation for the latter, e.g. by a Gaussian function in time, and are able to calculate all the static quantities, we will obtain a good approximation for the time dependence of $\mathbf{C}(t)$. In practice, however, one often limits oneself to truncate the hierarchy at the second level, i.e. to approximate $\mathbf{M}_1(t)$, since otherwise the calculations become too involved.

Applications of the above described formalism include, e.g., the case of the velocity autocorrelation function for a Brownian particle in which the memory function is approximated by a $\delta-$function in time, or for moderately dense liquids in which it can be modeled by an exponential decay (Balucani and Zoppi 1994, Barrat 1999).

5.2.2 *The Mode Coupling Approximations*

In the discussion of the Zwanzig-Mori formalism we had the idea in our mind that the variables $\mathbf{A}$ which we chose were the *slow* ones. Although the resulting equation of motion for $\mathbf{A}$ is exact, it is somewhat useless if we are not able to come up with a good approximation for the memory function $\mathbf{M}(t)$. In order to make such approximations it would of course be good if this function were simple, i.e. if it could be approximated by a rapidly decaying function in time. However, although there are cases in which such an approximation is quite accurate, e.g. for the velocity-autocorrelation function of a Brownian particle mentioned at the end of the previous subsection, this is generally not the case, as can be see as follows. Recall that the fluctuating force $\mathbf{f(t)}$ is orthogonal to $\mathbf{A}$ for all times, see Eq. (5.20). However, this property does not imply that it is also orthogonal to products of the form $A_m A_n$ or higher powers. If the components of $\mathbf{A}$ are slow it must, however, be expected that also such products are slowly varying variables and that hence the operator $\mathcal{P}$ has not projected out all the slow variables, i.e. $\mathbf{f(t)}$ will still contain slow components. This implies that also the memory function $\mathbf{M}(t)$ decays only slowly and hence it is probably not a good idea to approximate it by a rapidly decaying function.

In order to overcome this problem one can simply decide to include these higher products in the set of slow variables as well (Kawasaki 1966). For this we define a new operator $\mathcal{P}_2$ as follows:[11]

$$\mathcal{P}_2 g = (AA|g) \cdot (AA|AA)^{-1} AA \,. \tag{5.35}$$

The full memory function, which is given by Eq. (5.22),

$$M(t) = (f|\exp[i(1-\mathcal{P})\mathcal{L}t]f)(A|A)^{-1} \,, \tag{5.36}$$

must contain a *slowly varying* part that is due to $\mathcal{P}_2 f$ and which is given by substituting in Eq. (5.36) f by $\mathcal{P}_2 f$. If we assume that this piece is the only part in the memory function that is slow, we can approximate the full memory function by this term:

$$M_{MC}(t) = (\mathcal{P}_2 f|\exp[i(1-\mathcal{P})\mathcal{L}t]\mathcal{P}_2 f)(A|A)^{-1} \,, \tag{5.37}$$

[11] Note that in order to keep the nomenclature simple we will restrict ourselves to the case of only one slow variable which we will denote by A. The generalization to the case of several variables is straightforward.

which, by using Eq. (5.35), can be rewritten as

$$M_{MC}(t) = \tag{5.38}$$
$$|(AA|f)|^2(AA|AA)^{-1} \cdot (AA|AA)^{-1} \cdot (AA|\exp[i(1-\mathcal{P})\mathcal{L}t]AA) \cdot (A|A)^{-1}.$$

Note that since in many applications the variable A is a "mode", such as a density fluctuation for a given wave-vector, the memory function $M_{MC}(t)$ contains the part of the memory in which these modes are coupled together. Therefore the described procedure to obtain this memory function is referred to as "mode coupling theory" and hence we have labeled the memory function with the subscript "MC".

The function $M_{MC}(t)$ contains static four-point correlations that are usually very difficult to calculate. In addition it also depends on a time correlation function whose time dependence is given by the highly non-trivial propagator $\exp[i(1-\mathcal{P})\mathcal{L}t]$. In order to obtain a somewhat more tractable expression for M_{MC}, one usually makes the approximation to factor the four-point correlation into a product of two-point correlations and to replace the mentioned propagator by the usual one:

$$(AA|AA) \approx (A|A)^2 \tag{5.39}$$
$$(AA|\exp[i(1-\mathcal{P})\mathcal{L}t]AA) \approx (A|\exp(i\mathcal{L}t)A) \cdot (A|\exp(i\mathcal{L}t)A)\,. \tag{5.40}$$

Although these two approximations might seem very arbitrary and uncontrolled, there exist a substantial number of situations for which the resulting (approximative) equation of motion for the observable A gives a very good description of the real relaxation dynamics. For example in the case of critical phenomena the slow dynamics of density fluctuations at small wave-vectors are related to conservation laws (Hohenberg and Halperin 1977). Thus it is in fact quite reasonable to express the memory function as a product of two modes (Kawasaki 1966). Others examples are long time tails in the velocity auto-correlation function in dense liquids or the slow dynamics in magnetic systems (Hansen and McDonald 1986, Balucani *et al.* 2003).

With the mentioned approximations the memory function $M_{MC}(t)$ is transformed into the simple expression

$$M_{MC}(t) = |V(AA,f)|^2(A|A(t)) \cdot (A|A(t))\,, \tag{5.41}$$

where the (time independent) quantity $|V(AA,f)|^2$ can be obtained from Eqs. (5.39) – (5.40) and is usually called a "vertex". Note that if we have more than one observable, we can collect them again in the form of a column vector and the mode coupling function will be a bi-linear form that couples the various modes to each other.

5.2.3 *The Mode Coupling Theory of the Glass Transition*

In this section we will discuss the application of the mode coupling approximation to the case of supercooled liquids and the formulation of the so-called mode coupling theory of the glass transition (MCT), which is often just called mode coupling theory. Starting point of the theory is the physically motivated choice that the position of the particles (or equivalently the density fluctuations) are the slow variables, since in glass-forming systems these positions are changing only slowly with time. Let us consider the density fluctuations for wave-vector $\mathbf{q}$, see Eq. (2.11), in Sec. 2.1:

$$\rho_{\mathbf{q}}(t) = \sum_{j=1}^{N} \exp[i\mathbf{q} \cdot \mathbf{r}_j(t)] \tag{5.42}$$

and the corresponding time correlation function $F(q,t)$, i.e. the coherent intermediate scattering function:

$$F(q,t) = \frac{1}{N} \langle \rho_{\mathbf{q}}(t) \rho^*_{\mathbf{q}}(0) \rangle \,. \tag{5.43}$$

Also relevant is the relaxation dynamics of a *tagged* particle which can be characterized by the time dependence of the single particle density correlation function, or incoherent intermediate scattering function, given by

$$F_s(q,t) = \langle \rho^{(s)}_{\mathbf{q}}(t) \rho^{(s)*}_{\mathbf{q}}(0) \rangle \quad \text{with} \quad \rho^{(s)}_{\mathbf{q}}(t) = \exp[i\mathbf{q} \cdot \mathbf{r}_j(t)] \,. \tag{5.44}$$

Using the density fluctuations as slow variables, the Zwanzig-Mori formalism can be used to obtain an equation of motion for $F(q,t)$. In particular we are interested in the form of the theory in which the memory function is expressed itself as a (second order) memory function, see Eq. (5.33), which thus in turn leads to the expression for the Laplace transform of $F(q,t)$ as given in Eq. (5.34). If in addition one makes the mode coupling approximations discussed in the previous section one obtains the expression

$$\Phi(q,z) = \cfrac{-1}{z - \cfrac{\Omega_q^2}{z + \Omega_q^2 \{\widehat{M}^{\text{reg}}(q,z) + \widehat{M}(q,z)\}}} \,, \tag{5.45}$$

where $\Phi(q,z)$ is the Laplace transform of the normalized correlator

$F(q,t)/F(q,0) = F(q,t)/S(q)$. The squared frequency Ω_q^2 is given by

$$\Omega_q^2 = q^2 k_B T/(mS(q)), \tag{5.46}$$

where m is the mass of the particles. If one transforms this equation back into the time domain one obtains:

$$\ddot{\Phi}(q,t) + \Omega_q^2 \Phi(q,t) + \Omega_q^2 \int_0^t \left[M^{\text{reg}}(q,t-t') + M(q,t-t')\right] \dot{\Phi}(q,t')dt' = 0. \tag{5.47}$$

Note that we have split the memory function into two parts: The function $M^{\text{reg}}(q,t)$, often also called the "regular part" of the memory function, describes the time dependence of $\Phi(q,t)$ at short times whereas the memory function $M(q,t)$ is responsible for the relaxation dynamics at long times. We mention that $M^{\text{reg}}(q,t)$ is also present in the case of normal liquids, i.e. if the relaxation dynamics is not slow. To a first approximation it can be considered to be a Gaussian function in time, but more involved approximations have also been discussed (Tankeshwar *et al.* 1987, 1995, Casas *et al.* 2000, Singh *et al.* 2003). Obtaining an accurate description for $M^{\text{reg}}(q,t)$ can therefore be considered as a problem of the dynamics of standard liquids (although it is not a simple one!), i.e. it has nothing to do with glass-forming liquids. This is in contrast to the case of the memory function $M(q,t)$ which is responsible for the relaxation dynamics of the system at long times and which is therefore itself a slowly decaying function. As discussed in the previous subsection, see Eq. (5.41), the mode coupling approximation allows to express $M(q,t)$ as a bi-linear product of the correlation functions with coefficients that can be computed from static quantities. Thus one obtains

$$M(q,t) = \frac{1}{2(2\pi)^3} \int d\mathbf{k} V^{(2)}(q,k,|\mathbf{q}-\mathbf{k}|)\Phi(k,t)\Phi(|\mathbf{q}-\mathbf{k}|,t), \tag{5.48}$$

where the vertex $V^{(2)}$ is given by

$$V^{(2)}(q,k,|\mathbf{q}-\mathbf{k}|) = \frac{n}{q^2} S(q)S(k)S(|\mathbf{q}-\mathbf{k}|) \left(\frac{\mathbf{q}}{q}\left[\mathbf{k}c(k) + (\mathbf{q}-\mathbf{k})c(|\mathbf{q}-\mathbf{k}|)\right]\right)^2. \tag{5.49}$$

Here $n = V/N$ is the particles density and $c(k)$ is the so-called "direct correlation function" that is related to the structure factor via $c(k) = n(1 - 1/S(q))$ (Hansen and McDonald 1986, Barrat and Hansen 2003).[12]

[12] Note that in order to obtain Eq. (5.49) it is necessary to make also the so-called convolution approximation in which a static three-point correlation function is expressed

Since the memory function $M(q,t)$ can be calculated from the correlators and static quantities, Eqs. (5.47)-(5.49) form a closed set of equations for the time dependence of the correlation functions. This set of equations are called the "mode coupling equations" and they can be considered as the starting point of the mode coupling theory of the glass transition. Thus in principle this theory is nothing else than the mathematical investigation of the solution of these equations and subsequently the interpretation of these solutions in terms of physical phenomena, some of which are discussed in the following. However, before we start this discussion we make some general remarks on these equations:

- The form of Eq. (5.47) reminds the equation of motion of a damped harmonic oscillator with a damping term whose amplitude depends not only on temperature (via Eq. (5.49)) but also on the correlators. For the case of simple liquids a decrease of temperature will give rise to a static structure factor that is more peaked and hence vertices $V^{(2)}(q,k,|\mathbf{q}-\mathbf{k}|)$ that are larger. Therefore also the memory function at $t=0$ increases, i.e. the damping becomes larger. This increased damping will result in a slower decay of the time correlation function and, due to Eq. (5.48), to a slower decay of the memory function, i.e. the time span over which the damping is noticeable becomes larger. Thus within MCT this non-linear feedback mechanism is the reason why a relatively small change in the structure can lead to a very strong change in the relaxation dynamics (Götze 1978, Geszti 1983).
- In Sec. 5.2.1 we mentioned that one typical application of the Zwanzig-Mori approach is the case where the memory function can be approximated by a function that decays quickly in time, i.e. its time dependence is faster than the one of the observable of interest. Although from a mathematical point of view Eq. (5.47) is of course also a Zwanzig-Mori memory equation, the memory function does *not* show a time dependence that is faster than the one of the correlators since it is a bi-linear product of the latter. From a physical point of view this is very reasonable, since the memory function is just the correlation function of the forces, see

as the product of two-point correlation function, i.e. the static structure factor (Götze 1991). Although for simple systems like soft spheres this approximation seems to be quite reliable (Barrat *et al.* 1989), this is not the case for network-forming liquids like silica for which the three-point terms have to be taken into account (Sciortino and Kob 2001).

Eq. (5.30). In contrast to the case of, e.g., Brownian dynamics, these forces are directly related to the motion of the particles, i.e. they are not given by some sort of external bath which has at most a short memory. Thus if the motion of the particles is slow, it is almost inevitable that also the forces vary only slowly with time.

- Due to the above mentioned increase of the feedback effect, the dynamics of the system slows down with decreasing temperature. Surprisingly this slowing down is not necessarily a smooth function of T. Instead there is, e.g., for the case of a simple liquid, a *finite* temperature T_c at which the solution does not decay to zero anymore even at infinitely long times (and likewise for $T < T_c$) (Götze 1978, Bengtzelius *et al.* 1984, Leutheusser 1984). Thus at T_c the system undergoes a transition from an ergodic dynamics for $T > T_c$ to a nonergodic one for $T < T_c$. This transition is usually called "ideal glass transition". (Below we will discuss the relation of this transition with the experimental glass transition or the transition at the hypothetical Kauzmann temperature.) Because of this property the mode coupling equations as given in Eq. (5.47) are also often called "ideal mode coupling equations" in contrast to the "generalized mode coupling equations" discussed below.
- The feedback mechanism described above is related to the fact that with decreasing temperature the structure factor becomes more peaked which in turn increases the vertices and hence the damping. From this it follows immediately that any change of external parameters that will give rise to an increase of $S(q)$ will also lead to a slow dynamics. For example an increase in pressure will also lead to a more pronounced structure factor and also the onset of nematic ordering can have the same effect (Letz *et al.* 2000).
- The memory function in the mode coupling equations (5.47) includes the term M^{reg} that is responsible for the relaxation dynamics at short times. Since, however, this memory function decays (by definition) quickly, the long time dynamics is only governed by the MCT-memory function $M(q,t)$. As the latter depends only on static quantities, the long time dynamics will not depend on the details of the microscopic dynamics (apart from an overall shift in the time scale due to the change of M^{reg}). Thus this implies the interesting prediction of the theory that the details of the relaxation dynamics at long times (height of plateau of the correlators at intermediate times, stretching of the correlators, etc.) will be

independent of whether one uses, e.g., a Newtonian dynamics or a Brownian dynamics. Computer simulations have given strong evidence that this very remarkable prediction of the theory is indeed correct (Gleim *et al.* 1998, Szamel and Flenner 2004, Berthier 2007, Berthier and Kob 2007).

- So far we have only discussed the MCT-equations for the coherent intermediate scattering function $F(q,t)$, i.e. the dynamics of the collective modes. Using the same approach it is, however, also possible to derive the MCT-equations for the incoherent intermediate scattering function $F_s(q,t)$:

$$\ddot{F}_s(q,t) + \frac{q^2 k_B T}{m} F_s(q,t) + \int_0^t \left[M^{\mathrm{reg,s}}(q,t-t') + M^s(q,t-t')\right] \dot{F}_s(q,t')dt' = 0\,. \quad (5.50)$$

Here the memory function is given by

$$M^s(q,t) = \frac{nk_BT}{(2\pi)^3 m} \int d\mathbf{q}' \left(\frac{\mathbf{k}'\cdot\mathbf{q}}{q}\right)^2 c(q')S(q')\Phi(q',t)F_s(|\mathbf{q}-\mathbf{q}'|,t)\,, \quad (5.51)$$

where $\Phi(q,t)$ is again $F(q,t)/S(q)$. Note that there is a noticeable difference between the memory function $M^s(q,t)$ and the one for the collective variable, Eq. (5.48), in that the former is not a quadratic function in the correlators $F_s(q,t)$, as one might have naively guessed, but instead a bi-linear function of $F_s(q,t)$ and $F(q,t)$. Thus this result shows that in order to obtain the relaxation dynamics of $F_s(q,t)$ one must first know the one of the collective variables. This is in fact quite reasonable since $F_s(q,t)$ describes the relaxation dynamics of a tagged particle in an disordered environment that is itself time dependent. Thus it is evident that this latter dependence must be used as crucial input in order to obtain the former dynamics.

- The equations discussed so far are valid for a one component system. Using an appropriate matrix notation, it is quite straightforward to obtain also the corresponding MCT-equations for a multicomponent system (Barrat and Latz 1990). This fact is of course very important since usually one-component systems are not good glass-formers since they crystallize too easily. Furthermore it is also possible to generalize the theory to the case of molecular systems such as small molecules or polymers (Franosch *et al.* 1997a, Kawasaki 1997, Schilling and Scheidsteger 1997, Chong and Hirata 1998a, 1998b, Fabbian *et al.* 1999a, 1999b, Götze *et al.* 2000, Letz *et al.* 2000, Chong *et al.* 2001b, Theenhaus *et al.* 2001, Chong and Götze 2002a), a feature that allows to use the theory to predict the relaxation dynamics of a very large class of glass-forming systems such as polymers, ortho-terphenyl, glycerol, etc.

- The derivation we have given for the mode coupling equations made use of the Zwanzig-Mori formalism and the realization that for glass-forming systems the memory function is a slowly varying function of time. This approach is, however, not the only one that leads to the MCT-equations. Das *et al.* and Schmitz *et al.* have used an approach called "fluctuating hydrodynamics" to obtain very similar equations (Das *et al.* 1985a, 1985b, Das and Mazenko 1986, Schmitz *et al.* 1993), Kawasaki has used density functional theory (Kawasaki 1994, 1995), Zaccarelli *et al.* (2002a) have shown that a formal rewriting of Newton's equations of motion in terms of density fluctuations and a subsequent Gaussian approximation of these fluctuations also leads to the MCT-equations, and recentlty Andreanov *et al.* (2009) have shown that the MCT equations can be understood within a Landau theory. A further interesting point is that Biroli and coworkers have shown that fluctuations will change the mean field character of the MCT equations if the dimension is smaller than $d_u = 6$ (Biroli and Bouchaud 2004), a value that is increased to $d_u = 8$ if the coupling between dynamical fluctuations and slow conserved degrees of freedom is taken into account (Biroli *et al.* 2006, Biroli and Bouchaud 2007). Thus in summary we see that this type of equations of motion for the density correlation functions is quite generic and not directly linked to one particular type of approximation.

The equations discussed so far have been obtained by using the density fluctuations as relevant slow variables. Of course this is only an approximation, although, as we will see in the next section, a quite good one, and hence attempts have been made to include also other slow variables that are considered to be relevant for getting a reliable description of the relaxation dynamics at low temperatures. Das and Mazenko (1986) and independently Götze and Sjögren (1987, 1988) included fluctuations of the current density in the projection operator scheme and found that the equations of motions for $\Phi(q,t)$ are modified in the following way:

$$\Phi(q,z) = \frac{-1}{z - \cfrac{\Omega_q^2}{z + \cfrac{\Omega_q^2\{\widehat{M}^{\text{reg}}(q,z) + \widehat{M}(q,z)\}}{1 - \widehat{\delta}(q,z)[\widehat{M}^{\text{reg}}(q,z) + \widehat{M}(q,z)]}}} \,. \tag{5.52}$$

Thus the only difference between this equation and Eq. (5.45) is the presence of the term $\widehat{\delta}(q,z)$ which, in the time domain, can be approximated as a quadratic form in the current densities $\dot{\Phi}(q,t)$:

$$\delta(q,t) = \sum_{q_1,q_2} \tilde{V}^{(2)}(q,q_1,q_2)\dot{\Phi}(q_1,t)\dot{\Phi}(q_2,t)\,. \tag{5.53}$$

Although it is presently not possible to give a tractable expression for the vertices $\tilde{V}^{(2)}(q,q_1,q_2)$, it can be assumed that they are a smooth function of the external control parameters (temperature, density, ...).[13] As we will see in the next subsection, the solutions of the MCT equations as given by Eqs. (5.45)-(5.49) show at sufficiently strong coupling, i.e. large $V^{(2)}(q,k,|\mathbf{q}-\mathbf{k}|)$, a singular dependence on the control parameters in that they do not decay to zero anymore, which implies that the system has reached a structural arrest. As mentioned above, this state is sometimes also called "ideal glass state" and hence the corresponding equations are also often called "ideal mode coupling equations". The presence of the hopping term $\delta(q,t)$ in the generalized equations of motion do no longer allow for these nonergodic solutions, see next section, and hence there is no structural arrest. Therefore this set of equations is called "extended mode coupling theory", "mode coupling theory with hopping terms", or "generalized mode coupling theory".

The MCT-equations are quite complex and therefore it is not surprising that no explicit analytical solution has been found so far. Nevertheless it is possible to obtain at least some general features of these solution, without having to make very restrictive assumptions on the form of the structure factors and hence the vertices $V^{(2)}$ (Götze 1991). The above mentioned existence of a critical temperature T_c below which the (physically relevant) solution of the equations does not decay to zero anymore is one of these features and in the next subsection we will discuss it, and many other ones, in more detail. Since these features are generic, i.e., they are observed in a large class of systems, they can also be studied in very simplified models.

---

[13] Within kinetic theory it is possible to derive approximative expressions for $\tilde{V}^{(2)}(q,q_1,q_2)$. See Sjögren and Sjölander (1979), Sjögren (1980, 1990) for details.

Consider, e.g., the case that we approximate the static structure factor of a one component glass-forming liquid by a single $\delta-$function that is located at the main peak, i.e. $S(q) = \delta(q - q_0)$. Within this approximation the MCT-equations (5.47)-(5.49) reduce to one single equation that is given by

$$\ddot{\phi}(t) + \Omega^2\phi(t) + \zeta\Omega^2 \int_0^t \phi^2(t-t')\dot{\phi}(t')dt' = 0\,, \qquad (5.54)$$

with $\phi(t) = F(q_0, t)/S(q_0)$ and we have neglected the regular part of the memory function $M^{\rm reg}$, since, as discussed above, it does not influence the solution of the equations at long times. Here ζ is a parameter that characterizes the amplitude of the memory function, i.e. it will depend in a smooth way on temperature, pressure etc. Since the memory function is just a quadratic function of the correlator, this equation is also called "F_2-model". The approximation $S(q) = \delta(q - q_0)$ that led to Eq. (5.54) for the relaxation dynamics for the coherent intermediate scattering function, can also be used to obtain a simple model for the incoherent function, see Eq. (5.50), which reads (Sjögren 1986)

$$\ddot{F}_s(q_0, t) + \frac{q_0^2 k_B T}{m} F_s(q_0, t) + \lambda \int_0^t \dot{F}_s(q_0, t')\phi(t-t')dt' = 0\,. \qquad (5.55)$$

Here λ is the strength of the coupling vertex and thus depends on temperature and we have again set to zero the regular part of the memory function. The set of equations (5.54) and (5.55) is called "$F_{12}-$model" or "Sjögren-model". It constitutes the simplest model that describes within MCT the relaxation dynamics of the coherent and incoherent scattering function of a glass-forming liquid. More general models are of course possible in that, e.g., one can approximate the structure factor by the sum of two $\delta-$functions, and hence a memory function that is a quadratic polynomial in two correlators. The set of equations obtained in this way are called "schematic models". Note that such schematic models can also be introduced for the MCT equations that include the hopping terms by setting the function $\delta(q, t)$ from Eq. (5.53) equal to $\delta(t) = \tilde{V}\dot{\phi}^2(t)$ (Götze and Sjögren 1988) and at the end of this chapter we will discuss such models in more detail.

Although the approximation that $S(q)$ is just the sum of a small number of $\delta-$functions is of course very crude, the so obtained schematic models are very useful since, despite their simplicity, they have the same *universal* features (discussed below) as the MCT-equations that have the full wave-vector dependence. In addition it is possible to devise schematic models that have the same *generic* features as the full MCT-equation (e.g. height of the plateau, transition temperature, etc.) and therefore such models have been found to be very useful to make fits to experimental data in order to see to what extend MCT is able to reproduce the relaxation dynamics of real systems (Alba-Simionesco and Krauzman 1995, Franosch *et al.* 1997b, Rufflé *et al.* 1999, Götze and Mayr (2000), Götze and Voigtmann 2000, Brodin *et al.* 2002, Krakoviack and Alba-Simionesco 2002).

In the derivation of the MCT equations as given by Eqs. (5.47)-(5.49) we had to make various approximations, see Sec. 5.2.2, the accuracy of which is difficult to grasp. Therefore only the comparison of the solution of the MCT equations with real experimental data or the data from computer simulations can really tell whether or not MCT is a reliable theory. It is, however, of interest to point out that there exist systems for which the MCT equations are an *exact* description of the relaxation dynamics. These systems are the mean-field spin glasses discussed in Sec. 4.5. As has been shown in a series of seminal papers by Kirkpatrick, Thirumalai, and Wolynes the equation of motion for the spin-autocorrelation function has the same mathematical form as the schematic models of MCT (Kirkpatrick and Wolynes 1987b, Kirkpatrick and Thirumalai 1988, Thirumalai and Kirkpatrick 1988). This result establishes a link between spin glasses, i.e. systems in which the disorder is put in by hand in the Hamiltonian, and structural glasses, systems in which the Hamiltonian does not contain any random elements. How far this connection really goes is presently not quite understood. Nevertheless it has motivated a large number of studies in which concepts from spin glass physics (pure states, replicas, calculation of Kauzmann temperature, violation of the fluctuation dissipation theorem, etc.) are used to investigate the properties of structural glasses and in turn to apply concepts from structural glasses (shearing, tapping, landscapes, aging, ...) to spin glasses (Bouchaud *et al.* 1998, Mézard and Parisi 1999, 2000, Coluzzi *et al.* 2000, Dean and Lefevre 2001, Berthier and Barrat 2002b, Coluzzi and Verrocchio 2002, Cugliandolo 2003, Parisi 2003, Mézard and Parisi 2009).

5.2.4 *Predictions of Mode Coupling Theory*

In this subsection we will discuss some of the salient predictions that MCT makes for the relaxation dynamics of glass-forming systems. Since MCT is just the set of equations given by (5.47)-(5.49), we thus discuss in principle only the mathematical properties of the solutions of these equations and the physical interpretation of these properties. For the sake of simplicity of language we will assume in the following that temperature is the external parameter that is changed and hence drives the slowing down of the dynamics. However, the discussed results are unchanged if instead, e.g., one changes the pressure or the concentration of the particles since at the end the only relevant part is that the vertices $V^{(2)}$ in the MCT equations become larger. All the predictions discussed in this section are *generic*, e.g., they hold for a large class of glass-forming systems (but not all). If one wants to study the specific properties of a *particular* system, one has to take the corresponding static structure factor to calculate the vertices and to solve the resulting MCT-equations. Although it is not possible to do this analytically there are highly efficient numerical algorithms that allow to obtain the solutions of the MCT equations over 15 decades in time (Fuchs *et al.* 1991). Obtaining these solutions is important since they allow to test to what extend the analytical predictions of MCT, which hold only asymptotically close to T_c, are modified by the finite distance from T_c, which in turn permits to make a more thorough comparison of the theory with experimental data.

The most important feature of the MCT equations is that there exists a critical temperature T_c at which the nature of the solutions at long times changes qualitatively. To see this we make use of the relation $\lim_{t\to\infty} C(t) = -\lim_{z\to 0} z\widehat{C}(z)$, an equation that is just a generic property of the Laplace transform.

If we apply this relation to Eq. (5.45) and use that the long time limit of $M^{\text{reg}}(q,t)$ is zero, we obtain immediately

$$f(q) := \lim_{t\to\infty} F(q,t)/F(q,0) = -\lim_{z\to 0} z\Phi(q,z) = \frac{\mathcal{F}(q)}{\mathcal{F}(q)+1}, \tag{5.56}$$

where we have introduced $\mathcal{F}(q) = \lim_{t\to\infty} M(q,t) = -\lim_{z\to 0} z\widehat{M}(q,z)$, the long time limit of the memory function. This equation can be rewritten as

$$\begin{aligned}\frac{f(q)}{1-f(q)} &= \mathcal{F}(q) &(5.57)\\ &= \frac{1}{2(2\pi)^3}\int d\mathbf{k}V^{(2)}(q,k,|\mathbf{q}-\mathbf{k}|)f(k)f(|\mathbf{q}-\mathbf{k}|), &(5.58)\end{aligned}$$

where in the second line we have made use of the definition of the memory function given in Eq. (5.48). Thus this set of (implicit) equations for $f(q)$ can be used to obtain the value of the long time limit of $\Phi(q,t)$. We see immediately that $f(q) = 0$ is always a solution. However, if the vertices are sufficiently large there exist also solutions $f(q) > 0$ and it can be shown that in that case the long time limit of $\Phi(q,t)$ is indeed given by the *largest* of these non-zero solutions (Götze 1991). If such a non-zero solution exists, it implies that the system is no longer ergodic, since the correlation functions do not decay to zero even at infinitely long times. Instead the long time value is given by $f(q) > 0$ and therefore this value is also called "nonergodicity parameter" or Debye-Waller factor. (The latter name comes from the analogous quantity in the context of vibrations in crystals.[14]) The highest temperature at which these positive solutions occur defines thus the critical temperature T_c of mode coupling theory. Note that Eq. (5.58) depends neither on the frequency Ω_q nor on $M^{\rm reg}(q,t)$, i.e. the regular part of the memory function, and therefore also T_c and $f(q)$ are independent of these quantities. This fact implies that the predicted transition is independent of the microscopic dynamics, and computer simulations have shown that this is indeed the case (see Sec. 5.2.5 and Gleim *et al.* (1998), Szamel and Flenner 2004, Berthier and Kob 2007).

In the previous subsection we have seen that the idealized mode coupling equations given by Eq. (5.45) can be generalized by including additional terms, the coupling to the currents, Eqs. (5.52) and (5.53). As we will show now, the presence of these hopping terms has an important consequence for the nature of the solutions in that there is no longer a nonergodic phase. To see this we consider again $f(q) = \lim_{t\to\infty} F(q,t)/F(q,0) = -\lim_{z\to 0} z\Phi(q,z)$ which, with Eq. (5.52), is found to be

$$f(q) = -\lim_{z\to 0} z\Phi(q,z) = \lim_{z\to 0} \frac{-z\widehat{M}(q,z)}{-z\widehat{M}(q,z) + 1 - \widehat{\delta}(q,z)\widehat{M}(q,z)}. \tag{5.59}$$

(Here $\widehat{\delta}(q,z)$ is the Laplace transform of $\delta(q,t)$ and we have neglected the terms in z^2.) For $z \to 0$ we have that $-z\widehat{M}(q,z) \to \mathcal{F}(q)$ and thus we obtain

$$f(q) = \lim_{z\to 0} \frac{\mathcal{F}(q)}{\mathcal{F}(q) + 1 - \widehat{\delta}(q,z)\widehat{M}(q,z)}. \tag{5.60}$$

Since $\lim_{z\to 0} \widehat{\delta}(q,z)$ is just the time integral under $\delta(q,t)$ (times i), and likewise for $\widehat{M}$, and both of these integrals are positive, a solution $f(q) \neq 0$ is no longer possible. This can be seen by assuming that $f(q) \neq 0$ from

[14] Note that in the dynamic structure factor $S(q,\omega)$, accessible in inelastic neutron scattering, the presence of such a nonergodic component shows up in the elastic part, i.e. as a contribution at $\omega = 0$, since the time-Fourier transform of a constant in time is a $\delta-$function in frequency.

which it follows that also the memory function $M(q,t)$ does not decay to zero even at infinitely long times. Thus the time integral of $M(q,t)$ diverges and, since $\lim_{z\to 0} \widehat{\delta}(q,z) \neq 0$, the fraction (5.60) goes to zero, from which it follows that also $f(q)$ must be zero. Therefore we find that the long times limit of $\Phi(q,t)$ is always zero, i.e. the system is always ergodic and hence there is no glass state. Nevertheless it is possible to define for such a system a critical temperature T_c by setting the hopping terms $\delta(q,t)$ to zero and to use Eq. (5.56) in order to find the corresponding T_c, i.e. the highest temperature at which the equation has a non-vanishing solution for $f(q)$. As long as $\delta(q,t)$ is small, it can be expected that, for $T > T_c$!, the solutions of the extended MCT equations are qualitatively similar to the ones of the idealized solution and that therefore also the former show a rapid change in their quantitative behavior, i.e. a rapid change in the relaxation times. Below we will come back to this point in more detail.

The existence of a critical temperature T_c allows to introduce the reduced distance $\epsilon = (T_c - T)/T_c$ and many of the following results are valid only in the case of small ϵ. (Therefore these predictions are often termed "asymptotic"). Using this definition we can now discuss the $T-$dependence of the nonergodicity parameters $f(q)$. Two cases can be distinguished: The function $f(q)$ starts to increase at T_c in a *continuous* way, i.e. $f(q,\epsilon) \propto \epsilon$. A transition that has this behavior is usually called "type A transition" (There is no rational reason for the choice of this name.) This type of transition is found, e.g., in certain type of spin glasses and in binary mixtures in which the particles have a very different size (Bosse and Thakur 1987, Bosse and Kaneko 1995, Götze and Voigtmann 2003). In the second case the nonergodicity parameters jumps discontinuously to a finite value $f^c(q)$, the so-called "critical nonergodicity parameter", and we have $f(q,\epsilon) - f^c(q) \propto \sqrt{\epsilon}$ (here we have $\epsilon > 0$). This type of transition is called "type B transition" and is encountered in most structural glasses and also in some spin glasses.

If one approaches T_c from above, MCT predicts that the $\alpha-$relaxation time increases in the form of a power-law (see the discussion after Eq. (5.77) for the derivation of this result):

$$\tau_x(T) = C_x(T - T_c)^{-\gamma} . \tag{5.61}$$

Here $\tau_x(T)$ denotes the $\alpha-$relaxation time of a correlator x (e.g. this might be the coherent or incoherent intermediate scattering function for a certain wave-vector, a given species, etc.). The prefactor C_x will depend in general on x and show a $T-$dependence which, close to T_c, can be neglected. The exponent γ is a *system universal* constant, i.e. is independent of x. In addition it can be shown that $\gamma > 1.0$. Thus the implication of Eq. (5.61) is that the relaxation times for all correlation functions will show, close to T_c, the same $T-$dependence. (Note that by "all" we mean that the overlap of the observable of interest and $\rho_q(t)$, defined by Eq. (5.11), does not vanish. Hence, e.g., the velocities of the particles do not show this critical slowing down.) This property is also often called "$\alpha-$scale universality". This property should also hold for the incoherent intermediate scattering function which, for $q \to 0$, is related to the mean squared displacement via $F_s(q,t) = 1 - \langle q^2(r_j(t) - r_j(0))^2\rangle/6 + O(q^4)$. Since the latter is in turn connected directly to the diffusion constant D, it follows immediately that also D shows a power-law singularity:

$$D^{-1}(T) \propto \tau_x(T) \propto (T - T_c)^{-\gamma}. \tag{5.62}$$

Note that Eq. (5.62) implies that, if we assume that the viscosity is proportional to $\tau_x(T)$, the Stokes-Einstein relation

$$D(T) \cdot \tau_x(T) = \text{const.} \tag{5.63}$$

is valid. As we will see later, this relation does not hold at low T due to the presence of the dynamical heterogeneities mantioned above and which will be discussed in more detail in Chap. 6.

We now discuss the full time dependence of the solutions of the MCT equations. In Fig. 5.8 we show such solutions for the case of a hard sphere system (Fuchs *al.* 1998). The different curves correspond to different packing fractions η and we see that if η approaches the critical value $\eta_c = 0.5159$ the relaxation becomes indeed very slow. Thus from a qualitative point of view these curves are very similar to the ones found for the relaxation dynamics of real supercooled liquids in that they show a transition from a fast relaxation at weak coupling (high temperature) to a slow dynamics at high coupling (low temperature). In addition we see that for low T (or in this case high density) the correlators show at intermediate times the plateau discussed already in Sec. 5.1 and which is due to the cage effect, i.e. the temporary trapping of the particles in the cage formed by their neighbors. It can be shown that the existence of such a plateau is generic for systems showing a type B transition, i.e. for which the nonergodicity parameter makes at T_c a discontinuous jump (Götze 1991), and that it should be observable for all correlation functions that have a non-zero overlap with density fluctuations.

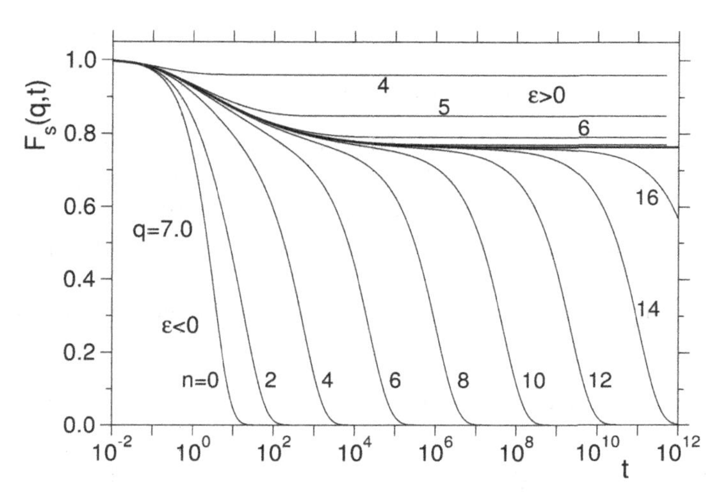

Fig. 5.8 Time dependence of the incoherent intermediate scattering function for a hard sphere system as predicted by mode coupling theory. The wave-vector is $q = 7.0$ (in units of the diameter of the spheres). The critical packing fraction is $\eta_c = 0.5159$. The different curves correspond to different packing fractions η with $\eta = \eta_c \pm 10^{-n/3}$, and the values of n are stated in the figure. Adapted from Fuchs *et al.* (1998).

From a physical point of view the existence of a plateau at intermediate times in a glass-forming liquid is quite plausible and one can envisage many simple theoretical approaches that will indeed give rise to such a feature. It is one of the merits of MCT that it makes very detailed predictions on the time dependence of the correlation functions close to the plateau which in turn can easily be tested in computer simulations or experimental data. One of the most remarkable predictions is the so-called "factorization property" which states that *in the $\beta-$relaxation regime* an arbitrary correlator $\Phi_x(t)$ can be written as follows:

$$\Phi_x(t) = \Phi_x^c + h_x G(t) . \tag{5.64}$$

Here Φ_x^c is the height of the plateau at T_c, i.e. the nonergodicity parameter, and h_x is an amplitude (often also called "critical amplitude"). The important prediction of MCT is that the time dependence of $\Phi_x(t)$ is given by the *system universal* function $G(t)$, i.e. it is independent of x. Thus we see that $\Phi_x(t) - \Phi_x^c$ is just the product of an $x-$dependent amplitude and a $x-$independent function $G(t)$, and this is the reason why Eq. (5.64) is called the "factorization property". (We emphasize that this property can be expected to hold only very close to the plateau, i.e. if $\Phi_x(t) - \Phi_x^c$ can be considered as small.)

If one makes a Fourier transform of Eq. (5.64) one obtains

$$\chi''_x(\omega) = h_x \chi''(\omega)\,, \tag{5.65}$$

where $\chi''_x(\omega)$ and $\chi''(\omega)$ are the susceptibilities associated to $\Phi_x(t)$ and $G(t)$, respectively (see Eq. (5.5)), and we have suppressed the $\delta-$function at $\omega = 0$. Equation (5.65) thus states that, close to T_c, all the dynamic susceptibilities have in the $\beta-$relaxation regime the same frequency dependence.

MCT also predicts that $G(t)$ has a very specific dependence on temperature which can be written as follows:

$$G(t) = \sqrt{|\sigma|}\, g_\pm(t/t_\sigma) \quad \text{with} \quad \sigma = C(T_c - T)/T_c\,, \tag{5.66}$$

where C is a constant and the $T-$independent functions $g_\pm$ correspond to $T < T_c$ and $T > T_c$, respectively. The time scale t_σ is the location of the plateau and is predicted to show a power-law divergence of the form

$$t_\sigma = t_0/|\sigma|^{1/2a}\,, \tag{5.67}$$

where t_0 is a microscopic time scale. (See Eq. (5.72) for the calculation of the exponent a.)

In order to determine the function $G(t)$, and hence $g_\pm$, one uses Eq. (5.64) as an Ansatz for the time dependence of the correlation functions. Using the relation $z\Phi(q,z) = -\Phi^c(q) + zh(q)G(z)$ and considering $zG(z)$ as a small parameter, the equation of motion (5.45) can be transformed, under the assumption to be close to the critical point, into (Götze 1985):

$$\sigma + \lambda G(t)^2 = \frac{d}{dt}\int_0^t G(t-t')G(t')dt' \tag{5.68}$$

or equivalently

$$\mp\zeta^{-1} + \zeta\hat{g}^2_\pm(\zeta) + i\lambda\int_0^\infty d\tau \exp(i\zeta\tau) g^2_\pm(\tau) = 0\,. \tag{5.69}$$

Here ζ is a rescaled frequency, $\zeta = z/t_\sigma$, that is adapted to the time scale of the $\beta-$relaxation, and the constant λ, which is also often called "exponent parameter", can be calculated from the vertices $V^{(2)}$ (Götze 1991). Since the functions $g_\pm(t)$ give the time dependence of the correlation functions in the $\beta-$relaxation regime, they are called "$\beta-$correlator". Note that the equation for $G(t)$, or equivalently $g_\pm(t)$, are independent of temperature or any other control parameter. This fact implies the remarkable prediction of MCT that the shape of the relaxation function in the $\beta-$regime is independent of T. The only $T-$dependence that is predicted to exist is the amplitude of the correlator and the time scale on which it relaxes, i.e. $\sqrt{|\sigma|}$ and t_σ, respectively.

Although the solution of Eq. (5.69) is not known analytically, it can be shown that it has the following asymptotic form:

$$g_\pm(\hat{t} \ll 1) = \hat{t}^{-a}\,. \tag{5.70}$$

In terms of the function $G(t)$ this corresponds to $G(t) = (t_0/t)^a$, which is valid for times much larger than t_0. Thus we have from Eq. (5.64) that

$$\Phi_x(t) = \Phi_x^c + h_x(t_0/t)^a\,, \tag{5.71}$$

i.e. that the correlation functions approach the plateau via a power-law. This time dependence is called "critical decay". The exponent a is related to the exponent parameter λ via the equation:

$$\frac{\Gamma(1-a)^2}{\Gamma(1-2a)} = \lambda\,, \tag{5.72}$$

where Γ is the $\Gamma-$function and a obeys the inequality $0 < a < 0.5$ (Götze 1999).

For times that are long with respect to t_σ but still short with respect to the $\alpha-$relaxation time τ of the system, the functions $g_\pm$ can be shown to have the following behavior:

$$g_+(\tilde{t}) = 1/\sqrt{1-\lambda} \tag{5.73}$$

$$g_-(\tilde{t}) = -B\tilde{t}^b\,, \tag{5.74}$$

with the rescaled time $\tilde{t} = t/\tau$. Equations (5.66) and (5.73) imply that in the glass phase, $\sigma > 0$!, the correlation functions do not decay to zero even at long times and, together with Eq. (5.64), that the height of their plateau increases as $\sqrt{\sigma} \propto \sqrt{T_c - T}$. In contrast to this we have for the liquid side, $\sigma < 0$, a time dependence that is given by a power-law, a functional form that in this context is often called the "von Schweidler-law" (von Schweidler 1907). The exponent b is directly related to λ, and hence to a, via

$$\frac{\Gamma(1+b)^2}{\Gamma(1+2b)} = \lambda\,. \tag{5.75}$$

Thus once one of the three quantities a, b, or λ is known, one can calculate immediately the two other ones.

Plugging Eq. (5.74) into Eq. (5.66) and then (5.64) we obtain

$$\Phi_x(t) = \Phi_x^c - h_x B(t/\tau)^b \tag{5.76}$$

with the time scale τ that is given by

$$\tau = t_0/|\sigma|^\gamma \qquad \text{with} \quad \gamma = 1/(2a) + 1/(2b) \,. \tag{5.77}$$

Hence we see that the time scale for leaving the plateau, and hence the time scale for the $\alpha-$relaxation, shows the power-law dependence discussed already in Eq. (5.61). This temperature dependence is a direct consequence of the two power-laws in the $\beta-$relaxation, i.e. the critical decay, Eq. (5.71), and the von Schweidler-law of Eq. (5.76).

For the derivation of these results we have made use of the equation of motion for the *ideal* version of MCT, Eqs. (5.45)-(5.48). As we have already mentioned before, this version of the theory can be extended by taking into account the hopping terms : Eqs. (5.52) and (5.53). If one uses the Ansatz (5.64) for this extended version, one finds instead of Eq. (5.68) the relation:

$$\sigma - \delta t + \lambda G(t)^2 = \frac{d}{dt}\int_0^t G(t-t')G(t')dt' \,, \tag{5.78}$$

where δ is the strength of the hopping which is here taken into account not in a $q-$dependent way, but as an effective, temperature dependent, parameter (Götze and Sjögren 1987). From this equation it is evident that we now have a new time scale $1/\delta$ and that therefore the nature of the $\beta-$relaxation is different from the one that corresponds to $\delta = 0$. More details on this can be found in the paper by Fuchs *et al.* (1992). Thus the main point that we want to emphasize is that from Eq. (5.78) it follows that even in the presence of hopping processes the factorization property holds, i.e. at a given temperature (close to T_c) the shape of the correlators is still independent of the observable considered, a result that has been confirmed in computer simulations of Horbach and Kob (2001, 2002) for the case of silicate glasses.[15] However, since in general δ will depend on temperature, also the function $G(t)$ becomes $T-$dependent. Thus in contrast to the ideal version of the theory where the shape of the relaxation functions are predicted to be independent of T, apart from a $T-$dependent amplitude and time scale, we now find that this shape depends on T.

[15] It is also remarkable that it is found that this factorization property seems to hold also for the case of glass-forming systems that are sheared (Berthier and Barrat 2002a, 2002b) or that are aging (Kob and Barrat 2000), which is evidence that this property is indeed a very robust feature of supercooled liquids.

The results discussed so far concerned the $\beta-$relaxation, i.e. the dynamics of the system on time scales at which the correlation functions are close to the plateau. We now turn our attention to the $\alpha-$relaxation, i.e. the time scale on which the correlators show the final relaxation to zero. One of the important predictions of MCT is that in this regime the so-called "time-temperature superposition principle" (TTSP) holds, which means that a correlator $\Phi_x(t)$ at temperature T can be written as

$$\Phi_x(t,T) = \tilde{\Phi}_x(t/\tau(T)), \tag{5.79}$$

where $\tau(T)$ is the $\alpha-$relaxation time of the system. (Note that due to Eq. (5.61) this time scale is, apart from a trivial prefactor, independent of x and can hence be absorbed in the definition of $\tilde{\Phi}_x$.) Thus the TTSP says that in the $\alpha-$relaxation regime the shape of the correlator is independent of T. In Sec. 5.1 we have seen that this shape can often be approximated very well by a Kohlrausch-Williams-Watts function, see Eq. (5.4). Thus the TTSP implies that the KWW-exponent β as well as the prefactor A are independent of T. Note that in general the KWW-law is *not* a solution of the MCT equations, as can be seen by plugging the KWW-law into the MCT equation.[16] However, if one determines these solutions numerically one finds that they can indeed be approximated very well by KWW-functions, in agreement with the findings for experimental data. It has to be pointed out that there is presently no real evidence that the KWW-law should indeed be an exact description for the $\alpha-$relaxation. Instead, it is much more likely that in most cases it is just a very good fitting function with no physical meaning. Thus within the framework of MCT one would argue that a KWW-law fits the experimental data because the former is similar to the solution of the MCT equations.

It is important to realize that within MCT the $\beta-$process is *not* the short time behavior of the $\alpha-$process, but that it is a *independent* process. One way to see this is to consider the short time expansion for the KWW-law which is given by $A(1-(t/\tau)^{\beta})$, see Eq. (5.4). Thus one might be tempted to identify this expression with the von Schweidler-law given by Eq. (5.76). However, this relation does *not* hold, since the von Schweidler exponent b is system universal, i.e. independent of the correlator considered, whereas the

[16] One remarkable exception to this is a result due to Fuchs (1994) who showed that in the limit of large q the solutions of the MCT equations are given by a KWW-law with an exponent β that is given by the von Schweidler exponent: $\beta = b$. Computer simulations have given strong evidence that this prediction of the theory is indeed true (Sciortino *et al.* 1997, Horbach *et al.* 2002, Puertas *et al.* 2003, Foffi *et al.* 2004).

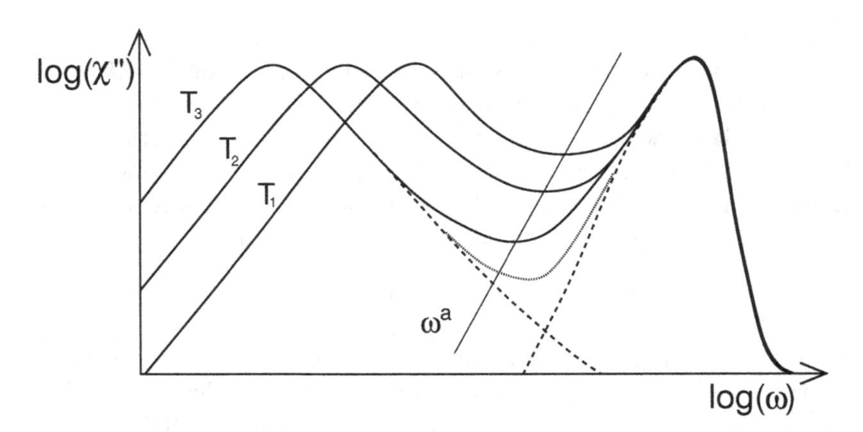

Fig. 5.9 Schematic plot of the frequency dependence of the imaginary part of the susceptibility as predicted by MCT. The three bold curves correspond to three different temperatures with $T_1 > T_2 > T_3$. With decreasing temperature the $\alpha-$peak moves quickly to lower frequencies without changing its shape, whereas the microscopic peak is basically independent of T. The dashed curves are the $\alpha-$peak and the microscopic peak at T_3. The dotted line is the sum of these two processes. Note that, due to the presence of the $\beta-$process, the height of the minimum between the $\alpha-$peak and the microscopic peak is higher than the sum of the two processes in this frequency range. The location of the minimum between the $\alpha-$peak and the microscopic peak is predicted to be given by a power-law (straight line).

KWW-exponent β does depend on the correlation function. Thus one must indeed conclude that from this point of view the $\beta-$process is independent from the $\alpha-$process.

So far we have mainly discussed the behavior of the relaxation functions in the time domain. However, since many experiments are done in the frequency domain, it is useful to see what the predictions of MCT mean in the frequency domain. The TTSP implies that for the imaginary part of the susceptibility $\chi''_x(\omega)$, see Eq. (5.5), the shape as well the height of the $\alpha-$peak is independent of T (see Fig. 5.4). Thus, according to MCT a decrease in T will make that the $\alpha-$peak moves quickly to lower frequencies, whereas the microscopic peak will basically show no $T-$dependence. Hence it is clear that at sufficiently low temperatures there will be a minimum between the microscopic peak and the $\alpha-$peak. However, it is one of the important predictions of MCT that close to the minimum, i.e. in the $\beta-$relaxation regime, the susceptibility is *not* just the sum of the $\alpha-$peak and the microscopic peak, but that instead there is an additional contribution, namely the $\beta-$process of MCT given by Eq. (5.65), see Fig. 5.9.

The time scale of the latter, given by t_σ from Eq. (5.67), has a *different* T−dependence than the one of the α−process, given by τ from Eq. (5.77), and must therefore indeed be taken as a new and independent process. Note that the predicted time dependence of the β−process, given by Eqs. (5.71) and (5.76), implies that the high frequency wing of the α−peak is given by $\chi''(\omega) \propto \omega^{-b}$ and the low frequency wing of the microscopic peak should show a dependence of the form $\chi''(\omega) \propto \omega^{a}$.

5.2.5 *The Relaxation Dynamics of Glass-Forming Liquids and Test of the Predictions of MCT*

In this subsection we will discuss some results from experiments and computer simulations that on one hand illustrate some features of glassy liquids and on the other hand show how the predictions of MCT can be tested. According to the theory the only observables that show a strong T−dependence are time correlation functions, in agreement with most results from experiments and computer simulations. (This holds of course only for the case of structural glasses. For spin glasses the situation is very different, as we have seen in Chap. 4.) Therefore we will focus in the following only on these dynamical observables and leave out completely the static quantities. For the sake of simplicity most of the results discussed here will concern simple liquids or network forming liquids, thus neglecting important systems like molecular and polymeric liquids. For an extensive discussion of the latter systems we refer the reader to the review articles by Binder *et al.* (2003) and Baschnagel and Varnik (2005).

One of the simplest time correlation functions is the mean squared displacement (MSD) $\langle \Delta r^2(t)\rangle = \langle |\vec{r}_i(t) - \vec{r}_i(0)|^2\rangle$ of a tagged particle. An example for its time and T−dependence is shown in Fig. 5.10. The system considered is a binary Lennard-Jones mixture, with 80% of the particles of type "A" and 20% of type "B", which in the past has been shown to be an excellent glass former (Kob 1999).[17] The curves shown correspond to the MSD for the A particles. At high temperatures two regimes can be distinguished: In the first one, which corresponds to short times, $t \leq 0.2$, the MSD is a quadratic function in time since the particles fly just ballistically. This behavior follows directly from the Taylor

[17]The advances of computer power of the last few years have allowed to do equilibrium simulation of this system at such low temperatures that this composition shows some trends to cristallization (Toxvaerd *et al.* 2009). However, other compositions seem still to be stable against crystallization on the time scale of present day computer simulations (Brüning *et al.* 2009).

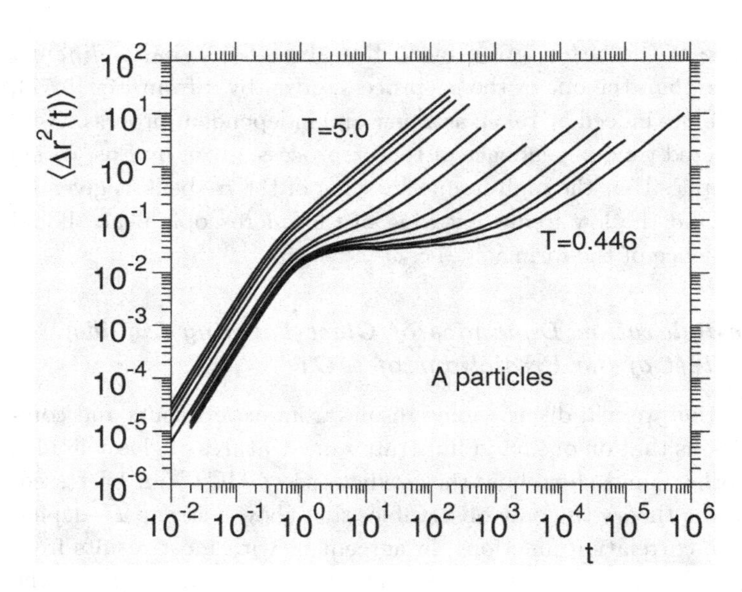

Fig. 5.10 Time dependence of the mean squared displacement of the A particles in a binary Lennard-Jones mixture. The curves correspond to the temperatures $T = 5.0$, 4.0, 3.0, 2.0, 1.0, 0.8, 0.6, 0.55, 0.50, 0.475, 0.466, and 0.446 (from left to right). For short times we find the ballistic regime, $\langle \Delta r^2(t) \rangle \propto t^2$, and at long times the diffusive one, $\langle \Delta r^2(t) \rangle \propto t^1$. At low T these two regimes are separated by the $\beta-$relaxation during which the particles are trapped (cage effect). From Gleim (1998).

expansion in time of the trajectory: $\vec{r}_i(t) = \vec{r}_i(0) + \vec{v}_i(0)t + O(t^2)$ and hence $\langle \Delta r^2(t) \rangle = \langle |\vec{v}_i(0)|^2 \rangle t^2 + O(t^4)$. Taking into account that the distribution of the velocity is given by a Maxwell-Boltzmann distribution one obtains immediately that $\langle |\vec{v}_i(0)|^2 \rangle = 3k_BT/(2m)$. Once the particles have moved for a time beyond which the quadratic Taylor expansion is no longer valid, i.e. the particles start to interact with each other and therefore higher order terms start to become important, the MSD enters the second regime. In this regime the particles collide with their neighbors and therefore their motion is no longer ballistic. At high temperatures these collisions give quickly rise to a complete randomization of the velocities and hence the particles start to show a diffusive motion and consequently the MSD becomes a linear function in time.

At low temperatures the MSD shows *three* regimes: Apart from the ballistic and diffusive one that we have just discussed, we have at intermediate times the "cage regime" at which the MSD is basically constant. As already mentioned in Sec. 5.1, in this time window, the $\beta-$relaxation, the particles are temporarily trapped by their neighbors and hence cannot move beyond a length scale that is related to the nearest neighbor distance of the particles and the range of interaction. From the figure we infer that this distance is of the order $\sqrt{0.03} \approx 0.17$, which has to be compared with the typical distance between two particles, which in this system is 1.1 (Kob and Andersen 1995a). Thus we see that the size of the cage is significantly smaller than the typical interparticle distance.[18] Only at relatively long times the particles are able to leave their cage and hence to become diffusive. The figure shows that with decreasing temperature the duration of this caging regime increases quickly and that in fact it is the only time regime which shows a noticeable T−dependence. Hence it is clear that in order to understand the slowing down of the relaxation dynamics of the system it is necessary to understand this caging regime, and MCT is a theoretical approach that makes precise predictions for the correlation function in this time window.

The MSD discussed so far is the one found in simple liquids, i.e. systems for which the interaction is short ranged and isotropic. For more complex liquids, such as polymers or systems that have an open network structure, the time dependence of the MSD is a bit more complex. For the case of the polymers this is demonstrated in Fig. 5.11 where we show $g_1(t)$, the MSD of the *central* monomer of a short polymer chain of length 10. From this figure we recognize that at short times we find again the ballistic motion and at long times the linear time dependence corresponding to diffusive motion. In contrast to the case of simple liquids, see Fig. 5.10, we have, however, at intermediate times not only a plateau but a more complex time dependence: At the end of the ballistic regime the particles are again caged and thus $g_1(t)$ shows a plateau. However, this plateau is not very long and the monomers try to leave their cage. Since they are attached to the rest of the polymer, this escape is hindered as they have to drag along the other monomers as well. This cooperative motion gives rise to a sub-diffusive time dependence, $g_1(t) \propto t^{0.63}$, and only once the central monomer has moved on the order of R_e, the end-to-end distance of the polymer (see Sec. 3.1), it starts to show

[18]We emphasize that the phenomenon of the cage effect is not necessarily related to the fact that the trapped particle is surrounded in *all* directions by other particles. For the case of systems with covalent bonds, such as SiO_2, the motion of the bridging oxygen atom is hindered because it is strongly linked to, only two!, silicon atoms (see Fig. 3.13). Thus, although there are no steric hindrance effects that impede the oxygen atom to move far away from its two silicon neighbors (in a direction perpendicular to the Si-O-Si line), the directional interactions between the three atoms do not allow for this motion. A similar mechanism can be expected to be relevant in gel-forming systems which form mechanically stable structures although the particle density is very small.

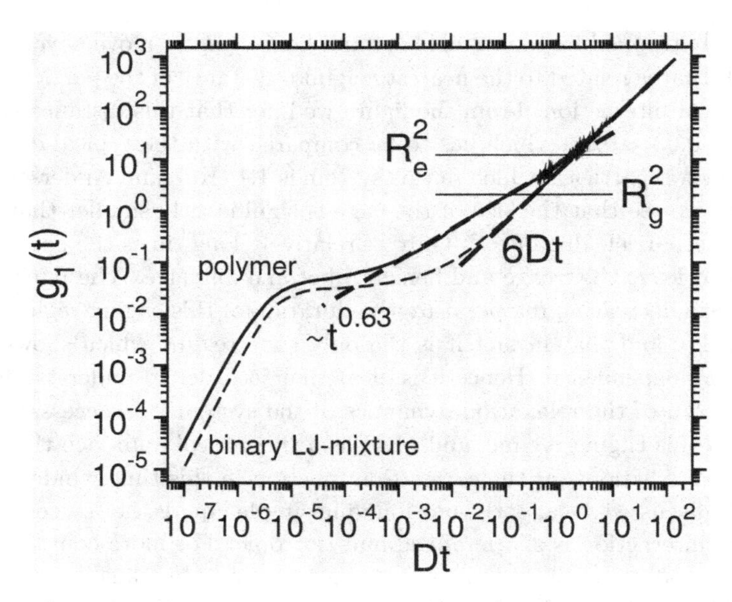

Fig. 5.11 Time dependence of the mean squared displacement of a central monomer of a polymer in a dense melt (bold solid curve). We find at short times a ballistic regime, followed by a short plateau. Subsequently we have a sub-diffusive regime, $g_1(t) \propto t^{0.63}$, before the MSD crosses over to the diffusive regime (dashed lines). The horizontal lines correspond to the square of the radius of gyration and the end-to-end distance. The dashed curve is the MSD of the binary Lennard-Jones system at $T = 0.466$ from Fig. 5.10. Adapted from Baschnagel *et al.* (2000).

a diffusive behavior. In order to show the qualitative difference between the relaxation dynamics of the polymer system and the one of a simple liquid, we have included in Fig. 5.11 also the MSD for the binary Lennard-Jones mixture discussed in Fig. 5.10. By comparing the two curves we thus can conclude that in a dense polymer melt the slow dynamics is not only due to the cage effect for the individual monomers, i.e. the mechanism that is responsible for the slowing down of the dynamics in simple liquids, but that there are also important effects due to the connectivity of the polymer.

Although it is in principle possible to compare the predictions of MCT regarding the $\beta-$relaxation with the time dependence of the MSD, we will restrict ourselves to do this for the case of the intermediate scattering function. In Fig. 5.12 we show the time dependence of the incoherent intermediate scattering function, defined in Eq. (5.44), for different temperatures. The system is the same binary Lennard-Jones mixture from Fig. 5.10. The wave-vector is $q = 7.25$, which corresponds to the location of the

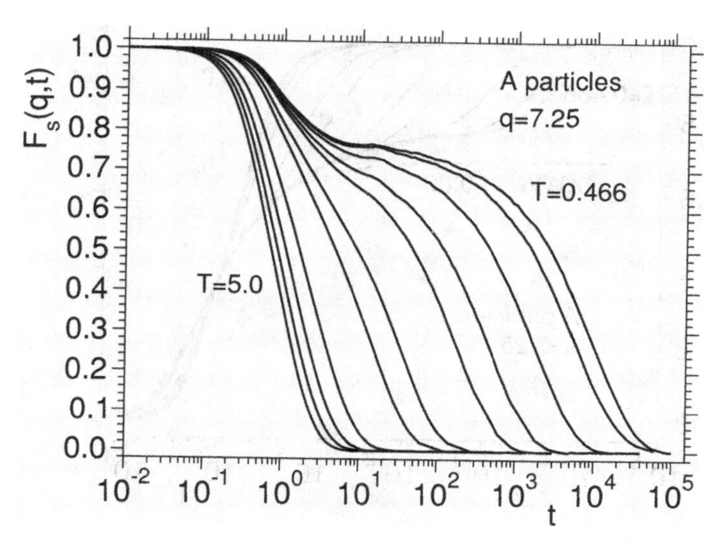

Fig. 5.12 Time dependence of the incoherent intermediate scattering function in a binary Lennard-Jones mixture. The curves correspond to the temperatures $T = 5.0$, 4.0, 3.0, 2.0, 1.0, 0.8, 0.6, 0.55, 0.50, 0.475, 0.466 (from left to right). The wave-vector corresponds to the location of the peak in the static structure factor for the A particles. From Kob and Andersen (1995b).

maximum in the partial structure factor, $S_{\rm AA}(q)$, for the A particles (Kob and Andersen 1995b). As can be recognized from the figure, at high temperatures $F_s(q,t)$ decays very quickly to zero and an analysis of the curve shows that this decay is exponential. (Note that at very short times, i.e. when the correlator is still larger than ≈ 0.95, the decay is just quadratic in time, since it corresponds to the ballistic motion of the particles.) With decreasing temperature the correlator starts to become non-exponential in that it shows a shoulder at intermediate times and in addition also the final relaxation is no longer exponential but can instead be described by a KWW-function, see Eq. (5.4), with an exponent $\beta < 1$. At even lower temperatures the shoulder has become a plateau which thus corresponds to the caging regime, i.e. to the $\beta-$relaxation.

According to MCT the shape of the correlation functions should, at temperatures close to T_c, be independent of T (time-temperature superposition principle), see Eq. (5.79). To what extend this prediction is correct can be checked by defining an $\alpha-$relaxation time $\tau(T)$ and by plotting the correlators as a function of $t/\tau(T)$. This is done in Fig. 5.13 where we show

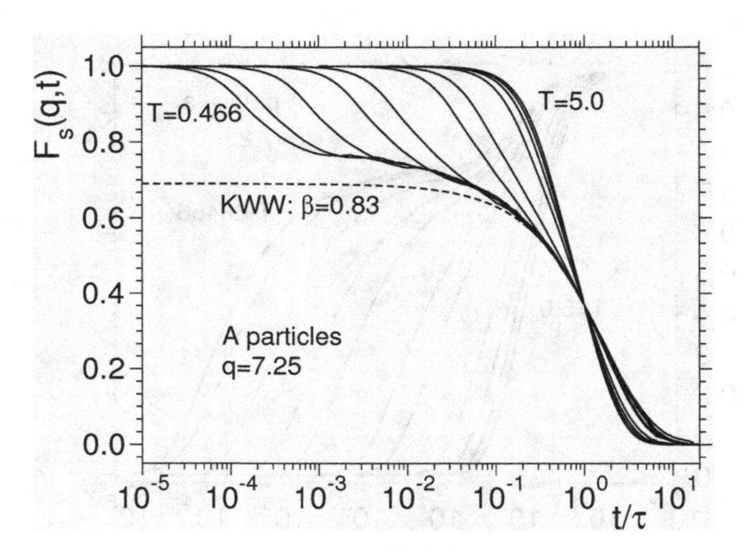

Fig. 5.13 The same data as in Fig. 5.12 as a function of t/τ with the $\alpha-$relaxation times defined as $F_s(q,\tau) = e^{-1}$. Also included is a fit with a Kohlrausch-Williams-Watts function to the low temperature data at long times (dashed curve). Adapted from Kob and Andersen (1995b).

the correlators from Fig. 5.12 and by using $F_s(q,\tau) = e^{-1}$ to define the $\alpha-$relaxation time. From the figure one recognizes that, roughly speaking, the correlators at high temperatures fall on top of each other and that this master curve is an exponential. Also at low temperatures we find a data collapse. However, in this case the master curve can be fitted well with a stretched exponential, included in the plot as well, with a KWW-exponent of $\beta = 0.83$. Thus we can conclude that the TTSP predicted by MCT does indeed hold. Note that the amplitude A for the KWW-law had to be chosen to be somewhat smaller than the height of the plateau in order to avoid small but systematic deviations at times at which the correlator falls just below the plateau. This is evidence that the KWW-exponent β and the von Schweidler exponent b, which for the present system is around 0.62 (Nauroth and Kob 1997), are indeed not the same. More systematic investigations on this point have shown that β does indeed depend on the correlator investigated (type of particle, wave-vector, etc.) whereas in the $\beta-$relaxation regime the relaxation dynamics can be described by *one* system universal function, the $\beta-$correlator from Eq. (5.69), and hence a unique value for the exponent parameter λ, in agreement with the theoretical prediction and the results from other simulations or experiments

(Signorini *et al.* 1990, van Megen and Underwood 1993a, Sciortino *et al.* 1996, Meyer 2002, Meyer *et al.* 1998, Tölle *et al.* 1997, 1998, Gleim and Kob 2000, Wuttke *et al.* 2000, Aichele and Baschnagel 2001a, Horbach and Kob 2001, Henrich *et al.* 2009).

The results shown in Figs. 5.12 and 5.13 are for the incoherent function of the A particles for the wave-vector at the maximum of the structure factor. It has been found, however, that other correlation functions (the coherent function, other wave-vectors, etc.) show qualitatively the same behavior. More important, the $\alpha-$relaxation time of these correlation functions shows at low temperatures all the same $T-$dependence (Kob and Andersen 1995b), in agreement with the power-law predicted by MCT, see Eq. (5.61), and with the findings in other systems (Sciortino *et al.* 1996, Bennemann *et al.* 1999, Aichele and Baschnagel 2001b, Puertas *et al.* 2003). Thus this is evidence that the $\alpha-$scale universality predicted by the theory, see Eq. (5.61), does indeed hold. Note that according to MCT the value of the exponent γ in this power-law should be independent of the way the critical point is approached: E.g. if the critical point corresponds to a temperature and pressure (T_c, P_c) this point can be approached by increasing P at constant temperature $(= T_c)$ or by decreasing T at constant pressure $(= P_c)$, or an arbitrary combination of these two paths. Since the value of γ is a property of the system and the critical point only, it is hence independent on how the latter is approached. Computer simulations of a glass-forming polymer melt by Bennemann *et al.* (1999) have shown that this remarkable property seems indeed to hold which is thus evidence that the critical line $T_c(P)$ does indeed exist.

Using the Einstein relation $D = \lim_{t\to\infty}\langle \Delta r^2(t)\rangle/6t$ we can calculate from the mean squared displacement the diffusion constant D. According to MCT this quantity as well as the $\alpha-$relaxation time τ should show close to T_c a power-law dependence, see Eq. (5.62). That for the binary Lennard-Jones system this is indeed the case is demonstrated in Fig. 5.14 where we show D and τ as a function of $T - T_c$ in a double logarithmic representation. (Here the value of T_c has been chosen such that *all* the data is rectified over the largest possible range.) We see that for intermediate and low temperatures the data is indeed compatible with the power-law given in Eq. (5.62), with a *unique* value of T_c. (The deviations from the power-law seen at the lowest temperatures will be discussed in more detail below.) Also included in the figure are fits with such power-laws, solid straight lines, and their slopes give the exponent γ. For the case of the relaxation time, τ, these slopes are basically independent of the type of

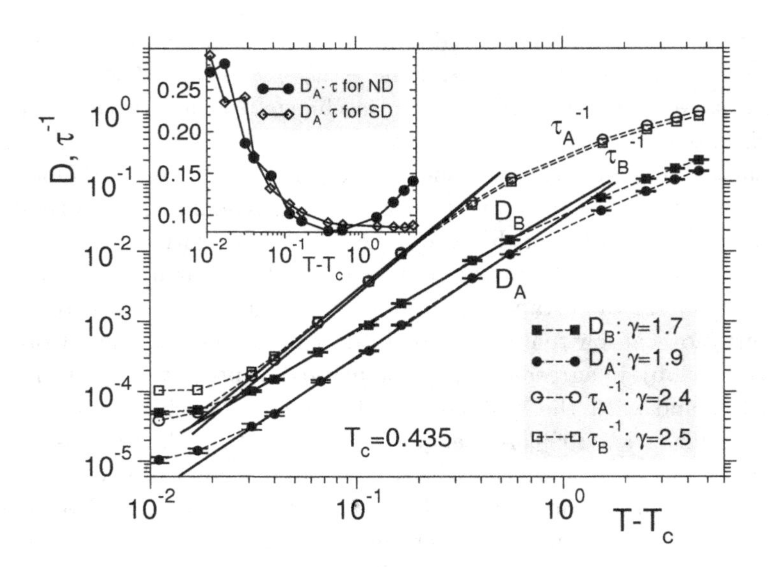

Fig. 5.14 The diffusion constant D and the inverse of the α−relaxation time τ as a function of $T - T_c$ for the case of a binary Lennard-Jones system. The straight lines are fits to the data with power-laws, see Eq. (5.62). The values of the exponents γ are given in the figure as well. Inset: T−dependence of the product $D_{\rm A} \cdot \tau_{\rm A}$. The full symbols correspond to a Newtonian dynamics (ND) whereas the open ones correspond to a stochastic dynamics (SD). Adapted from Gleim (1998).

particles and also of the wave-vector considered (Kob and Andersen 1995b, Gleim *et al.* 1998). This is, however, not the case for the diffusion constant in that we find that the value of γ for $D_{\rm B}$ is smaller than the one for $D_{\rm A}$ and that both of them are significantly smaller than the ones for τ. This finding is thus in contrast to the prediction of the theory.[19] This can also be directly seen by plotting the product $D \cdot \tau$ as a function of temperature, see inset of Fig. 5.14. (Note that for this representation of the data it is not necessary to know T_c and therefore there is no uncertainly about this.) According to MCT this product should, close to T_c be independent of T, see Eq. (5.63). However, as can be recognized from the figure this is not the case since at intermediate and low T the product increases by about a factor of three. The reason for this discrepancy with the prediction of the theory is likely the presence of the dynamical heterogeneities mentioned in

[19] Note that in principle it is possible to rectify the various data sets with the *same* value of γ if one allows that T_c depends on the observable, i.e. on D_α and τ. However, such an approach cannot be justified from a theoretical point of view and hence should be avoided.

Sec. 5.1 and which will be discussed in more detail in Chap. 6. As discussed at that place these dynamical heterogeneities, in the form of a few percent of fast moving particles, can have a very strong influence on the diffusion constant and it is not a surprise, and also not a big flaw, that MCT is not able to catch the relaxation dynamics of the system on this level of detail. We mention, however, that also hopping processes can make that the product is not a constant (Chong 2008).

Note that the presence of these dynamical heterogeneities do not only affect the T−dependence of the diffusion constant D, and hence the product $D \cdot \tau$, but also the time dependence of the incoherent intermediate scattering function $F_s(q,t)$ for small wave-vectors, since the latter is directly related to the mean squared displacement and hence to D, see the discussion in context of Eq. (5.62). Hence it must be expected that the T−dependence of these functions will show appreciable deviations from the power-law predicted by MCT, see Eq. (5.62) and, e.g., computer simulations of SiO_2 have indeed shown this to be true (Horbach and Kob 2001).

So far we have discussed the relaxation dynamics of a binary Lennard-Jones mixture, i.e. of a so-called "simple liquid" (Hansen and McDonald 1986, Barrat and Hansen 2003), since its structure and relaxation dynamics is very similar to the one of a hard sphere system. Another important class of glass-forming systems are liquids that have a network-like structure. The paradigm of such a system is silica in which the structure is given by tetrahedra that are connected to each other at their corners, see Chap. 3 and Fig. 3.13, but other glass-formers, such as sodium-silicates, alumino-sodo-silicates, phosphor-oxides, etc. are important examples as well (Zarzycki 1991, Varshneya 1993). It is therefore important to see to what extend MCT is able to rationalize the relaxation dynamics of these glass-forming systems. In Fig. 5.15 we show an Arrhenius plot of the diffusion constant as obtained from a computer simulation of SiO_2 (Horbach and Kob 1999). As expected for a system that is a strong glass-former, see Fig. 5.2, we find at low temperatures an Arrhenius law, solid straight lines. The activation energies are in good quantitative agreement with the experimental data thus giving evidence that the accuracy of the interaction potential used in the simulation is quite high (Horbach and Kob 1999). Surprisingly the figure shows however, that at sufficiently high temperatures the data from the simulation starts to deviate from the Arrhenius law in that it has a weaker T−dependence. Within the framework of MCT this crossover can be rationalized by the presence of the critical point T_c close to which the relaxation dynamics of the system changes from a flow-like dynamics above

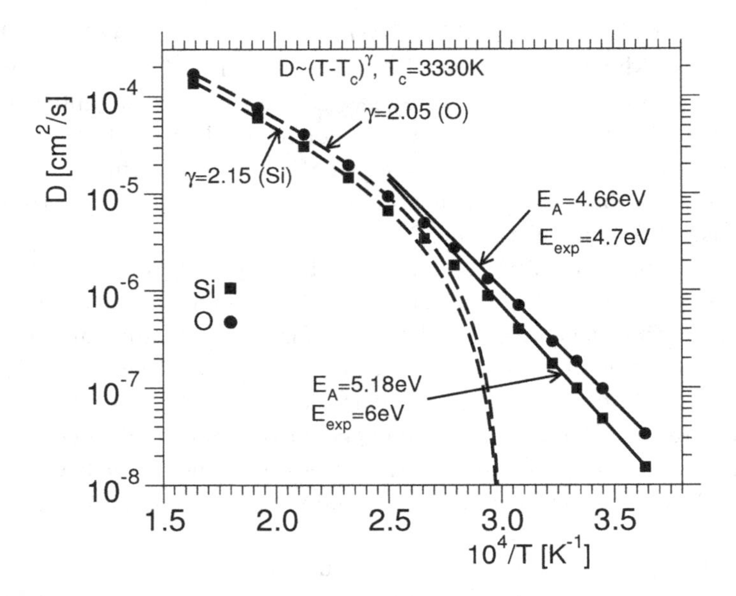

Fig. 5.15 Arrhenius plot of the diffusion constants as determined from a computer simulation of SiO_2. The solid lines are fits with Arrhenius laws with the given activation energies. Also given are the experimental values for the activation energies from Brébec *et al.* (1980) and Mikkelsen (1984). The dashed lines are power-law fits to the diffusion constants with a critical temperature T_c=3330 K. Adapted from Horbach and Kob (1999).

T_c to a hopping like motion below T_c, see discussion in Sec. 5.2.3. Also included in the figure are fits to the diffusion constants with the power-laws predicted by MCT, dashed lines, which shows that the data is indeed compatible with such a functional form. It is evident, however, that these fits give a good description of the data over only about 1.5 decades in D, thus significantly less than the 3 decades we found for the Lennard-Jones system, see Fig. 5.14. Thus within MCT one can argue that this rather limited range of validity of the *ideal version* of MCT is due to the fact that in silica the hopping processes, see Sec. 5.2.3, are much more pronounced than in the simple glass-former.

The fact that the strong glass-former SiO_2 shows at high temperatures strong deviations from the Arrhenius law found at low temperatures might look somewhat circumspect, since it seems to be in contradiction to the experimental findings, see Fig. 5.2. Nevertheless, the results of the simulation concern a temperature range in which so far no experimental data is available and hence there is for the moment no disagreement. However, the result from the simulation makes one wonder whether at the end not *all* glass-formers have a temperature range in which their relaxation time shows a non-Arrhenius behavior. Rössler, Hess *et al.* and later on Elmatad *et al.* investigated this point by using the viscosity data to make a new type of scaling plot (Hess *et al.* 1996, Rössler *et al.* 1998, Elmatad *et al.* 2009). The goal of such a plot was to see whether the relaxation dynamics of all strongly supercooled liquids does show the same temperature dependence and hence to give evidence for the existence of a universal mechanism for the slowing down. Note that the Angell-plot, shown in Fig. 5.2, is not appropriate to see this since the local slopes of the curves, i.e. the activation energies, will of course depend on the system considered. (E.g. two glass-formers that have both an Arrhenius dependence will in general not fall onto a master curve, since the respective activation energy and the prefactor is different.) This problem can be avoided by scaling the temperature axis by a system dependent factor. In practice Rössler, Hess *et al.* proceeded as follows: Denoting by T_g the calorimetric glass transition temperature, they chose as a scaling factor $F = T_x/(T_x - T_g)$ and adjusted for each liquid the temperature T_x such that the viscosity data coincided at low temperatures as best as possible with the viscosity data of ortho-terphenyl. (The fact that these authors chose ortho-terphenyl as reference is irrelevant, since the resulting master curve will not strongly depend on this choice.) In the upper panel of Fig. 5.16 we show the original viscosity data for the liquids considered, which include SiO_2, B_2O_3, salol, propylene carbonate, CKN ($= 40Ca(NO_3)_2 60KNO_3$) and many other glass-formers. As can be seen from the figure these substances range from very fragile to very strong ones. In the lower panel of Fig. 5.16 we show that the scaling of the $T-$axis does indeed lead to a nice collapse of the data onto a master curve, at least for intermediate and high viscosities, whereas at small viscosities there is no longer a master curve. Hence we have evidence that there is indeed a special temperature T_x at which the relaxation dynamics changes qualitatively. Surprisingly the analysis of Rössler, Hess *et al.* has shown that the value of T_x coincides within the experimental error bar with the literature value of T_c, which have been determined by making the appropriate MCT analysis of the time correlation functions, i.e. without using viscosity data. Thus in summary one can conclude that the viscosity data gives strong evidence that *all* glass-forming liquids show a change in the transport mechanism and that there is good experimental evidence that this change occurs at the critical temperature of MCT.

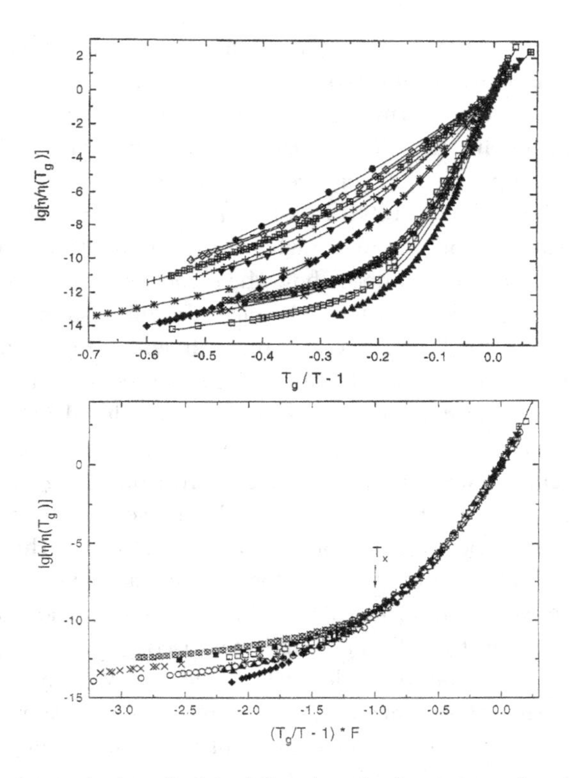

Fig. 5.16 Upper panel: Angell-plot of the viscosity for various glass-forming systems (silica, B_2O_3, salol, propylene carbonate, CKN ($40Ca(NO_3)_2 60KNO_3$), etc.). Here T_g is the calorimetric glass transition temperature. Lower panel: Same data plotted as a function of $(T_g/T - 1) \cdot F$ with $F = T_x/(T_x - T_g)$. T_x is a temperature that has been adjusted for each liquid such that the viscosity data falls at low T on a master curve. Adapted from Rössler *et al.* (1998).

One of the most interesting predictions of MCT is that the relaxation dynamics of a system at intermediate and long times can be calculated from its static properties. This follows directly from the MCT equations, Eqs. (5.47)-(5.49), since the vertices, and hence the memory function, depends only on static quantities such as the particle density or the static structure factor. Only the short time dynamics will depend on quantities like the mass of the particles, the nature of the microscopic dynamics (Newtonian, Brownian, ...) etc. whereas the $\beta-$and $\alpha-$relaxation are predicted to be independent of them. In a real experiment it is unfortunately difficult to change these parameters. In contrast to this computer simulations allow easily to investigate this point. Roux *et al.* (1989) have found, e.g., that for the case of a binary mixture of soft spheres the glass transition temperature is independent of the mass ratio between the particles. In a different study Gleim *et al.* (1998) investigated to what extend the relaxation dynamics depends on the microscopic dynamics. For this they compared the time dependence of the intermediate scattering function as determined for the above mentioned binary Lennard-Jones system following a standard Newtonian dynamics (ND), see Fig. 5.12, with the one obtained for the same system but now with a "stochastic dynamics". Here "stochastic dynamics" (SD) means that the equations of motion of the particles are given by

$$m\ddot{\mathbf{r}}_i + \nabla_i \sum_l V_{il}(|\mathbf{r}_l - \mathbf{r}_i|) = -\zeta\dot{\mathbf{r}}_i + \eta_i(t)\,. \tag{5.80}$$

Here V_{il} is the potential between particles i and l, $\eta_i(t)$ are Gaussian distributed random variables with zero mean, i.e., $\langle\eta_i(t)\rangle = 0$, and ζ is a damping constant. The fluctuation dissipation theorem relates ζ to the second moment of η_i, and thus we have $\langle\eta_i(t)\cdot\eta_l(t')\rangle = 6k_BT\zeta\delta(t-t')\delta_{il}$ (Pathria 1986). Gleim *et al.* used a value of ζ of 10, which is large enough to ensure that the relaxation dynamics at intermediate and long times is independent of ζ, apart from a trivial scaling of the time scale by a factor ζ. Note that since the interactions between the particles in the Newtonian and the stochastic system are exactly the same, all equilibrium properties must be identical.

In Fig. 5.17 we show the time dependence of the incoherent intermediate scattering function as obtained by this stochastic dynamics (solid lines). We see that qualitatively the correlators look very similar as the ones found for the ND, see Fig. 5.12, in that at high temperatures we find a rapid decay whereas at low T the correlation functions show again an extended plateau. In order to allow a more detailed comparison between the two type of dynamics we have included in the figure also some of the correlators which were obtained from the ND (bold dashed lines). We see that at high T the relaxation dynamics for the ND system is significantly faster in that $F_s(q,t)$ has a relaxation time that is about 7 times smaller than the one for the SD system. At these temperatures also the shape of the curves depends on the nature of the microscopic dynamics. This is in contrast to the case of low temperatures in that we see that the SD and ND curves have exactly the same shape. (Note that since in both systems the TTSP is fulfilled one can compare correlators at different (low!) temperatures.). Since we see that also the height of the plateau at intermediate times is independent of the microscopic dynamics, we can conclude that

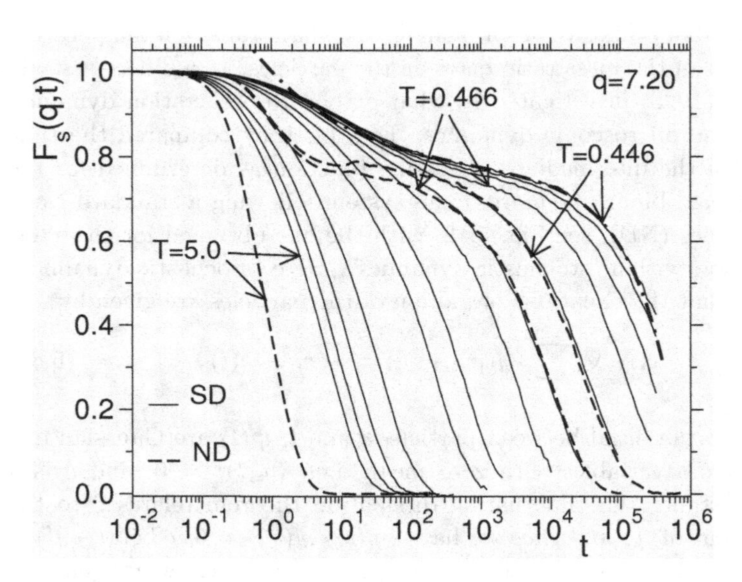

Fig. 5.17 Time dependence of the incoherent intermediate scattering function for the A particles in a binary Lennard-Jones mixture. The solid lines are for the case of the stochastic dynamics at the temperatures $T = 5.0$, 4.0, 3.0, 2.0, 1.0, 0.8, 0.6, 0.55, 0.5, 0.475, 0.466, 0.452, and 0.446 and the bold dashed lines are for the Newtonian dynamics at $T = 5.0$, 0.466, and 0.446. The dashed-dotted lines is a fit to the SD curve at $T = 0.446$ with the $\beta-$correlator from the MCT. From Gleim *et al.* (1998).

indeed the whole $\alpha-$process does not depend on the details of this dynamics, in agreement with the prediction of the theory. The only relevant difference between the relaxation dynamics for the SD and ND systems is the time dependence of the correlators at short times. Whereas for the ND the correlation function approaches the plateau very quickly, the one of the SD make this approach much more smoothly. This slow approach is in agreement with MCT which predicts that it should be given by the $\beta-$correlator (see Eqs. (5.64), (5.66), and (5.69)). A fit with this functional form to the SD data in the $\beta-$regime is included in the figure as well, bold dashed-dotted curve. We see that this theoretical curve gives a very good description of the data for about 4 decades in time thus giving strong evidence that this functional form is indeed a correct description of the dynamics.[20] Since a similar good agreement has been found for the case of

[20]Note that in order to make such a fit one had to adjust three parameters: The time scale t_σ, the height of the plateau $f^c(q)$ and the amplitude $h(q)$. The exponent parameter λ that determines the shape of the $\beta-$correlator, see Eq. (5.69), was taken

soft spheres, colloidal suspensions and simple molecular systems, (Barrat and Latz 1990, Götze and Sjögren 1991, Li *et al.* 1992, van Megen and Underwood 1993b, Cummins *et al.* 1997, Meyer *et al.* 1998, Rinaldi *et al.* 2001, Chong and Sciortino 2004, Flenner and Szamel 2005) we can conclude that this good agreement is not just a particularity of the Lennard-Jones system but that the theoretically predicted time dependence can indeed be found in many glass-forming systems.

It is evident from the figure that the correlators for the ND show no sign of the critical decay, i.e. the approach to the plateau in form of a power-law, see Eq. (5.71). The reason for this is the presence of phonons that influence the dynamics of the system even at intermediate time scales. These phonons are taken into account in the regular part of the memory function, i.e. $M^{\rm reg}$ in Eqs. (5.47) and (5.50). By making a simple Ansatz for this regular term and plugging it into the MCT equations it has been demonstrated that the solution of these equations do not necessarily show the critical decay *if one is not extremely close to the critical point* (Franosch *et al.* 1997c, Fuchs *et al.* 1998). Thus the results of the simulations and the analytical calculations demonstrate that it is not sufficient to focus only on the predictions of the theory that hold asymptotically close to the transition point, such as the prediction on the shape of the $\beta-$correlator, etc., but instead to consider also the corrections to these asymptotic results. Such calculations have been done by Franosch *et al.* (1997c) and Fuchs *et al.* (1998), and subsequently Gleim and Kob (2000) have given evidence that the predicted corrections can indeed be found in numerical simulations. Thus the absence of the critical decay in real data, such as in the ND system shown here but also observed in many other atomic systems, is not necessarily in contradiction to the theory. For the present case it has in fact been found that a reasonable description of the regular part of the memory function gives solutions of the MCT equations that are in good *quantitative* agreement with the data from the simulation, and that therefore the critical decay is not even observed in the exact MCT equations if one is not exceedingly close to T_c (Kob *et al.* 2002, Kob 2003).

In the previous section we have shown that according to MCT "all" correlation functions have in the $\beta-$relaxation regime the same shape, i.e. that the factorization property holds: $\Phi_x(t) = \Phi_x^c + h_x G(t)$, see Eq. (5.64). Recall that this prediction is not only valid for the ideal version of MCT but also for the extended one, i.e. in the case that hopping processes are

from a theoretical calculation, i.e. it was not a fit parameter. See Gleim and Kob (2000) for more details.

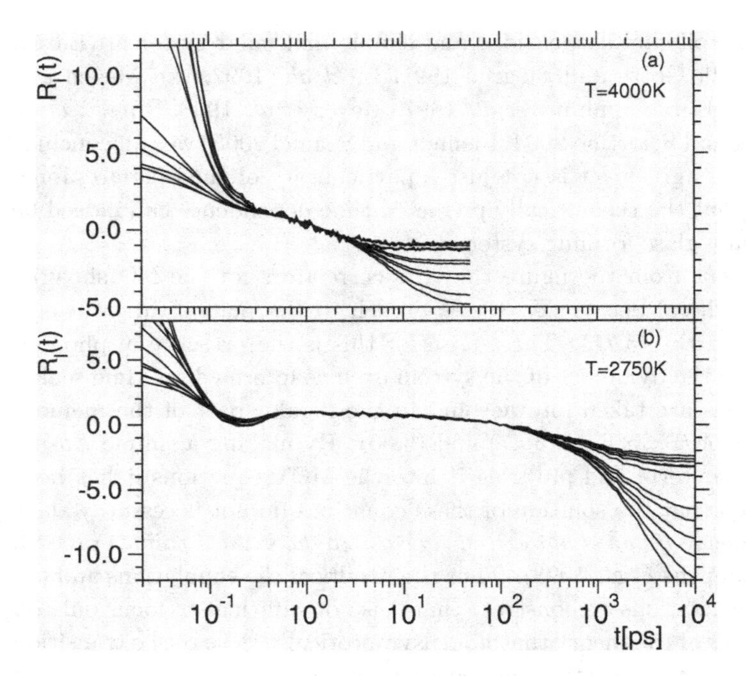

Fig. 5.18 Time dependence of the functions $R_l(t)$ as defined in Eq. (5.81) for the case of silica. The upper and lower panel corresponds to $T = 4000$ K and $T = 2750$ K, respectively. The correlators used are $F_s(q,t)$ for silicon and oxygen at $q = 1.7$, 2.2, 2.8, 4.43, 5.02, and 5.31 Å^{-1}. The times t'' and t' are 0.4 ps and 1.6 ps for $T = 4000$ K and 11 ps and 106 ps for $T = 2750$K, respectively. Adapted from Horbach and Kob (2001).

present. One simple possibility to check whether or not this prediction is correct has been proposed by Signorini *et al.* (1990): Consider an arbitrary correlator $\Phi_l(t)$. If the factorization property holds, it follows immediately that in the $\beta-$relaxation regime the ratio

$$R_l(t) = \frac{\Phi_l(t) - \Phi_l(t')}{\Phi_l(t'') - \Phi_l(t')} \tag{5.81}$$

is independent of l since it is just $(G(t) - G(t'))/(G(t'') - G(t'))$. Here t' and t'' are two arbitrary times in the $\beta-$regime. In Fig. 5.18 we show the time dependence of $R_l(t)$ for the case of the SiO_2 system discussed above. The correlators used to calculate these curves are the incoherent intermediate scattering function for Si and O at $q = 1.7$, 2.2, 2.8, 4.43, 5.02, and 5.31 Å^{-1} (see the structure factor in Fig. 2.5), but other correlators can be included as well (Kob *et al.* 1999). The upper panel shows that for times in which the correlators are close to the plateau, the functions

$R_l(t)$ do indeed fall on a master curve, in agreement with the prediction of the theory. That this scaling is far from trivial can be recognized from the fact that for times that are shorter than the $\beta-$regime and times that are significantly longer than this regime, the various curves do not at all fall onto a master curve. If the temperature is decreased, lower panel, the time range over which the $R_l(t)$ coincide extends rapidly, also this in agreement with the expectation of the theory. Thus we can conclude that for the SiO_2 system investigated, the factorization property predicted by MCT does indeed hold. Qualitatively similar results have been found for other systems, such as the Lennard-Jones mixture discussed above, water, ortho-terphenyl, polymers, and others (Signorini *et al.* 1990, Li *et al.* 1992, van Megen and Underwood 1993b, Wuttke *et al.* 1994, Sciortino *et al.* 1997, Tölle *et al.* 1997, Kämmerer *et al.* 1998, Wuttke *et al.* 1998, Aichele and Baschnagel 2001a), thus showing the relevance of this theoretical prediction. We mention, however, that for the case of systems with orientational degrees of freedom the situation is somewhat more complicated. Since also these degrees of freedom couple to the translational degrees of freedom, one would expect that the factorization property extends also to the orientational correlation function. A comparison of neutron scattering data, in which one measures the translational correlation function $F(q,t)$, with results from dielectric spectroscopy, where the orientational degrees of freedom are probed, shows, however, that the location of the minimum in the corresponding susceptibilities, see Fig. 5.4, differs by a factor of around ten in frequency (Lunkenheimer *et al.* 1996b, Kämmerer *et al.* 1998, Goldammer *et al.* 2001). This has sometimes been taken as evidence that the time scale for the $\beta-$relaxation depends on the correlator, in contrast to the prediction of MCT. It has subsequently been shown, however, that this conclusion is not correct, since the amplitude of the $\alpha-$process as well as the corrections to the asymptotic shape of $\chi(\omega)$, both of which depend strongly on the observable considered, can easily be used to rationalize the *apparent* discrepancy with the *asymptotic* prediction which holds only for $T \to T_c$ (Theenhaus *et al.* 2001, Chong and Götze 2002b). Although for temperatures extremely close to T_c the theory does indeed predict that the location of the minimum in $\chi''(\omega)$ is independent of the observable, in reality the hopping processes become so important that MCT can no longer be considered to be a reliable theory from this point of view.

So far we have discussed some features of the relaxation dynamics of glass-forming systems and have found that these results are in *qualitative* agreement with the predictions of MCT. In view of the approximations that are needed to derive the MCT equations it is, however, also important to check to what extend the theory is also able to make reliable *quantitative* predictions. Since the only input to the theoretical calculation are static quantities (density of particles, static structure factor, etc.) the knowledge of these quantities allows to make quantitative predictions for the relaxation dynamics of the system. This has been done for the case of colloidal suspensions which can be described very well by a hard sphere model (Barrat *et al.* 1989, Götze and Sjögren 1991, van Megen and Underwood 1993a). Using as theoretical input the static structure factor calculated within the Verlet-Weiss approximation (Hansen and McDonald 1986), Götze and Sjögren (1991) determined numerically the solutions of the MCT equations and compared them with the dynamic light scattering measurements of van Megen and Pusey (1991), a work that was later extended by van Megen and Underwood (1993a, 1994). As an example for the nice agreement between theory and experiment we show in Fig. 5.19 the coherent intermediate scattering function for the colloidal system and the theoretical predictions of the theory. The different curves correspond to different packing fractions and the upper and lower panel to a wave-vector at the location of the maximum in the static structure factor and the first minimum, respectively. The figure shows that the theory is indeed able to give a quantitatively correct description of the relaxation dynamics for this system. In particular MCT is able to reproduce the approach of the correlators to the plateau, the height of this plateau, and the shape and the relaxation time of the curves in the $\alpha-$relaxation regime.

It should be noted that for the calculation of the theoretical curves of Fig. 5.19 it was necessary to adjust two fit parameters: The first one is the time t_0 from Eq. (5.67) which makes the connection between the time scale of the $\beta-$relaxation with the time scale of the microscopic dynamics of the system, and which here depends on the nature of the solvent, the mass of the particles, etc. The second parameter is the critical packing fraction ϕ_c. In the experiments this latter has been determined to be 0.560 ± 0.005 whereas MCT predicts 0.52 ± 0.01. (The error bar in theory stems from the error made by using a structure factor that is not exact.) Thus we find that the theory *overestimates* the tendency of the system to form a glass. This inaccuracy of the theory is not a particularity of the hard sphere system discussed here, but has been found to be present also in many others, such as Lennard-Jones, soft spheres, water, silica, etc. (Bengtzelius 1986, Nauroth and Kob 1997, Theis *et al.* 2000, Winkler *et al.* 2000, Rinaldi *et al.* 2001, Sciortino and Kob 2001, Sciortino *et al.* 2003, Voigtmann 2003, Chong and

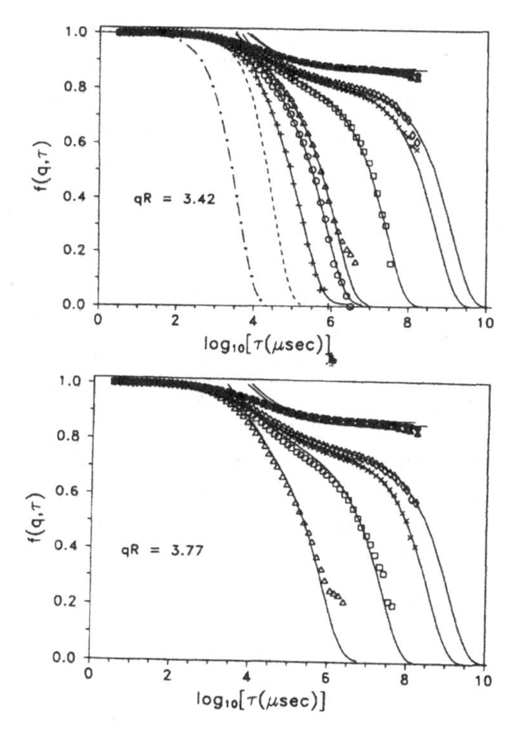

Fig. 5.19 Time dependence of the coherent scattering function as measured in a dynamic light scattering experiment of a colloidal system at different packing fractions (symbols). The solid lines are the predictions of MCT for a hard sphere system. Upper panel: $qR = 3.42$, where R is the radius of the colloidal particles. Lower panel: $qR = 3.77$. Adapted from van Megen and Underwood (1994).

Sciortino 2004). Thus this is a hint that some of the relaxation channels that exist in reality are not properly taken into account by the theory. Note that we are here *not* referring to the hopping processes discussed in the previous section, since these do not change the value of T_c but only smear out the dynamical singularity.

For the case of the hard sphere system the static structure factor can be calculated with quite high accuracy, using the approximations of Percus-Yevick or Verlet-Weiss (Hansen and McDonald, 1986), and hence one has reliable input data for the MCT calculations. This is not the case for more complex systems, like multi-component simple liquids, network-forming liquids, or molecular liquids. Furthermore there is the additional problem that in an K-component systems all the $K(K+1)/2$ partial structure factors are needed and for the molecular system also the distribution functions for the orientational degrees of freedom. Obtaining all this information from experiments is for most cases impossible and the available analytical approximations for these functions are presently not sufficiently reliable. The only possibility one thus has for the moment is to use computer simulations in order to generate the necessary input data for the MCT calculations. It is evident that these simulations can as well be used to determine the relaxation dynamics of the system and therefore provide the reference data against which the theory can be tested. For the case of the binary Lennard-Jones system described above this has been done by Nauroth and Kob (Nauroth 1999, Nauroth and Kob 1997). Among other quantities these authors used Eq. (5.58) to calculate the wave-vector dependence of the non-ergodicity parameters at the critical temperature.[21] In Fig. 5.20 we show these three functions and compare the results from the simulation, dots, with the prediction from the theory. In the derivation of the MCT equations we made the approximation that a three-point correlation function was factorized into a product of two-point correlation functions, see discussion in footnote 12. If one makes this approximation the theory predicts the curves labeled by c_2. If one avoids this approximation and instead extracts from the simulation data also the static three-point correlation function c_3, the theory predicts the curves labeled by c_3. The figure demonstrates that the agreement between the theoretical prediction and the results from the simulation is very good even if one uses only the c_2 theory and that the theoretical description of the simulation data becomes even better if this approximation is avoided. Using the theory it is also possible to calculate the value of the exponent γ which is found to be 2.34 (Nauroth and Kob 1997). This value is in very good agreement with the one found in the simulation data for the relaxation times which are around 2.4, see Fig. 5.14. Thus we can conclude that MCT is indeed able to make reliable *quantitative* predictions for the relaxation dynamics of this system. However, also in this case the theory overestimates the critical temperature significantly in that it predicts a $T_c = 0.92$ whereas one finds in the simulations $T_c = 0.435$ (Kob and Andersen 1994). (The inclusion of the c_3 terms change this value only by 1%, see Sciortino and Kob (2001), in agreement with the result by Barrat *et al.* (1989) for soft spheres.) Thus we find that with respect to this quantity the theory is indeed not reliable. The reason for this flaw,

[21] Note that since this is a binary system one has to use the generalization of this equation to the multi-component case, see Barrat and Latz (1990) and Nauroth and Kob (1997), and one obtains $K(K+1)/2$ (=3 for $K = 2$) nonergodicity parameters.

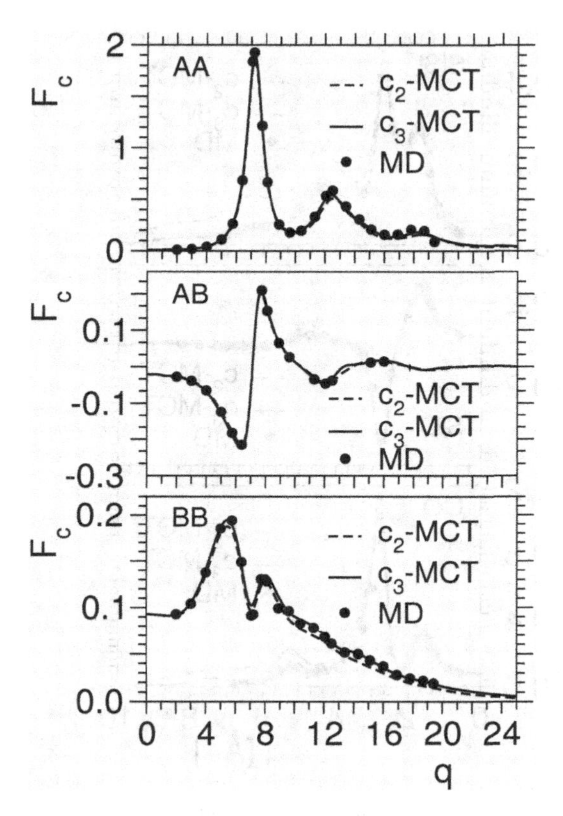

Fig. 5.20 Wave-vector dependence of the nonergodicity parameters of $F(q,t)$ for the case of a binary Lennard-Jones system. The dots are the results from the molecular dynamics simulations by Gleim *et al.* (1998). The dashed lines are the prediction of MCT if the contribution of the c_3 terms in the vertices are neglected. The full lines are the prediction if these terms are taken into account. From Sciortino and Kob (2001).

which as mentioned above is found also in most other systems, is presently not known.

The local structure of the colloidal system and of the binary Lennard-Jones model discussed so far is very similar to the one of a hard sphere system. Thus one must wonder whether the good agreement between the relaxation dynamics of these systems and the theoretical prediction of MCT is just related to this fact or whether the theory is also able to make reliable quantitative predictions for liquids in which the local structure is very different. One system for which this is the case is silica, whose structure is given by a open network of interconnected tetrahedra, see Sec. 3.6. Although for

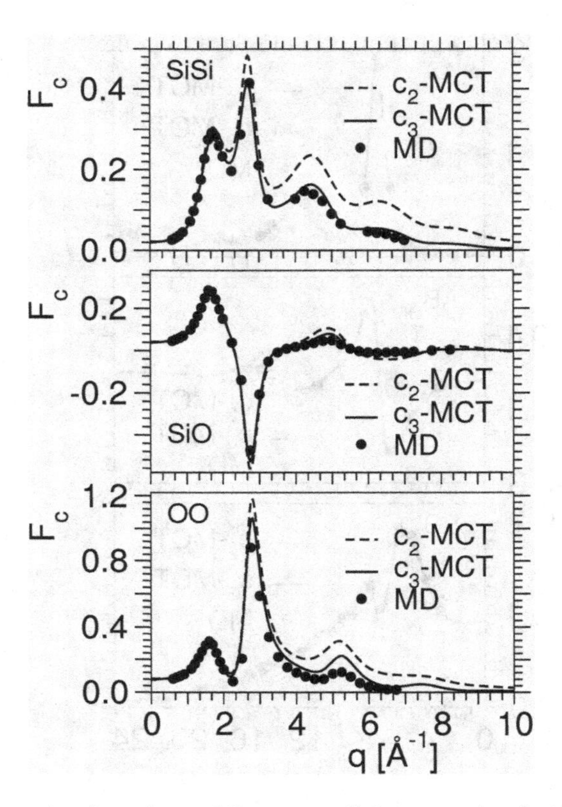

Fig. 5.21 Wave-vector dependence of the nonergodicity parameters for the case of silica. The dots are the results from the molecular dynamics simulations by Horbach and Kob (2001). The dashed lines are the prediction of MCT if the contribution of the c_3 terms in the vertices are neglected. The full lines are the prediction if these terms are taken into account. From Sciortino and Kob (2001).

SiO_2 it is not possible to measure in a real experiment the partial structure factors, due to the lack of appropriate isotopes, the interaction potential by van Beest *et al.* 1990 discussed above, see Eq. (3.132), gives a sufficiently reliable description of the local structure (Horbach and Kob 1999). Thus Sciortino and Kob (2001) used these structure factors as well as the three point correlation functions c_3 to determine the wave-vector dependence of the nonergodicity parameters. These functions were then compared with the corresponding data determined from a molecular dynamics computer simulation of the same model by Horbach and Kob (2001), see Fig. 5.21. The dots are the results from the simulations, the dashed line is the version of the theory in which c_3 is set to zero, and the solid line is the prediction of MCT by taking into account the c_3 terms. From the figure we can conclude that in this case the contribution of the c_3 terms are important in that the two theoretical curves differ significantly. This result is not a surprise since the presence of an open network structure will make that in this case the function c_3 is larger than the one for a more dense structure that resembles the one of a closed packed hard sphere system and hence it must be expected that the vertices of the memory functions change significantly.[22] Figure 5.21 shows that this change in the vertices leads to a significant improvement of the accuracy of the theoretical prediction in that the curves for the c_3-theory agrees now very well with the simulation data for the Si-Si- and Si-O-correlation. The description of the O-O-data is still fair, but certainly of inferior quality than the two other correlations. The reason why here the discrepancy is larger is presently not known. The inclusion of the c_3 terms also leads to a significant change in the theoretical prediction for the critical temperature in that it is calculated to be 4680 K, whereas the prediction with $c_3 = 0$ gives $T_c = 3960$ K. Since the value from the simulation has been estimated to be 3330 K (Horbach and Kob 2001), both values are significantly different from the correct one, a result that is coherent with the findings for other systems, see discussion above, and which shows the inaccuracy of the theoretical prediction regarding this quantity.

The results discussed in the previous paragraphs give strong evidence that MCT is able to predict reliably quantities like the wave-vector dependence of the nonergodicity parameter, the critical exponent and others, once the structure of the liquid, characterized by $S(q)$ and c_3, is known. One might thus wonder whether the theory is also able to give an accurate description of the *full* time-and wave-vector dependence of the correlation functions. That this is indeed the case has been demonstrated in Fig. 5.19 for the case of the colloids which were described as a system of hard spheres. Qualitatively similar results have been obtained for the binary Lennard-Jones mixture discussed above, Kob *et al.* (2002), for a binary mixture of hard sphere particles in which the theoretical predictions have been compared with light scattering data, Voigtmann (2003), for a simple model for ortho-terphenyl, Chong and Sciortino (2004), and for

[22] Recall that the function c_3 is defined by the residual of the product of the two-point correlation function: $\langle \delta\rho(\mathbf{q})\delta\rho(\mathbf{k})\delta\rho(\mathbf{q}+\mathbf{k})\rangle = S(q)S(k)S(|\mathbf{q}+\mathbf{k}|)[1+n^2c_3(\mathbf{q},\mathbf{k})]$, where n is the particle density. Thus in a dense hard sphere like system the arrangement of the particles is made in such a way that the local structure is basically given by the two-point correlation function since it is very similar to a closed packed structure, and hence $c_3 \approx 0$, whereas in a open network structure this function contains relevant information.

sodosilicate systems, Voigtmann and Horbach (2006). For the case of systems in which the microscopic dynamics is given by a Brownian one, such as it is the case for colloidal systems, the theoretical curves give a good description for the *whole* time dependence of the correlation functions. However, if the microscopic dynamics is a Newtonian one, only the $\beta-$and $\alpha-$relaxation are described well, whereas for times that corresponds to the microscopic time scale of the dynamics the agreement between the theoretical curves and the ones from the simulation is not very satisfactory (Nauroth and Kob 1999, Kob *et al.* 2002). The reason for the observed discrepancies is believed to be related to an inaccurate description of the regular part of the memory function $M^{\rm reg}(q,t)$, see Eq. (5.47), i.e. a quantity that has nothing to do with mode coupling theory. It thus can be hoped that once a better description for $M^{\rm reg}$ is found, it is possible to obtain a quantitatively good description of the time dependence of the correlators over the whole time domain also for systems with a Newtonian dynamics.

All the results discussed so far in this section concern the *ideal* mode coupling theory, i.e. the version of MCT in which the hopping terms $\delta(q,z)$ are neglected, see Eq. (5.52). As we have mentioned already earlier, the presence of these terms changes qualitatively the nature of the time dependence of the correlators close to T_c in that the correlation functions decay to zero even for $T < T_c$ and hence makes that the relaxation times stay finite for all temperatures. Due to a lack of knowledge of the details of the hopping terms it has so far not been possible to understand the details of the solutions of the extended MCT equations. However, it is possible to study at least some of the generic features of these solutions by investigating schematic models, i.e. equations that have the same structure as the full $q-$dependent equations but are stripped by all the details that are irrelevant for the generic behavior at intermediate and long times, see Eqs. (5.54) and (5.55). This approach has been used by Götze and Sjögren (1988) who solved Eq. (5.52) numerically for the case that the $q-$dependence is completely neglected and by setting $M^{\rm reg}(q,z) = i\nu$, where ν is a damping constant. The hopping term $\delta(t)$ was approximated by

$$\delta(t) = \lambda_3 \dot{\Phi}^2(t) C(t) , \qquad (5.82)$$

where λ_3 is the strength of the hopping and $C(t) = \Theta(t - t_0)$ is a cut-off function which makes that $\delta(t)$ does not influence the relaxation dynamics at times shorter than t_0, where t_0 is a time that is large with respect to the microscopic time scale. (Recall that this latter dynamics is governed by

$M^{\rm reg}(q,z) = i\nu$.) In addition Götze and Sjögren made the approximation that $\delta(z)$ is approximated by the zero-frequency limit of Eq. (5.82), i.e.

$$\widehat{\delta}(z) = i\lambda_3 \int_{t_0}^{\infty} \dot{\Phi}^2(t)dt \equiv i\delta\,. \tag{5.83}$$

Thus the resulting equation has the form:

$$\Phi(z) = \frac{-1}{z - \cfrac{\Omega^2}{z + \cfrac{\Omega^2\{i\nu + \widehat{M}(z)\}}{1 - i\delta\{i\nu + \widehat{M}(z)\}}}}\,, \tag{5.84}$$

where δ is obtained from Eq. (5.83). As memory function the authors used

$$M(t) = \lambda_1\Phi(t) + \lambda_2\Phi^2(t)\,, \tag{5.85}$$

where λ_1 and λ_2 are coupling constants. For the case $\delta = 0$ this model has been discussed by Götze (1984) who showed that it has an (ideal) transition for coupling constants $\lambda_1^c = (2\lambda - 1)/\lambda$ and $\lambda_2^c = 1/\lambda^2$, where the number $1/2 < \lambda < 1$ parametrizes the transition line.

In Fig. 5.22a we show, for the case $\delta = 0$, the time dependence of Φ for various values of ϵ, a small parameter that characterizes the distance from the transition line (Götze 1984):

$$\lambda_1 = \lambda_1^c + \epsilon\lambda/[1 + (1-\lambda)^2] \tag{5.86}$$

$$\lambda_2 = \lambda_2^c + \epsilon\lambda(1-\lambda)/[1 + (1-\lambda)^2]\,. \tag{5.87}$$

The values used are $\lambda = 0.7$, $\nu = 5\Omega$, and $t_0 = 1000/\Omega$. (Note that in Fig. 5.22 the time scale is expressed in Ω^{-1}.) From the figure we see that qualitatively the curves look indeed similar to the ones obtained from the solutions of the MCT-equation of a hard sphere system, i.e. a calculation in which the full q−dependence has been taken into account, see Fig. 5.8. In particular we find that upon approach to the critical point the relaxation times increase rapidly and that for sufficiently strong coupling the correlator does not decay to zero any longer, thus showing that such schematic models do indeed capture the generic behavior of the full MCT equations.

In Fig. 5.22b we show the solutions of Eq. (5.84) for the same values of ϵ but with $\delta \neq 0$, namely $\lambda_3 = 1$. A comparison of these curves with the ones from panel (a) shows immediately that for weak couplings, i.e. $\epsilon < 0$ and $|\epsilon|$ large, the corresponding curves look very similar. This result is reasonable since the correlators decay to zero on a time scale that is on the order of t_0

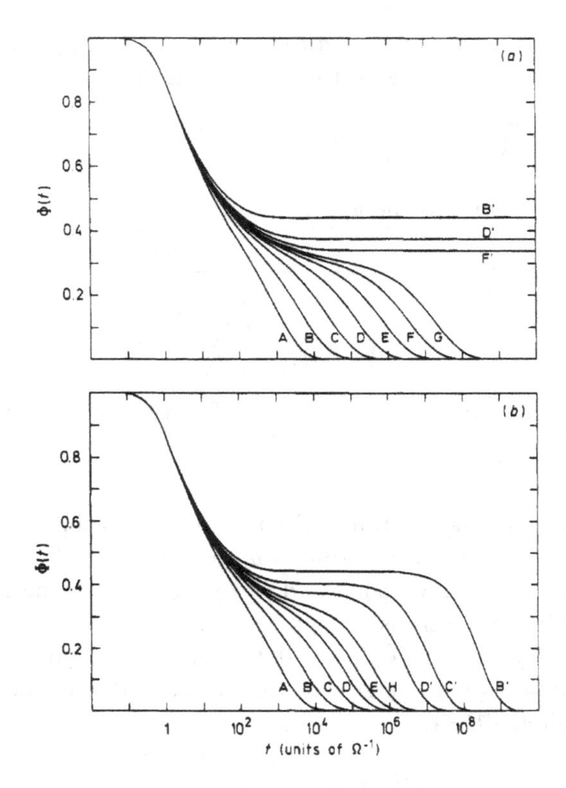

Fig. 5.22 Time dependence of the solution of Eqs. (5.84) for the case $\delta = 0$, upper panel, and $\delta \neq 0$, lower panel. The different curves correspond to value of $\epsilon = -0.2/2^n$ with $n = 0 \ldots 6$ and $\epsilon = 0.1, 0.05$, and 0.025. (See Eqs. (5.86) and (5.87) for a definition of ϵ.) From Götze and Sjögren (1988).

and hence they are not strongly influenced by the presence of the hopping term. If the coupling is increased, see curve E, we start to see the effect of the hopping term in that the relaxation time is about a factor 10 smaller than the one for the same coupling in the case $\delta = 0$. Furthermore also the shape of the correlators starts to change in that the curve for $\delta > 0$ is less stretched than the one for $\delta = 0$. If one goes to the critical point, defined by $\epsilon = 0$ (curve H), or even stronger coupling (curves D', C', B') the effects of the hopping terms become even more pronounced in that there is no longer a nonergodic state since the correlators always decay to zero at long times. This final decay, the α–process, is not stretched but given by a simple exponential with a typical time scale $O(\delta^{-1})$. Note, however, that for short and intermediate times the form of the correlators, height of the plateau and the approach to it, is independent of δ, i.e. the hopping terms affect only the long time dynamics of the system. Although the results presented here are for a schematic model, it can be shown that they hold also if the full wave-vector dependence is taken into account and hence can be viewed as generic (Sjögren 1990).

Thus in summary we can conclude that according to MCT the relaxation dynamics of a real system, i.e. a system in which the hopping terms are present, will be quite similar to the ones of the solutions of the ideal version of MCT *for couplings that are weak or moderate* in that an increase of the coupling toward the critical value will lead to a rapid increase of the relaxation times and in that the correlators will obey approximately the time-temperature superposition principle and be stretched. This holds for couplings such that the relaxation time of the ideal system is smaller than $O(\delta^{-1})$, defined by Eq. (5.83). Once the relaxation time of the ideal system is larger than this time scale, the correlators of the real system will be different from the ones of the ideal system in that the relaxation times increase more slowly than a power-law and the stretching decreases again. It can be shown (Sjögren 1990) that if the exponent parameter λ is larger than $\pi/4$, this temperature dependence is given by an Arrhenius law, i.e. for this case MCT predicts indeed the experimentally observed crossover from a super-Arrhenius dependence at high temperatures to an Arrhenius dependence at low T, see Fig. 5.2.

The predicted temperature dependence of the shape of the correlation functions is in agreement with the one found in the computer simulations of Horbach and Kob (2001) on viscous SiO_2, who found that the stretching does indeed show a non-monotonic dependence on T if T_c is crossed. Also the fact that for this system it has been possible to use the ideal version of MCT to calculate reliably the $q-$dependence of the various nonergodicity parameters, i.e. the height of the plateau, is strong evidence that for short and intermediate times the real system can be approximated well by the ideal MCT equations. Hence we conclude that for the case of SiO_2 the scenario proposed by the extended version of MCT is indeed able to rationalize the relaxation dynamics. To what extend this is also true for systems for which the hopping terms are less pronounced is presently not clear since the corresponding computer simulations have not yet been done due to the difficulty to equilibrate liquids that have only weak hopping processes. It is, however, evident that it is only a matter of time until this point can be addressed.

5.2.6 *Some General Remarks on Mode Coupling Theory*

When the equations that today are known as "mode coupling theory", Eqs. (5.45)-(5.51), were proposed for the first time, it was believed that they describe the dynamics of a glass-forming liquid close to the (experimental) glass transition, i.e. in the temperature range where the viscosity is on the order of $10^9 - 10^{13}$Pa·s, since the divergence of the relaxation time in the (ideal) MCT was thought to be closely connected to the corresponding divergence of τ at the Kauzmann temperature. This possibility and the multitude of predictions of the theory triggered an extremely high activity in the experimental and computational communities in order to test these predictions (Götze 1991, 1999, 2009, Cummins *et al.* 1994, Götze and Sjögren 1992). The result of this effort is that we now have very good evidence that MCT does not describe the relaxation dynamics of glass-forming systems close to T_g but instead the dynamics in a temperature range at which the relaxation dynamics crosses over from the one found at high temperatures, and that is related to normal liquid dynamics, to the relaxation dynamics that is found for temperatures close to T_g, i.e. where the viscosity is on the order of 10^{13}Pa·s. For most liquids this crossover falls in the temperature range for which the viscosity is between $10^{-1} - 10^4$Pa·s, i.e. where the relaxation times are on the order of ps to 100 ns.[23] This is also the range where most glass formers show the (more or less marked) upward turn in the Angell-plot, see Fig. 5.2. Thus MCT is able to give a microscopic reason for this crossover phenomenon and in addition makes many predictions on the time and temperature dependence of other observables (time correlation functions, susceptibilities, ...). As we have seen in the previous subsection, these predictions are often found to hold not only on a qualitative level but also on a quantitative one. This is strong evidence that the theory does not just propose a convenient and flexible set of fitting functions but instead allows to make a real calculation within the framework of statistical mechanics.

Note that for most real systems it is not possible to make such an "*ab initio*" approach, since usually neither the necessary microscopic

[23] An exception to this are colloidal suspensions of hard spheres whose typical relaxation times are between μs and many days. It has been found that for these systems the range of validity of MCT extends over more decades in time than for atomic systems which is evidence that for the former systems the hopping processes are not very relevant. Only recent work has given evidence that even colloidal systems can be equilibrated below the critical packing fraction ϕ_c of MCT, i.e. that also they show hopping processes (Brambilla *et al.* 2009).

information (all the partial structure factors) nor the Hamiltonian are known with sufficient high accuracy to permit to calculate the vertices and hence the memory function. In this situation all one can do is thus to check to what extend the *generic* predictions of the theory hold. In most cases it is, however, not a simple matter to give compelling evidence that the theory does indeed give an accurate description of the relaxation dynamics, since most predictions of the theory involve several unknown parameters that can be adjusted (critical temperature, exponent parameter, height of plateau, critical amplitude, ...). E.g. it is evident that in a restricted temperature range it will always be possible to fit the relaxation time with a power-law and thus this observation *alone* is certainly not a proof for the validity of the theory. However, most of the different predictions involve the *same* parameters, or the theory predicts certain relations between them, and hence it becomes indeed possible to investigate whether or not MCT gives a reliable description of the relaxation dynamics. For this it is thus necessary to check for the same system and under the same conditions as many of these predictions as possible and to see whether it is possible to describe within MCT the relaxation dynamics within *one* consistent set of parameters.

As discussed at the end of the previous subsection, the inclusion of the hopping processes restores the ergodicity of the system and hence the singularity in the relaxation times is avoided, in agreement with the findings from experiments. Thus from a phenomenological point of view the extended version of MCT seems to be correct. However, the hopping terms considered in Eq. (5.53) contain only one type of hopping, the one related to the longitudinal current, and it is expected that there are also other contributions such as a coupling to the transverse current and multi-phonon processes (Götze and Sjögren 1987, 1988). Unfortunately there is presently no good understanding how exactly these additional terms can be included in the theory and hence progress on this matter has been rather slow in the last few years (Saltzman and Schweizer 2006, Chong 2008). Therefore it is presently not clear whether or not the approach proposed by MCT can be extended to the temperature range around the experimental glass transition.

Despite this difficulty one has to emphasize that MCT has been able to make an impressive number of predictions on the dynamics of glass forming liquids (factorization property, existence of the β−process, shape of β−correlator, wave-vector dependence of nonergodicity parameters, ...), most of them have been verified in experiments and computer simulations, or to rationalize results that have been known since a long time, without

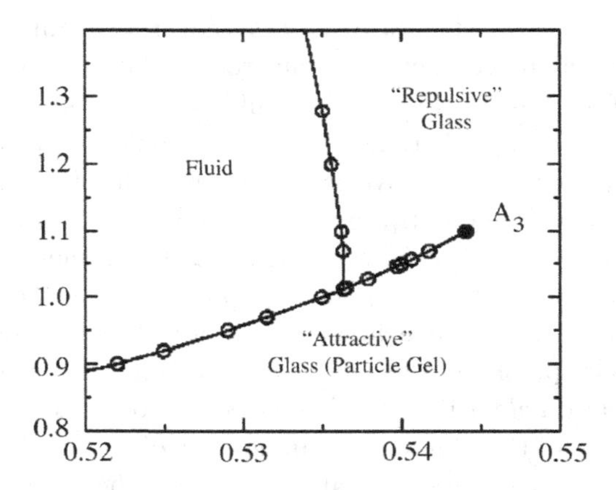

Fig. 5.23 Location of the glass transition lines in a system of particles that have a hard sphere interaction and an attractive well (of width 3% of the sphere diameter). The abscissa is the packing fraction ϕ of the particles and the ordinate is temperature. Note the presence of three different phases: The fluid at low ϕ, high T. A "repulsive glass" at high ϕ, high T, and an "attractive glass" at low T, high ϕ. From Dawson (2002).

making *ad hoc* assumptions (such as the crossover temperature in the Angell plot or the stretching of the time correlation functions). In the last few years the theory has also been successfully applied to increasingly complex systems, such as molecules, polymers, and gels, thus showing that it is a useful theoretical framework also for such systems and not only for simple liquids (Franosch *et al.* 1997a, 1997b, 1997c, Schilling and Scheidsteger 1997, Bergenholtz and Fuchs 1999, Fabbian *et al.* 1999b, Götze *et al.* 2000, Chong *et al.* 2001a, Chong and Götze 2002b, Chong and Fuchs 2002, Miyazaki and Yethiraj 2002, Kroy *et al.* 2004). Recently the theory has even predicted a new type of glass, the so-called "attractive glass", in which the relaxation dynamics is very different from the usually studied hard sphere type systems ("repulsive glasses"). The first example of such a type of glass-former was a system of particles in which the interaction is a hard sphere plus a narrow attractive well (Dawson *et al.* 2001). For these systems at intermediate densities the dynamics slows down with increasing density like in a standard hard sphere system. However, once the typical distance between the particles becomes comparable with the width of the well, some of the particles get (temporarily) trapped in this well, thus freeing up configurations space for the other particles. As a consequence the other particles can move more easily and the dynamics is (on avarage) *accelerated*. At even higher densities all the particles feel the attractive well, and hence the system slows (again) down with increasing ϕ. The result of this theoretical prediction is shown in Fig. 5.23 in which the location of the glass transition is plotted as a function of temperature (verical axis) and packing fraction ϕ (horizontal axis). We see the standard hard sphere glass line at high density has a negative curvature, thus showing the re-entrant behavior. At low temperatures and intermediate and high densities most of the particles are trapped inside the well, thus forming a relatively weakly connected structure, i.e. a gel. (In Sec. 6.3 we will discuss this in more detail.) It is remakable that the theory predics the existence of a glass-glass transition, i.e. a line in parameter space at which two different types of glasses (that have a different structure) coexist and this line ends at a special point, a so-called A_3- singularity, a special higher order singularity of MCT (Götze 2009).

In addition to this reentrant behavior MCT predicts that the time correlation functions show over a large time window a logarithmic time dependence (Dawson *et al.* 2001, Götze and Sperl 2002, Sperl 2003). This new form of glass has subsequently been found in computer simulations and experiments of colloidal systems (Mallamace *et al.* 2000, Foffi *et al.* 2002, Pham *et al.* 2002, Zaccarelli *et al.* 2002b, Chen *et al.* 2002, Chen *et al.* 2003a, 2003b, Sciortino *et al.* 2003) and later on also in telechelic gels (Chaudhuri *et al.* 2010). Thus this shows that the theory has the virtue of not only being able to rationalize results that have been known beforehand but also to predict new and unexpected phenomena. Finally we mention that recently the theory has also be generalized to the situation that the system is inhomogeneous (e.g. in which the liquid is confined in a pore) and new predictions have been made (Krakoviack 2005, Biroli *et al.* 2006) and in Chap. 6 we will discuss these results in more detail.

Last not least we point out the fact that there are glass forming systems that have a structure that is strictly independent of the control parameter that drives the transition. One such system is an array of infinitely thin rods of length ℓ whose center of mass is fixed on the vertices of a given lattice (simple cubic, fcc, ...) that has a lattice spacing a. The rods can rotate freely but are not allowed to intersect each other. This system can thus be considered as a simple model to describe the steric hindrance effect of the orientational degrees of freedom in a plastic crystal, discussed in Sec. 4.6. Since the rods have thickness zero, there are no static correlations at any temperature, or for any length ℓ/a of the rods, the relevant parameter of the system. Therefore a theory like MCT will not predict any dynamical transition for this system. However, computer simulations have shown that there is indeed such a transition in that with increasing ℓ/a the orientational dynamics of the rods seem to become frozen (Renner *et al.* 1995, Obukhov *et al.* 1997). The reason for this slowing down is that the dynamics of the rods is hindered by the presence of the nearest, next nearest neighbor, ... rods who form a dynamic cage around it. Thus also here the slow dynamics is due to a cage effect, but in contrast to the one found in usual glass forming systems where the cage is related to a static structure that becomes more and more rigid with increasing coupling, in this case the cage is of purely dynamic origin.[24]

Due to the absence of any static correlations in such systems it is clear that theories like MCT will not be able to give an accurate description of the relaxation dynamics and therefore other approaches have to be used (Schilling and Szamel 2003, Ricker and Schilling 2005). This simple example shows that although the theory of glass forming systems has now reached a certain level of maturity, it is by far not complete and thus will be the focus of research for quite a few more years.

[24]There is an apparent similarity of this hindrance to the one present in the kinetically constrained lattice gas models that will be discussed in Chap. 6 where the imposed kinetic rules lead to a slowing down in the dynamics. However, it is presently not clear how far the analogy really goes.

References

Adam, G., and Gibbs, J.H. (1965) *J. Chem. Phys.* **43**, 139.

Aichele, M., and Baschnagel, J. (2001a) *Eur. Phys. J.* **E5**, 229.

Aichele, M., and Baschnagel, J. (2001b) *Eur. Phys. J.* **E5**, 245.

Alba-Simionesco, C., and Krauzman, M. (1995) *J. Chem. Phys.* **102**, 6574.

Andreanov, A., Biroli, G., and Bouchaud, J.-P. (2009) *Europhys. Lett.* **88**, 16001.

Angell, C.A. (1985) *Relaxation in Complex Systems* (K.L. Ngai and G.B. Wright, eds.) p. 1 (US Dept. Commerce, Springfield).

Angell, C.A., and Tucker, J.C. (1974) *J. Phys. Chem.* **78**, 278.

Angell, C.A., Poole P.H., and Shao J. (1994) *Nuovo Cimento* **D16**, 993.

Ashcroft, N.W., and Mermin, N.D. (1976) *Solid State Physics* (Holt-Saunders International Editors, Philadelphia).

Balucani, U., and Zoppi, M. (1994) *Dynamics of the Liquid State* (Oxford University Press, Oxford).

Balucani, U., Lee, M.H., and Tognetti, V. (2003) *Phys. Rep.* **373**, 409.

Barrat, J.-L. (1999) in *Physics of Glasses* (P. Jund and R. Jullien, eds.) p. 47, (APS, New York).

Barrat, J.-L., and Hansen, J.-P. (2003) *Basic Concepts for Simple and Complex Liquids* (Cambridge University Press, Cambridge).

Barrat, J.-L., and Kob, W. (1999) *Europhys. Lett.* **46**, 637.

Barrat, J.-L., and Latz, A. (1990) *J. Phys.: Condens. Matter* **2**, 4289.

Barrat, J.-L., Götze, W., and Latz, A. (1989) *J. Phys.: Condens. Matter* **1**, 7163.

Baschnagel, J., and Binder, K. (1995) *Macromolecules* **28**, 6808.

Baschnagel, J., and Varnik, F. (2005) *J. Phys.: Condens. Matter* **17**, R851.

Baschnagel, J., Bennemann, C., Paul, W., and Binder, K. (2000) *J. Phys.: Condens. Matter* **12**, 6365.

Bässler, H. (1987) *Phys. Rev. Lett.* **58**, 767.

Benassi, P., Krisch, M., Masciovecchio, C., Mazzacurati, V., Monaco, G., Ruocco, G., Sette, F., and Verbeni, R. (1996) *Phys. Rev. Lett.* **77**, 3835.

Bengtzelius, U. (1986) *Phys. Rev.* **A33**, 3433.

Bengtzelius, U., Götze, W., and Sjölander, A. (1984) *J. Phys.* **C17**, 5915.

Bennemann, C., Paul, W., Baschnagel, J., and Binder, K. (1999) *J. Phys.: Condens. Matter* **11**, 2179.

Bergenholtz, J., and Fuchs, M. (1999) *Phys. Rev.* **E59**, 5706.

Berne, B.J., and Pecora, R. (2000) *Dynamic Light Scattering: With Applications to Chemistry, Biology, and Physics* (Dover, New York).

Berthier, L. (2004) *Phys. Rev.* **E69**, 020201.

Berthier, L. (2007) *Phys. Rev.* **E76**, 011507.

Berthier, L., and Barrat, J.-L. (2002a) *J. Chem. Phys.* **116**, 6228.

Berthier, L., and Barrat, J.-L. (2002b) *Phys. Rev. Lett.* **89**, 095702.

Berthier, L., and Garrahan, J.P. (2003a) *Phys. Rev.* **E68**, 041201.

Berthier, L., and Garrahan, J.P. (2003b) *J. Chem. Phys.* **119**, 4367.

Berthier, L., and Kob, W. (2007) *J. Phys.: Condens. Matter* **19**, 205130.

Berthier, L., and Witten, T.A. (2009) *Europhys. Lett.* **86**, 10001.

Binder, K., and Young, A.P. (1986) *Rev. Mod. Phys.* **58**, 801.

Berthier, L., Biroli, G., Bouchaud, J.-P., and Jack, R.L. (2010) arxiv:1009.4765.

Binder, K., Baschnagel, J., and Paul, W. (2003) *Prog. Polym. Sci.* **28**, 115.

Birge, N.O., and Nagel, S.R. (1985) *Phys. Rev. Lett.* **54**, 2674.

Biroli, G., and Bouchaud, J.-P. (2009) arxiv:cond-mat 0912.2542.

Biroli, G., and Bouchaud, J.-P. (2004) *Europhys. Lett.* **67**, 21.

Biroli, G., and Bouchaud, J.-P. (2007) *J. Phys.: Condens. Matter* **19**, 205101.

Biroli, G., and Monasson, R. (2000) *Europhys. Lett.* **50**, 155.

Biroli, G., and Bouchaud, J.-P., Miyazaki, K., Reichman, D.R. (2006) *Phys. Rev. Lett.* **97**, 195701.

Biroli, G., Bouchaud, J.-P., Cavagna, A., Grigera, T.S., and Verrocchio, P. (2008) *Nature Phys.* **4**, 771.

Böhmer, R., Ngai, K.L., Angell, C.A., and Plazek, D.J. (1993) *J. Chem. Phys.* **99**, 4201.

Bosse, J., and Kaneko, Y. (1995) *Phys. Rev. Lett.* **74**, 4023.

Bosse, J., and Thakur, J.S. (1987) *Phys. Rev. Lett.* **59**, 998.

Bouchaud, J.-P., Cugliandolo, L.F., Kurchan, J., and Mézard, M. (1996)

Physica **A226**, 243.

Bouchaud, J.-P., Cugliandolo, L.F., Kurchan, J., and Mézard, M. (1998) in *Spin Glasses and Random Fields* (A.P. Young, ed.) p. 161, (World Scientific, Singapore).

Brambilla, G., El Masri, D., Pierno, M., Berthier, L., Cipelletti, L., Petekidis, G., and Schofield, A.B. (2009) *Phys. Rev. Lett.* **102**, 085703.

Brébec, G., Seguin, R., Sella, C., Bevenot, J., and Martin, J.C. (1980) *Acta Metall.* **28**, 327.

Brodin, A., Frank, M., Wiebel, S., Shen, G.Q., Wuttke, J., and Cummins, H.Z. (2002) *Phys. Rev.* **E65**, 051503.

Brüning, R., and Samwer, K. (1992) *Phys. Rev.* **B46**, 11318

Brüning, R., St-Onge, D.A., Patterson, S., and Kob, W. (2009) *J. Phys.: Condens. Matter* **21**, 035117.

Buchenau, U., Prager, M., Nücker, N., Dianoux, A.J., Ahmad, N., and Phillips, W.A. (1986) *Phys. Rev.* **B34**, 5665.

Büchner, S., and Heuer, A. (2000) *Phys. Rev. Lett.* **84**, 2168.

Butler, S., and Harrowell, P. (1991) *J. Chem. Phys.* **95**, 4454.

Casas, J., González, D.J., González, L.E., Alemany, M.M.G., and Gallego, L.J. (2000) *Phys. Rev.* **B62**, 12095.

Cavagna, A. (2009) *Phys. Rep.* **476**, 51.

Chandler, D. (1987) *Introduction to Modern Statistical Mechanics* (Oxford University Press, Oxford).

Chaudhuri, P., Berthier, L., Hurtado, P.I., and Kob, W. (2010) *Phys. Rev.* **E81**, 040502.

Chen, S.H., Mallamace, F., Faraone, A., Gambadauro, P., Lombardo, D., and Chen, W.R. (2002) *Eur. Phys. J.* **E9**, 283.

Chen, S.H., Chen, W.R., and Mallamace, F. (2003a) *Science* **300**, 619.

Chen, W.R., Mallamace, F., Glinka, C.J., Fratini, E., and Chen, S.H. (2003b) *Phys. Rev.* **E68**, 041402.

Chong, S.H. (2008) *Phys. Rev.* **E78**, 041501.

Chong, S.H., and Fuchs, M. (2002) *Phys. Rev. Lett.* **88**, 185702.

Chong, S.-H., and Götze, W. (2002a) *Phys. Rev.* **E65**, 041503.

Chong, S.-H., and Götze, W. (2002b) *Phys. Rev.* **E65**, 051201.

Chong, S.-H., and Hirata, F. (1998a) *Phys. Rev.* **E57**, 1691.

Chong, S.-H., and Hirata, F. (1998b) *Phys. Rev.* **E58**, 6188.

Chong, S.-H., and Sciortino, F. (2004) *Phys. Rev.* **E69**, 051202.

Chong, S.-H., Götze, W., and Singh, A.P. (2001a) *Phys. Rev.* **E63**, 011206.

Chong, S.-H., Götze, W., and Mayr, M.R. (2001b) *Phys. Rev.* **E64**, 011503.

Cicerone, M.T., Blackburn, F.R., and Ediger, M.D. (1995) *J. Chem. Phys.* **102**, 471.

Cohen, M.H., and Grest, G.S. (1979) *Phys. Rev.* **B20**, 1077.

Coluzzi, B., and Verrocchio, P. (2002) *J. Chem. Phys.* **116**, 3789.

Coluzzi, B., Parisi, G., and Verrocchio, P. (2000) *Phys. Rev. Lett.* **84**, 306.

Coslovich, D., and Pastore, G. (2007) *J. Chem. Phys.* **127**, 124504.

Courtens, E., Foret, M., Hehlen, B., and Vacher, R. (2001) *Solid State Commun.* **117**, 187.

Courtens, E., Foret, M., Hehlen, B., Rufflé, B., and Vacher, R. (2003) *J. Phys.: Condens. Matter* **15**, S1279.

Cugliandolo, L.F. (2003) in *Lecture Notes for "Slow relaxations and nonequilibrium dynamics in condensed matter", Les Houches July, 1–25, 2002; Les Houches Session LXXVII* (J.-L. Barrat, M. Feigelman, J. Kurchan, and J. Dalibard, eds.) p. 367 (Springer, Berlin).

Cugliandolo, L.F., and Kurchan, J. (1993) *Phys. Rev. Lett.* **71**, 173.

Cugliandolo, L.F., Kurchan, J., and Le Doussal (1996) *Phys. Rev. Lett.* **76**, 2390.

Cugliandolo, L.F., Kurchan, J., and Peliti, L. (1997) *Phys. Rev.* **E55**, 3898.

Cummins, H.Z., Li, G., Du, W.M., and Herandez, J. (1994) *Physica* **A204**, 169.

Cummins, H.Z., Li, G., Hwang, Y.H., Shen, G.Q., Du, W.M., Herandez, J., and Tao, N.J. (1997) *Z. Phys. B: Condens. Matter* **103**, 501.

Das, S.P. (2004) *Rev. Mod. Phys.* **76**, 785.

Das, S.P., and Mazenko, G.F. (1986) *Phys. Rev* **A34**, 2265.

Das, S.P., Mazenko, G.F., Ramaswamy, S., and Toner, J.J. (1985a) *Phys. Rev. Lett.* **54**, 118.

Das, S.P., Mazenko, G.F., Ramaswamy, S., and Toner, J.J. (1985b) *Phys. Rev.* **A32**, 3139.

Dasgupta, C., Indrani, A.V., Ramaswamy, S., and Phani, M.K. (1991) *Europhys. Lett.* **15**, 307.

Dawson, K.A. (2002) *Curr. Opin. Coll. Inter. Sci.* **7**, 218.

Dawson, K.A., Foffi, G., Fuchs, M., Götze, W., Sciortino, F., Sperl, M., Tartaglia, P., Voigtmann, T., and Zaccarelli, E. (2001) *Phys. Rev.* **E63**, 011401.

Dean, D.S., and Lefevre, A. (2001) *Phys. Rev. Lett.* **86**, 5639.

Debenedetti, P.G. (1997) *Metastable Liquids* (Princeton University Press, Princeton).

Debenedetti, P.G., Stillinger, F.H., Truskett, T.M., and Lewis, C.P. (2001) *Adv. Chem. Eng.* **28**, 21.

Di Leonardo, R., Angelani, L., Parisi, G., and Ruocco, G. (2000) *Phys. Rev. Lett.* **84**, 6054.

Doliwa, B., and Heuer, A. (1998) *Phys. Rev. Lett.* **80**, 4915.

Doliwa, B., and Heuer, A. (2003) *Phys. Rev.* **E67**, 030506.

Donth, E. (2002) *The Glass Transition* (Springer, Berlin).

Dyre, J.C. (2006) *Rev. Mod. Phys.* **78**, 953.

Ediger, M.D. (2000) *Ann. Rev. Phys. Chem.* **51**, 99.

Edwards, S.F., and Anderson, P.W. (1975) *J. Phys.* **F5**, 965.

Elliott, S.R. (1987) *Adv. Phys.* **36** 135.

Elmatad, Y.S., Chandler, D., and Garrahan, J.P. (2009) *J. Phys. Chem.* **B113**, 5563.

Ernst, R.M., Nagel, S.R., and Grest, G.S. (1991) *Phys. Rev.* **B43**, 8070.

Fabbian, L., Götze, W., Sciortino, F., Tartaglia, P., and Thiery, F. (1999a) *Phys. Rev.* **E59**, R1347.

Fabbian, L., Latz, A., Schilling, R., Sciortino, F., Tartaglia, P., and Theis, C. (1999b) *Phys. Rev.* **E60**, 5768.

Feltz, A. (1993) *Amorphous Inorganic Materials and Glasses* (VCH, Weinheim).

Flenner, E., and Szamel G. (2005) *Phys. Rev.* **E72**, 031508.

Foffi, G., Dawson, K.A., Buldyrev, S.V., Sciortino, F., Zaccarelli, E., and Tartaglia, P. (2002) *Phys. Rev.* **E65**, 050802.

Foffi, G., Götze, W., Sciortino, F., Tartaglia, P., and Voigtmann, T. (2004) *Phys. Rev.* **E69**, 011505.

Franosch, T., Fuchs, M., Götze, W., Mayr, M.R., and Singh, A.P. (1997a) *Phys. Rev.* **E56**, 5659.

Franosch, T., Götze, W., Mayr, M.R., and Singh, A.P. (1997b) *Phys. Rev.* **E55**, 3183.

Franosch, T., Fuchs, M., Götze, W., Mayr, M.R., and Singh, A.P. (1997c) *Phys. Rev.* **E55**, 7153.

Franz, S., and Montanari, A. (2007) *J. Phys. A: Math. Theor.* **40**, F251.

Fuchs, M. (1994) *J. Non-Cryst. Sol.* **172–174**, 241.

Fuchs, M., Götze, W., Hofacker, I., and Latz, A. (1991) *J. Phys.: Condens. Matter* **3**, 5047.

Fuchs, M., Götze, W., Hildebrand, S., and Latz, A. (1992) *J. Phys.: Condens. Matter* **4**, 7709.

Fuchs, M., Götze, W., and Mayr, M.R. (1998) *Phys. Rev.* **E58**, 3384.

Fulcher, G.S. (1925) *J. Amer. Ceram. Soc.* **8**, 339.

Geszti, T. (1983) *J. Phys. C: Solid State Phys.* **16**, 5805.

Gleim, T. (1998) *Ph.D. Thesis* (Johannes Gutenberg Universität Mainz).

Gleim, T., Kob, W., and Binder, K. (1998) *Phys. Rev. Lett.* **81**, 4404.

Gleim, T., and Kob, W. (2000) *Eur. Phys. J.* **B13**, 83.

Glotzer, S.C. (2000) *J. Non-Cryst. Solids* **274**, 342.

Goldammer, M., Losert, C., Wuttke, J., Petry, W., Terki, F., Schober, H., and Lunkenheimer, P. (2001) *Phys. Rev.* **E64**, 021303.

Goldstein, M. (1969) *J. Chem. Phys.* **51**, 3728.

Götze, W. (1978) *Sol. State Comm.* **27**, 1393.

Götze, W. (1984) *Z. Phys.* **B56**, 139.

Götze, W. (1985) *Z. Phys.* **B60**, 195.

Götze, W. (1991) in *Liquids, Freezing and the Glass Transition, Les Houches. Session LI, 1989* (J.P. Hansen, D. Levesque, and J. Zinn-Justin, eds.) p. 287 (North-Holland, Amsterdam).

Götze, W. 1999 *J. Phys.: Condens. Matter* **10**, A1.

Götze, W. (2009) *Complex Dynamics of Glass-Forming Liquids. A Mode Coupling Theory* (Oxford University Press, Oxford).

Götze, W., and Mayr, M.R. (2000) *Phys. Rev.* **E61**, 587.

Götze, W., and Sjögren, L. (1987) *Z. Phys.* **B65**, 415.

Götze, W., and Sjögren, L. (1988) *J. Phys.* **C21**, 3407.

Götze, W., and Sjögren, L. (1991) *Phys. Rev.* **A43**, 5442.

Götze, W., and Sjögren, L. (1992) *Rep. Prog. Phys.* **55**, 241.

Götze, W., and Sperl, M. (2002) *Phys. Rev.* **E66**, 011405.

Götze, W., and Voigtmann, T. (2000) *Phys. Rev.* **E61**, 4133.

Götze, W., and Voigtmann, T. (2003) *Phys. Rev.* **E67**, 021502.

Götze, W., Singh, A.P., and Voigtmann, T. (2000) *Phys. Rev.* **E61**, 6934.

Grigera, T.S., Martin-Mayor, V., Parisi, G., and Verrocchio, P. (2001) *Phys. Rev. Lett.* **87**, 085502.

Hansen, J.-P., and McDonald, I. R. (1986) *Theory of Simple Liquids* (Academic, London).

Hehlen, B., Courtens, E., Vacher, R., Yamanaka, A., Kataoka, M., and Inoue, K. (2000) *Phys. Rev. Lett.* **84**, 5355.

Henrich, O., Weysser, F., Cates, M.E., and Fuchs, M. (2009) *Phil. Trans. Roy. Soc.* **A367**, 5033.

Hess, K.-U., Dingwell, D.B., and Rössler, E. (1996) *Chem. Geol.* **128**, 155.

Heuer, A. (2008) *J. Phys.: Condens. Matter* **20**, 373101.

Hohenberg, P.C., and Halperin, B.I. (1977) *Rev. Mod. Phys.* **49**, 435.

Horbach, J., and Kob, W. (1999) *Phys. Rev.* **B60**, 3169.

Horbach, J., and Kob, W. (2001) *Phys. Rev.* **E64**, 041503.

Horbach, J., and Kob, W. (2002) *J. Phys.: Condens. Matter* **14**, 9237.

Horbach, J., Kob, W., and Binder, K. (2001) *Eur. Phys. J.* **B19**, 531.

Horbach, J., Kob, W., and Binder, K. (2002) *Phys. Rev. Lett.* **88**, 125502.

Johari, G.P., and Goldstein, M. (1971) *J. Chem. Phys.* **53**, 2372.

Jung, Y., Garrahan, J.P., and Chandler, D. (2004) *Phys. Rev.* **E69**, 061205.

Kämmerer, S., Kob, W., and Schilling, R. (1998) *Phys. Rev.* **E58**, 2141.

Kauzmann, W. (1948) *Chem. Rev.* **43**, 219.

Kawasaki, K. (1966) *Phys. Rev.* **150**, 291.

Kawasaki, K. (1994) *Physica* **A208**, 35.

Kawasaki, K. (1995) *Trans. Theory Stat. Phys.* **24**, 755.

Kawasaki, K. (1997) *Physica* **A243**, 25.

Kegel, W.K., and van Blaaderen, A. (2000) *Science* **287**, 290.

Kirkpatrick, T.R., and Thirumalai, D. (1987) *Phys. Rev. Lett.* **58**, 2091.

Kirkpatrick, T.R., and Thirumalai, D. (1988) *Phys. Rev.* **B37** 5342.

Kirkpatrick, T.R., and Wolynes, P.G. (1987a) *Phys. Rev.* **A35**, 3072.

Kirkpatrick, T.R., and Wolynes, P.G. (1987b) *Phys. Rev.* **B36**, 8552.

Kirkpatrick, T.R., Thirumalai, D., and Wolynes, P.G. (1989) *Phys. Rev.* **A40**, 1045.

Kob, W. (1999) *J. Phys.: Condens. Matter* **11**, R85.

Kob, W. (2003) in *Lecture Notes for "Slow relaxations and nonequilibrium dynamics in condensed matter", Les Houches July, 1-25, 2002; Les Houches Session LXXVII* (J.-L. Barrat, M. Feigelman, J. Kurchan, and J. Dalibard, eds.) p. 199 (Springer, Berlin).

Kob, W., and Andersen, H.C. (1994) *Phys. Rev. Lett.* **73**, 1376.

Kob, W., and Andersen, H.C. (1995a) *Phys. Rev.* **E51**, 4626.

Kob, W., and Andersen, H.C. (1995b) *Phys. Rev.* **E52**, 4134.

Kob, W., and Barrat J.-L. (2000) *Eur. Phys. J.* **B13**, 319.

Kob, W., Donati, C., Plimpton, S.J., Glotzer, S.C., and Poole, P.H. (1997) *Phys. Rev. Lett.* **79**, 2827.

Kob, W., Horbach, J., and Binder, K. (1999), in AIP Conference Proceedings Vol. 469, *Slow Dynamics in Complex Systems, Tohwa University 1998* (M. Tokuyama, ed.) p. 441 (AIP, Woodbury).

Kob, W., Nauroth, M., and Sciortino, F. (2002) *J. Non-Cryst. Solids* **307-310**, 181.

Kohlrausch, R. (1847) *Ann. Phys. (Leipzig)* **12**, 393.

Krakoviack, V. (2005) *Phys. Rev. Lett.* **94**, 065703.

Krakoviack, V., and Alba-Simionesco, C. (2002) *J. Chem. Phys.* **117**, 2161.

Kroy, K., Cates, M.E., and Poon W.C.K. (2004) *Phys. Rev. Lett.* **92**, 148302.

La Nave, E., Mossa, S., and Sciortino, F. (2002) *Phys. Rev. Lett.* **88**, 225701.

Latz, A. (2000) *J. Phys.: Condens. Matter* **12**, 6353.

Latz, A. (2001) preprint cond-mat/0106086.

Laughlin, W.T., and Uhlmann, D.R. (1972) *J. Phys. Chem.* **76**, 2317.

Letz, M., Schilling, R., and Latz, A. (2000) *Phys. Rev.* **E62**, 5173.

Leutheusser, E. (1984) *Phys. Rev.* **A29**, 2765.

Li, G., Du, W.M., Chen, X.K., Cummins, H.Z., and Tao, N.J. (1992) *Phys. Rev.* **A45**, 3867.

Lovesey, S.W. (1994) *Theory of Neutron Scattering from Condensed Matter* (Oxford University Press, Oxford).

Lunkenheimer, P., Pimenow, A., Dressel, M., Gonchunov, Y.G., Böhmer, R., and Loidl, A. (1996a) *Phys. Rev. Lett.* **77**, 318.

Lunkenheimer, P., Pimenov, A., Schiener, B., Böhmer R., and Loidl, A. (1996b) *Europhys. Lett.* **33**, 611.

Magill, J.H. (1967) *J. Chem. Phys.* **47**, 2802.

Mallamace, F., Gambadauro, P., Micali, N., Tartaglia, P., Liao, C., and Chen, S.H. (2000) *Phys. Rev. Lett.* **84**, 5431.

Menon, N. (1996) *J. Chem. Phys.* **105**, 5246.

Meyer, A. (2002) *Phys. Rev.* **B66**, 134205.

Meyer, A., Wuttke, J., Petry, W., Randl, O.G., and Schober, H. (1998) *Phys. Rev. Lett.* **80**, 4454.

Mézard, M., and Parisi, G. (1999) *Phys. Rev. Lett.* **82**, 747.

Mézard, M., and Parisi, G. (2000) *J. Phys.: Condens. Matter* **12**, 6655.

Mézard, M., and Parisi, G. (2009) arXiv:cond-mat 0910.2838.

Mikkelsen, J.C. (1984) *Appl. Phys. Lett.* **45**, 1187.

Miyazaki, K., and Yethiraj, A. (2002) *J. Chem. Phys.* **117**, 10448.

Montanari, A., and Semerjian, G. (2006) *J. Stat. Phys.* **125**, 23.

Mori, H. (1965) *Prog. Theor. Phys.* **33**, 423.

Nauroth, M. (1999) *Ph.D. Thesis* (Technical University of Munich).

Nauroth, M., and Kob, W. (1997) *Phys. Rev.* **E55**, 657.

Ngai, K.L., and Paluch M. (2004) *J. Chem. Phys.* **120**, 857.

Nieuwenhuizen, Th.M. (1998) *Phys. Rev. Lett.* **80**, 5580.

Obukhov, S.P., Kobsev, D., Perchak, A., and Rubinstein, M. (1997) *J. Phys. I* **7**, 563.

Ogielski, A.T., and Stein, D.L. (1985) *Phys. Rev. Lett.* **55**, 1634.

Parisi, G. (1997) *Phys. Rev. Lett.* **79**, 3660.

Parisi, G. (2003) in *Lecture Notes for "Slow relaxations and nonequilibrium dynamics in condensed matter", Les Houches July, 1-25, 2002; Les Houches Session LXXVII* (J.-L. Barrat, M. Feigelman, J. Kurchan, and J. Dalibard, eds.) p. 271 (Springer, Berlin).

Pathria, R.K. (1986) *Statistical Mechanics* (Pergamon Press, Oxford).

Perera, D.N., and Harrowell, P. (1996) *J. Chem. Phys.* **104**, 2369.

Pham, K.N., Puertas, A.M., Bergenholtz, J., Egelhaaf, S.U., Moussaid, A., Pusey, P.N., Schofield, A.B., Cates, M.E., Fuchs, M., and Poon, W.C.K. (2002) *Science* **296**, 104.

Puertas, A.M., Fuchs, M., and Cates, M.E. (2003) *Phys. Rev.* **E67**, 031406.

Rat, E., Foret, M., Courtens, E., Vacher, R., and Arai, M. (1999) *Phys. Rev. Lett.* **83**, 1355.

Renner, C., Löwen, H., and Barrat, J.-L. (1995) *Phys. Rev.* **E52**, 5091.

Richert, R. (1994) *J. Non-Cryst. Solids* **172-174**, 209.

Richert, R. (2002) *J. Phys.: Condens. Matter* **14**, R703.

Richert, R., and Bässler, H. (1990) *J. Phys.: Condens. Matter* **2**, 2273.

Ricker, M., and Schilling R. (2005) *Phys. Rev.* **E72**, 011508.

Rinaldi, A., Sciortino, F., and Tartaglia, P. (2001) *Phys. Rev.* **E63**, 061210.

Rössler, E., Hess, K.-U., and Novikov, V.N. (1998) *J. Non-Cryst. Solids* **223**, 207.

Roux, J.-N., Barrat, J.-L., and Hansen, J.-P. (1989). *J. Phys.: Condens. Matter* **1**, 7171.

Rufflé, B., Ecolivet, C., and Toudic, B. (1999) *Europhys. Lett.* **45**, 591.

Rufflé, B., Parshin, D.A., Courtens, E., and Vacher, R. (2008) *Phys. Rev. Lett.* **100**, 015501.

Ruocco, G., Sciortino, F., Zamponi, F., de Michele, C., and Scopigno, T. (2004) *J. Chem. Phys.* **120**, 10666.

Saltzman, E.J., and Schweizer K.S. (2006) *J. Chem. Phys.* **125**, 044509.

Sastry, S. (2004) *J. Phys. Chem.* **B108**, 19698.

Sastry, S., Debenedetti, P.G., and Stillinger, F.H. (1998) *Nature* **393**, 554.

Scheidler, P., Kob, W., Latz, A., Horbach, J., and Binder, K. (2001) *Phys. Rev.* **B63**, 104204.

Scheidler, P., Kob, W., and Binder, K. (2002) *Europhys. Lett.* **59**, 701.

Schilling, R., and Scheidsteger, T. (1997) *Phys. Rev.* **E56**, 2932.

Schilling, R., and Szamel, G. (2003) *Europhys. Lett.* **61**, 207.

Schmid, B., and Schirmacher, W. (2008) *Phys. Rev. Lett.* **100**, 137402.

Schirmacher, W., Diezemann, G., and Ganter, G. (1998) *Phys. Rev. Lett.* **81**, 136.

Schmidt-Rohr, K., and Spiess, H.W. (1991) *Phys. Rev. Lett.* **66**, 3020.

Schmitz, R., Dufty, J.W., and De P. (1993) *Phys. Rev. Lett.* **71**, 2066.

Schrøder, T.B., Sastry, S., Dyre, J.C., and Glotzer, S.C. (2000) *J. Chem. Phys.* **112**, 9834.

Sciortino, F., and Kob, W. (2001) *Phys. Rev. Lett.* **86**, 648.

Sciortino, F., and Tartaglia, P. (2001) *Phys. Rev. Lett.* **86**, 107.

Sciortino, F., Gallo, P., Tartaglia, P., and Chen, S.H. (1996) *Phys. Rev.* **E54**, 6331.

Sciortino, F., Fabbian, L., Chen, S.H., and Tartaglia, P. (1997) *Phys. Rev.* **E56**, 5397.

Sciortino, F., Kob, W., and Tartaglia, P. (1999) *Phys. Rev. Lett.* **83**, 3214.

Sciortino, F., Tartaglia, P., and Zaccarelli, E. (2003) *Phys. Rev. Lett.* **91**, 268301.

Shell, M.S., Debenedetti, P.G., La Nave, E., and Sciortino, F. (2003) *J. Chem. Phys.* **118**, 8821.

Sherrington, D., and Kirkpatrick, S. (1975) *Phys. Rev. Lett.* **35**, 1972.

Signorini, G.F., Barrat, J.-L., and Klein, M.L. (1990) *J. Chem. Phys.* **92**, 1294.

Sillescu, H. (1999) *J. Non-Cryst. Solids* **243**, 81.

Simon, F. (1931) *Z. Anorg. Allg. Chem.* **203**, 219.

Singh, S., Srivastava, S., Kumar, C.N., and Tankeshwar, K. (2003) *J. Phys. Chem. Liq.* **41**, 567.

Sjögren, L. (1980) *Phys. Rev* **A22**, 2866.

Sjögren, L. (1986) *Phys. Rev* **A33**, 1254.

Sjögren, L. (1990) *Z. Phys.* **B79**, 5.

Sjögren, L., and Sjölander, A. (1979) *J. Phys.* **C12**, 4369.

Sokolov, A.P. (1999) *J. Phys.: Condens. Matter* **11**, A213.

Sperl, M. (2003) *Phys. Rev.* **E68**, 031405.

Stickel, F., Fischer, E.W., and Richert, R. (1995) *J. Chem. Phys.* **102**, 6251.

Stillinger, F.H. (1988) *J. Chem. Phys.* **88**, 7818.

Stillinger, F.H. (1999) *Phys. Rev.* **E59**, 48.

Stillinger, F.H., and Weber, T.A. (1982) *Phys. Rev.* **A25**, 978.

Szamel, G., and Flenner E. (2004) *Europhys. Lett.* **67**, 779.

Tammann, G., and Hesse, W.Z. (1926) *Anorg. Allg. Chem.* **156**, 245.

Tankeshwar, K., Pathak, K.N., and Ranganathan, S. (1987) *J. Phys.* **C20**, 5749.

Tankeshwar, K., Pathak, K.N., and Ranganathan, S. (1995) *J. Phys.: Condens. Matter* **7**, 5729.

Theenhaus, T., Schilling, R., Latz, A., and Letz, M. (2001) *Phys. Rev.* **E64**, 051505.

Theis, C., Sciortino, F., Latz, A., Schilling, R., Tartaglia, P. (2000) *Phys. Rev.* **E62**, 1856.

Thirumalai, D., and Kirkpatrick, T.R. (1988) *Phys. Rev.* **B38**, 4881.

Tölle, A., Schober, H., Wuttke, J., and Fujara, F. (1997) *Phys. Rev.* **E56**, 809.

Tölle, A., Schober, H., Wuttke, J., Randl, O.G., and Fujara, F. (1998) *Phys. Rev. Lett.* **80**, 2374.

Toninelli, C., Biroli, G., and Fisher, D.S. (2004) *Phys. Rev. Lett.* **92**, 185504.

Tool, A.Q., and Eichlin, C.G. (1931) *J. Opt. Soc. Amer.* **14**, 276.

Toxvaerd, S., Pedersen, U.R., Schrøder, T.B., and Dyre, J.C. (2009) *J. Chem. Phys.* **130**, 224501.

Uhlmann, D.R. (1972) *J. Non-Cryst. Solids* **7**, 337.

van Beest B.W.H., Kramer G.J., and van Santen R.A. (1990) *Phys. Rev. Lett.* **64**, 1955.

van Megen, W., and Pusey, P.N. (1991) *Phys. Rev.* **A42**, 5429.

van Megen, W., and Underwood, S.M. (1993a) *Phys. Rev. Lett.* **70**, 2766.

van Megen, W., and Underwood, S.M. (1993b) *Phys. Rev.* **E47**, 248.

van Megen, W., and Underwood, S.M. (1994) *Phys. Rev.* **E49**, 4206.

van Zon R., and Schofield J. (2005) *J. Chem. Phys.* **122**, 194502.

Vogel, H. (1921) *Phys. Z.* **22**, 645.

Voigtmann, T. (2003) *Phys. Rev.* **E68**, 051401.

Voigtmann, T., and Horbach J. (2006) *Europhys. Lett.* **74**, 459.

Vollmayr, K., Kob, W., and Binder, K. (1996) *J. Chem. Phys.* **105**, 4714.

von Schweidler, E. (1907) *Ann. Phys.* **24**, 711.

Wales, D.J. (2003) *Energy Landscapes* (Cambridge University Press, Cambridge).

Weeks, E.R, Crocker, J.C., Levitt, A.C., Schofield, A., and Weitz, D.A. (2000) *Science* **287**, 627.

Williams, G., and Watts, D.C. (1980) *Trans. Faraday Soc.* **66**, 80.

Winkler, A., Latz, A., Schilling, R., and Theis, C. (2000) *Phys. Rev.* **E62**, 8004.

Winterling, G. (1975) *Phys. Rev.* **B12**, 2432.

Wolfgardt, M., Baschnagel, J., Paul, W., and Binder, K. (1996) *Phys. Rev.* **E54**, 1535.

Wuttke, J., Hernandez, J., Li, G., Coddens, G., Cummins, H.Z., Fujara, F., Petry, W., and Sillescu, H. (1994) *Phys. Rev. Lett.* **72**, 3052.

Wuttke, J., Seidl, M., Hinze, G., Tölle, A., Petry, W., and Coddens, G. (1998) *Eur. Phys. J.* **B1**, 169.

Wuttke, J., Ohl, M., Goldammer, M., Roth, S., Schneider, U., Lunkenheimer, P., Kahn, R., Rufflé, B., Lechner, R., and Berg, M.A. (2000) *Phys. Rev.* **B61**, 2730.

Yamamoto, R., and Onuki, A. (1998) *Phys. Rev.* **E58**, 3515.

Zaccarelli, E., Foffi, G., De Gregorio, P., Sciortino, F., Tartaglia, P., and Dawson, K.A. (2002a) *J. Phys.: Condens. Matter* **14**, 2413.

Zaccarelli, E., Foffi, G., Dawson, K.A., Buldyrev, S.V., Sciortino, F., and Tartaglia, P. (2002b) *Phys. Rev.* **E66**, 041402.

Zarzycki J. (Ed.) (1991) *Materials Science and Technology, Vol. 9*, (VCH Publ., Weinheim).

Zwanzig, R. (1960) *J. Chem. Phys.* **33**, 1338.

Zwanzig, R. (1961) *Phys. Rev.* **124**, 983.

Chapter 6

Further Models for Glassy Dynamics; Dynamical Heterogeneities; Gels; Driven Systems

6.1 Models for Slow Relaxation

Having reviewed in the previous chapter some of the salient features of glass-forming liquids we now will discuss some of the theoretical approaches that have been proposed to rationalize this slow dynamics. Roughly speaking one can distinguish two large classes of models: Phenomenological models that make assumptions on the mechanism for the slowing down of the dynamics (existence of traps, importance of cooperative motion, etc.) and then deduce some expressions that describe the relaxation time as a function of external parameters such as temperature, density etc. Examples of such models are the theories of Adam and Gibbs (1965), the trap model by Bouchaud (Monthus and Bouchaud 1996, Bouchaud *et al.* 1998), the approach on frustrated limited domains by Kivelson and coworkers (Tarjus *et al.* 2000, Sausset *et al.* 2008, Sausset and Tarjus 2010), or the free volume theory by Cohen and Turnbull (1959) which has later been extended by Cohen and Grest (1979, 1982). In the second class of models one is more ambitious in that one determines, sometimes by making strong approximations, the relaxation dynamics of many body models and compares this dynamics with the one found in real systems. Due to the difficulty to calculate this dynamics, these models are often rather simple and seem to be somewhat remote from real systems. However, they have the important merit that they do allow to understand the slow dynamics and the reason that gives rise to it in a unbiased way, i.e. without referring to a supposed mechanism. In addition these models can sometimes be used to test the relevance of the phenomenological approaches mentioned above. Examples of such models are the facilitated spin models, as the one proposed by Fredrickson and Andersen (1984), the theory of Gibbs and

DiMarzio (1958), or the mode coupling theory of the glass transition (Götze 1991, 2009, Götze and Sjögren 1992) discussed in the previous chapter. In the present section we will discuss some of these models in more detail. Due to its importance and large number of predictions, the presentation of the mode coupling theory has already been given in a separate chapter (Chap. 5).

6.1.1 *The Theory of Adam and Gibbs*

Starting point of the approach by Adam and Gibbs is the physical idea that in a glass-forming liquid the relaxation dynamics at low temperatures is the result of a sequence of individual events in which a subregion of the system relaxes to a new local configuration (Adam and Gibbs 1965). These relaxation events are supposed to occur because a local fluctuation of the enthalpy has allowed the particles in the subregion to make a *collective* motion. Therefore these subsystems are also called "cooperatively rearranging regions". Adam and Gibbs made the assumption that the relaxation dynamics of these subregions are *independent* of each other and therefore it is possible to describe each of them as an independent many body system that obeys the rules of statistical mechanics. (Note that for this one has to assume that the number of particles in each region is sufficiently large.) Consider one of these regions and let z be the number of particles in it.[1] The isothermal-isobaric partition function for the subsystem is then given by

$$\Delta(z,P,T) = \sum_{E,V} w(z,E,V)\exp(-\beta H)\,, \tag{6.1}$$

where $w(z,E,V)$ is the degeneracy of the subsystem at energy level E and volume V, and H is its enthalpy. If we restrict in Eq. (6.1) the sum to only those states that allow the system to undergo a rearrangement, we can define a partition function Δ'. Thus the fraction of systems that can undergo a relaxation event is given by

$$f(z,T) = \Delta'/\Delta = \exp[-\beta(G' - G')] \tag{6.2}$$

[1]In the following we will assume that for each value of z there is only one type of region, i.e. collection of arrangements of particles that are related to each other by means of simple relaxation events. Although for most systems this assumption is not correct, the following calculation can be easily generalized to the case that for each z there is a finite number of regions. The only change in the result is a different pre-exponential factor in the final expression, Eq. (6.7).

where we have introduced the Gibbs free energies $G = -k_BT \ln \Delta$ and $G' = -k_BT \ln \Delta'$. Introducing $\delta\mu = (G' - G)/z$, the difference in chemical potential per particle, and making use of the fact that $W(z,T)$, the probability that a system of size z makes a cooperative rearrangement within one unit of time, is proportional to $f(z,T)$, we thus obtain

$$W(z,T) = A \exp(-\beta z \delta\mu) \,. \tag{6.3}$$

Here A is a prefactor and in the following we will assume that its dependence on z and T is very weak with respect to the exponential function, $z\delta\mu \gg 1$, and hence can be neglected. Note that $\delta\mu$ is related to the *effective* energy barrier that a configuration has to overcome in order to make a relaxation event and hence its $T-$dependence can be assumed to be weak as well.

We now calculate the average transition rate per particle of the *whole* system. For this we imagine that the latter is (on average) the collection of $n(1,T), n(2,T), \ldots$ subsystems of size $z = 1, 2, \ldots$. Since the subsystems are assumed to be independent we have

$$\sum_{z=1}^{N} z n(z,T) = N \,, \tag{6.4}$$

where N is the number of particles in the whole system. The average transition probability per particle per unit of time is then given by

$$\overline{W(T)} = N^{-1} \sum_{z=1}^{N} z n(z,T) W(z,T) \,. \tag{6.5}$$

Due to the cooperative nature of the motion for the relaxation dynamics in each cluster, the particles in very small clusters will not be able to make such a movement and therefore the sum in Eq. (6.5) will only start at z^*, the smallest cluster that allows for a cooperative rearrangement and, using Eq. (6.3), $\overline{W(T)}$ can thus be written as

$$\overline{W(T)} = \frac{z^* n(z^*,T) \exp[-\beta z^* \delta\mu]}{N} \sum_{z=z^*}^{N} \frac{z n(z,T)}{z^* n(z^*,T)} A \exp[-\beta(z - z^*)\delta\mu] \,. \tag{6.6}$$

Under the assumption that for $z > z^*$ the number of clusters $n(z,T)$ does not depend very strongly on z, we thus find, since $\beta\delta\mu \gg 1$, that we can approximate the right hand side by the first term in the sum and hence obtain

$$\overline{W(T)} = \overline{A} \exp[-\beta z^* \delta\mu] \,, \tag{6.7}$$

where $\overline{A}$ can be considered as a constant. Thus this equation implies that the relaxation dynamics of the whole system is dominated by the cooperative events occurring in the *smallest* possible cooperatively rearranging regions since these relax by a factor of $O(\exp[-\beta\delta\mu])$ faster than the larger regions.

The final task is now to determine z^*. In principle this quantity could be measured directly in computer simulations by investigating, e.g., how many particles participate in a typical local relaxation event. However, in a real experiment this is usually not possible and hence one would like to express z^* in terms of quantities that are directly measurable. One such possibility is the configurational entropy $S_{\rm conf}$: In a seminal paper Goldstein (1969) pointed out that at low temperatures it should be possible to decompose the dynamics of a glass-forming system into two parts since the particles will oscillate for a long time around some equilibrium positions and only rarely undergo a hopping motion to a new local minimum (and this latter motion is most likely of cooperative nature). Hence it should be a good approximation to write the partition function of the system as a term that describes the *vibrational* degrees of freedom and a term that describes the *configurational* degrees of freedom. Hence also the entropy of the system should be given by two terms: a vibrational one, $S_{\rm vib}$, and a configurational one, $S_{\rm conf}$.[2]

Consider now a system with N particles and assume that it can be decomposed into cooperatively rearranging regions of size z^*. Thus the number of such regions is given by $n(z^*, T) = N/z^*$. If the configurational entropy of the whole system is $S_{\rm conf}$, the one of the subsystems are given by $s_{\rm conf} = S_{\rm conf}/n(z^*, T)$. (Note that $s_{\rm conf}$ must be of order $k_B \ln 2$, since a subsystem must have at least two distinct configurations. This value is, however, not an upper bound, since there can be excitations that have a degeneracy higher than two, e.g. if particles make a cooperative ring-like motion (Donati *et al.* 1998).) Thus we find that the size of a cooperatively rearranging region is given by

$$z^* = \frac{N}{n(z^*, T)} = \frac{N s_{\rm conf}}{S_{\rm conf}} \,. \tag{6.8}$$

[2]Note that this type of splitting is also the reason why one can decompose the specific heat into two parts, as we have done in the context of the Kauzmann temperature.

Together with Eq. (6.7) we thus obtain

$$\overline{W(T)} = \overline{A} \exp\left[-\frac{\beta N s_{\text{conf}} \delta\mu}{S_{\text{conf}}}\right] = \overline{A} \exp\left[-\frac{C}{T S_{\text{conf}}}\right] . \qquad (6.9)$$

If we assume that the relaxation time $\tau(T)$ or the viscosity $\eta(T)$ is proportional to $\overline{W(T)}^{-1}$, we thus obtain immediately the final result of the theory of Adam and Gibbs:

$$\tau(T) \propto \eta(T) \propto \exp\left[\frac{C}{T S_{\text{conf}}}\right] , \qquad (6.10)$$

i.e. that the vanishing of the configurational entropy leads to a divergence of the relaxation time. In order to check to what extend this relation is indeed able to describe the slowing down of the dynamics of glass-forming liquids it is necessary to know the temperature dependence of the configurational entropy. As mentioned in Sec. 5.1, this dependence can, at least in principle, be obtained by the thermodynamic integration of the specific heat. For this we approximate $S_{\text{conf}}(T)$ by $\Delta S(T)$, the difference between the specific heat of the liquid and the one of the crystal, see Eq. (5.6), and thus a measurement of the specific heat will give us the configurational entropy. (See, however, Johari (2000) and Angell and Borick (2002), for possible pitfalls in this approach.) Experimentally it is found that certain simple glass-forming liquids show a $\Delta C_p(T)$ that is, to a good approximation, proportional to the inverse of the temperature: $\Delta C_p(T) = K/T$, where K is a constant (Alba *et al.* 1990). Thus in this case the configurational entropy can be obtained immediately from Eq. (5.6) as $\Delta S(T) = K(T_K^{-1} - T^{-1})$. (Note that we have used here that $\Delta S(T_K) = 0$.) With the identification $S_{\text{conf}}(T)$ by $\Delta S(T)$ we thus obtain from Eq. (6.10) immediately that

$$\tau(T) \propto \exp\left[\frac{C T_K / K}{T - T_K}\right] , \qquad (6.11)$$

i.e. the Vogel-Fulcher-Tammann law, Eq. (5.2), which, as we have mentioned in Sec.5.1, gives a quite good description of the temperature dependence of the relaxation times. Note that the Adam-Gibbs theory implies that the Vogel temperature T_0 at which the relaxation time diverges, see Eq. (5.2), is the *same* as the Kauzmann temperature T_K at which the configurational entropy vanishes. Hence the theory establishes a direct link between the relaxation dynamics of a system and its thermodynamics. This predicted equality is also often used as a first check in order to see whether or not the theory is reasonable and the experimental data is indeed often

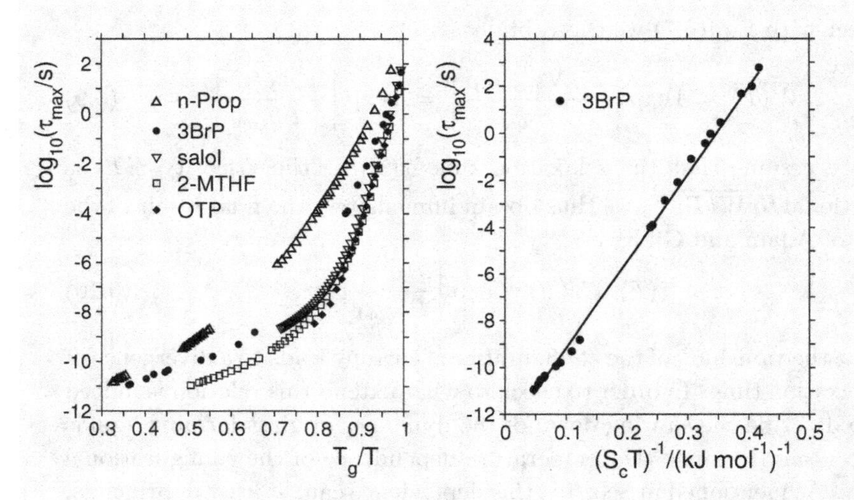

Fig. 6.1 Left panel: Arrhenius plot of $\tau_{\max}$ for 3-bromopentane, n-propanol, salol, 2-methyltetrahydrofuran, and ortho-terphenyl. $\tau_{\max}$ is defined as $\omega_{\max}^{-1}$, where $\omega_{\max}$ is the location of the $\alpha-$peak in a dielectric measurement. Right panel: Logarithm of the relaxation time as a function of $(S_{\text{conf}}T)^{-1}$ for 3-bromopentane to demonstrate the validity of the Adam-Gibbs equation, Eq. (6.10). The solid line is a fit with a Vogel-Fulcher law, Eq. (5.2) with a Vogel temperature T_0 that is identical to the Kauzmann temperature T_K. Adapted from Richert and Angell (1998).

compatible with this relation, at least for fragile glass-formers (Angell 1997, Richert and Angell 1998, Tanaka 2003). A more stringent test is the comparison of the $T-$dependence of the relaxation times and the one of S_{conf}, i.e. to test directly the Adam-Gibbs relation given by Eq. (6.10). This has indeed been done for various glass-forming systems. As an example for such a test we show in Fig. 6.1 the relaxation time $\tau_{\max}$ for the molecular glass former 3-BrP (3-bromopentane) as a function of $1/(S_{\text{conf}}T)$ (Magill 1967, Richert and Angell 1998). Here $\tau_{\max}$ is defined by $1/\omega_{\max}$, where $\omega_{\max}$ is the location of the $\alpha-$peak in a dielectric measurement. From this figure we recognize that at sufficiently low temperatures this presentation does indeed lead to a straight line over about 13 decades in dynamics, and hence gives evidence for the validity of the Adam-Gibbs theory. (Note that there is no adjustable parameter that could be used to rectify the data!).[3] Further support for the theory comes from other experiments and computer

[3] One also has to mention that 3-BrP is a glass-former with intermediate fragility, i.e. an Arrhenius plot of the relaxation time is *not* a straight line, see left panel of Fig. 6.1. Hence we conclude that the presence of the factor S_{conf}^{-1} in the abscissa of Fig. 6.1 is important for the rectification of the data.

simulations in which one attempted to estimate the configurational entropy directly from the inherent structures, hence avoiding the error-prone procedure of calculating S_{conf} from specific heat data (Saika-Voivod *et al.* 2004, Roland *et al.* 2004, and references therein).

Despite the success of the theory for some glass-forming systems, there are also a significant number of other systems for which it fails in that, e.g., the value of the Kauzmann temperature T_K is above the one of the Vogel temperature T_0 (Debenedetti 1997, Tanaka 2003). In view of the approximations made, this is certainly not a surprise. In addition the theory has the drawback that the nature of the cooperatively rearranging region is not defined in a precise way and hence it is not possible within the theory to make any statement about their size, the number of particles they contain etc. The main merit of the theory is thus to propose a connection between the relaxation dynamics of a glass-former and its thermodynamic quantities. Furthermore the theory predicts that this relaxation is related to the cooperative motion of particles whose number increases with decreasing T. This prediction has thus triggered a substantial number of investigations in which such a divergent dynamical (!) length scale has been looked for. Although such an increasing length scale has been reported (Donati *et al.* 1998, 1999, Weeks *et al.* 2000, Kegel and van Blaaderen (2000), Donth *et al.* 2001, Scheidler *et al.* 2002a,b, Berthier 2004, 2005a, Dalle-Ferrier *et al.* 2007, Crauste-Thibierge *et al.* 2010), the typical scales are usually only of the order of a few, 3-8, particle diameters, since the accessible temperatures are all relatively far away from T_K. Furthermore one has to mention that the Adam-Gibbs theory is not the only theoretical approach that predicts the presence of a growing length scale (see, e.g., the so-called "random first order transition" proposed by Wolynes and coworkers, Xia and Wolynes 2001, Lubchenko and Wolynes 2007, Biroli and Bouchaud 2009) and hence the observed growth of a length scale cannot be taken as strong evidence for the validity of the Adam-Gibbs theory. We will discuss the problem of growing length scales in more detail in Sec. 6.2.

6.1.2 *The Free Volume Theory*

The starting point for the theory of Adam and Gibbs was the idea that at low temperatures particles can move only if they do this in a cooperative way. In contrast to this the main ingredient of the so-called "free-volume theory" is the view that a particle can only move if it has the space to do so (Cohen and Turnbull 1959, Turnbull and Cohen 1961, Turnbull and Cohen

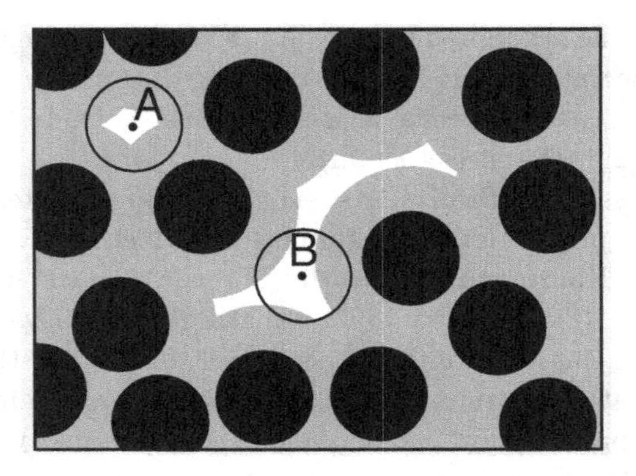

Fig. 6.2 Accessible volume of two atoms in a system of hard disks (white area). Atom A is confined by its neighbors and thus is not able to leave its cage. Atom B is able to leave its environment and hence to make a relaxation motion.

1970). This idea can be explained best by considering a hard sphere system, see Fig. 6.2. Consider the particle labeled A. If none of the other atoms move, the space accessible to that particle is given by the white area. Since this region is confined within the cage formed by the nearest neighbors of particle A, the latter is not able to make a movement that corresponds to relaxation, i.e. to leave its cage. The situation is different for the particle labeled B. Here the accessible space includes also a region in which the particle has changed its local environment, i.e. the atom can undergo a relaxation movement. The free volume theory is an attempt to relate the size of these "allowed regions" to the diffusivity of the particles and hence to the relaxation dynamics of the system. To this aim we define the free volume v_f of the system by

$$v_f = v - v_0 \,, \tag{6.12}$$

where v is the volume per particle and v_0 is the volume per particle that is excluded to all the other particles, i.e. in a hard sphere system this is the volume of the hard sphere whereas in a system with finite potentials it corresponds to that region for which the Boltzmann weight is very small. Using the geometrical construction shown in Fig. 6.2 one can associate to each particle i its free volume v_{fi}. (Note that this association is not one to one since a given free volume element will usually be shared by several particles.) If we discretize v_{fk}, a given configuration of particles will have

N_k particles that have a free volume v_{fk}. Thus we have

$$\gamma \sum_k N_k v_{fk} = N v_f \qquad \text{with} \qquad \sum_k N_k = N\,, \tag{6.13}$$

where γ is a geometrical factor that takes into account the overlap between the free volume of neighboring particles and N is the total number of particles.

We now make the assumption that the free volume of the system can be exchanged from one particle to another without any energetic cost. Hence the total available free volume will be distributed in such a way that the number of possible configurations is maximized. Thus we have to determine which distribution of the numbers N_k maximizes the quantity

$$N!/\prod_k N_k! \tag{6.14}$$

under the constraints given by Eqs. (6.13). Making the variable v_{fk} continuous, one finds that this distribution is just an exponential (Cohen and Turnbull 1959):

$$p(v') = \frac{\gamma}{v_f} \exp\left(-\frac{\gamma v'}{v_f}\right). \tag{6.15}$$

In order to make a connection between the free volume and the diffusion constant of the system we first note that a particle that makes a random walk with steps a has a diffusion constant

$$D = \frac{1}{3} u \int_0^\infty p(a) a da\,, \tag{6.16}$$

where $p(a)$ is the probability that the particle makes a displacement of length a and u is the average thermal velocity of the particle, i.e. $u = (3k_B T/m)^{1/2}$, where m is the mass of the particle. (This exact relation can be easily obtained from the mean end-to-end distance of a random walk, see Eq. (3.4)). In order to apply this expression to our case of the glass-forming liquid we make the assumption that to each free volume v' we can associate a typical distance $a(v')$ for an elementary diffusion step. Thus in this case the integral of Eq. (6.16) can be written as an integral over v' and we obtain for the average diffusion constant of a particle

$$D = \frac{1}{3}u \int_0^\infty p(v')f(v')a(v')dv' , \tag{6.17}$$

where we have introduced the so-called correlation factor $f(v')$, i.e. the probability that a diffusion step can actually be made instead of being rejected. In their first papers Cohen and Turnbull made the approximation that $f(v')$ is just zero if $v' < v^*$, i.e. there is no motion if the available free volume is below a certain threshold v^* (Cohen and Turnbull 1959, Turnbull and Cohen 1961). The value of v^* is on the order of the hard-core volume v_0 of the particles and thus the mentioned cutoff expresses the idea that in order for a particle to move to a new "independent" place, i.e. to make a relaxation motion, it needs to move at least by a distance that allows it to occupy a new region in space that has only a small overlap with the initial space it occupied. In these papers it was also assumed, somewhat arbitrarily, that the distance $a(v)$ is given by a constant $a_0/2$, the radius of the particles. Under these assumptions the diffusion constant becomes

$$D = \frac{ua_0}{6} \int_{v^*}^\infty p(v')dv' = \frac{ua_0}{6} \exp\left(-\frac{\gamma v^*}{v_f}\right) . \tag{6.18}$$

Thus we find that the diffusion constant goes to zero if the free volume vanishes. This predicted connection between D and v_f is very similar to the one by Doolittle who studied the viscosity of hydrocarbon liquids and who proposed $\eta^{-1} = A\exp(-bv_0/v_f)$, where b is a constant of order unity (Doolittle 1951). The result given by Eq. (6.18) holds for hard spheres, i.e. an athermal system. As we have mentioned above, it is reasonable to assume that the concept of free volume can also be extended to thermal systems and thus v_f becomes a function of temperature. If we denote by T_0 the temperature at which the free volume becomes zero, we can expect that close to T_0 the free volume is just a linear function of temperature: $v_f = \alpha(T - T_0)$, where α is the coefficient for thermal expansion. Thus this $T-$dependence of the free volume leads, together with Eq. (6.18), immediately to a Vogel-Fulcher law for the diffusion constant and hence, under the assumption that $D^{-1} \propto \eta \propto \tau$, for the viscosity and the relaxation time.

In a subsequent paper Turnbull and Cohen have refined their approach by assuming that the distance $a(v)$ for the displacement is not a constant but proportional to v: $a(v) = \alpha v$ (Turnbull and Cohen 1970). The resulting expression for the diffusion constant is then

$$D = \frac{u\alpha}{3}(v^* + v_f/\gamma)\exp(-\gamma v^*/v_f) \tag{6.19}$$

and does indeed give a good description of molecular dynamics simulation data for hard sphere systems as well as simple liquids (Turnbull and Cohen 1970, Angell *et al.* 1981, Woodcock and Angell 1981, Fox and Andersen 1984). w In the free volume theory that we have just discussed one makes the assumption that the free volume can be redistributed without cost in the free energy. Nothing is said, however, about the local free energy of a particle that has a certain free volume at disposition. Cohen and Grest (1979, 1982) showed that if this local free energy is taken into account the predicted $T-$dependence for the diffusion constant changes significantly. For this these authors divided the system into "cells", where each cell is a region that encloses one particle, that has a free volume v, as well as its cage. A cell is now denoted "liquid-like" or "solid-like" depending whether its free volume is larger or smaller than a critical value v_c, respectively. The idea of Cohen and Grest was now that the system will be a liquid if the liquid-like cells form a percolating cluster and a glass if this is not the case. Hence the slow dynamics of the system is not only due to the decreased mobility of the particles within each cell, because with decreasing T or increasing density the average free volume decreases, but also due to the increasing dearth of the liquid-like cells that can make a real transport of the particles. (Note that according to this model a particle in a liquid-like cell can only move into an adjacent cell if the latter is also liquid-like.)

Under the assumption that the cells are independent and that their free energy is just a function of their free volume, we thus can write the free energy of the system as

$$F = N\int [p(v)f(v) + k_BTp(v)\ln p(v)]dv - TS_{\text{com}}\,, \tag{6.20}$$

where $p(v)$ is the probability that a cell has a free volume v and the second term in the integral is just the entropic contribution to the free energy due to the distribution $p(v)$. The quantity S_{com} is the so-called "communal entropy" and is related to the fact that the particles that are in a connected cluster of liquid-like cells have a larger accessible volume and hence more possibilities to distribute themselves in this volume (Kirkwood 1950). The value of S_{com} thus depends on the distribution of liquid-like clusters, which in turn is a function of the distribution $p(v)$. Using a simple functional form for $f(v)$, quadratic for $v < v_c$ and quadratic plus a linear term for $v > v_c$, and by making a mean-field approximation for the connectivity of liquid-like cells, Cohen and Grest determined the dependence of the free energy F on the probability that a cell is liquid-like, i.e. on $W = \int_{v_c}^{\infty} p(v)dv$. Surprisingly they found that this free energy shows a behavior that corresponds to a system with a first order transition in that there is a temperature T_p at which the most probable value of W jumps discontinuously from a small to large value if T is increased. This change is related to the percolation phenomena occurring for the liquid-like cells. As a consequence of this transition the specific heat has a spike at T_p, i.e. there is a latent heat, and also the volume shows a discontinuity. Since in real experiments *in equilibrium* (!) such singularities have never been observed, Cohen and Grest argued that T_p is somewhat below the experimental glass transition temperature and hence it will be difficult to see the effect of the percolation transition. Although in principle this might be a valuable argument for the lack of experimental evidence for the percolation transition, it is nevertheless bothersome that so far no indication exists for this transition.

The theory seems to be more reliable regarding the T−dependence of the relaxation dynamics. The average free volume can be calculated by averaging the local free volume over the liquid-like fraction of the material (recall that solid-like cells do not have free volume):

$$\overline{v}_f = \int_{v_c}^{\infty} (v - v_c)p(v)dv / \int_{v_c}^{\infty} p(v)dv \,, \tag{6.21}$$

and is found to be

$$\overline{v}_f = \frac{k_B}{2\zeta_0} \left\{ T - T_0 + [(T - T_0)^2 + 4v_a\zeta_0 T/k_B]^{1/2} \right\} , \tag{6.22}$$

where ζ_0, T_0, and v_a are constants that characterize the shape of the local free energy $f(v)$. Using the same arguments that Cohen and Turnbull (1959) used to express the diffusion constant as a function of free volume, see Eq. (6.18), it is now possible to obtain the temperature dependence of the viscosity:

$$\log_{10} \eta = A + \frac{2B}{T - T_0 + [(T - T_0)^2 + 4v_a\zeta_0 T/k_B]^{1/2}} , \tag{6.23}$$

where A and B are constants. Note that, in contrast to the free volume theory by Cohen and Turnbull, here the free volume vanishes only at $T = 0$ and hence the relaxation time stays finite for all $T > 0$. As documented by

Cohen and Grest 1979, the expression given by Eq. (6.23) is able to give a very good description of the viscosity data of glass forming liquids, and this over a dynamic range of 14 decades in η, a result that has subsequently been confirmed in other investigations (Cummins *et al.* 1997, Schneider *et al.* 1999, Paluch *et al.* 2003, Comez *et al.* 2004).[4] Despite this success one must mention that Cohen and Grest (1979) also made a very plausible Ansatz for the pressure dependence of the local free energy $f(v)$ and thus obtained an expression on how the viscosity should depend on temperature and pressure. Comez *et al.* (2004) have found that this predicted dependence does not agree with the experimental data and thus that the theory is not reliable with respect to this. Hence one must conclude that the idea of free volume, although very intuitive and thus appealing, is probably not a reliable approach to describe the phenomenon of the glass transition.

Before we conclude this subsection we mention that before Cohen and Turnbull proposed their free volume theory, which was subsequently extended by Cohen and Grest, Gibbs and DiMarzio have advanced a somewhat related approach for the case of polymers (Gibbs and DiMarzio 1958, DiMarzio and Gibbs 1958). These authors considered a lattice model for a polymer melt and calculated, by making a simple mean-field approximation, the entropy of this system. Under the assumption that at low temperatures the vibrational entropy of this melt is the same as the one of the corresponding crystalline system, they found that the configurational entropy vanishes at a given packing density and that the melt undergoes a true second order phase transition. The critical slowing down of this thermodynamic transition is hence responsible for the slow relaxation of the melt at high packing fraction. This model predicted thus the dependence of the critical temperature (and hence of the glass transition temperature) on the length and stiffness of the polymer chains. Subsequently the model was extended to take also into account the effect of "plasticizers", i.e. the phenomena that the addition of small molecules decreases T_g, effects of mixing various polymers etc. (see DiMarzio (1997) for a review). Many of these predictions have been confirmed in experiments and computer simulations and therefore the theory is quite popular.

[4]In view of the four free parameter of the theory its success to describe the experimental data might not be surprising. Despite this success one must mention, however, that most other functional forms, such as the Vogel-Fulcher-law or the power-law proposed by mode coupling theory, have also three parameters and describe the relaxation dynamics only on a much smaller range in η (see, e.g., Stickel *et al.* 1995).

Surprisingly it turns out that this success of the theory is not related to the predicted result that the entropy vanishes at the critical point, but to the fact that the theory is able to give a reasonably good description of the dependence of the entropy on the experimentally relevant parameters like length of the polymers, their stiffness etc. This was demonstrated by Wolfgardt *et al.* who used computer simulations to calculated the configurational entropy of a simple lattice polymer (Wolfgardt *et al.* 1996). These authors found that the mean-field approximation made by Gibbs and DiMarzio to calculate this entropy at *low* density, *underestimates* the real configurational entropy, and this error is propagated to the high density state. Therefore the vanishing of the entropy, and hence the existence of the thermodynamic transition, at a finite packing fraction is due to a inaccurate approximation in the low viscosity regime. If this approximation is improved, using e.g. a simple expression proposed by Milchev (1983), the configurational entropy remains finite for all densities (Wolfgardt *et al.* 1996). However, there exists indeed a range of density in which $S_{\rm conf}(\rho)$ shows a strong variation in that it decreases very rapidly. Thus the success of the theory is related to the fact that it is able to predict this rapid decrease and how the density at which this happens depends on parameters like chain length etc. Since the relevant ingredient for the theory are packing considerations, expressed in terms of configurational entropy, this shows again that free volume is indeed a quite useful concept. Nevertheless since it is not possible to calculate this quantity within a precise statistical mechanics framework, the success of theories based on this concept should always be looked at with a good amount of skepticism.

At this point we again wish to warn the reader about the fact that in different contexts (and thus in different papers in the original literature) the same term "configurational entropy" has different meanings. The "configurational entropy" approximately computed by Gibbs and DiMarzio (1958) and estimated by numerically exact Monte Carlo methods by Wolfgardt *et al.* (1996) for *lattice models* of polymers is, in fact, the total entropy of these models (which do not possess vibrational entropy, of course). This entropy counts states as different states which differ only by local changes in the occupation of small regions of sites on the lattice. In accord with the arguments of Stillinger (1988) discussed in Sec. 5.1., this entropy must be nonzero at all temperatures (for systems with short-range forces), and this is what the calculation of Wolfgardt *et al.* (1996) bears out.

Such "states" which differ only by rearrangements of a small number of degrees of freedom are not different "states" in the sense of the "ergodic components" of thermodynamics (Chap. 1), of course. The latter must be separated from each other by infinitely high barriers, cf. upper part of Fig. 1.16, requiring rearrangements of an infinite number of degrees of freedom, in the thermodynamic limit. The entropy associated with this number of ergodic components, which we have called "complexity" (Chap. 1, and Sec. 4.6) is often also referred to as "configurational entropy", but it has nothing to do with the configurational entropy estimated in the theory of Gibbs and DiMarzio (1958), contrary to many inaccurate citations in the literature. Likewise, the "configurational entropy" estimated from the total entropy of the liquid (with the entropy of the crystal subtracted, see Secs. 1.2.4 and 5.1, to approximately subtract the vibrational entropy) has nothing to do with the configurational entropy in the sense of complexity, which should count the thermodynamically distinct metastable states of the glass or supercooled liquid. Obviously, this experimentally accessible entropy difference includes many configurations which differ from each other only by minor rearrangements of small groups of particles, or even single particles (cf. the two states of particle B in Fig. (6.2). If for real glasses a true glass transition with a relaxation time diverging at ia temperature T_0 exists, one expects that for T slightly above T_0 some states are separated from each other by very high free energy barriers, and for all practical purposes such states can then be treated like ergodic components and thus a configurational entropy in the sense of complexity can be associated with them, at least approximately. However, as T is raised, a "configurational entropy" or "complexity" introduced in this way becomes more and more fuzzy (Mézard and Parisi 2009). Using the analogy with metastable states in standard first-order transitions (Binder 1984), where metastable states (in systems with short-range forces) become fuzzy before the (mean field) spinodal is reached (i.e. there is no longer a barrier), the critical temperature T_c of mode coupling theory (Chap. 5) can analogously be interpreted as a spinodal, and so for $T > T_c$ the notion of configurational entropy (complexity) becomes meaningless (Mézard and Parisi 2009). This consideration seems to imply that the apparent success of the Adam-Gibbs theory (Sec. 6.1) in which apparently different states are present at even relatively high temperatures (certainly above T_c is merely accidental, since the configurational entropy which is used there is the total entropy difference (relative to the crystal) of the fluid.

An instructive example for these considerations is the mean field Potts glass, see Fig. 1.15 and, Sec. 4.6.2: The total entropy has a kink at T_0, but the Kauzmann temperature where the (extrapolated) entropy of the high temperature phase vanishes exists but is irrelevant. A complexity is well-defined and nonzero only for $T_0 < T < T_c$, cf. Fig. 4.36.

In view of the importance to make the connection between real thermodynamic states (i.e. states that have in mean field an infinite lifetime) and a local minimum in the free energy landscape (i.e. states that have a finite lifetime), it is not surprising that quite a few efforts have been made to investigate the latter landscape. However, this is not an easy task since, in lack of better ideas, the latter is often approximated by the potential energy landscape, an approximation whose quality is often doubtful. (Recall that, e.g. in the case of a hard sphere system the potential energy landscape is flat.) A review on these landscape approaches can be found in Heuer (2008).

6.1.3 *Kinetically Constrained Models*

Although the approaches discussed in the two previous subsections have the merit that they allow to rationalize to some extend the slowing down of the dynamics of glass-forming systems, and also predict the temperature/density dependence of the relaxation times, these theories have a certain number of free parameters that need to be adjusted to experimental data. Despite the fact that the underlying mechanism(s) on which these theories are based upon (cooperativity, free volume, etc.) are certainly quite reasonable, they do not permit to make a real statistical mechanics calculation in which these free parameters are directly computed. Since for realistic systems such full calculations are extremely difficult, it is natural to consider rather simple models for which an analytical approach seems more feasible and in this subsection we will discuss some of these models.

Many of the difficulties of devising a theory of glass-forming liquids is related to the fact that the degrees of freedom are continuous and hence it is necessary to calculate *integrals* in order to obtain quantities like the partition function or time-correlation functions. In the theory of liquids one has realized long time ago that some of the relevant questions can also be addressed with lattice models, the so-called "lattice-gases", in which the degrees of freedom are discrete and hence much simpler to tackle (Gaunt and Fisher 1965, Gaunt 1967). It was therefore quite natural when in the eighties lattice models were proposed that included some of the features that are thought to be relevant for glassy dynamics, such as cooperativity. One of the seminal papers was the one of Fredrickson and Andersen who proposed a class of models that today are known as "spin facilitated kinetic

Ising models", which turned out to be very useful to gain insight into the dynamics of glassy systems (Fredrickson and Andersen 1984, 1985). This work was the origin of an area of research on such models which is presently still very active. An extensive review on the work done until 2002 can be found in the article of Ritort and Sollich (2003) and in the proceedings of a conference on these systems (Ritort and Sollich 2002). A more recent discussion is presented in Léonard *et al.* (2007) and Chandler and Garrahan (2010).

In their model Fredrickson and Andersen considered a collection of N Ising spins, $\sigma_i \in \{-1, +1\}$, on a (regular) lattice. The spins do not interact with one another and are in an external field h. Thus the total energy E is given by

$$E = h \sum_{i=1}^{N} \sigma_i \,. \tag{6.24}$$

This Hamiltonian is very simple and it is obvious that from a thermodynamic point of view one does not expect any singular behavior for any temperature $T > 0$.[5] The nontrivial part of this model lies in its *dynamics* which is defined as follows: Consider a spin σ_i and denote by m_i the number of its nearest neighbor spins that are in the up-state. The spin σ_i can flip only if $m_i \geq n$, where n is a given integer. In order to take account the condition of detailed balance, the rates for up and down flips will differ by a factor of $\exp(-2h\beta)$, where $\beta = 1/k_B T$. Thus we have for the transition probabilities for a spin flip

$$w(\sigma_i \to -\sigma_i) = \frac{\alpha}{m_i!} \exp[\beta h(\sigma_i - 1)] \prod_{k=0}^{n-1} (m_i - k) \,. \tag{6.25}$$

Here α is a constant that sets the time scale. Hence a spin can only flip if a sufficiently large number of its neighbors are in the up-state ($\sigma = +1$), i.e. facilitate its change, a feature that lead to the name "spin facilitated kinetic Ising models" or "n−SFM". Thus since the flip rate of a spin depends on the state of the neighboring spins, it can be expected that the relaxation dynamics is cooperative. This property is thus in common with the ones of

[5] Note that in the original paper of Fredrickson and Andersen the Hamiltonian included also a nearest neighbor interaction between the spins (Fredrickson and Andersen 1984). However, such a coupling opens the possibility that the system has a thermodynamic phase transition and hence the glassy dynamics related to the dynamic rules for the spin flips might be obscured by this transition. Hence most of the studies done for the model have been done for systems in which this coupling has been set to zero.

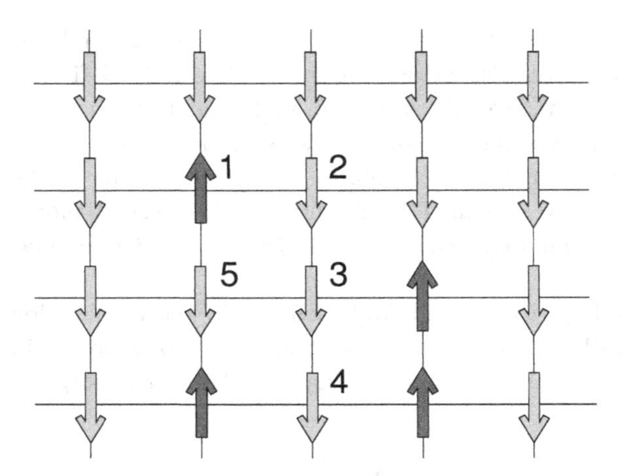

Fig. 6.3 Example of a spin configuration. For the case of a 1-SFM the presence of the up-spin at vertex 1 allows the spin at 2 to flip. Hence the "defect" $\sigma = 1$ has been propagated to the right. This type of dynamics is not possible for n−SFM with $n > 1$. For a 2−SFM one possibility to flip spin number 2 is to flip spin number 4, then spin 3 and finally spin 2. Thus the dynamics is more cooperative than for the case $n = 1$.

glass-forming liquids and which is one of the main motivations for having introduced these specific transition probabilities.

Since the Hamiltonian of the system is so simple, one can easily calculate some of its equilibrium properties. E.g. the concentration c of up spins is found to be

$$c = \frac{1}{1 + \exp(2\beta h)} \tag{6.26}$$

and the entropy s per spin is

$$s = k_B \ln[2\cosh(\beta h)] - \beta h \tanh(\beta h) . \tag{6.27}$$

Thus we see that with decreasing temperature the concentration of up-spins decreases rapidly, basically like $\exp(-2\beta h)$, see Eq. (6.26). Since a spin can only flip if it has a sufficiently large number of up-spins in its neighborhood, it is evident that at low temperatures the relaxation dynamics of the system becomes very slow.

At this point it is useful to distinguish two cases of models: Those for which $n = 1$ and those with $n > 1$, since their dynamical behavior is quite different as we will see below.

We start the discussion with the case $n = 1$. Consider the spin configuration given in Fig. 6.3 and imagine that we are at low T. Since spin number 1 is in the up-state it allows spin number 2 to flip upward also. Since $\sigma_1 = 1$ corresponds to a high energy state, it is likely that it will flip down quite quickly, and since it now has a neighbor that is in the up-state, it can do so. Thus the effect of these flips is that the "defect", i.e. the up-spin, has moved by one lattice spacing. It is hence evident that at low temperatures the relaxation dynamics of the system will be ruled by the diffusive motion of these defects (also called "excitations"). The time scale for this motion is clearly given by the inverse of the concentration of these defects, i.e. by c^{-1}.[6] Due to this diffusive motion it is thus clear that the 1-SFM will never show a dynamical transition at a finite temperature, since each spin will relax on the time scale it takes for one of these defects to find the spin and to flip it. This time scale can be estimated quite simply as follows (Fredrickson and Andersen 1985, Garrahan and Chandler 2002, Ritort and Sollich 2003). In a $d-$dimensional space the typical distance ℓ between two defects scales like $c^{-1/d} \approx \exp(2\beta h/d)$. For a given spin to relax it has to wait until one of these defects comes by. Since the latter make random walks this will typically take a time ℓ^2/D, where D is the diffusion constant for a defect. As we have seen above the latter is proportional to c and hence we find that the relaxation time τ of the system is given by

$$\tau \sim c^{-2/d}/c \approx \exp[2\beta h(1 + 2/d)]\,. \tag{6.28}$$

Thus the 1-SFM will show at low temperatures an Arrhenius behavior, and hence can be classified as a strong glass-former. Since the relaxation dynamics of the system is tightly related to the diffusive motion of the defects discussed above, it can be expected that the time correlation functions, such as the spin-autocorrelation function $\langle \sigma_i(t)\sigma_i(0)\rangle$, show basically a simple exponential behavior, which is in agreement with the results from simulations (Whitelam and Garrahan 2004). One noticeable exception is the model for $d = 1$ for which subtle correlations between the random trajectories of neighboring defects, induced by the low dimensionality, make that the time correlation function decays much slower. Garrahan and Chandler (2002) have put forward arguments that predict that due to these correlations the decay should be given by a stretched exponential with exponent 1/2, a result that has indeed been confirmed in simulations (Berthier and Garrahan 2003). These correlation also make that the system has a very

[6]Note, that the local magnetization of the system is not a conserved quantity, since one defect can destroy a neighboring one but also can create one on a nearest neighbor site.

heterogeneous dynamics, in agreement with the strong stretching of the correlation function (Garrahan and Chandler 2002).

The reason that 1-SFM show at low temperatures an Arrhenius law is that the defects can move independently. Thus from this point of view the local dynamics is not very cooperative. This changes if we consider the case $n > 1$. From Fig. 6.3 it becomes clear that spin number 2 cannot relax. Instead it has to wait until, e.g., spin number 3 has flipped which in turn can only flip if spin number 4 has flipped. Thus it is evident that an isolated defect is immobile and does not lead by itself to a relaxation of the neighboring spins.[7] Instead this local configuration has to wait until other defects arrive in order to relax. Thus in this case the relaxation dynamics is much more cooperative than for the 1-SFM and hence it can be expected that the relaxation times are non-Arrhenius, an expectation that has indeed been confirmed in computer simulations (Fredrickson and Brawer 1986, Leutheusser and De Raedt 1986, Butler and Harrowell 1991, Graham *et al.* 1997). These simulations have also shown that the relaxation functions are non-exponential and in a seminal paper Butler and Harrowell (1991) have shown that this non-Debye relaxation is related to the presence of strong dynamical heterogeneities with a length scale that increases upon cooling. These simulations have also demonstrated that in these models there is no evidence for the existence of a dynamical transition at finite c, i.e. $T > 0$, a result that is in contrast with the prediction of the diagrammatic expansions of Fredrickson and Andersen (1984, 1985).

Although there is no firm theoretical results that demonstrate the absence of a dynamical transition at finite T, there are some plausible arguments for an ergodic behavior at any finite temperature (Jäckle *et al.* 1991, Reiter 1991, Ritort and Sollich 2003). To see this consider an arbitrary configuration of spins in an n-SFM in d dimensions. Let's assume that it is possible to find a dynamically allowed path from this configuration of spins to the configuration in which all spins are up. All configurations that have such a path are also connected to each other and we denote the collection of all these configurations the "high temperature partition". Thus the question is whether at a given concentration c of up spins the number of configurations that are *not* in the high temperature partition is negligibly small, or finite. In the latter case the system is nonergodic and thus has

[7] The same is also true on a hypercubic lattice for two neighboring defects since also they are immobile. However, for a triangular lattice this no longer holds since a pair of defects are able to move around in a diffusive motion (Fredrickson and Andersen 1985). Thus there are also $n-$SFM with $n > 1$ that will show an Arrhenius behavior at low temperatures.

a dynamic transition at a finite temperature. For the sake of argument we consider now the 2-SFM in two dimensions, but the generalization to higher dimensions is simple.

Imagine that we have an $\ell \times \ell$ square of spins that are pointing upward. It is easy to see that if each side of the four borders of this square has at least one spin that is pointing upward, all the nearest neighbor spins of the square can be flipped upward and the same is true for the four spins at the corners. Thus we have expanded the original square to a new one having size $(\ell+2) \times (\ell+2)$. The probability to have on each side of the square at least one spin pointing upwards is larger than $p_\ell = [1-(1-c)^\ell]^4$. If the expanded square has again at least one upward spin on each of its sides, this expansion process can be repeated until all the spins of the system are in the up state. The probability that this process does not break down due to a lack of appropriate spins pointing upward is higher than

$$P_\ell^\infty = \prod_{k=0}^{\infty} p_{\ell+k} = \exp\{4\sum_{k=0}^{\infty} \ln[1-(1-c)^{\ell+k}]\}\,. \tag{6.29}$$

For the case that ℓ is large we can approximate this expression by

$$P_\ell^\infty \approx \exp\{-4\sum_{k=0}^{\infty}(1-c)^{\ell+k}\} = \exp\left[-4(1-c)^\ell/c\right]\,, \tag{6.30}$$

and thus we see that this probability becomes close to one if ℓ is large. We hence can conclude that once a given spin configuration has a cluster of up spins of linear size

$$\ell(c) \approx \ln(c)/\ln(1-c) \approx -\ln(c)/c \tag{6.31}$$

this cluster can most likely be expanded by flipping the spins on its boarders upwards and hence the configuration of spins will belong to the high temperature partition. The probability to have such a cluster in a given system is on the order of $\exp(-\ell^2) \approx \exp(-1/c^2)$. Thus if we have a system that has a linear dimension smaller than

$$\mathcal{L}(c) \approx \exp(1/c)\,, \tag{6.32}$$

there is only a very small probability to find a cluster that can be expanded and hence the system will *not* be ergodic.

To calculate the probability that a single up spin is the origin of such an expandable cluster we return to Eq. (6.29). For the case of $c \ll 1$ we can convert the sum into an integral and obtain

$$P_1^\infty \approx \exp\left\{4\int_0^\infty du \ln[1-(1-c)^u]\right\} = \exp\left\{-\frac{4}{\ln(1-c)}\int_0^\infty dv \ln(1-e^{-v})\right\}, \tag{6.33}$$

from which we conclude

$$P_1^\infty \approx \exp(-\text{const}/c)\,. \tag{6.34}$$

Since this is a finite number and since in an infinite system we have for each $c > 0$ infinitely many up spins, this result implies that in the thermodynamic limit the probability that a given configuration belongs to the high temperature partition is one. This result is in agreement with simulation studies by Nakanishi and Takano (1986) and theoretical calculations by Aizenman and Lebowitz (1988) who considered the equivalent problem within the framework of the so-called "bootstrap percolation".[8]

We thus can conclude that there are two different important length scales in the spin facilitated problem: The size of the critical droplet $\ell(c)$, and $\mathcal{L}(c)$, the size of the system needed in order to find such a droplet with high probability. In order that a cluster of up spins is mobile, i.e. can move around through the whole system, it is necessary that its size is larger than $\ell(c)$. If it is smaller it will be (most probably) be confined in a finite area. If the system is smaller than $\mathcal{L}(c)$ it is likely to be nonergodic whereas it is ergodic if it is larger than this threshold. In the latter case it will consist of domains of up spins that are localized if they have a size smaller than $\ell(c)$ whereas the domains that are larger than this value will diffuse around and hence relax the system at long times. Thus we see that the question regarding the existence of a dynamic transition is closely related to the size of the system, a phenomenon that has already been realized very early (Fredrickson and Brawer 1986). The given argument implies, however, that in the thermodynamic limit the system is indeed ergodic for all temperatures and hence there is no dynamic transition at finite T. These results are believed to hold for all n-SFM in d dimensions if $n \leq d$ (Reiter 1991) in agreement with the known results from the simulations.

[8] In bootstrap percolation (Chalupa *et al.* 1979) particles are placed randomly on a lattice with probability p. Subsequently all the particles that have less than m nearest neighbors are removed, and this procedure is repeated until no further particles can be removed (because there are no more particles left or the remaining ones have all m or more nearest neighbors). The question of interest is then to know how the number of remaining particles depends on the system size and the initial density p. See Adler (1991) for a review and Balogh *et al.* (2009) for a more recent discussion on the subject.

Note that the time scale for the diffusive motion is expected to scale like $\exp(c^{-1})$, since this is the size of the critical droplets. From Eq. (6.26) we see that at low temperature the concentration of up spins is approximately given by $c \approx \exp(-2\beta h)$ and we thus expect the typical relaxation time to scale like $\tau \sim \exp(\exp(2\beta h))$, i.e. the $T-$dependence is extremely strong, in agreement with some heuristic arguments and results from simulations (Butler and Harrowell 1991).

Before we end the discussion on these models we mention that for all these models the entropy is given by Eq. (6.27). Hence *if* one equates this entropy with the configurational entropy of the Adam-Gibbs theory, it is possible to test directly the Adam-Gibbs relation, $\log(\tau) \sim h/TS_{\text{conf}}(T)$, see Eq. (6.10). For the two-dimensional 2-SFM, Fredrickon and Brawer have indeed found that this relation holds for the accessible dynamic range, i.e. six decades in τ (Fredrickon and Brawer 1986). However, subsequently Butler and Harrowell (1991) found small deviations from this law at low T. Furthermore one must realize that the Adam-Gibbs relation predicts that all models of this type have the *same* slowing down, since the entropy is independent of n and d. In view of the very different dynamics of the various models, i.e. different values of n and dimensionality d, such a connection seems somewhat improbable and therefore it is likely that the result of Fredrickson and Brawer is just a fortuitous coincidence, and that the Adam-Gibbs relation does not hold for all SFMs.

An interesting modification of the spin facilitated model are systems in which the kinetic constraint is no longer isotropic but directed. The so-called "East model", proposed by Jäckle and Eisinger (1991), has the same Hamiltonian as a one-dimensional 1-SFM, but now the constraint is that a spin can flip only if the spin to its left is in the up state (whereas in the Fredrickson-Andersen model the up spin could be on the left or the right). The "North-East model" is a two-dimensional system, with the usual trivial Hamiltonian but with the constraint that a spin can only flip if the nearest neighbor spins to its left and below it are pointing up (Reiter *et al.* 1992).[9] Although the change of the dynamical rules seem at first sight minor, it has in fact a strong influence on the relaxation dynamics of the system as can easily be seen for the East model: First of all we recognize immediately that an isolated defect, i.e. an up spin, is pinned. Although such a defect

[9] In the original version of the models the constraints were "to the right" and "to the right and on the top", which explains the name of the models. However, for the analytical description it is simpler to change the orientation of the constraints and therefore we adapt this nomenclature here also.

can relax the spins that are on its right, the inverse is not true. Instead the defect has to wait until another defect comes in from the left and only then it can relax. Having recognized this it is indeed possible to calculate the typical relaxation time of the model (Mauch and Jäckle 1999, Sollich and Evans 1999) and in the following we give the main idea to arrive at this result.

We consider a part of the chain in which we have a domain of spins that are all pointing downward with the exception of the first spin, σ_0, and the last spin, σ_d, that we assume to point upward. We now have to find the most economical path to relax the spin σ_d, i.e. the sequence of spin configurations that will allow to flip spin σ_d by watching that $u(d)$, the maximum number of up spins not counting σ_0, in this sequence, stays as low as possible, since this number will be the energy barrier that has to be overcome in the process. One possible path is of course to flip first σ_1, then σ_2 until all spins are in the up state, then flip σ_d down, and subsequently flip all the remaining spins downward. The maximum energy for this sequence is obviously dh and corresponds to the state where all the spins are pointing upward. A more economical path can be found by a simple recursive procedure: Consider the case that $d = 2^n$, where n is a positive integer. In order to relax spin σ_{2^n} we will first construct a sequence that makes that the central spin of the domain, $\sigma_{2^{n-1}}$, is pointing upward. For this we leave the spins $2^{n-1}+1, \dots, 2^n$ as they are, and flip only the spins $1, \dots, 2^{n-1}$. Thus this procedure will cost a maximum energy of $u(2^{n-1})$ and we are left with a spin configuration in which the first, the central, and the final spins look upward, whereas all the other spins point downward. Working now on the domain made of $\sigma_{2^{n-1}}, \sigma_{2^{n-1}+1}, \dots, \sigma_{2^n}$, we can flip spin σ_{2^n} and this procedure will cost us also not more than an energy $u(2^{n-1})$. Hence we have the relation $u(2^n) = u(2^{n-1}) + h$, where the $+h$ arises because in order to flip the last spin the central spin has to point upward. Iterating this procedure we thus obtain $u(2^n) = u(2) + h(n-1)$. For $u(2)$ we find immediately $u(2) = 2h$, since it corresponds to the sequence $101 \to 111 \to 110 \to 100$. Hence we find the final result $u(2^n) = h(n+1)$. This line of reasoning can be generalized to arbitrary integer values of d (Sollich and Evans 1999) and hence we find that the energy barrier for a domain of length d is given by $u(d) = \ln d / \ln 2$.

Using this result and the fact that the typical size of a domain is given by $1/c \approx \exp(2\beta h)$ we thus can estimate the relaxation time τ:

$$\tau \sim e^{\beta u(d)} = e^{\beta \ln d / \ln 2} = e^{2\beta^2 h / \ln 2} . \qquad (6.35)$$

Hence we find that for this model the logarithm of the relaxation time increases like $1/T^2$, i.e. like the Bässler law discussed in Sec. 5.1 and which describes some of the experimental data quite well (Bässler 1987, Richert and Bässler 1990).[10]

Although it is quite easy to calculate the relaxation time for this model, there is no simple way to obtain also the time dependence of the spin-autocorrelation functions. Due to the simplicity of the model it is not very difficult to come up with various plausible phenomenological approaches to describe this relaxation dynamics, to make diagrammatic expansions or to write down mode coupling equations for the time dependence of the correlator (see Sec. 5.2 for a discussion of the mode coupling theory). However, it turns out that all these approaches are not very reliable at low temperatures in that they predict, e.g., a blocking transition at finite c, even if some of them are very accurate at high temperatures (Jäckle and Eisinger 1991, Reiter *et al.* 1992, Eisinger and Jäckle 1993, Pitts *et al.* 2000, Pitts and Andersen 2001). This flaw is not present in an approach taken by Sollich and Evans who used some phenomenological arguments and ideas from coarsening dynamics to obtain expressions for the spin correlation functions in the case $T \to 0$ (Sollich and Evans 2003). They predicted that this correlator becomes very stretched with an exponent that decreases like T, a result that agrees with the calculations by Buhot and Garrahan (2001). More results on this model, its out of equilibrium dynamics, as well as its various generalizations can be found in the review articles of Ritort and Sollich (2003) and Chandler and Garrahan (2010).

The spin facilitated models we have discussed so far are by no means the only lattice models that have been proposed to describe the relaxation dynamics of supercooled liquids. Some of the other models and the known results on them are discussed in Ritort and Sollich (2003). Here we will only mention one other class of models that have been studied extensively, namely lattice models in which, in contrast to the Fredrickson-Andersen model, the number of spins is conserved. In this case it is thus more appropriate to think of particles that move on a lattice and that obey some rules for their dynamics instead of the defects/excitations that are relevant for the dynamics of the $n-$facilitated spin models. One of the simplest models of this kind is the one proposed by Ajay and Palmer (1990) who considered a square lattice with particles on its vertices, leaving however some of the

[10]Although we have made here the assumption that it is sufficient to consider the average quantities (mean domain size, activated processes), this line of argument can be made more rigorous. See Aldous and Diaconis (2002) for details.

vertices empty, the so-called vacancies. A particle can move to its nearest neighbor site if it is empty, i.e. if there is a vacancy. It is obvious that this model is always ergodic as long as there is a finite concentration of vacancies. Nevertheless, it is surprising that even this simple model shows a non-Debye relaxation at intermediate times, i.e. the time scale on which the particles explore their local environment. The reason for this is that each particle will have an environment that is somewhat different regarding the number of vacancies and hence will thus relax with its own relaxation time, thus giving rise to the stretched exponential. For larger times the relaxation becomes, however, exponential.

A somewhat more complex model has been proposed by Ertel *et al.* (1988) who considered a hard-square lattice gas with nearest neighbor exclusion. Here a particle can move only to a new nearest neighbor site if the latter has only nearest neighbor sites that are vacant (apart from the one on which the particle is). The maximum occupation is thus $c = 0.5$ and there are two ordered states, the two sublattices of a checker board. Thus if one increases the density of particles from low c, there will be a disorder-order transition at around $c^* = 0.37$ at which the particles on one of the two sublattices percolate and hence the system decides (randomly!) on which one of the sublattices it will have the majority of particles. Associated with this transition, which is of second order, is a critical slowing down. However, here we are interested at concentrations somewhat higher than c^* and hence the relaxation times are finite.

It is easy to see that in this model a diagonal that is occupied with particles is blocked, see Fig. 6.4. Hence the relevant question is whether or not at sufficiently high density these blocking diagonals form a percolating cluster and hence lead to a vanishing diffusion constant. The answer is negative, since it can be shown that the probability for having a completely filled diagonal vanishes in the thermodynamic limit (for reasons that are basically the same as the ones that prevent the existence of an ordered phase in a one-dimensional system of particles with short range interactions). Nevertheless the diffusion constant shows a very strong dependence on the concentration and is approximated well by $D(c) \propto \exp[-1/(0.5-c)]$ (Jäckle *et al.* 1991). Although this expression looks formally very similar to the one from the free volume theory, see Eq. (6.18), the underlying mechanism is very different. Recall that within this theory the free volume is defined as the mean local free space that is accessible to a particle in order to move. For the hard square lattice gas at high density this volume will be proportional to $(0.5-c)^3$, since it is just given by the condition that the three neighboring

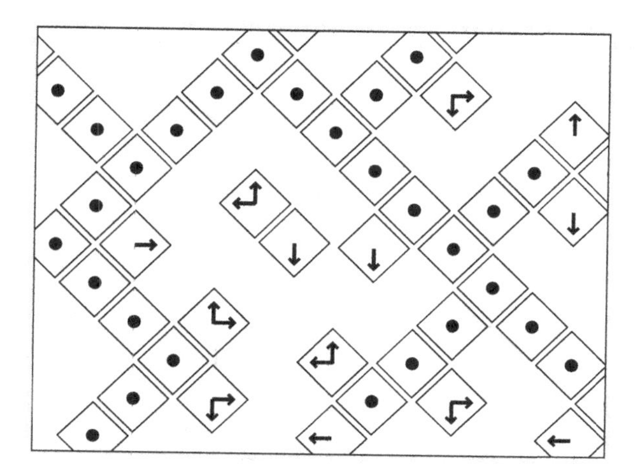

Fig. 6.4 Possible moves in the hard square lattice gas model. The particles that have a dot at their center are, at least temporarily, blocked. The particles with arrows can move in the indicated directions. Adapted from Ertel *et al.* (1988).

sites of the target site of the jump are empty. For concentrations close to 0.5 the free volume theory predicts thus $D \propto \exp(-\text{const.}/(0.5 - c)^3)$, in contrast to the finding of the simulation. From the setup of the model, it is clear that in this case the slow dynamics is related to the cooperative dynamics of the particle and not to the local free volume. Hence it is not very surprising that the free volume theory is not able to give a reliable description of the relaxation dynamics at high density. More results on these type of lattice gases can be found in Jäckle (2002) and Ritort and Sollich (2003).

The relaxation dynamics of the hard-square lattice model that we have just discussed shows at high concentration a very strong dependence on the size of the system (Jäckle *et al.* 1991), a feature that does not find its correspondence in real glass-forming liquids. The reason for these finite size effects are the mentioned percolating clusters of diagonals that are responsible for the blocking transition. In order to avoid these effects, Kob and Andersen (1993) have proposed a somewhat different lattice gas, which turned out to be affected much less by these effects and that does indeed have quite a few properties that are in common with real glass-forming systems (Sellitto 2002). In this model particles occupy a simple cubic lattice and the dynamic rules are as follows: A particle can move

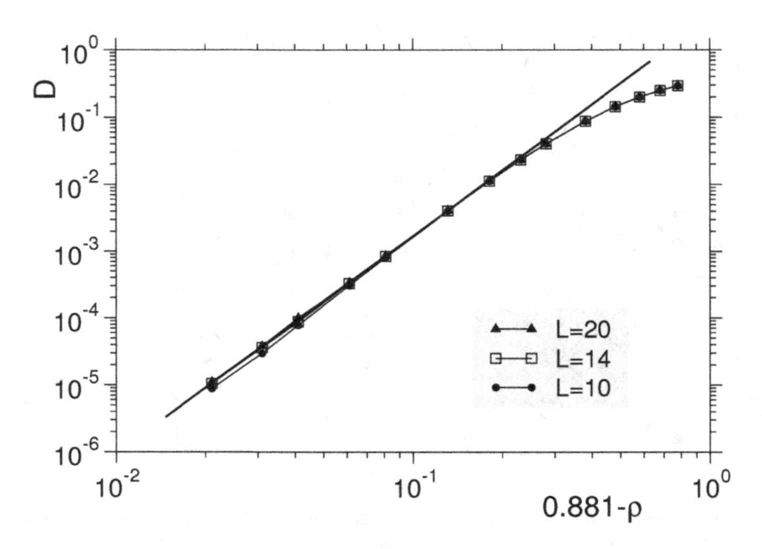

Fig. 6.5 Density dependence of the diffusion constant for the Kob-Andersen lattice gas. The different symbols correspond to different system sizes. The straight line is a power-law fit to the high-density data and has a slope of 3.1. Adapted from Kob and Andersen (1993).

to a nearest neighbor site if and only if 1) the target site is empty, 2) the particle has less than m nearest neighbors, and 3) once the particle has moved to the target site it also has less than m nearest neighbors. It is evident that the value of the integer m plays an important role for the dynamics. If $m = 2$ or 3 a cube of volume 2^3 is not able to dissociate even at low densities, and therefore the model is not ergodic. The smallest value for which such trivial nonergodic configurations are avoided is $m = 4$. Note that since the particles do not interact, apart from the hard core condition, the Hamiltonian is trivial, i.e. it is just a constant. Therefore any slow relaxation will not be related to the presence of a thermodynamic transition.

Computer simulations have shown that this model shows indeed an apparent dynamical transition at a finite concentration of particles in that the diffusion constant as well as the relaxation time for the (incoherent) intermediate scattering function, which are stretched, show a singular behavior that is described well by a power-law. This is demonstrated in Fig. 6.5 where we show a double-logarithmic plot of the diffusion constant as a function of $\rho_c - \rho$, where $\rho_c = 0.881$ is the critical density which has been obtained from a fit with a power-law to the high density data. The straight line in the graph is a power-law with exponent 3.1 and we see that it described the data very well for more than 3 decades in D. The α−relaxation times show the same critical behavior, however the exponent is around 5.0 (Kob and Andersen 1993). Thus the product $D \cdot \tau$ depends strongly on density, i.e. the Stokes-Einstein relation (see Sec. 2.3) is not fulfilled. The reason for the breakdown of this relation are again the presence of strong dynamical heterogeneities that have been found for this system (Berthier 2003, Chaudhuri *et al.* 2008). What is meant by "dynamical heterogeneities" will be discussed in the next section.

From Fig. 6.5 one can also conclude that for the densities investigated the diffusion constant does not depend on L, the size of the system. It must be expected, however, that for densities even closer to ρ_c such system size effects are present. To see this one should realize that also in this model there can be structures that are not able to relax since they are stable under the dynamical rules. One example is a 2 times 2 column that spans the system and which, due to the periodic boundary conditions, blocks itself. Thus one has to wonder to what extend such permanent structures will induce finite size effects by affecting the relaxation dynamics. Due to the nature of the dynamic rules we can make again a connection with the bootstrap percolation mentioned above. If we use a given particle configuration as a starting point of the bootstrap procedure (remove repeatedly all the particles that have less than $m = 4$ nearest neighbors), the final state will have particles that have at least m nearest neighbors and hence are permanently blocked or all the particles have been removed. In the latter case the starting configuration *might* belong to the ergodic component of the system, although we have here only a necessary and not a sufficient condition for this. The probability that a configuration contains such blocked particles (and hence has a relaxation time that is infinite) depends of course on L and it is found that at ρ_c one has to go to a value of L around 50 in order to have only a negligibly small probability to have such states (Kob and Andersen 1993). Thus it can be expected that these types of finite size effects will no longer be present for $\rho = \rho_c$ if $L > 50$.

Although the data for the diffusion constant seems to show a singularity at $\rho = \rho_c$, see Fig. 6.5, there are arguments that in reality the dynamic transition is avoided (Kob and Andersen 1993, Toninelli *et al.* 2004, 2005, Cancrini *et al.* 2010). One way to see this goes back to the problem of the bootstrap percolation (van Enter *et al.* 1990). Imagine an empty slab of size N^2 and width 1 in a system of size L^3. Within the repeated removal sequence of the particles the square forming one of the sides of the slab becomes unstable, i.e. all the particles can be removed, if there is a percolating cluster of vacancies within this square (since within the square many particles have two vacancies as neighbors, and since they are also neighbor of the empty slab, they cannot have more than three particles as neighbors, and hence they can be removed). Repeating the arguments that we used to derive Eq. (6.32), we conclude that the probability to have such a percolating cluster is high if $p = 1 - \rho$ is on the order of $1/\ln N$. Thus if N is larger than $\exp(O(1/p))$ the square will most likely be unstable within the bootstrap dynamics. The probability that such a large slab is indeed empty is given by

$$p^{N^2} = p^{[\exp(O(1/p))]^2} \approx p^{\exp(O(1/p))} = \exp[\ln p \exp(O(1/p))] . \qquad (6.36)$$

In order to have a reasonable probability to find such an empty slab one thus needs a system that has a size of at least the inverse of this probability, i.e.

$$L^3 \approx \exp[-\ln p \exp(O(1/p))] . \qquad (6.37)$$

Inverting this relation we thus obtain a "critical density"

$$p_c(L) = 1 - \rho_c(L) = O\left(\frac{1}{\ln(\ln L)}\right) . \qquad (6.38)$$

This result applies to the bootstrap dynamics. However, Toninelli *et al.* (2004, 2005), see also Cancrini *et al.* (2010), have shown that once a square has become unstable, it is possible, by making appropriate moves of particle from the square into the slab that is (at the beginning) empty, to shift the whole empty slab by one unit (in all three directions) and that hence the system is not blocked, a result that has also been refined by De Gregorio *et al.* (2004).

Thus if for a given L the density of particles is lower than $\rho_c(L)$, the system will be ergodic, whereas it will have blocked states if $\rho > \rho_c(L)$. However, since in the thermodynamic limit $\rho_c(L) \to 1$, we have that in this case the system is indeed ergodic. Nevertheless one must realize that the convergence is extremely slow. If we use, e.g. a value of $L = 10^8$, i.e. the system is macroscopically large, we have $1 - 1/\ln(\ln 10^8) = 0.65$, thus still significantly different from 1.0. Hence from a practical point of view the system will have a dynamic transition at a finite concentration of vacancies, in qualitative agreement with real glass-formers. Finally we mention that there are facilitated model that do have a transition at a finite concentration even in the thermodynamic limit (Toninelli *et al.* 2006), but here we do not this discuss these models in more detail.

In the present section we have studied only the equilibrium dynamics of the spin facilitated models. However, due to their simplicity it is also possible to investigate analytically their out of equilibrium dynamics and it is found that this dynamics if very rich. For details on these results and references to the literature we refer the reader to the review articles of Sellitto (2002) and Ritort and Sollich (2003).

6.2 Dynamical Heterogeneities and Growing Length Scales

If a disordered system such as a supercooled fluid would exhibit some sort of static inhomogeneity, it would be possible to find a correlation function of some physical quantity to probe this inhomogeneity (just as in a fluid close to the vapor-liquid critical point in which the density is inhomogeneous, and this is seen in the density correlation). However, even in the absence of such a static inhomogeneity on the nanoscale or larger scales, the disorder can lead to pronounced differences in the local dynamics of the system since, as we have discussed above, the relaxation dynamics of a glassy system depends very sensitively on density, composition etc. As a consequence, near the glass transition, particles in one region of the fluid can relax several orders of magnitude faster than in a region only a few nanometers away. This spatially inhomogeneous dynamic behavior of the particles in a fluid is called "dynamical heterogeneity" and has been the focus of interest of many recent investigations (Richert 1994, Cicerone et al,. 1995, Cicerone and Ediger 1996; for early reviews of experimental work see Sillescu (1999), Ediger (2000) and Richert (2002), and for a recent comprehensive review see Berthier *et al.* 2010).

One can expect that the length scale associated with these dynamic heterogeneities is related to a growing *static* "glass correlation length" but so far no direct way to measure such a length is known. Therefore it is important to obtain a good understanding on the nature of the dynamical heterogeneities, since they are experimentally more readily accessible, and to establish a link how the observed dynamic length scale is related to this static length scale. E.g. the kinetic Ising models discussed in the previous section do not have any particular intrinsic length scale, since the spins to not interact. Thus the only static length scale is the mean distance between the defects and it is indeed possible to express the relaxation dynamics as a function of the concentration of these defects (and hence of their distance), see below for more details. (Note that this does not exclude the presence of dynamical heterogeneities, since even these simple models allow for local fluctuations in the concentrations of the defects.) However, whether in real structural glasses this is true as well is currently a matter of debate in that there are theoretical approaches, like the random first order theory discussed below, that imply the existence of a increasing static length scale.

How can such dynamic length scales be identified? One (indirect) possibility to obtain information on a correlation length ξ comes from finite size effects and surface effects. For ordinary critical phenomena at standard second-order phase transitions, the critical slowing down (Hohenberg and Halperin 1977) can be expressed as a power-law relation between ξ and the relaxation time τ,

$$\tau \propto \xi^z \,, \tag{6.39}$$

z being the dynamic exponent. In a finite system (such as a hypercube of linear dimension L in d dimensions, with periodic boundary conditions in all spatial directions, as normally used in computer simulations) the growth of ξ is limited by L, and hence one expects a "finite size scaling" behavior (Privman 1990, Binder 1992; see also Sec. 3.2.3)

$$\tau(T, L) \propto L^z \tilde{\tau}(\xi/L) \tag{6.40}$$

where for $T > T_c$ the scaling function $\tilde{\tau}(\zeta) \propto \zeta^z$ if ζ is small, to recover Eq. (6.39). If one would assume the validity of Eq. (6.40) also for a glass-forming fluid, as it should make sense at least for T exceeding the critical temperature T_c, one could expect similar finite size scaling for the self-diffusion constant of the particles (which is related to τ via a Stokes-Einstein relation, Eq. (6.42), or a fractional Stokes-Einstein relation, Eq. (6.47), discussed below).

However, attempts to see finite size effects for the self-diffusion constant of a model for a glassforming polymer melt in $d = 2$ dimensions did reveal only very weak effects (Ray and Binder 1994). Various other studies applying different methods and different models (e.g. Ernst *et al.* 1991, Dasgupta *et al.* 1991, Ghosh and Dasgupta 1996) also failed to obtain evidence for a growth of a correlation length as $T \to T_c$, so at that time the belief was strengthened that no such growing length scale exists. However, as we will see below, this conclusion was premature.

Before we discuss the experiments that have provoked to postulate the existence of dynamical heterogeneities, we briefly discuss a "Gedankenexperiment" that is instructive for the comprehension of the phenomenon: Let us compare two snapshots of a system for which one can see all the individual molecules, which are all thought to be labelled, and let Δt be the time interval between these two snapshots. If Δt is of the order of the timescale of the β-relaxation or smaller (see Fig. 5.3), the particles are still confined in their cages, and so the mean-squared displacements of the particles are all comparable; thus in this sense the dynamics of the system is still homogeneous, on this timescale. The same is true if Δt is taken much larger than the structural relaxation time τ of the α-relaxation (defined, e.g., from the viscosity) since then all particles have escaped from their cages and diffused over a distance of at least a few nanometers and all particles behave in the same way, in a statistical sense: For such a large time window the time average resembles the ensemble average over the particles (disregarding crystallization, the supercooled fluid is assumed to be in equilibrium). However, a very different picture emerges if Δt and τ are of the same order: Some particles have already escaped their cages early and moved a lot, others are still caught in the cages formed by their neighbors. Thus we see that the presence of dynamical heterogeneities is not really surprising since they are related to the fact that we have a separation of time scales. What we will subsequently find, however, is that the spacial distribution of the dynamical heterogeneities is far from trivial and that they might reveal the mecanism for the relaxation dynamics of glass-forming liquids.

Of course, this "Gedankenexperiment" can easily be realized in a computer simulation, and much insight into dynamical heterogeneity has in fact been gained from such investigations (Muranaka and Hiwatari 1995, Hurley and Harrowell 1995, Perera and Harrowell 1996, 1999, Yamamoto and Onuki 1997, 1998, Kob *et al.* 1997, Donati *et al.* 1998, Doliwa and Heuer 1998, Bennemann *et al.* 1999, Glotzer 2000, Gebremichael *et al.* 2001, 2004, Aichele *et al.* 2003, Vogel *et al.* 2004, Appignanesi *et al.* 2006, Vallée *et al.* 2006, 2007ab, Chaudhuri *et al.* 2008, Appignanesi and Rodriguez-Fris 2009, Chong and Kob 2009). However, it must be said that most of the simulation evidence refers to temperatures rather far above the glass transition temperature T_g (most studies are done at temperatures even higher than where one would locate the critical temperature of mode coupling theory), while the (indirect) experimental evidence comes from the temperature region rather close to T_g. Gratifyingly, also direct real-space observations of particles near the glass transition have become possible for colloid dispersions (Kegel and van Blaaderen 2000, Weeks *et al.* 2000). A rather direct evidence for heterogenous dynamics has also been possible from following the trajectories of the fluorescence lifetime of probe molecules in glassforming polymer melts (Vallée *et al.* 2006), since these trajectories give direct evidence of the fact that sometimes the probe molecule stays in a fast relaxing region, sometimes in a slowly relaxing region. A recent overview of these investigations can be found in the collection of articles edited by Berthier *et al.* (2010).

As will be justified below, this dynamical heterogeneity has important consequences for the relaxation in glassforming fluids in general, such as pronounced nonexponential relaxation in the α-relaxation regime, a breakdown of the Stokes-Einstein relation between self-diffusion and viscosity, and a breakdown of the Stokes-Einstein-Debye relation between translational and rotational diffusion (Sillescu 1999, Ediger 2000, Richert 2002). It was these last phenomena that gave the first clear evidence for the existence of dynamic heterogeneity. Hence we briefly discuss some of this experimental evidence.

One argument is based on the details of the time dependence of the various relaxation functions $\phi(t)$ at intermediate and long times (see Fig. 5.3). At high temperatures (melting temperature T_m or higher), where relaxation functions do not show a plateau and relax to zero in a single step, the relaxation is exponential, $\phi(t) \propto \exp(-t/\tau)$. In contrast to this, for deeply supercooled fluids a stretched exponential decay occurs, described by the Kohlrausch-Williams-Watts function ("KWW law"), cf. Eq. (5.4), $\phi(t) \propto \exp[-(t/\tau)^{\beta}]$. Close to T_g the stretching exponent β is often independent of temperature (then the "time-temperature superposition principle" holds), but the value of β will depend on the physical quantity the relaxation of which is considered (and of course on the material considered).

Since it is always possible to formally associate with $\phi(t)$ an inverse Laplace transform $G(\tau)$,

$$\phi(t) = \int_0^\infty G(\tau') \exp(-t/\tau') d\tau' \,, \qquad (6.41)$$

where $G(\tau)$ can be interpreted as a "distribution of relaxation times", one can attribute a nonexponential decay to equilibrium to dynamic heterogeneity: One can argue (Richert 1994, 2002, Sillescu 1999, Ediger 2000) that different particles i (whose relaxations $\phi_i(t)$ contribute to $\phi(t)$ such that $\phi(t) = (1/N)\sum\phi_i(t))$ relax exponentially towards equilibrium with relaxation times τ', and it is the integration over a broad distribution of the relaxation times τ' of the different particles in different regions of space which causes the stretched exponential decay.

Of course, *a priori* one is not forced to make such an hypothesis, since one could also argue that stretched relaxation is an intrinsic phenomenon already at a local level, and all particles everywhere in the system relax in the same way. In fact, stretched exponential relaxation is also found in mean-field systems such as the p-state Potts glass (where every spin interacts with every other spin with interactions drawn from the same distribution, see Sec. 4.6.) or the p-spin interaction spin glass. As discussed in Sec. 4.6, there is no space that one could associate with such models, but nevertheless they exhibit an ergodic-to-nonergodic transition at some dynamical transition temperature T_D, and at temperatures slightly higher than T_D a pronounced two-step relaxation, similar to Fig. 5.3, occurs: The second step describing the final decay to equilibrium is distinctly non-exponential and can be well approximated by the KWW expression, Eq. (5.4). According to such a homogeneous interpretation of nonexponential relaxation, $G(\tau)$ for glass-forming fluids would not have an immediate physical interpretation, while in the heterogeneous case $G(\tau)$ just describes the distribution of slowly relaxing and fast relaxing particles, where one assumes that the asymptotic decay of $\phi_i(t)$ for each particle is a single exponential, $\phi_i(t) \propto \exp(-t/\tau_i)$. Of course, also intermediate cases where each particle already exhibits a nonexponential decay, described like Eq. (6.41) by a distribution of relaxation times $G_i(\tau)$, but the global relaxation (described by $G(\tau)$) is further broadened by dynamical heterogeneity, also is conceivable.

The first indirect experimental evidence for dynamical heterogeneity came from the finding that near the glass transition both the Stokes-Einstein (SE) relation (Tyrell and Harris 1984; see also Sec. 2.3) for the selfdiffusion constant of a particle in a liquid

$$D_t = k_B T/(6\pi\eta R) \tag{6.42}$$

and the Stokes-Einstein-Debye relation (Böttcher and Bordewijk 1978)

$$D_r = k_B T/(8\pi\eta R^3) \tag{6.43}$$

are violated. In Eq. (6.42), D_t is the translational diffusion coefficient of a particle (with hydrodynamic radius R) diffusing in a fluid with shear viscosity η, while Eq. (6.43) yields its rotational diffusion coefficient (D_r). Equations (6.42) and (6.43) derive from a phenomenological treatment (using hydrodynamics where one supposes that R is much larger than the molecules of the fluid). But there is broad evidence, for ordinary liquids, far away from any glass transition, that Eqs. (6.42), (6.43) hold (apart from numerical factors of order unity) even if R is of the same order as the size of the molecules in the fluid, so these equations can be applied to describe tracer diffusion for any of these fluid molecules. Note that Eqs. (6.42) and (6.43) imply that the ratio of D_r and D_t should yield a simple constant,

$$D_r/D_t = (3/4R^2)\,. \tag{6.44}$$

Now Ehlich and Sillescu (1990) have used various photoreactive dye molecules which were dissolved in glass-forming polymer melts and have measured D_t via forced Rayleigh scattering from the decay of holographic gratings. They found that Eq. (6.42) fails near T_g (or one must assume a rather unphysical temperature variation of the Stokes radius R of the tracer particle). Of course, one could also argue that the dye molecule motions get increasingly decoupled from the dynamics of the polymer matrix, but computer simulations of coarse-grained models (Vallée *et al.* 2006, 2007ab) suggest that no such decoupling takes place, small molecules at low concentration in polymer matrices are "faithful reporters" of the matrix dynamics.

Subsequently Fujara *et al.* (1992), Chang *et al.* (1994), and Chang and Sillescu (1997) discovered that actually Eq. (6.42) is strongly violated also for temperatures well above T_g (Fig. 6.6). Rotational diffusion in o-terphenyl were measured by deuteron-NMR methods over the entire temperature range of interest, while self-diffusion coefficients were obtained from field-gradient NMR methods. More precise and complete data were obtained for both D_r and D_t of tetracene probe molecules in o-terphenyl (Cicerone and Ediger 1996). Cicerone and Ediger (1996) also put forward a nice qualitative argument to explain why $D_t \gg D_r$ should be a consequence of dynamical heterogeneity. For simplicity, let us assume that there are just two kinds of regions, slow (s) and fast (f), and that their volume fractions ϕ_s, ϕ_f are equal, $\phi_s = \phi_f = 1/2$. Defining the rotational relaxation time as an integral of the decay of the orientational correlation function, one concludes that the average rotational relaxation time is

$$\langle \tau_r \rangle = \phi_f\, \tau_{rf} + \phi_s \tau_{rs} \approx \phi_s \tau_{rs} = \frac{1}{2} \tau_{rs}\,, \tag{6.45}$$

where we have assumed that $\tau_{rs} \gg \tau_{rf}$. The rotational diffusion constant scales inversely as $\langle \tau_r \rangle$, and hence $D_r \propto \tau_{rs}^{-1}$. On the other hand, the translational displacement of a molecule occurs mostly when it is in a "fast" region, and taking a suitable average of the Einstein relation along a particle trajectory then yields

$$\langle [\vec{r}_i(t) - \vec{r}_i(0)]^2 \rangle = 6t[\phi_f D_{tf} + \phi_s D_{ts}] \approx 6t\phi_f D_{tf} \,. \qquad (6.46)$$

In this way, it is possible to have $D_t \gg D_r$ (Cicerone and Ediger 1996). However, this qualitative argument does not explain why the data (Fig. 6.6) seems to follow a fractional Stokes-Einstein relation

$$D_r \propto (\eta/T)^{-\zeta} \,. \qquad (6.47)$$

The exponent $\zeta \approx 0.75$ is not universal, however, and Eq. (6.47) is no more than an empirical fitting function. In addition one has to realize that the translation-rotation decoupling certainly depends on the shape of the molecules. E.g. if they are almost spherical the coupling is weak whereas it is strong in the case of elongated molecules.[11]

In order to test these ideas on the violation of the SE relation Kumar *et al.* (2006) did simulations of a (polydisperse) hard sphere fluid and considered the diffusion constants of "sedentary particles" (i.e., particles residing in slow regions). They found that these particles still do satisfy the ordinary SE relation, and that indeed, in the spirit of the above argument, the subset of fast diffusing particles are responsible for the violation of the SE relation. An interesting framework for the theoretical discussion of Eq. (6.47) was also provided by Jung *at al.* (2004) and Berthier *et al.* (2005b) in terms of "excitation lines" in space-time structures that one can define precisely for kinetically constrained lattice models (see Sec. 6.1.3). We will briefly come back to these ideas at the end of this section.

Another question concerns the temperature at which this "rotation-translation-decoupling" occurs: Figure 6.6 suggests that in o-terphenyl this occurs at a "cross-over temperature " $T_x \approx 290$ K, while for $T > T_x$ both D_r and D_T are inversely proportional to the viscosity, so the Stokes-Einstein and Stokes-Einstein Debye relations, Eqs. (6.42), (6.43), both hold. Since in o-terphenyl the critical temperature T_c of mode coupling theory (MCT), according to the analysis of neutron scattering data (Tölle 2001) coincides

[11]As a side remark we mention that within the framework of mode coupling theory it is possible to determine a "critical" elongation of the molecules and to predict different types of glass transition scenarios in which the rotational and translational degrees of freedom occur at the same/at different temperatures (Chong and Götze 2002ab, Chong *et al.* 2005, Moreno *et al.* 2005).

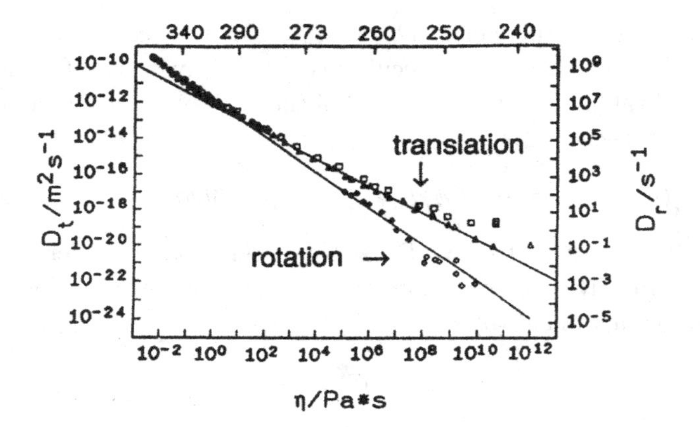

Fig. 6.6 Translational, D_t, and rotational, D_r, diffusion coefficients in ortho-terphenyl as a function of viscosity (abscissa) and temperature (labels on top of the figure). Self-diffusion constants are shown by solid circles, tracer diffusion constants for two probes by open squares and open triangles. Rotational diffusion constants for deuterated o-terphenyl are shown by filled and open diamonds. Full lines shown $\eta^{-0.75}$ and η^{-1} dependencies. Note that the translational diffusion has a weaker dependence on viscosity below $T \approx 290$ K than above this cross-over temperature. From Chang *et al.* (1994)

with T_x, $T_c = 290$ K, it has been often suggested that supercooled fluids do not have a dynamic heterogeneity for $T > T_c$, and relaxation should be homogeneous above this temperature. Since the breakdown of MCT traditionally is associated with the neglect of thermally activated relaxation ("hopping processes"), often reference is made to the energy landscape picture of Goldstein (1969), in which the system at low enough temperatures explores its phase space through activated jumps over (free) energy barriers, from one local minimum via a saddle point to the next one. It is then argued that these barriers are only felt for $T < T_x$, while for $T > T_x$ entropic effects essentially remove these barriers. Tentatively, the dynamical heterogeneity of the system is thus associated with the rugged free energy landscape.

While on a qualitative level such arguments seem to make sense, one cannot take them too literally: In fact, from simulations there exists clear evidence that dynamic heterogeneity sets in not at T_c *but already at temperatures distinctly above* T_c (Kämmerer *et al.* 1997, 1998a,b, De Michele and Leporini 2001, Moreno *et al.* 2005, Jose *et al.* 2006, Lombardo *et al.* 2006, Vallée *et al.* 2007b). While at high enough temperatures, where the structural relaxation time is of the order 10^{-12} sec, $\phi(t)$ decays essentially as single exponential function, and Eqs. (6.42)-(6.44) apply, a different

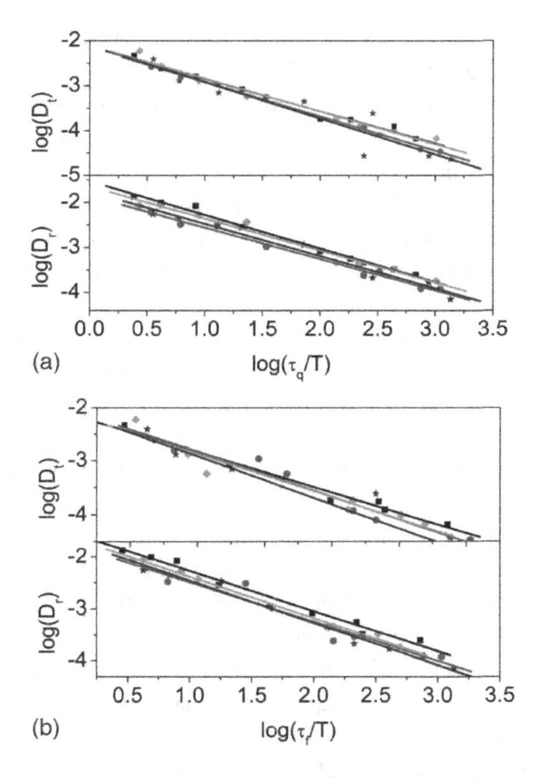

Fig. 6.7 Power-law fits of translational (a) and rotational (b) diffusivities D_t and D_r as function of τ_q/T (upper part) and τ_f/T (lower part), for four different choices of a model of a dumbbell in a bead-spring model of a polymer melt (chain length $N = 10$, density $\rho = 1$, in Lennard-Jones units). Here τ_q is the structural relaxation time of the intermediate incoherent scattering function at wavenumber $q = 6.9$ where the static structure factor has its peak, while τ_f is an estimate for the fluorescence lifetime (the dumbbell mimicks a fluorescent molecule as used in single molecule spectroscopy experiments). The straight lines illustrate relations $D_t \propto (\tau_{q,f}/T)^{-\zeta_t}$ and $D_r \propto (\tau_{q,f}/T)^{-\zeta_r}$ for various values of the mass M and Lennard-Jones diameter σ of the beads of the dumbbell (relative to those of the polymer): $M = 1$, $\sigma = 1$ (squares); $M = 2.25$, $\sigma = 1$ (diamonds); $M = 1$, $\sigma = 1.22$ (circles); $M = 2.25$, $\sigma = 1.22$ (stars). The effective exponents ζ_t, ζ_r (slopes of the straight lines in the fits) lie in the range $0.71 \leq \zeta_t, \zeta_r \leq 0.89$. From Vallée *et al.* (2007b).

behavior already sets in as soon as the structural relaxation time significantly increases. Figure 6.7 shows, as an example, simulation results of a probe molecule in a polymer melt, giving evidence for the presence of the fractional Stokes-Einstein as well as the fractional Stokes-Einstein-Debye relation. Since in this model $T_c = 0.45$ and only temperatures above T_c

are included ($0.47 \leq T \leq 0.7$; all temperatures being quoted in Lennard-Jones units) it is clear that the fractional SE law can be observed already above T_c. The reason why experiments seem to see evidence for dynamical heterogeneity only close to T_g, and not also above T_c like simulations do (Ediger 2000) is not clear, however.

While the breakdown of the Stokes-Einstein relation can hence be taken at least as indirect evidence for the presence of dynamical heterogeneity (although many aspects, such as whether Eq. (6.47) is more than an empirical fitting formula, are not yet understood), also direct evidence for dynamical heterogeneity exists from methods whose measurement signal comes from "sub-ensembles" of the molecules, rather than taking an average over all molecules.

One such technique to do that is the four-dimensional nuclear magnetic resonance (NMR) (Schmidt-Rohr and Spiess 1991), where "four-dimensional" means that one measures some parts of a four-time correlation function. Within this technique one first selects from the polymer melt a group of C-H vectors which have reoriented an unusually small angle during the time interval $t_2 - t_1$. After some waiting time the same subset of the system is probed again (between the times t_3 and t_4, choosing $t_4 - t_3 = t_2 - t_1$) in order to see whether now it has reoriented by similarly small angles or rather shows an average reorientation behavior. Thus one checks over what time (here $t_3 - t_2$) a subensemble of slow particles stays slow. For poly(vinylacetate) at $T_g + 10$ K it was found that the subset of vectors that was initially slow did remain slow if the waiting time $t_3 - t_2$ was short, hence providing for these conditions heterogeneous dynamics. After a very long waiting time, the slow subset had evolved to a behavior identical to the average behavior of such subsets. This is expected if the system is ergodic, and hence time averages and ensemble averages are equivalent. In this way, also the lifetime of the dynamical heterogeneities is probed. Similar behavior was also established for glassforming fluids consisting of small molecules such as o-terphenyl, for instance (Boehmer *et al.* 1996).

Apart from NMR, also optical experiments studying the reorientation of dilute probe molecules such as tetracene in o-terphenyl (Cicerone and Ediger 1995) yield information on subensemble dynamics. The sample is illuminated for a few milliseconds with polarized laser light, causing photobleaching of some of the probe molecules. Subsequent illumination of the sample with a weak "reading beam" whose polarization is modulated is used to study the decay of the probe anisotropy. Particularly elegant extensions of such optical techniques are the single molecule spectroscopy experiments

(Vallée *et al.* 2006), where trajectories of the fluorescence lifetime of "rotor probes" are analyzed. While some experiments have indicated that also subensembles of slow degrees of freedom exist whose lifetime is much larger than the α-relaxation time (see Ediger 2000 for a discussion) we maintain that the typical lifetime of dynamical heterogeneities indeed is of the same order as the α-relaxation time, and the character of dynamical heterogeneity must imply that some subsets of degrees of freedom exist whose lifetime is much larger than the average, as well as other subsets whose lifetime is much smaller than the average.

Of course, such large differences in relaxation time between slow and fast degrees of freedom can only exist, if the slow particles are not randomly mixed with the fast ones. Instead one can expect that both slow and fast particles form some sort of clusters, i.e. when a "slow" molecule is considered, it is reasonable to assume that most of its neighbors also are slow. Therefore we can conclude that there must be some spatial length scale associated with these dynamic heterogeneities. The experiments described above give only a very limited and indirect information on this spatial length scale. An attempt in this direction due to Tracht *et al.* (1999), also based on 4-dimensional NMR, yielded an estimate of 2 to 3 nm for polyvinylacetate at $T_g + 10$ K. However, there are no convincing experiments that could give precise information of the dependence of this length scale on temperature (and/or density or pressure, respectively), although the attempts to find this length scale are numerous (see e.g. Arndt *et al.* 1997, Donth *et al.* 2001, and the articles in Berthier *et al.* 2010).

Simulations (e.g. Yamamoto and Onuki 1997, Kob *et al.* 1997) are nicely suited to follow up on this distinction of subsets of slow and fast particles. Considering subsets such as the 5% fastest and 5% slowest particles, one typically finds that the fast particles do indeed form clusters of somewhat stringlike, ramified shape, while the slow particles form clusters that are more compact. As expected, the size of these clusters increases with decreasing temperature — although the conclusions on the temperature dependence close to the glass transition must remain speculative, since most of present simulation data on these clusters are only for temperatures $T \geq T_c$. In addition, the results depend somewhat on the question whether one considers clusters of the 5% or 7% or 10% slowest particles — clearly there is some undesirable arbitrariness in this percentage.

However, the existence of clusters of rather immobile (or mobile) particles of some linear dimension ℓ suggests that the motion of a particle i in such a cluster during a time interval $t < \tau$ is correlated with the motion of a particle j from the same cluster during the same time interval. Hence it is reasonable to assume that it must be possible to detect such correlations from a direct analysis of suitable dynamic correlation functions. A first successful attempt in this direction is due to Donati *et al.* (1999) and Bennemann *et al.* (1999) who introduced a displacement-displacement correlation function $G_u(\vec{r}, \Delta t)$,

$$G_u(\vec{r}, \Delta t) = \int d\vec{r}' \langle [u(\vec{r}' + \vec{r}, t, \Delta t) - \langle u \rangle][u(\vec{r}', t, \Delta t) - \langle u \rangle] \rangle \tag{6.48}$$

where

$$u(\vec{r}', t, \Delta t) = \sum_{i=1}^{N} |\vec{r}_i(t + \Delta t) - \vec{r}_i(t)| \delta(\vec{r} - \vec{r}_i(t)) \,. \tag{6.49}$$

Thus $G_u(\vec{r}, \Delta t)$ measures the space-time correlation function of the "mobility" u.

A further useful quantity is the "total displacement" U and its mean-square fluctuation,

$$U(t, \Delta t) = \int d\vec{r} u(\vec{r}, t, \Delta t) \,, \tag{6.50}$$

$$\langle [U(t, \Delta t) - \langle U \rangle]^2 \rangle = \int d\vec{r} G_u(\vec{r}, \Delta t) \equiv \langle u \rangle \langle U \rangle k_B T \kappa_u(\Delta t) \tag{6.51}$$

where the second equality defines the "susceptibility" $\kappa_u(\Delta t)$, in analogy to the fluctuation relation for the particle number N in terms of the isothermal compressibility κ_T (see Eq. (2.25)). One finds that $\kappa_u(\Delta t)$ exhibits a maximum at a time $\Delta t = \Delta t^*$, a time which is of the same order as the α-relaxation time (Donati *et al.* 1999). Figure 6.8 shows plots of the temperature dependence of Δt^* and $\kappa_u(\Delta t^*)$ and we see that both quantities increase with decreasing temperature and are compatible with a power-law behavior as the critical temperature T_c of mode coupling theory is approached, i.e.

$$\kappa_u(\Delta t^*) \propto (T - T_c)^{-\gamma_\kappa} \,, \quad \Delta t^* \propto (T - T_c)^{-\gamma} \,, \tag{6.52}$$

where for the considered Kob-Andersen (1994) model $T_c = 0.435$ (in Lennard-Jones units) was used, and $\gamma_\kappa \approx 0.84$, $\gamma \approx 2.3 \pm 0.2$ was found (Donati *et al.* 1999). Taking the space Fourier transform of $G_u(\vec{r}, \Delta t)$ it was found that the resulting structure factor exhibits at small wavenumbers a peak which, if one assumes that it can be described by the standard

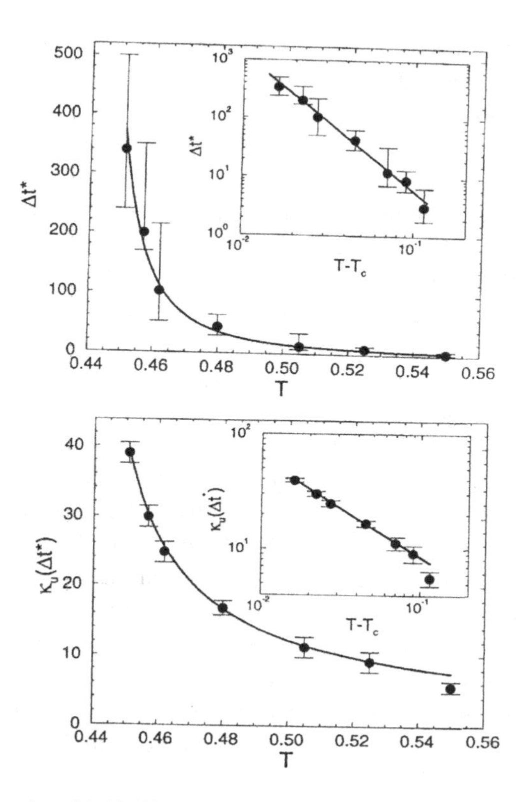

Fig. 6.8 Δt^* (a) and $\kappa_u(\Delta t^*)$ (b) plotted vs. temperature (all quantities measured in standard Lennard-Jones units, for the 80:20 binary Lennard-Jones mixture at pressure $p = 3.03$) where the mode coupling critical temperature is estimated to be $T_c = 0.435$. Full curves show power-law fits. These power-laws are also displayed in log-log form in the inserts. From Donati *et al.* (1999).

Ornstein-Zernike form

$$S_u(q, \Delta t^*) \propto \kappa_u(\Delta t^*)/(1 + \xi^2(\Delta t^*)q^2)\,, \tag{6.53}$$

can be used to determine a dynamic correlation length $\xi(\Delta t^*)$. Although Donati *et al.* saw an increase of $\xi(\Delta t^*)$ with decreasing T, the quality of the data was insufficient to make strong claims on a possible divergence of this length scale for $T \to T_c$. (For more recent attempts to characterize this growth see Stein and Andersen 2008 and Karmakar *et al.* 2009.)

When these results were discovered by Donati *et al.* (1999) for the Kob-Andersen (1994) model (and similar results were found by Bennemann *et al.* (1999) for a model of a glassforming polymer melt), this finding was an unexpected surprise since the prevailing prejudice was that mode coupling theory (MCT), present in these results via the critical temperature T_c, describes a purely local "dynamical arrest" of particles in their cages, driven by a nonlinear feedback mechanism of the corresponding equations (see Chap. 5), with no growing length scale whatsoever.

However, this opinion about MCT has more recently been found to be erroneous since also this theory predicts the divergence of a dynamic correlation length (Biroli and Bouchaud 2004, 2007, Biroli *et al.* 2006). These results are based on the insight that for the case of $p-$spin glasses it is possible to show that the susceptibility does indeed diverge at the MCT temperature T_c and that the generalisation of this susceptibility to the case of glass-forming liquids is a four-point correlation function, similar to the one that is given in Eq. (6.48) and discussed in more detail below (Franz and Parisi 2000). Also recall, see end of Sec. 5.2.3 and Kirkpatrick and Wolynes (1987b) and Kirkpatrick and Thirumalai 1988a,b, that the equations of motion for the spin correlation function of these systems are formally identical to the ones of MCT for glass-forming liquids, and thus the connection put forward by Franz and Parisi is not that strange.

Let us now discuss some of the details of the mentioned four-point correlation function. Since we want to investigate the geometric arrangement of the domains of particles that move fast or slow, it is useful to introduce a "mobility" $c_j(t,0)$ which gives a measure how much particle j moves between time $t=0$ and t. (An explicit definition for this quantity will be given below.) Now we can introduce a "mobility field" $c(\vec{r};t,0)$ via

$$c(\vec{r};t,0) = \sum_j c_j(t,0)\delta(\vec{r}-\vec{r}_j)\,. \tag{6.54}$$

The spatial correlation of mobility can now be characterized by the correlation function

$$G_4(\vec{r};t) = \langle c(\vec{r};t,0)c(\vec{0};t,0)\rangle - \langle c(\vec{0};t,0)\rangle^2\,. \tag{6.55}$$

Thus the function $G_4(\vec{r};t)$ measures the correlation of mobility (as determined over a time interval t) between two points in space that are a distance $\vec{r}$ apart.

How can one characterize the mobility $c_j(t,0)$ of a particle? The choice $c_j(t,0) = |\vec{r}_j(t) - \vec{r}_j(0)|$, used in the context of Eq. (6.48), is one possibility. It is, however, more useful to consider mobilities that tell us something on how fast a particle is moving on a given length scale ℓ which, e.g., we can write as $\ell = 2\pi/q$, where q is a wave-vector. For this we can introduce the observable $o_j(q,t) = \exp[i\vec{q}\cdot\vec{r}_j(t)]$ and define the mobility as $c_j(t,0) = o_j(q,t)o_j(-q,0)$. Thus with this choice the average of the mobility field is directly related to $F_s(q,t)$, the self-part of the intermediate scattering function, see Eq. (2.75). Plugging in these expressions into the definition for the function $G_4(\vec{r};t)$ and passing from the particle based observable $o_j(q,t)$ to a field $o(\vec{r};q,t)$ we obtain

$$G_4(\vec{r};t) = \langle o(\vec{r};q,t)o(\vec{r};-q,0)o(\vec{0};q,t)o(\vec{0};-q,0)\rangle - \langle o(\vec{0};q,t)o(\vec{0};-q,0)\rangle^2 . \tag{6.56}$$

(Note that here the function G_4 depends also on the parameter q.) We see that G_4 depends on four points in space-time and thus is indeed a four-point function. It measures the correlation of the mobility on a length scale r on a time scale t, the mobility being probled on a length scale $2\pi/q$.

Independently of the precise choice of the definition of mobility one can make an analogy to critical phenomena and thus can expect that for large r the function $G_4(\vec{r};t)$ decays like

$$G_4(\vec{r};t) \sim \frac{A(t)}{r^p}\exp[-r/\xi_4(t)] \tag{6.57}$$

with an exponent $p > 0$ and a dynamical correlation length $\xi_4(t)$.

From $G_4(\vec{r};t)$ one can also define a susceptibility $\chi_4(t)$ via

$$\chi_4(t) = \int d\vec{r} G_4(\vec{r};t) \tag{6.58}$$

which we will discuss in more detail below. By introducing the quantity

$$C(t,0) = \int d\vec{r}\; c(\vec{r};t,0)\,, \tag{6.59}$$

this susceptibility can be expressed as

$$\chi_4(t) = N[\langle C(t,0)^2\rangle - \langle C(t,0)\rangle^2]\,. \tag{6.60}$$

Thus $\chi_4(t)$ is just the variance of the average mobility $C(t,0)$, i.e. can be interpreted as a measure the number of overlapping particles in two configurations that are separated a time t (Franz *et al.* 1999, Donati *et al.* 2002). Equation (6.60) is the analog of the "spin glass susceptibility" which also was the sum of a special four-spin correlation function, cf. Sec. 4.5.2 but there the presence of quenched disorder allowed to consider a purely static correlation, while in the supercooled fluid the glass-type correlations only show up on the time-scale of the structural relaxation time, but not on much longer time scales.

Instead of considering the function $G_4(\vec{r}, t)$ one can also study its space Fourier transform

$$S_4(\vec{k}, t) = \int d\vec{r} e^{i\vec{k}\cdot\vec{r}} G_4(\vec{r}; t) \tag{6.61}$$

since for analytical calculation this function is more accessible. (Note that one should not confuse the wave-vector q, used to define the mobility, with the wave-vector k, used here for the Fourier transform.) Of course one has $\chi_4(t) = \lim_{k\to 0} S_4(\vec{k}, t)$. As shown by Franz *et al.* (1999), Donati *et al.* (2002) and Lacevic *et al.* (2003), the function $S_4(\vec{q}, t)$ (for precise definitions of this function see the quoted literature since it depends on the details of the definition of the mobility) is well-suited to extract a growing (dynamic) glass correlation length, which is consistent with a power-law divergence at T_c, see Fig. 6.9. It is interesting to note that the estimated exponent ν of this dynamical correlation length, $\nu \approx 0.82$, while the MCT result of Biroli and Bouchaud (2004, 2007) and Biroli *et al.* 2006 predict an exponent $\nu = 1/4$. However, since the upper critical dimension $d_u = 6$ (estimated from a Ginzburg-type criterion, see Biroli and Bouchaud 2004, a value that has later been conjectured to be $d_u = 8$, see Biroli and Bouchaud 2007), one does not expect that the MCT prediction should be accurate in $d = 3$ dimensions. Biroli and Bouchaud (2007) then argue that "nonclassical critical fluctuations" are responsible for the breakdown of the SE relation, discussed above. However, an explicit theory supporting this interesting suggestion to our knowledge is still lacking.

Unfortunately, so far no direct experiment that could measure $S_4(q, t)$ or at least $\chi_4(t)$ for an atomic system has been possible. However, one can show (Berthier *et al.* 2005a) that $\chi_4(t)$ satisfies the inequality

$$\chi_4(t) \geq \frac{k_B T^2}{c_p} \chi_T^2(t) \tag{6.62}$$

where c_p is the specific heat per particle at constant pressure, and $\chi_T(t)$ is defined by

$$\chi_T(t) = \frac{N}{k_B T^2} \langle \delta H(0)\, \delta C(t, 0) \rangle \,. \tag{6.63}$$

Here $\delta H(0)$ is the fluctuation of the enthalpy per particle at time $t = 0$ and $\delta C(t, 0)$ the fluctuating part of the correlator defined in Eq. (6.59). Although the inequality (6.62) gives only a lower bound to $\chi_4(t)$, it has

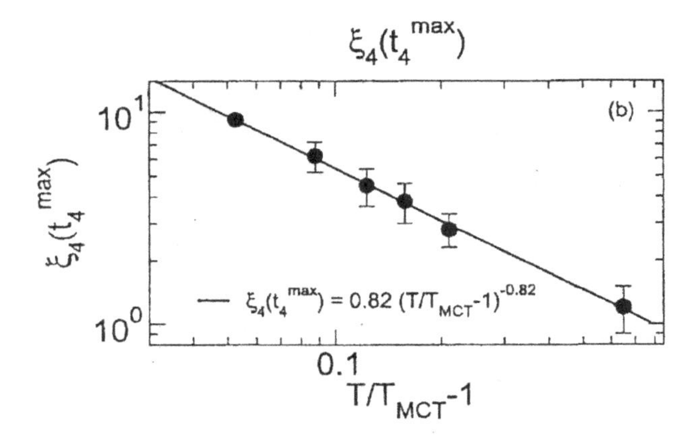

Fig. 6.9 Log-log plot of the dynamic glass correlation length $\xi_4(t_4^{\max})$ versus $T/T_{\rm MCT}-1$ ($T_{\rm MCT}$ is denoted as T_c in the text), for a 50:50 binary Lennard-Jones mixture. $t_4^{\max}$ is defined as the time at which $\chi_4(t)$ has its maximum. From Lacevic *et al.* (2003).

been found in simulations that at low temperatures the right hand side gives a good estimate for $\chi_4(t)$ and therefore the knowledge of $\chi_T(t)$ allows to make a good estimate of $\chi_4(t)$ (Berthier *et al.* 2005a, 2007ab). This latter quantity can be expressed as

$$\chi_T(t) = \left.\frac{\partial\langle C(t,0)\rangle}{\partial T}\right|_{N,P}, \qquad (6.64)$$

and thus can be measured easily by in an experiment, since $\langle C(t,0)\rangle$ is just the coherent intermediate scattering function.[12]

Berthier *et al.* (2005a) have extracted $\chi_T(t)$ from dielectric susceptibility measurements of the glassforming fluid glycerol and thus studying the right hand side of the inequality, Eq. (6.62). In this way they obtain an estimate for ξ near the glass transition of glycerol is found and find that ξ increases from 0.9 nm at $T = 232$ K to $\xi = 1.5$ nm at $T = 192$ K, compatible with the estimates extracted indirectly from multidimensional NMR measurements for the size of dynamical heterogeneities, discussed above. Subsequently Dalle-Ferrier *et al.* (2007) have extended this work to a multitude of glass-formers and therefor measured the maximum of $\chi_T(t)$ as a function of temperature. The result of this analysis is shown in Fig. 6.10 in which is, apart from a prefactor, the value of this maximum. Note that

[12]Note that similar experessions can be obtained for the case that density instead of temperature is the control variable (Berthier *et al.* 2007ab).

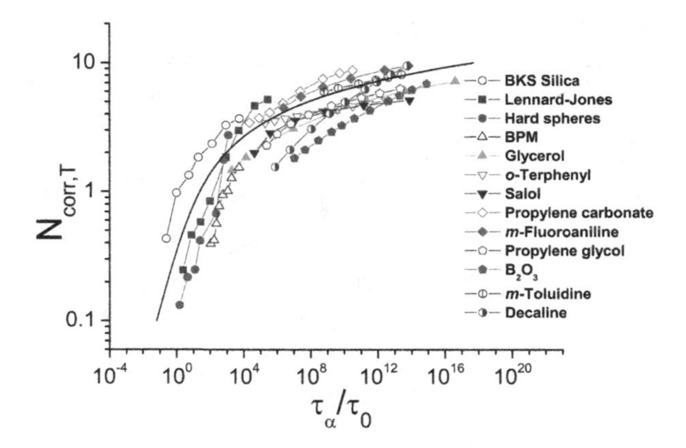

Fig. 6.10 Dependence of $N_{\text{corr},4}$ on the $\alpha-$relaxation time $\tau_\alpha(T)$ for a variety of glass-formers (experiments and simulations). The time τ_0 is a micoscopic time scale. From Dalle-Ferrier *et al.* (2007).

this quantity, denoted by $N_{\text{corr},4}$ is *not* the number of particles that are in a cluster of dynamically correlated particles, since a (unknown) prefactor is missing (basically the prefactor $A(t)$ in Eq. (6.57).

Thus although in recent years there has certainly been made a lot of progress on the characterization of the dynamical heterogeneities, important issues, such as the real size of the clusters, their shape, etc., still remain open and thus the focus of current research.

We now return to the possibility to extract estimates for the glass correlation length from suitable finite size scaling analyses, mentioned at the beginning of this subsection, see Eq. (6.40). Striking evidence supporting the validity of Eq. (6.40) is due to the work of Berthier (2003) on the one-dimensional version of the spin-facilitated Ising model introduced by Fredrickson and Andersen (1984). He starts from the well-known observation for static phase transitions that finite size scaling is most convincingly demonstrated by an analysis of moments of the order parameter distribution (Binder 1981). E.g., for an Ising ferromagnet a convenient order parameter is the density of the magnetization, $m = L^{-d}\sum_i S_i$, $S_i = \pm 1$ being the spin variables at lattice site i. Then a useful quantity is the reduced fourth order cumulant

$$U(T, L) = 1 - \langle m^4 \rangle_L / (3\langle m^2 \rangle_L^2) , \tag{6.65}$$

since it satisfies finite size scaling without a power-law prefactor of L such as in Eq. (6.40), i.e. $U(T, L)$ can be expressed as

$$U(T, L) = \tilde{U}(\xi/L)\,, \tag{6.66}$$

where $\tilde{U}$ is a scaling function and hence criticality is easily found by studying $U(T, L)$ as function of T for various L, and since $\xi \to \infty$ as the temperature approaches the critical temperature T_c, the latter is found from common intersection point $\tilde{U}(\infty)$ of these curves.

Now for a glass-forming system the analog of the order parameter density m is the two-time density correlation function

$$\varphi(\tau) = L^{-d} \int d^d\vec{r}\, F(\vec{r}, \tau), \qquad F(\vec{r}, \tau) = \rho(\vec{r}, \tau + t_0)\rho(\vec{r}, t_0)\,, \tag{6.67}$$

and then the function G_4 considered above is just an averaged correlation function of the local time-displaced quantity $F(r, \tau)$. In the spirit of Eq. (6.65), it then makes sense to consider moment ratios, but unlike the Ising model (where there is a symmetry between m and $-m$ in the absence of external fields, $F(r, \tau)$ as defined above is a non-negative quantity lacking any particular symmetries. The generalization of Eq. (6.65) to this case without symmetries is

$$B(T, L) = 1 - \frac{\langle\varphi^4\rangle_L - 4\langle\varphi^3\rangle_L\langle\varphi\rangle_L + 6\langle\varphi^2\rangle_L\langle\varphi\rangle_L^2 - 3\langle\varphi\rangle_L^4}{3[\langle\varphi^2\rangle_L^2 - 2\langle\varphi^2\rangle_L\langle\varphi\rangle_L^2 + \langle\varphi\rangle_L^4]}\,. \tag{6.68}$$

Berthier (2003) studied this function for the $d = 1$ Fredrickson-Andersen model, which does not have a finite temperature transition temperature T_c, but nevertheless exhibits a characteristic length $\ell(T)$ and a relaxation time $\tau(T)$ that both diverge exponentially as $T \to 0$ (Butler and Harrowell 1991, Garrahan and Chandler 2002),

$$\ell(T) = \exp(1/T)\,, \quad \tau = \exp(3/T), \tag{6.69}$$

where the energy scale in the Hamiltonian simply has been chosen to be unity. Note that Eq. (6.69) is compatible with Eq. (6.39) with $z = 3$ if we assume $\ell(T) \propto \xi$. Since $T_c = 0$ one does not find any intersection point when $B(T, L)$ is plotted vs. T, see Fig. 6.11, but one finds that a finite size scaling of the type of Eq. (6.66) can be established very convincingly, choosing $\log\,(\xi/L) = \log(\ell(T)/L) = 1/T - \log L$ as an abscissa variable (Fig. 6.11). Similar results were also obtained (Berthier 2003) for a second model, the kinetically constrained lattice gas on cubic lattices (Kob and

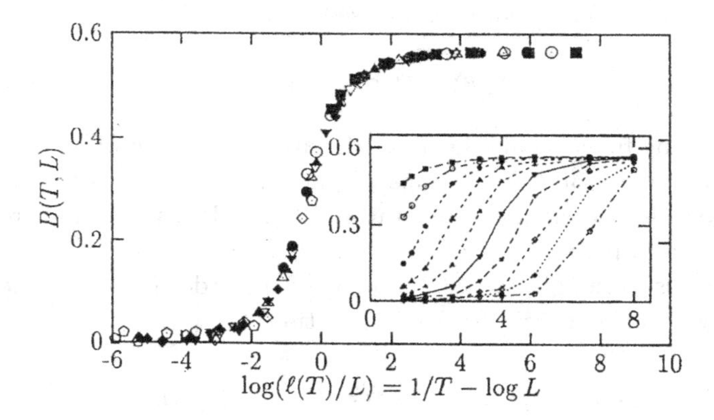

Fig. 6.11 The parameter $B(T, L)$ as defined in Eq. (6.68), plotted versus the variable $1/T$ - log L, for a one-dimensional spin-facilitated kinetic Ising model, including various system sizes $L = 2, 4, \cdots, 1024$, distinguished by different symbols. The inset shows the same data as a function of $1/T$ ($L = 2, 4, \cdots, 1024$, from top to bottom). From Berthier (2003).

Andersen 1993) and very recently by Karmakar *et al.* 2009 for the Kob-Andersen binary Lennard-Jones mixture.

An alternative method to learn about a growing correlation length ξ is to study how far the disturbing effect of a surface extends away from this surface into the bulk (Binder and Hohenberg 1972, 1974, Binder 1983). Let us suppose that at the surface of an Ising magnet a local magnetic field H_1 acts while there is no magnetic field in the bulk: then the local magnetization $m(z)$ decays with distance z from the surface proportional to $\exp(-z/\xi)$. A case more closely related to glassforming systems is a fluid (e.g. hard spheres) confined by a hard wall: the hard wall acts like a local ordering field both with respect to density (leading to the characteristic oscillatory density variation $\rho(z)$, the so-called "layering effect", which is quite similar to the oscillations in the radial density distribution $g(r)$ with r, and similarly such a hard wall also induces orientational order (Ricci *et al.* 2007).

An important general feature is that the correlation length, which in the bulk can only be obtained by investigating the order parameter correlation function, can already be determined from the order parameter profile $\varphi(z)$ itself, for any order parameter φ that couples linearly to an ordering field present at the surface at $z = 0$ (if one is in the disordered phase and thus $\varphi = 0$ in the bulk). This consideration suggests that near a surface the average $\varphi(z, \Delta t)$ of the local order parameter, $\varphi(z, \Delta t) = \langle \rho(\vec{r}, t)\rho(\vec{r} + \vec{\delta}, t + \Delta t)\rangle$, evaluating the thermal average at a given value of z, contains the same information on the glass-type correlation $\xi(\Delta t)$ as the four-particle correlation function, $G_4(\Delta t)$, Eq. 6.55. In addition, a study of surface effects on the relaxation dynamics and hence on the glass transition is interesting because it should be useful for the interpretation of experiments on the glass transition of fluids confined into pores or in thin film geometry (see e.g. Jones 1999, Forrest and Dalnoki-Verress 2001, Frick *et al.* 2000, 2003, Alcoutlabi and McKenna 2005, Baschnagel and Varnik 2005, and Koza *et al.* 2007 for overviews over a vast literature on the subject). However, such experiments are difficult, since in most cases no spatially resolved information is obtained, but only an average information integrated over the pore volume (or integrated across the thin film). Therefore one tries to extract information on the characteristic length by comparing data for several pore diameters or several film thicknesses, respectively. But also this approach has many problems since it is hard to measure the density of a fluid inside a nanopore very precisely, and to make sure that the density is the same, irrespective of the pore diameter. In addition, the roughness of the pore surface and other defects may introduce elements of randomly quenched disorder into the problem. For thin films, the typical geometry is a film with a "free" surface on the top, while the bottom surface is provided by a substrate. While one expects that the "free" surface (i.e., a surface against air) enhances the mobility of fluid particles, the bottom surface (due to attractive interactions from the substrate) typically has the opposite effect. Simulations by Torres *et al.* (2000) of thin films confined between two surfaces that attract the particles suggest that then the glass transition temperature T_g gets enhanced with decreasing thickness. For films with one free surface and one supported surface other simulations (Peter *et al.* 2006) suggest that a net reduction of the critical temperature T_c of MCT remains, but in any case any conclusion from the proposed decrease of $T_c(h)$ with film thickness h, such as the dependence

$$T_c(h) = T_c(\infty)/(1 + h_0/h)\,, \tag{6.70}$$

on the correlation length ξ, encoded in h_0, is very indirect. Again, Eq. (6.70) is a tentative phenomenological Ansatz, for which a fundamental justification is lacking. For long entangled polymers also free-standing films (which hence have two free surfaces) have been studied (e.g. Forrest and Dalneki-Verress 2001), but these films can only be prepared by heating them up from temperatures below T_g, and the possibility of artefacts due to the lack of equilibrium of some of the degrees of freedom is hard to rule out. Although there have also been interesting attempts to study the dynamics of molecules in the surface region of thin films (Erichsen *et al.* 2004) or inside of thin films spatially resolved by a fluorescence mulilayer method (Ellison and Torkelson 2003), in our view the most convincing studies of spatially resolved relaxation near surfaces in glassforming fluids again are due to simulation studies (Scheidler *et al.* 2000, 2002ab, 2004, Varnik *et al.* 2002ab, Baschnagel and Varnik 2005). We discuss here only some results for the binary Lennard-Jones mixture, a model that has already been introduced as a test of MCT in the previous chapter. Systems confined by two parallel walls a distance $D = 15\sigma_{AA}$ apart (σ_{AA} is the range of the LJ interaction between the A particles, henceforth used as unit of length, $\sigma_{AA} \equiv 1$) are studied. Two types of walls are used:

(i) A smooth repulsive wall, modelled by a potential

$$V_{\alpha w}(z) = (4/45)\pi\rho_w\sigma_{AB}^3\varepsilon_{\alpha w}(\sigma_{AB}/(z-z_w))^9\,, \quad \alpha \in A, B, \qquad (6.71)$$

where σ_{AB} is the range of the LJ potential for AB pairs, z_w is the position of the wall, and the energy parameters were chosen as $\varepsilon_{AW} = 1.0$, $\varepsilon_{BW} = 3.0$ (in units of $\varepsilon_{AA} \equiv 1.0$). By choosing $z_w = -0.65$ and $z_w = 15.65$ a film of width $D = 15$ was then realized. Note that the z^{-9} potential in Eq. (6.71) simply results when the Lennard-Jones r^{-12} repulsion is integrated over an halfspace.

(ii) A rough wall was realized by simulating a larger system with periodic boundary conditions in all directions, and then simply freezing a slice of thickness $2.5.\sigma_{AA}$ of the LJ liquid. The rigidly frozen particles interact with the freely moving particles in between these rough boundaries with the same LJ potentials as in equilibrium. The quantity of interest then is the spatially resolved intermediate incoherent scattering function.

$$F_s^\alpha(\vec{q}, z, t) = \frac{1}{N_\alpha}\sum_{j=1}^{N_\alpha}\langle \exp[i\vec{q}\cdot(\vec{r}_j(t) - \vec{r}_j(0))]\delta(z_j - z)\,, \alpha \in A, B\,. \quad (6.72)$$

Thus this correlator considers only particles that had a distance z from the wall at time $t = 0$. Scheidler *et al.* (2002a) oriented the wave vector $\vec{q}$ parallel to the wall, and chose $|\vec{q}| = 7.2$, the location of the maximum in the bulk static structure factor. At low temperatures $F_s^\alpha(\vec{q}, z, t)$ shows a two-step relaxation, and hence one can characterize the α-relaxation time

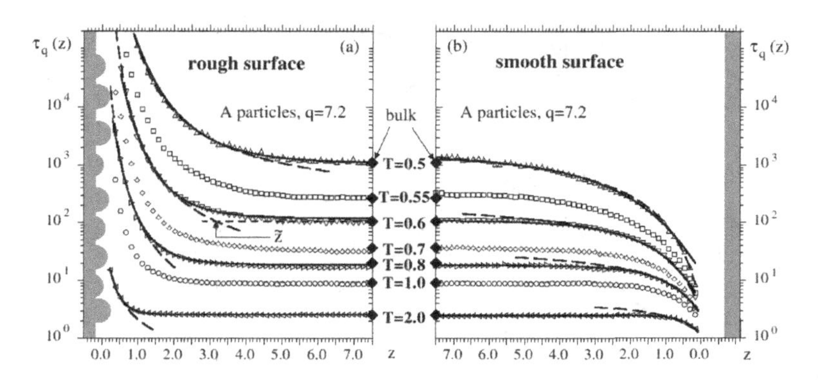

Fig. 6.12 Structural relaxation time $\tau_q(z)$ for the A particles in a binary (A, B) Lennard-Jones mixture at a wave-vector $q = 7.2$ as a function of particle distance z from the wall for (a) rough and (b) smooth surface at different temperatures. The large diamonds are the bulk values (note that $T_c = 0.435$ for the chosen density in the bulk) and the curves indicate various fits from which characteristic lengths were extracted. From Scheidler *et al.* (2002a).

$\tau_q(z)$ by requiring

$$F_s(\vec{q}, z, \tau_{\vec{q}}(z)) = 1/e\,. \tag{6.73}$$

Figure 6.12 shows these relaxation times both for rough and for smooth surfaces. One can see that for particles far from the wall $\tau_q(z)$ reduces to the relaxation time τ_q in the bulk. The range of z values near the wall over which $\tau_q(z)$ significantly differs from the bulk clearly increases as T decreases, thus indicating the presence of a growing correlation length in the fluid.

The most natural way to introduce a correlation length $\xi(t)$ is by making the Ansatz (Scheidler *et al.* 2002b)

$$F_s^{\alpha}(\vec{q}, z, t) = F_s^{\alpha,\mathrm{bulk}}(\vec{q}, t) + a(t)\exp[-(z/\xi(t))^{\beta(t)}] \tag{6.74}$$

where at each time t the parameters a, ξ and β are phenomenological fitting parameters. $F_s^{\alpha,\mathrm{bulk}}(\vec{q}, t)$ is known from independent simulations of bulk systems, of course. In comparison to standard phase transitions, this ansatz is more complicated since it allows for a stretched exponential z-dependence rather than a simple exponential. It turns out that in the regime of interest (where $\xi(t)$ reaches its maximum $\xi_{\max}$) $\beta(t) \approx 1.2 \pm 0.1$ not significantly depending on temperature.

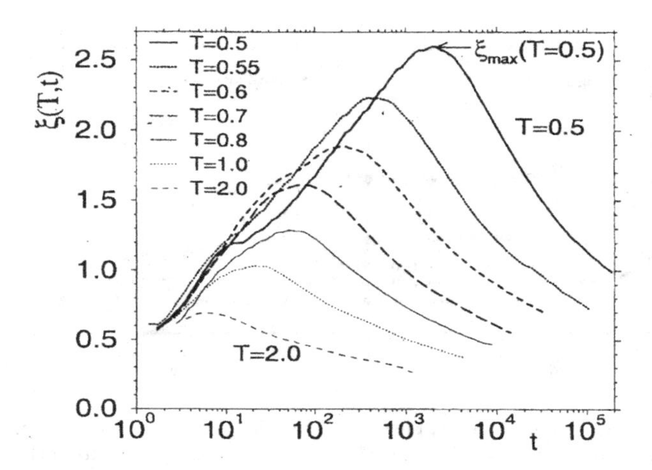

Fig. 6.13 Time dependence of the correlation length $\xi(T,t)$ extracted from a fit of Eq. (6.74) to the data for $F_s^A(\vec{q}, z, t)$ for binary Lennard-Jones mixture of Fig. 6.12 for the case of rough walls. The value $\xi_{\max}(T)$ at the maximum of the curves is used to define the length scale characterizing the glass transition. From Scheidler *et al.* (2002b).

Figure 6.13 shows that $\xi(T,t)$ increases with time, reaches a maximum at a time which is of the order of the structural relaxation time in the bulk, and then decreases again. Qualitatively, these data are just as expected from our discussion of the four-particle correlation functions in the bulk, Eqs. (6.55)-(6.60). However, it must be stressed that the data of Scheidler *et al.* (2002ab) refer to temperatures that are distinctly higher than the critical temperature of MCT; extracting from the data shown in Fig. 6.12, 6.13 a quantitatively reliable information on the temperature dependence of the glass correlation length is difficult. Moreover different ways to extract this correlation length from the simulation data yield somewhat different results (Scheidler 2002a).

What can then be concluded in general on the existence and behavior of a correlation length that grows when the glass transition is approached? There are several rather different interpretations. One very interesting concept emphasizes the fact that all these correlation lengths discussed above are not defined from *static* correlation functions but refer to an order parameter which has time t as an explicit variable, in addition to spatial coordinates. Based on extensive studies of spin-facilitated kinetic Ising models (Garrahan and Chandler 2002, 2003, Jung *et al.* 2004, Whitelam and Garrahan 2004), Merolle *et al.* (2005) and Jack *et al.* (2006), see also Chandler and Garrahan (2010), suggested that the glass transition can be viewed as a symmetry breaking (or phase coexistence) in trajectory space of the system. Thus the time evolution of the (coarse-grained) dynamical variables of the system, symbolically denoted as $\{x(t)\}$, can be described by a probability density functional $P\{x(t)\}$ and one can associate with this probability density an action $\mathcal{E}\{x(t)\}$,

$$P\{x(t)\} \equiv \exp(-\mathcal{E}\{x(t)\}) . \quad (6.75)$$

One expects that this action is extensive in N and proportional to t (for large t). Now one can also consider a distribution function for this action,

$$P(\mathcal{E}) \equiv \langle \delta(\mathcal{E} - \mathcal{E}\{x(t)\}) \rangle = \Omega(\mathcal{E}) \exp(-\mathcal{E}) . \quad (6.76)$$

Here the brackets $\langle \cdots \rangle$ denote an average over the ensemble of trajectories (which all extend over the range of time from zero to t), and $\Omega(\mathcal{E})$ then is the number of such trajectories with action $\mathcal{E}$. When N is much larger than any dynamically correlated volume in space, and t is much larger than any correlation time, one expects an extensivity property,

$$\ln \Omega(\mathcal{E}, N, t) = S(\mathcal{E}) N t , \quad (6.77)$$

where the entropy per space-time point, $S(\mathcal{E})$, has been introduced. It then is useful to define a susceptibility from the mean square fluctuation of the action,

$$\chi_{\mathcal{E}}(N, t) = \langle \mathcal{E}^2 \rangle - \langle \mathcal{E} \rangle^2 . \quad (6.78)$$

For spin-facilitated Ising models, Merolle *et al.* (2005) find that $\chi_{\mathcal{E}}(N, t)$ is extensive as well, and $\chi_{\varepsilon}/(Nt)$ develops a peak at low temperatures, with a height which strongly increases with increasing time t, if one varies T (but keeps t large and fixed). This peak means that for temperatures T less than the temperature of the maximum, the system has fallen out of equilibrium, i.e. if one interprets the temperature of the maximum as a glass transition temperature, it is a clearly observation-time dependent nonequilibrium phenomenon. A further interesting finding is the fact that $P(\mathcal{E})$ shows a tail at low $\mathcal{E}$ which means that for these low values of the action, the entropy of trajectory space is subextensive in time.

This interpretation of the glass transition as a kind of order-disorder phenomenon in trajectory space ("space-time thermodynamics", as termed by Merolle *et al.* 2005), has been made more precise recently by Garrahan *et al.* (2007, 2009), see also Chandler and Garrahan (2010). Defining by $K\{x(t)\}$ the number of configuration changes in that particular trajectory

$\{x(t)\}$ of the system (in practice K can be taken to be the mean squared displacement), a dynamical partition function (Eckmann and Ruelle 1985) can be defined as

$$Z_K(s,t) = \sum_{\{x(t)\}} P\{x(t)\} \exp[-sK\{x(t)\}] , \tag{6.79}$$

where the "field" s is a variable conjugate to K, such that sK has the dimension of an action. One then can show, when one considers the dynamical order parameter K as a function of s, that K/Nt is singular at $s = 0$: there occurs a first order transition, the dynamics of the system is characterized by two phases, an "active" one (where $K/Nt > 0$) for $s < 0$, and an inactive one ($K/Nt = 0$) for $s > 0$. Of course, the physical situation of the considered models corresponds to $s = 0$, and thus there one has coexistence between the active and inactive "phases". Physically this means that some trajectories have a very large number of configuration changes, and other trajectories a few or none. Thus, this finding can be viewed as a mathematical description of dynamical heterogeneity in these models.

Of course, while the work of Garrahan *et al.* (2007), (2009) clearly has considerably improved our insight into the behavior of the spin-facilitated kinetic Ising models, there is no consensus on the extent to which these models can describe the glass transition of real materials. Hedges *et al.* (2009) have made an interesting attempt to carry these concepts over to the binary Lennard-Jones mixture (Kob-Andersen 1994 model). For a system of $N = 150$ particles, Hedges *et al.* (2009) show that already at rather high temperatures ($k_BT/\varepsilon_{AA} = 0.7$ and 0.6) signatures of a first order phase transition in space-time are seen. However, this transition occurs not at $s = 0$ but at a nonzero value of s. Thus we see that although this line of research opens interesting perspectives to understand dynamical heterogeneities, the details still have to worked out.

Another very interesting hypothesis starts from the fact, that mode coupling theory, which is a quite successful description of the onset of slow relaxation of many glass-forming fluids (see Chap. 5 of this book), also describes the ergodic-to-nonergodic transition in the dynamic version of certain mean-field type spin glasses, such as the p-spin interaction spin glass or the p-state mean-field Potts glass with $p > 4$ (see Chap. 4). This has been shown by Kirkpatrick and Wolynes (1987ab), Kirkpatrick and Thirumalai (1988a,b), Kirkpatrick *et al.* (1989) and discussed further by many others (e.g. Bouchaud *et al.* 1996 and Castellani and Cavagna 2005). As was already discussed in Chap. 4, in the disordered phase for $T > T_D$

the free-energy "landscape" contains a single "valley" or "thermodynamic state", the system is ergodic, while for $T < T_D$ there occurs an infinite number of valleys (in the thermodynamic limit), separated from each other by infinitely high free energy barriers, so the system stays in one of these valleys with no chance to visit any of the other valleys, the system being nonergodic. This transition at T_D shows up via a divergence of the relaxation time, but in most static quantities (spin glass susceptibility, specific heat, etc.) this transition does not show up at all. An important exception is the free energy itself, since $F = \bar{F} - TI$, where $\bar{F}$ is the (average) free energy of a single valley. The quantity I is called "complexity" and its physical interpretation is in terms of the entropy due to the existence of a large number K^* of valleys ($I = k_B \ln K^*$) and thus it is also called "configurational entropy" in the literature.

Now a striking feature of the p-spin interaction spin glasses and $p > 4$ Potts glasses (in this mean-field limit only, of course) is that, see Fig. 4.36, I vanishes continuously at a lower, static transition temperature T_0: At this temperature, a nonzero glass order parameter shows up, and also the spin glass susceptibility jumps to infinity. Despite this first order character of the transition, there is no latent heat, the specific heat exhibits a kink singularity only. As was already discussed in Chap. 4, this static transition prevents an "entropy catastrophe" (i.e. a regime of temperatures where the entropy is negative) to occur (see Chap. 1). An elegant formulation of this mean field theory, which can be generalized to other glassforming systems, can be found in Mézard and Parisi (2009) in terms of the coupling between real replicas of the system (see Sec. 4.6).

Since the equations for the schematic models of MCT (i.e. where the wavenumber dependence of the structure factor is not taken into account) are formally identical to those derived for these mean-field spin glass models, it is tempting to speculate that this scenario of two successive transitions holds also for the problem of the glass transition in supercooled fluids. However, there is one important problem: In real systems we deal with systems where the interactions between the relevant degrees of freedom is short-range. As a result, when we wish to identify the critical temperature T_c of MCT with the analog of T_D but for a system with finite interaction range, one must accept that for $T < T_c$ (T_D) the free energy barriers separating the different "valleys" in the free energy "landscape" are not infinite but finite. These "valleys" hence are not really states in a sense of thermodynamics, but rather metastable states — the system will stay in one such "state" only for a finite lifetime; this lifetime then can only diverge at T_0 (if $T_0 > 0$), where the "configurational entropy" I (complexity) associated with the number of these valleys vanishes, and a static glass transition is postulated.

The static transition speculatively proposed by this so-called "mosaic theory" is often called "random first order transition" (RFOT) (see e.g. Bouchaud and Biroli 2004, Biroli *et al.* 2008). The state of the supercooled fluid for $T_0 < T < T_c$ is viewed as a mosaic-like patchwork of regions, each of which being in a different metastable state. By a nucleation-like event each region can move from the state in which it is in to a different metastable state. However, while in ordinary nucleation the driving force for nucleation is a gain in free energy in the bulk of the droplet (and the barrier that needs to be overcome it due to the surface tension of the droplet surface), here we have no gain in free energy when a domain changes from one metastable state to another, and thus unlike standard nucleation and growth phenomena there is no growth of a domain occurring after its transition. Thus one calls these droplets of the "mosaic state" sometimes "entropic droplets" (Xia and Wolynes 2001, Bouchaud and Biroli 2004), to avoid a too strong analogy with standard nucleation phenomena.

The task then is to estimate the size ξ of these droplets, and the barriers in free energy that need to be overcome to change their state. There are interesting speculations on this question in the literature (e.g. Kirkpatrick *et al.* 1989, Xia and Wolynes 2001, Bouchaud and Biroli 2004), but in our opinion a clear link to a more microscopic description (e.g. via model Hamiltonians) is still lacking, and the evidence from both simulation and experiment for this "mosaic theory" is, at best, indirect. As a further caveat, we mention that in the case of the 10-state Pott glass (described in Chap. 4) the mean-field scenario of the two subsequent transitions at T_c and T_0 was rather useless to understand the short-range version of this model, which had a transition (where relaxation times diverge to infinity according to Arrhenius laws, like in "strong glass-formers") at $T = 0$ only. If one assumes, however, that a static glass transition at some temperature T_0 exists where ξ diverges, $\xi \propto (T - T_0)^{-\nu}$, one can account for Vogel-Fulcher type laws. We recall that according to nucleation theory concepts, the barrier of a nucleation event where a spherical droplet of radius ξ is formed would be estimated from a decomposition of the droplet formation free energy $\Delta F_{\text{drop}}(\xi)$ in terms of volume and surface terms,

$$\Delta F_{\text{drop}}(\xi) = 4\pi\xi^2\Gamma - (4\pi/3)\xi^3\Delta f_{\text{bulk}}\,, \qquad (6.80)$$

where Γ is a surface free energy per unit area, and Δf_{bulk} the free energy gain (Kelton and Greer 2010). The free energy barrier against nucleation, ΔF^*_{drop}, occurs for ξ^* where $\partial(\Delta F_{\text{drop}}(\xi))/\partial\xi = 0$, yielding $\Delta F^* = (4\pi/3)\xi^{*2}\Gamma$, and then the time τ needed to form a nucleus of radius ξ^* is estimated from an Arrhenius relation, $\tau \propto \exp(\Delta F^*/k_BT)$.

Now Eq. (6.80) assumes compact spherical droplets and a scaling of the surface free energy (in d dimensions) with the surface area (ξ^{d-1}). Now in the sections on spin glasses and systems with random fields we have already seen, that in disordered systems interfaces are very rough, droplet shapes are very irregular (or even fractals), and surface free energies scale with a much smaller power θ of length ($\theta < d-1$). So the mosaic theory assumes that in a glass-forming system the free energy landscape for a domain of size ξ in the vicinity of the barrier where it can move from one of its locally ordered states to another one can be described as

$$\Delta F_{\text{domain}}(\xi) = \xi^\theta \Gamma - \Delta S \xi^d\,, \tag{6.81}$$

where geometric factors are disregarded, and Δf_{bulk} is replaced by an entropic term, ΔS, since it is entropy that drives the transitions between the many local minima in the free energy landscape, due to the different "ordered states" that the cooperative regions of size ξ can take. Note that Eq. (6.81) is assumed to hold only for ξ near ξ^*, but not for $\xi \gg \xi^*$. Now $\partial\Delta F_{\text{domain}}(\xi)/\partial\xi = 0$ yields $\xi^* \propto (\Gamma/\delta S)^{1/(d-\theta)}$ and $\Delta F^* \propto \Gamma\xi^{*\theta}$ and thus, assuming again an Arrhenius law for the relaxation time, we have

$$\log\tau \propto \Gamma(\Gamma/\Delta S)^{\theta/(d-\partial)}\,. \tag{6.82}$$

Assuming now that ΔS is related to the "complexity" of the mean-field model, one expects that $\Delta S \propto T - T_0 \to 0$ as $T \to T_0$. If $\theta/(d-\theta) = 1$, a choice for which there is not really a good theoretical justification, Eq. (6.82) gives the Vogel-Fulcher law but it is clear that also exponents $\theta/(d-\theta)$ different from unity would be compatible with experimental data. Finally we note that the mosaic theory makes quite a number of predictions regarding the correlations between various quantities of glass-forming systems (e.g. between height of jump in the specific heat and $T_c - T_K$, value of the stretching parameter β and fragility, etc.), some of which seem to be supported by experimental data (see Lubchenko and Wolynes 2007 for a review).

There exist several variations and extensions of this "mosaic picture" (e.g. Xia and Wolynes 2001, Bouchaud and Biroli 2004, and Leuzzi and Nieuwenhuizen 2008, and references therein). But so far there is still no microscopic theory that could serve as basis for this "mosaic picture". Thus, we will not discuss this "mosaic theory" further here, except stressing the implication that it predicts the existence of a static correlation length that should increase with decreasing temperature, for $T < T_c$. An interesting attempt to find such a static length was recently reported by Biroli *et al.* (2008), for a soft-sphere model of a glassforming fluid (Bernu *et al.* 1987), for which a very efficient Monte Carlo algorithm exists (Grigera and Parisi 2001) and thus allows to equilibrate the system below T_c (which is known in this model, see Roux *et al.* 1989). Biroli *et al.* (2008) apply the concept to look for this length via boundary effects, as described earlier in this section, but they consider a different quantity, namely the local overlap of configurations of particles inside a small box in the center of a sphere of radius R, freezing all particles outside of this sphere in the system (as done for the study of a binary LJ fluid confined between rough walls, cf. Fig. 6.12). The overlap of configurations is measured by dividing the box in many small cubic cells of linear dimension ℓ chosen such that the probability of finding more than one particle in a single cell is negligible. If r_i is the number of particles in cell i, the overlap is defined as

$$q_c(R) = \frac{1}{\ell^3 N_i} \sum_{i \in v} \langle n_i(t_0) n_i(t_0 + t) \rangle, \quad t \to \infty \tag{6.83}$$

where N_i is the number of cells (inside the central box) that is averaged over ($N_i = 125$ was chosen), and the brackets denote an average over 8 to 32 configurations of frozen particles outside the sphere. Choosing systems with N=2048 or N=16384 particles in total, the number of mobile particles inside the sphere could be varied from $M = 20$ to $M = 3200$. Subtracting the value q_0 that results for two completely uncorrelated configurations of particles in a sphere, one finds that $q_c(R) - q_0$ crosses over from an exponential decay as a function of R for $T \gg T_c$ to a compressed exponential

$$q_c(R) - q_0 \propto \exp[-(R/\xi)^\zeta], \quad \zeta > 1 \tag{6.84}$$

for $T < T_c$, but this crossover is rather smooth, contrary to what one finds for a one-dimensional Kac model (Franz and Montanari 2007). This work clearly suggests that the influence of boundary conditions propagates into the bulk on an increasingly large length scale upon cooling. However, as the authors point out, their results seem to imply that the effective interface tension Γ between neighboring domains does not have a sharp value but is distributed, thus imposing the presence of additional parameters. Thus while the results of Biroli *et al.* (2008) can be taken as a piece of evidence in favor of the concepts of the "mosaic theory", many open questions clearly remain.

6.3 Gels versus Glasses

In Section 4.3.3 we have already considered the sol-gel-transition for the case of "chemical gelation" in which molecular building blocks in a solution aggregate and form via an (irreversible) chemical reaction random clusters that are held together by chemical bonds. The concentration p of the bonds that are used to join together the clusters is a useful parameter to discuss these gelation processes, since then the problem can be mapped onto a percolation problem (Stauffer and Aharony 1992), see Sec. 3.2. For example, if p reaches a critical value p_c, the "percolation threshold", a cluster of infinite size appears. Then, for $p > p_c$, a finite nonzero fraction of the molecular constituents, which previously were contained in clusters that have a finite size, is part of the percolating cluster that has infinite size. This percolating cluster forms the "gel phase", while for $p < p_c$ where only finite clusters are present one speaks about the "sol phase". If the chemical reaction is stopped at some value of p (e.g., the reactions can be triggered by radiation that can be switched off, or by initiator molecules that are chemically deactivated or removed, etc.), one can consider the resulting system at a given value of p and the corresponding structure of random clusters (and/or a random percolation network) as a system in thermal equilibrium. For $p < p_c$ and small wavenumbers q the intermediate scattering functions $F(q,t)$ and $F_s(q,t)$ decay to zero as time $t \to \infty$, since in the sol phase the clusters are able to diffuse and thus the asymptotic decay for small q is dominated by a factor $\exp[-Dq^2t]$, D being an effective diffusion constant. This is not the case in the gel phase, however, since the gel can be considered as a kind of disordered solid phase, see Sec. 4.3.3. Thus, for $t \to \infty$ $F(q,t)/F(q,0)$ and $F_s(q,t)$ tend to some "frozen" parts, $f(q)$ and $f_s(q)$ respectively, similar to the ergodic-to nonergodic transitions discussed in Chap. 5. However, for p slightly above p_c, only the very large length scale structure of the gel is frozen and hence $f(q)$ and $f_s(q)$ are distinctly nonzero only for very small q. As p increases further one reaches a fairly compact amorphous structure, which for $p = p_{\max}$ is also frozen on small length scales, similar to a normal dense glass-forming system. (Note that in actual gelation processes the "chemical conversion" as p is then called (the number of covalent bonds between the mutually reactive groups relative to the total number of pairs of such groups) stops at a value $p_{\max} < 1$, even if the reaction time t_{react} tends to infinity, unlike the simplistic problem of bond percolation on a lattice, where one can increase p up to $p = 1$, the ideal perfect lattice.)

Thus we see that chemical gelation is a possible route to create nonergodic, i.e. glassy, structures, and this is, e.g., exploited for the production of thermoset resins (Pascault *et al.* 2002). Some features of the relaxation phenomena characteristic for glass-forming liquids can already be found at much smaller chemical conversions p, depending on the length scale that is studied. For example, Corezzi *et al.* (2006) have studied a reactive mixture of epoxy-amines by inelastic X-ray scattering (at wave vectors 1 $\text{nm}^{-1} \leq q \leq 15\ \text{nm}^{-1}$) and found an ergodic to nonergodic transition (compatible with a description in terms of mode coupling theory, as developed by Götze and Sjögren 1992) already for $\tilde{p}_c = 0.34$, while gelation occurs only at $p_c = 0.5$ and the whole network appears frozen at $\tilde{p}_g \approx 0.65$. As emphasized already in Sec. 4.3.3, for $p \to p_c$ the viscosity of the system diverges, i.e. for $p > p_c$ the system no longer flows and it has a solid-like response on macroscopic scales. However, on the scales studied in the scattering experiment by Correzi *et al.* (2006) no anomaly whatsoever could be detected near $p = p_c$. While the view has been advocated (Corezzi *et al.* 2006, Zaccarelli 2007) that in such reactive mixtures p can be considered as as control parameter to study glass transitions, similarly as one uses temperature and pressure as control parameters for the study of glassification in ordinary supercooled fluids, we caution the reader: For systems undergoing chemical gelation the fact that upon approaching the threshold p_c the correlation length of the corresponding percolation problem diverges (cf. Sec. 3.2.2) has the consequence that the length scale on which one studies glassification matters. Thus in our view, the quoted value $\tilde{p}_c$, $\tilde{p}_g$ have a meaning on the length scale of nm, but not on much larger length scales. Note that this is very different from the situation of normal glass-forming systems in which the liquid becomes nonergodic at the *same* tempeature at *all* length scales.

The above discussion of chemical gelation assumes that the lifetime of every bond that was formed is strictly infinite. However, as we have already mentioned in Sec. 3.6 and Chap. 5, even in network-forming glass melts such as molten SiO_2, GeO_2, B_2O_3 etc., the covalent bonds between the atoms that define the networks do not have an infinite lifetime, because infinite lifetime can only arise as a consequence of an infinite energy and the energy of such covalent bonds is large (in the range of several eV) but finite.

Now in experiments many aggregation processes are termed "physical gelation" or "reversible gelation" if the energy scale of bonds is not very high, a few tens of k_BT or even a few k_BT only.[13] So bonds can reversibly break and form again, many times during the course of experiments which are extended over macroscopic time-scales. One might expect that these physical gelation processes should have some similarity to the network-forming glass melts, if the latter are considered at very high temperature (T of the order of several 1000 K, see e.g. Horbach and Kob (1999, 2001) for simulations of SiO_2). However, it turn out that this analogy is not really very useful, because physical gels have very special features due to the supramolecular nature of the building blocks forming such gels (namely, bonds are being formed between associative polymers, colloidal systems, etc., see Zaccarelli (2007) for a more detailed discussion).

A very important aspect in gel-forming systems is the interplay between relaxation dynamics and the dynamics related to phase separation. In order to clarify ideas, it is useful to consider the extreme case of a simple lattice gas, i.e. a system in which lattice sites can either be occupied by a particle or stay empty, and an energy ε is gained if two neighboring sites are occupied. One can then ask under which conditions percolating clusters of particles connected by such nearest neighbor bonds occur (Fig. 6.14, Hayward *et al.* 1987). For $T/\varepsilon \to \infty$ this question just yields the standard problem of random site percolation (see section 3.2). For finite temperatures (but still in the one-phase region) the percolation concentration $c_p(T)$ bends over from its limit $c_p(T \to \infty)$ (≈ 0.312 for the simple cubic lattice) and hits the coexistence curve at $c_p(T_{\text{coex}}) \approx 0.23$ for the simple cubic lattice (Heermann and Stauffer 1981). The critical exponents α_p, β_p, γ_p, $\cdots$ (see Sec. 3.2) of this correlated percolation transition (Coniglio 1975, Coniglio *et al.* 1977) are the same as for random percolation, because the correlation length ξ describing the site occupancy correlations, $G(|\vec{r}_i - \vec{r}_j|) = \langle c_i c_j \rangle - \langle c_i \rangle \langle c_j \rangle \propto \exp(-|\vec{r}_i - \vec{r}_j|/\xi)$, stays finite along $c_p(T)$. As a result, in this correlation function no sign of any singularity is detected if $c_P(T)$ is crossed: This line of "transient percolation" does not have any significant physical consequences. Hayward *et al.* (1987) have shown that this line can be continued into the two-phase coexistence region, where it

[13] Recall that 1eV$\approx$11600k_BK, thus for room temperatures these energies are much smaller than the one stored in a typical covalent bond. This is the reason why these gels are soft and are often called "soft matter".

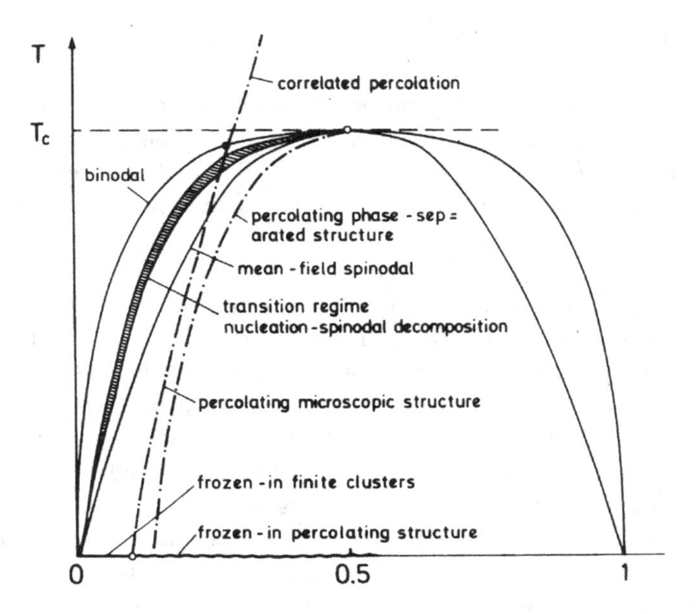

Fig. 6.14 Schematic phase diagram of a lattice gas system with short-range attractive interactions, indicating various regions in the plane of variables temperature T versus concentration c of occupied sites, as well as various lines of percolation transitions. The binodal (or coexistence curve) separates the one-phase region (outside of the binodal) from the region where in equilibrium macroscopic domains coexist. The mean-field spinodal as well as the actual transition region from nucleation to spinodal decomposition (see text) are indicated in the left side $c \leq 0.5$) of the phase diagram. The dash-dotted curves indicate percolation transitions (see text). From Hayward *et al.* (1987).

describes a percolation transition that can be observed as a time-dependent, transient phenomenon in "quenching experiments" in which one studies the kinetics of phase separation. The canonic "quenching experiment" for the study of phase separation kinetics (Binder and Fratzl 2001, Puri 2009) is a sudden decrease of temperature, ideally starting at $T \to \infty$ (i.e., random initial occupation of sites by the particles) and setting at time $t = 0$ the system to the desired temperature T (below the coexistence curve). The particles that diffuse on the lattice at conserved concentration c via hopping to neighboring empty sites cluster together, and if the quench started at a concentration c with $c_p(0) < c < c_p(T \to \infty)$, the system undergoes a percolation transition at some time t after the quench. Here $c_p(0)$ refers to quenches to $T = 0$ in which case the phase separation stops at some nonequilibrium structure of frozen-in finite clusters (for $c < c_p(0)$)

or with a frozen-in percolating structure (for $c > c_p(0)$), the situation resembling chemical gelation since no bond that was formed between two particles can be broken again by a process where one particle hops away if this process costs nonzero energy. The standard method (both in experiments and simulations, see e.g. Binder and Fratzl 2001) to study phase separation kinetics is, however, not through the connectivity properties of clusters that form. Rather one studies the time-dependence of the (equal-time) occupancy correlation function $G(|\vec{r}_i - \vec{r}_j|, t)$ or its Fourier transform, the equal-time structure factor $S(q,t)$. It turns out that this percolation transition does not lead to any observable anomalies in the structure factor $S(q,t)$ or in the real space correlations, however.

Nevertheless, the structure factor does develop a very nontrivial behavior, namely there occurs a peak at a characteristic inverse lengh scale $q_m(t) = 2\pi/\ell(t)$ where $\ell(t)$ is the length scale on which phase separation is essentially already complete at time t after the quench. One observes (Binder and Fratzl 2001, Puri 2009) that $\ell(t \to \infty)$ diverges ("coarsening behavior") and that typically there occur power-laws and scaling during the late stages of phase separation,

$$S(q,t) = [\ell(t)]^d \tilde{S}_\phi[q\ell(t)], \quad \ell(t) \propto t^a\,, \tag{6.85}$$

where d is the dimensionality of the system, $\tilde{S}_\phi$ a scaling function that depends on the volume fraction ϕ of the minority phase, and a is a (presumably universal, see Bray 1994) exponent ($a = 1/3$ for the above diffusive model).

Again one can ask the question whether the mesoscopically phase-separated structure described by Eq. (6.85) has an interconnected, percolating morphology, or whether separate droplets of the minority phase form on the background of the majority phase, like islands in the sea. This is another type of (correlated) percolation transition, which occurs at some critical volume fraction ϕ_c, and also is shown in Fig. 6.14.

Finally, we emphasize that these percolation transition lines in phase separation should not be confused with the concept of "spinodal curves" (or spinodals) (Binder 2009). The "spinodal curve" arises when one introduces a free energy $F(c)$ for the system for concentration c inside the coexistence curve, assuming that states with a globally homogeneous concentration can be defined there. In that case the spinodal curve acts as a limit of metastability: States in between coexistence curve and spinodal are metastable, i.e. they decay towards equilibrium, i.e. a macroscopically phase-separated state, by nucleation and growth (Zettlemoyer 1969, Kashchiev 2000, Binder 2009), whereas states inside the spinodal are unstable, i.e. long-wavelength fluctuations of the concentration grow with time spontaneously (one speaks about "spinodal decomposition", see Binder and Fratzl 2001 and Puri 2009). Thus the spinodal line can also be viewed as the curve in the phase diagram where the free energy barrier ΔF^* against homogeneous nucleation vanishes and the lifetime of metastable states becomes very small.

However, we remind the reader that spinodals are a mean-field concept (i.e. only well-defined for systems with long-range interactions, e.g. an Ising magnet in which every spin interacts with every other spin with the same energy, irrespective of the distance between the spins). In such mean-field systems, fluctuations are suppressed and thus the lifetime of metastable states is infinite right up to the spinodal. In systems with short-range interactions, spinodals cannot be uniquely defined, as a "Ginzburg criterion" for spinodals shows (Binder 1984), and anyway would not have a physical significance: If the nucleation barrier is only of the order of a few k_BT, many "critical clusters" can be formed more or less simultaneously throughout the system, and the sum of these many growing critical clusters resembles a wavepacket of growing concentration waves that also coarsen (Binder and Stauffer 1976). This region in the phase diagram, where the character of the initial stages of phase separation changes gradually from nucleation and growth to spinodal decomposition, is indicated by the shaded area in Fig. 6.14.

After this digression on the kinetics of phase separation, we recall the main point of this discussion, namely that for k_BT/ε of order unity the percolation transition of the particles has no consequences for most observable quantities, unlike chemical gelation, and such bonds that can reversibly form and break again normally give rise to phase separation, not to glass formation.

Of course, it is possible to have a glass transition in a phase-separating system by a completely independent mechanism, and this may give rise to phenomena which are very interesting in their own right. Consider, e.g., the phase diagram of polymer solutions, such as atactic polystyrene dissolved in cyclohexane (Hikmet *et al.* 1988), Fig. 6.15. As most polymers, melts of pure atactic polystyrene fail to crystallize and rather show a glass transition (at $T_g \approx 115°$C, for the molecular weight chosen by Hikmet *et al.* 1988). If the polymer is mixed with the solvent (cychohexane in this case), the relatively more mobile solvent molecules act as "plasticizer" on the polymer, i.e. $T_g(c)$ decreases if the polymer concentration c in the solution decreases. Since at the temperatures of interest the solvent "quality" for this polymer is not so good, the system tends to phase separate into a phase of almost pure solvent (the left branch of the coexistence curve is almost indistinguishable from the ordinate axis) and a polymer-rich phase. If one quenches the mixture to a temperature below $T_m \approx 50°C$, the temperature at which the coexistence curve intersects the glass transition curve $T_g(c)$, the polymer-rich phase that forms by spinodal decomposition (or by nucleation and growth of solvent fluid droplets) is at a concentration for which $T_g(c) > T$, and hence the phase separation process is "arrested" by glassification. The resulting gel structure is an irregular cellular network, with a characteristic length scale in the μm range formed in this thermoreversible gelation process. By varying the concentration and/or the speed of quenching one thus can vary the morphology of the gel in a wide range (see lower panels of Fig. 6.15).

Similar phenomena are exploited for the large-scale production of polymeric foam materials in chemical industry, mostly using supercritical CO_2 rather than cyclohexane as a solvent, since it is convenient to use also pressure rather than only temperature as a control parameter to produce microcellular polymer foams (Stafford *et al.* 1999, Krause *et al.* 2001, Cooper 2001). By a sudden reduction of pressure, the polymer plus solvent mixture is brought into the metastable or unstable region of the phase diagram, and as CO_2 vapor bubbles form, the remaining polymeric matrix glassifies. By such procedures it is possible to create bicontinuous nanoporous polymers (i.e., both the polymer matrix and the pore network form a percolating structure), as shown by Krause *et al.* (2001). In all these cases, the system is inhomogeneous on much larger scales than the length scales that one considers when one discusses the glass transition of bulk polymer melts (such as the size of cooperatively rearranging regions, etc.). Note that the phenomenon of a phase separation that is arrested by the nearby $T_g(c)$ line has more recently also been investigated for colloidal systems. Here the phase separation is induced by the addition of short polymers that lead to a depletion interaction (Lu *et al.* 2008).[14] In these systems it is possible to

[14]While polymer coils may overlap with small energy cost, polymers and colloids cannot overlap. Thus, each colloidal particle has a "forbidden zone" of width R_g (the redius of gyration of a polymer) around it, where no polymer can enter. However, when such "forbidden zones" of colloidal particles overlap, there is more room for the polymers left in the remaining volume of the solution, i.e. they have a larger entropy. The range of this attraction is of order R_g, while its strength can be controlled by the polymer concentration (Asakura and Oosawa 1954).

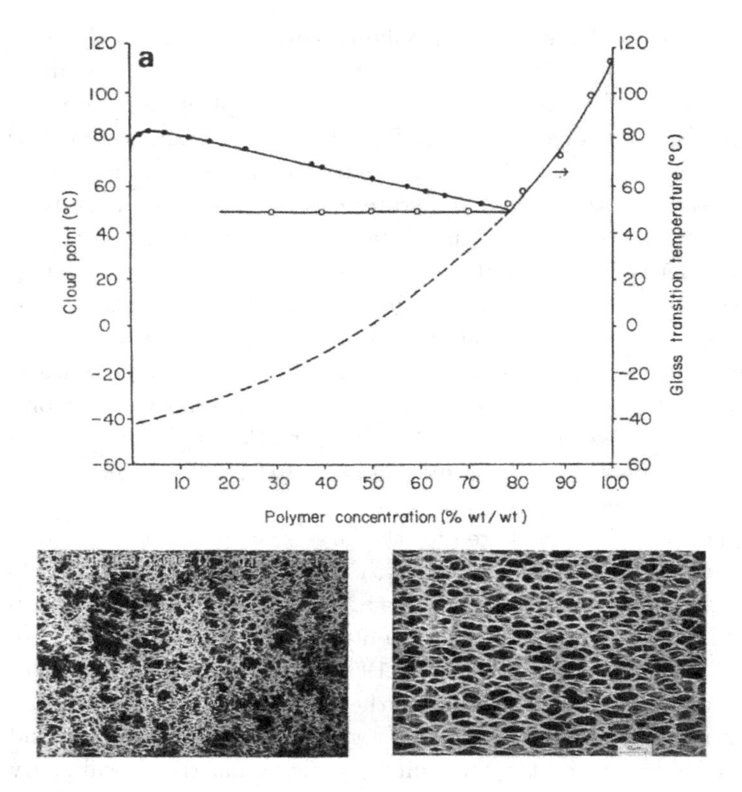

Fig. 6.15 Top panel: Phase diagram of the polymer atactic polystyrene (molecular weight $M_w = 2.75 \cdot 10^6$) dissolved in cyclohexane. This polymer solution has a critical point of unmixing at about $T_c = 83°$C and polymer concentration $c_{\text{crit}} = 4$ weight %. The polymer-rich part of the coexistence curve intersects the glass transition line $T_g(c)$ at about $T = 50°$C and $c \approx 78$ weight %. The broken curve is a tentative continuation of $T_g(c)$ into the two-phase coexistence region. If the system is quenched to a state below the horizontal line ($T_m = 50°$C), the system freezes into sponge-like polymeric foam structure as shown in the two lower panels. If the system is heated up above $T_m = 50°$C, the gel melts and phase separation (at state points underneath the coexistence curve) reaches completion. Lower panels: Scanning electron micrograph of a 2% gel sample (left panel) and a 16% gel sample (right panel). From Hikmet *et al.* (1988).

use confocal microscopy techniques to measure directly the positions of the colloids and hence to study the resulting percolating structure in a direct way.

In the phase diagram of Fig. 6.15, one can also ask where do polymer coils "percolate" in the sense that they strongly overlap. Of course, this concentration is to the left of the critical concentration, as in Fig. 6.14, because only if the polymer coils interact in an infinite percolating network, can a critical point of phase separation occur. Thus, Fig. 6.15 clearly illustrates that percolation of reversible interactions has nothing to do with glass formation, if the interaction strength and temperature are comparable.

However, if one could avoid the phase separation in Fig. 6.14 and consider a system where bonds of strength $\varepsilon \gg k_BT$ occur between constituent particles, the system would behave in many respects similar to a chemical gel, and nevertheless the concepts and methods of statistical mechanics would be fully applicable. There are several mechanisms by which the vapor-liquid type phase separation in Figs. 6.14 and 6.15 can be pushed to low concentrations/temperatures (Sciortino *et al.* 2005). One possibility is to add to the short-range attractive energy ε a longer range weak repulsive interaction. This case seems to be realized rather often in colloidal dispersions. For example, Campbell *et al.* (2005) have studied spherical colloidal particles with a radius around 0.8 μm in a mixture of cycloheptyl bromide and cis-decalin. As usual (Poon and Pusey 1995), the poly(methylmetacrylate) PMMA spheres are stabilized against aggregation by a thin (10 nm) polymer layer at their surfaces. These particles have then a long-range electrostatic repulsion (describable by a screened Coulomb potential, where the Debye screening length is of the same order as the particle size, in the 1 μm range). Now a short range attraction between these particles can be induced by adding a soluble polymer, polystyrene with a gyration radius R_g of about 0.1 μm, giving rise to a depletion interaction (Asakura and Oosawa 1954). In fact, for colloids where such Coulomb repulsions are absent, this depletion attraction is known to give rise to an Ising-type separation as in the case of the lattice gas model (Fig. 6.14) and in polymer solutions (Fig. 6.15); this phase separation is well documented both experimentally (Poon 2002) and by simulations (Vink *et al.* 2005). By confocal microscope imaging of the system one can show that at small volume fraction ϕ_c of colloidal particles the latter arrange themselves into clusters, which with increasing ϕ_c are not compact but more string-like, and form then at high enough ϕ_c an interconnected network. One estimates that in this case the interparticle attraction is about 9 k_BT (Campbell *et al.* 2005). Association of these clusters in the network then gives rise to dynamical arrest (in the sense that the lifetime of the percolating network exceeds the observation time, and notwithstanding the fact that at small scales the network can restructure itself). Fig. 6.16 reproduces some typical observations (Campbell *et al.* 2005). A corresponding simulation study

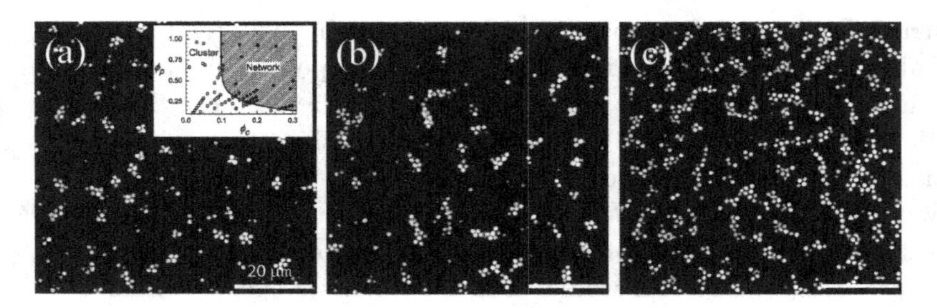

Fig. 6.16 Confocal microscopy images of colloid-polymer mixtures at different volume fractions. From left to right: $\phi_c = 0.080$ (a), 0.094 (b) and 0.156 (c). The attractive interactions are the same in all samples, of order $9k_BT$. Images (a) and (b) contain clusters while (c) is a two-dimensional cut through a three-dimensional network phase. The bars are 20 μm long. Inset shows phases observed as a function of the volume fractions of colloids (ϕ_c) and polymers (ϕ_p). From Campbell *et al.* (2005).

predicting such a behavior is due to Puertas *et al.* (2002, 2003) (see also Sciortion *et al.* 2004 and de Candia *et al.* 2005).

Another mechanism to stabilize a physical gel state consisting of a network of particles held together by bonds of finite strength ε but $\varepsilon \gg k_BT$ is the reduction of coordination number of the particles in the network. E.g., if a particle can be bound to only two neighbors, all what can happen is the formation of chains of particles, but no macroscopic phase separation can occur. If one allows only three-fold coordination (thus less than the 6 nearest neighbors of an Ising model on the simple cubic lattice and much less than that for off-lattice models for which the number of nearest neighbors would be distinctly higher), the region in the phase diagram in the (T,ϕ) plane where macroscopic phase separation occurs gets much reduced (Zaccarelli *et al.* 2005, 2006). Such a limited "valency", binding of a colloidal particle to only very few neighbors, can be realized with particles which can only bind at "sticky spots". For such "patchy particles" one finds very slow relaxation, attributable to physical gel formation, over a wide range of volume fraction ϕ at low temperatures, in computer simulations of suitable models (Zaccarelli *et al.* 2006, Russo *et al.* 2009), see Fig. 6.17.

Also for this system one can identify a transient percolation line, which characterizes whether in a "snapshot picture" of the configuration of the particles the colloids are arranged in finite clusters connected by "bonds" (determined in terms of a simple square well potential of a very short range Δ, $\Delta/(\sigma+\Delta) = 0.03$, $\sigma = 1$ being the hard core diameter of the particles, and the attraction strength $\varepsilon = 1$ fixes the scale of temperature) or form

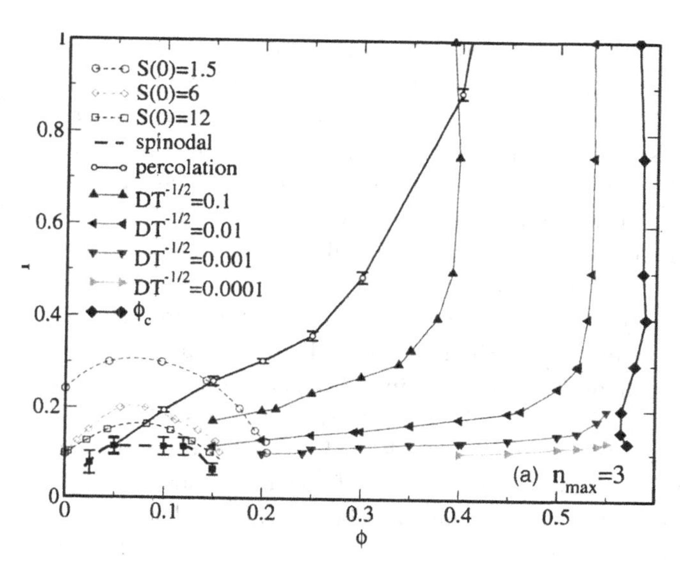

Fig. 6.17 Thermodynamic phase diagram in the plane of variables temperature T and volume fraction ϕ and various kinetic characterizations for a model where a colloidal particle may bind to at most $n_{\max} = 3$ neighbors. Broken curves $S(0) =$ const in the left part of the figure show contours of constant small angle scattering intensity (the spinodal shown is a tentative extrapolation of these data to $S(q \to \infty)$. Open circles connected by a full curve is the transient percolation transition (just as in Fig. 6.14, it has no significance for either static or time-dependent correlations in the system). Various triangles connected by thin full lines, $DT^{-1/2} =$ const, are isodiffusively lines (the selfdiffusion constant D is normalized by $\sqrt{T}$ to take into account the trivial contribution of the thermal velocity with T). The diamonds (right part) show estimates for the glass transition line $\phi_c(T)$ obtained from the assumption of a power-law vanishing, $D \propto (\phi - \phi_c(T))^{\gamma(T)}$. From Zaccarelli *et al.* (2006).

a percolating network. It is found that, as in Fig. 6.14, this percolation transition is irrelevant for the behavior of both static and time-dependent correlation functions of the particle positions. However, unlike the case shown in Fig. 6.14, the short range of the attraction relative to range of the repulsion, together with the small "valency" (coordination number) of at most $n_{\max} = 3$ neighbors has the effect that the phase coexistence region is very much reduced, as the analysis of the static structure factor $S(q \to 0)$ shows: Phase separation is expected only for small volume fractions, $\phi < 0.2$, and low T, while the standard colloidal glass transition as estimated from a MCT analysis (disregarding crystallization, of course) occurs near $\phi_c(T \gg 1) \approx 0.58$. At intermediate temperatures the line $\phi_c(T)$ moves to slightly higher ϕ but bends back to a minimum value $\phi_c(T \approx 0.15) \approx 0.57$.

Lines of constant diffusion constant D (normalized by $\sqrt{T}$) run more or less parallel to the abscissa for small ϕ but bend over to run parallel to $\phi_c(T)$ at larger ϕ. Thus, if one lowers T at $\phi = 0.2$ or $\phi = 0.3$, one finds clear signs of "dynamical arrest" as in dense glass-forming systems but the "localization length" (i.e. the length describing the range over which particles have moved in the region where the mean squared displacement $\langle \Delta r^2(t) \rangle$ of the particles exhibits its plateau, see Chap. 5) will be rather large, and increases with decreasing ϕ since this length will be related to the amplitude of the oscillation that a typical branch of the percolating structure will do, a scale that is normally much larger than the typical size of a particle. The glass transition of the model thus also distinctly depends on the length scale of observation, as for chemical gelation: On length scales smaller than the localization length the system resembles a fully ergodic system, and only on large scales one finds the signatures of dynamical arrest. Finally we also mention that at small and intermediate ϕ the $T-$dependence of the diffusion constant at low T is given by an Arrhenius-law. This result is related to the fact that in this $\phi - T$ regime the relaxation dynamics of the system is closely related to the breaking of a single bond, i.e. a well defined process with a single activation energy. This dependence is thus very different from the one found for the one in which T is constant (and one varies ϕ) and which is given by the power-law dependence proposed by MCT.

Regarding the simulations of such systems we mention that there are many variations to this theme: Models in which the lifetime of bonds is introduced as an ad-hoc parameter to interpolate between chemical and physical gelation (Del Gado *et al.* 2003), models for colloidal particles with directional interactions (Del Gado and Kob 2007, Saw *et al.* 2009), etc. However, we will not go into details here but refer the reader to the review of Zaccarelli (2007) on this suject.

In this section we have focussed on colloidal systems since on the one hand, because of the large size of colloidal particles, one can directly visualize the real-space structure of colloidal glasses (as long as the size of the particles is larger than around 1 μm), both in $d = 3$ (van Blaaderen and Wiltzius 1995) and $d = 2$ (König *et al.* 2005) dimensions, and can follow even the dynamics of these particle configurations in real time (König *et al.* 2005), and on the other hand because they are model systems with tunable interactions (Likos 2001). At low volume fractions, they can form both gels (as discussed above) and glasses (such as the "Wigner glass", proposed by Bosse and Wilke 1998, stabilized by strong electrostatic repulsions, and experimentally analyzed by Bonn *et al.* 1999a). However, not all colloidal glasses can be considered as "simple" systems since some of them have properties that are difficult to understand. For example in solutions containing laponite clay the distinction between gel and glassy behavior turns out to be rather subtle, see Bonn *et al.* (1999b) and Ruzicka *et al.* (2008) and also some observations, such as the "cluster fluids" of Lu *et al.* (2006) are difficult to rationalize.

At the end of this section, we emphasize that in soft matter systems there are in fact many examples for the formation of physical gels, and often very large relaxation times occur, systems fall out of equilibrium, qualitatively similar to the freezing of glass-forming fluids. One classic example are "associative polymers" in solution (Russo 1987, Semenov and Rubinstein 1998, Kumar and Douglas, 2001), i.e. heteropolymers which contain both soluble and insoluble monomers, the latter acting as "stickers" that form reversible junctions which in turn leads to the aggregation of chains. Depending on the fraction of stickers along a chain and the strength of these stickers, there is an interplay between gel formation and phase separation, qualitatively similar to the suspensions of sticky colloids. Other examples include the so-called "telechelic micelles" (Vlassopoulos *et al.* 1999, 2003, Michel *et al.* 2000) and "telechelic star polymers" (Lo Verso *et al.* 2006a,b, Lo Verso and Likos 2008). Micelles form when diblock copolymers $A_c B_{1-c}$, where a polymer of type A (chain length N_A) and of type B (chain length N_B; $c = N_A/(N_A + N_B)$) are covalently linked at one chain end and are dissolved in a selective solvent which is good for B but poor for A (we assume here $c \ll 1$). Then a number n_{AB} of such chains aggregate in a micelle, provided the concentration exceeds the "CMC" (critical micelle concentration), and then the insoluble A parts of the chain cluster together in a dense "core", while the soluble B-parts form a "corona", where each chain stretches away from the corona, in a self-avoiding walk like configuration, into the solvent. However, if there is another insoluble group at the free end of the B block this group acts as "sticker", similar to the "stickers" in the case of the associating polymers, and gel-like networks with a high functionality can form, where the cores of the micelles act as (bulky) junctions in the network.

In a micellar solution (at least under some circumstances) micelle formation is reversible, there is a "condensation/evaporation"-type equilibrium where chains are taken up by the micelle from the solution or get dissolved from the micelle into the solution, though this process typically has a free energy barrier of many k_BT. In contrast to this, f-arm star polymers are formed in irreversible chemical reactions, where f linear polymers are irreversibly bonded to a core. However, apart from the vicinity of the core, micelles and star polymers have a similar structure, if n_{AB} and f are of the same order. If these star polymers are "telechelic", i.e. have "sticky" endgroups, again large clusters or gels can be formed, and dynamical arrest is observed and can also be theoretically described, at least on an MCT level (Lo Verso *et al.* 2006b). While one has a good understanding of such systems at a mesoscopic level, on a more microscopic level these systems still pose many unsolved problems (e.g., the theoretical prediction of how many chains n_{AB} aggregate in a micelle is a problem, see Milchev *et al.* 2001 and Cavallo *et al.* 2006). Thus, these systems will not be considered further here.

6.4 Driven Systems

6.4.1 *Glasses under Shear*

The linear and nonlinear response of glassforming fluids and glasses to external pertubations such as shear is a topic of great importance and hence strong current activity. Due to this the subject has many facets and thus quite a few textbooks exist that describe the rheological behavior of glassy materials (Larson 1999, Coussot 2005). Thus here we do not attempt to give a comprehensive review of this rapidly evolving subject, but rather confine ourselves to the most basic salient features, and focus on some pertinent simulations (Varnik *et al.* 2003, 2004, Varnik 2006, Varnik and Henrich 2006) to provide instructive examples of the behavior which is quite complex even for very simple models.

Note that the investigation of rheological properties of glass-forming systems is not only important for applications, but also for obtaining a better understanding of the glassy dynamics. Take as an example a hard sphere system. In equilibrium the only relevant parameter is the density ϕ and not much can be said about the system at densities above ϕ_g, i.e. when it is in the glassy state. However, if we shear the system at a finite shear rate $\dot{\gamma}$, we can explore the properties of the fluid also at very high densities and, e.g., by varying $\dot{\gamma}$, investigate the structure etc. above ϕ_g (Berthier and Witten 2009).

Let us impose to a fluid a flow in the x-direction with a gradient in the y-direction, such that the velocity profile is given by $v_x(y) = \dot{\gamma}y$, $\dot{\gamma}$ being the shear rate. If linear response holds, i.e., for a so-called "Newtonian fluid", the stress tensor is then given by $\sigma_{xy} = \eta\dot{\gamma}$, where η is the zero-shear viscosity. Since in glassforming fluids the (zero shear) viscosity is extremely high when $T \to T_g$, the linearity of the response breaks down already for very small shear rates, and one finds a viscosity $\eta(\bar{\gamma})$ which decreases with increasing shear rate $\dot{\gamma}$; this effect is called shear thinning. Basically, the onset of shear thinning is controlled by the "Weissenberg number" $We = \dot{\gamma}\tau$, where τ is the structural relaxation time of the fluid (Larson 1999). As soon as $We > 1$, the response is nonlinear. This relation is plausible, since it implies that there is a competition between the structural relaxation time (of the quiescent system) and the timescale imposed by the external drive.

Qualitatively, the phenomenon of shear thinning is described by many different theoretical approaches, such as, e.g. the "soft glassy rheology" (SGR) model (Sollich 1998, Fielding *et al.* 2000), the driven p-spin-interaction spin glass (Berthier *et al.* 2000), and the non-equilibrium extensions of the mode coupling theory (MCT) of the glass transition (Fuchs and Cates 2002, Miyazaki *et al.* 2004). The latter approach has the distinctive advantage, that MCT makes many testable predictions on the relaxation behavior of glassforming fluids in equilibrium (see Chap. 5) and hence we will come back briefly to the predictions of this theory below. However, the main focus here will be on Molecular Dynamics (MD) simulations due to Varnik (2006) and Varnik *et al.* (2003, 2004) using the 80:20 binary (AB) Lennard-Jones (LJ) mixture which has before been extensively studied by Kob and Andersen (1994), (1995 a,b) in equilibrium.

Using the Lees-Edwards (1973) "sliding brick" boundary conditions and the so-called SLLOD algorithm (Evans and Morriss 1990), one can create the desired linear velocity profile in the finite $L \times L \times L$ simulation box, to simulate the flow of a macroscopic system (Varnik 2006). Figure 6.18 shows typical results for a system containing $N = 1200$ particles (see Chap. 5 for a description of this binary Lennard-Jones model and its properties in equilibrium), at a temperature $T = 0.45$ (in LJ units; remember $T_c = 0.435$ is the estimate for the critical temperature of MCT for this model), and various shear rates $\dot{\gamma}$. One sees that the pronounced two-step relaxation of the incoherent intermediate scattering function is more and more suppressed as $\dot{\gamma}$ increases, the relaxation of the fluid is sped up by the shear, i.e. particles can escape easier from their "cages". The effect of increasing $\dot{\gamma}$ is very similar to increasing T in equilibrium (see corresponding data in Chap. 5). There it was found for T near T_c the "time-temperature superposition" principle holds which suggests to test the present off-equilibrium

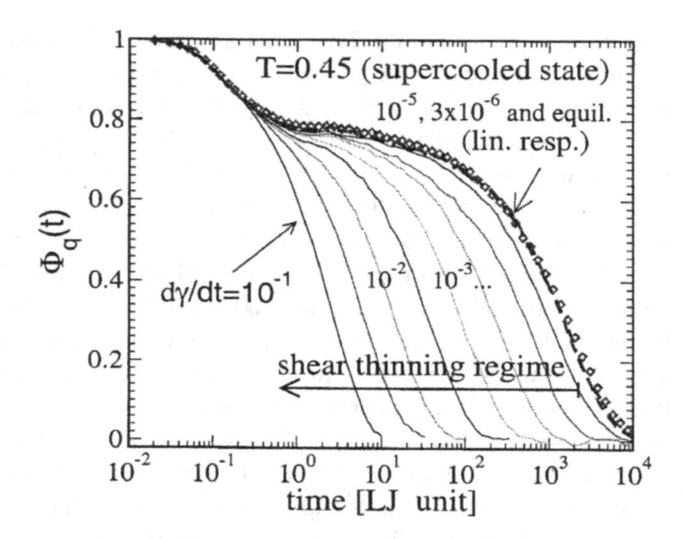

Fig. 6.18 Incoherent intermediate scattering function of the A particles in a binary 80:20 LJ mixture plotted vs. time (on a logarithmic scales), for a temperature $T = 0.45$ and choosing the wavenumber $q = 7.1$ where the static structure factor in equilibrium has its maximum. Shear rates $\dot{\gamma} = d\gamma/dt$ shown are 10^{-1}, 3×10^{-2}, 10^{-2}, $\cdots$, 3×10^{-6} (from left to right). From Varnik (2006).

data for a "time-shear superposition principle" (Fig. 6.19). One again finds that such a property holds for T slightly above T_c and small shear rates.

It is also of interest to measure the stress tensor in the simulation and define an apparent viscosity as $\eta_{\text{app}} = \sigma_{xy}/\dot{\gamma}$. For T slightly above T_c one finds that $\eta_{\text{app}} \propto \dot{\gamma}^{-1}$, i.e. there is an apparent divergence of the viscosity when one approaches the glassy state. For higher temperatures, however, the behavior $\eta_{\text{app}} \propto \dot{\gamma}^{-1}$ at small enough shear rates crosses over to a constant viscosity, independent of $\dot{\gamma}$, indicating that the regime of linear response has been reached.

While the shear thinning behavior caused by the interaction of the flow with the slow structural rearrangements in the supercooled fluid is compatible (at leat qualitatively) with all the theoretical concepts that were mentioned above, a difference is found when one studies the nonlinear response in the region of the frozen glassy state (Varnik *et al.* 2003, 2004, Varnik 2006). One finds that the system responds to the applied shear like an elastic solid, until a yield stress σ_y is exceeded (Fig. 6.20). Note that in this simulation a "Couette cell" geometry was used, in which a fluid is confined between two parallel plates, and the flow is created by moving

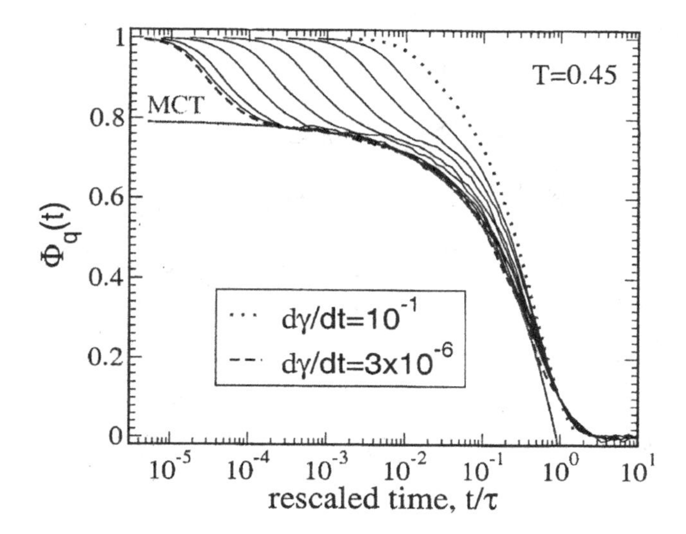

Fig. 6.19 A test of the time-shear superposition principle, using the data of Fig. 6.18, replotting them as function of a rescaled time t/τ, where τ was determined requiring $\Phi_q(t = \tau) = 0.1$. Curves follow for $\phi_q < 0.7$ a master curve for small $\dot{\gamma}$, while for larger $\dot{\gamma}$ systematic deviations occur. The thick solid line is a fit using the von Schweidler law, $\Phi_q(t) = f_q - A(t/\tau)^b$, with parameters $f_b = 0.79$, $b = 0.53$ and $A = 0.0922$. From Varnik (2006).

one wall against the other with a constant velocity U_{wall}. The walls are created by first simulating a large system in "equilibrium" with periodic boundary conditions in all three directions at $T = 0.2$, after a quench from a well-equilibrated system at $T = 0.5$, and then completely freezing the positions of all the particles outside the two parallel xy-planes at positions $z_{\text{wall}} = \pm L_z/2$. Of course, the properties of the glass prepared in this way somewhat depend on the "waiting time" t_w for which it was equilibrated (or rather annealed) at $T = 0.2$, e.g. $t_w = 4 \times 10^4$ MD time units ("aging phenomena" which can be observed when t_w is varied will be briefly mentioned in the next subsection). An interesting phenomenon that one finds when the yield stress is slightly exceeded is "shear banding" (Fig. 6.21). Thus only a certain part of the fluid is moving, showing a linear profile $u(z)$ from zero up to U_{wall}, and the remaining part of the fluid is still glassy. In the glassy part, $F_q(z, t)$ is not decaying to zero (on the timescale of the simulation), while in the fluid part $F_q(z, t)$ decays to zero, as in Fig. 6.18. When U_{wall} is increased enough, all the glass is "shear melted" and a linear velocity profile throughout the full Couette cell is established.

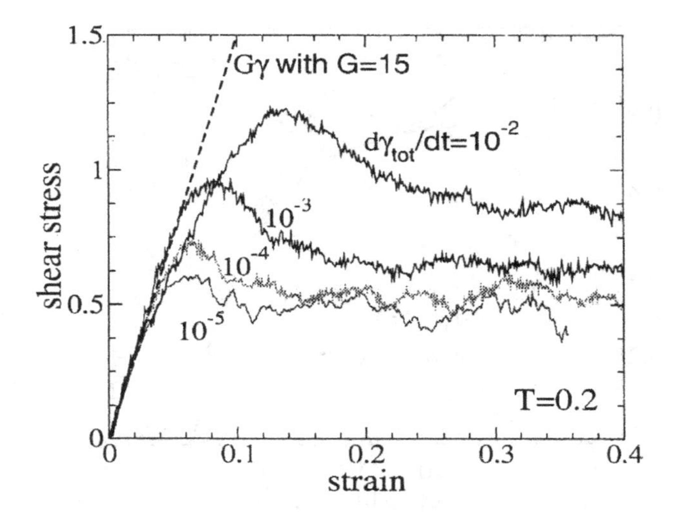

Fig. 6.20 Stress response vs. applied strain, $\gamma = \dot{\gamma}_{\rm tot} t$, for strain rates $U_{\rm wall}/L_z = \dot{\gamma}_{\rm tot} = 10^{-5}$, 10^{-4}, 10^{-3} and 10^{-2}, for the same system as in Figs. 6.18 and 6.19, but at $T = 0.2$ and under confinement between two parallel walls at $z = \pm L_z/2$, with $L_z = 40$, to simulate a Couette cell (linear dimensions $L_x = L_y = 10$, as in Figs. 6.18 and 6.19). From Varnik *et al.* (2004).

Figure 6.22 shows the resulting shear stress as a function of the imposed rate, $\dot{\gamma}_{\rm tot}$, under homogeneous flow conditions. Under imposed shear, if the corresponding steady state stress falls below the horizontal dotted line, a heterogeneous flow (i.e., "shear banding") can be expected, whereas for a stress above this threshold the flow will be homogeneous.[15] Note that the yield stress shown here is a lower bound (it increases with increasing t_w due to the "aging"), and so the boundary (vertical dotted line in Fig. 6.22) also slightly shifts to larger values of $d\gamma_{\rm tot}/dt$ (Varnik *et al.* 2004).

The existence of a yield stress can also be understood from an extension of MCT to nonequilibrium situations (Fuchs and Cates 2002). We recall (Chap. 5) that according to the "factorization theorem" in equilibrium (see Chap. 5 for a test of this theorem for the present model), close to T_c any

[15]In order to obtain the yield stress σ_y (shown in Fig. 6.22) accurately, Varnik *et al.* (2004) measured the maximum velocity, $U_{\rm max}$, which occurs in the simulation in the layer adjacent to the (not moving) left wall at imposed stress, increasing the stress slowly in steps of $d\sigma = 0.02$ once in 4000 LJ time units, and measuring $U_{\rm max}$ in between two subsequent stress increments. One finds that for applied stress below 0.6 this maximum velocity is very small (below 10^{-4}) but then increases rapidly by several orders of magnitude when 0.62 is reached. Once this flow sets in, a linear velocity profile is formed across the system.

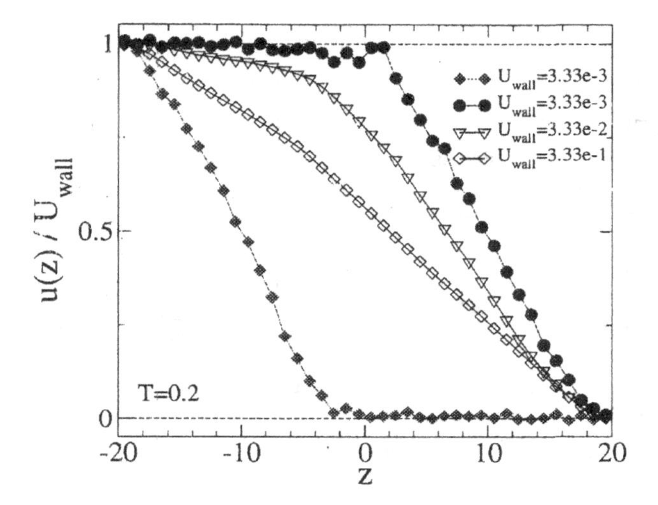

Fig. 6.21 Velocity profiles $u(z)$ normalized by the wall velocity U_{wall} for the binary LJ mixture in a Couette cell, at $T = 0.2$ for several choices of U_{wall}, as indicated in the figure. Note that these profiles do not possess any particular symmetry relative to the midplane $z = 0$. From Varnik *et al.* (2003).

time correlation function $\phi_q(t)$ can be written as $\phi_q(t) \approx f_q^c + h_q G(t)$. Here f_q^c is the non-ergodicity parameter (i.e. the height of the plateau) at the critical temperature where (in idealized MCT) glassy freezing sets in. In the presence of shear one finds that the function $G(t)$ obeys the following equation (Fuchs and Cates 2002):

$$\varepsilon - c_\gamma(\dot{\gamma}t)^2 + \lambda G^2(t) = \frac{d}{dt}\int_0^t dt' G(t-t')G(t') \,. \tag{6.86}$$

Here ε measures the distance from the transition, λ is the exponent parameter of MCT, and c_γ is a new parameter characterizing the effect of shear (Fuchs and Cates (2002) quote $c_\gamma \approx 3$ as a value appropriate for hard spheres). Equation (6.86) implies that for $\varepsilon < 0$, i.e. $T > T_c$, non-Newtonian flow behavior occurs, described by two timescales, the relaxation time τ in equilibrium and $\tau_{\dot{\gamma}} = \sqrt{(\lambda - 1/2)c_\gamma}/|\dot{\gamma}|$, which shows up in the long time decay of $G(t)$, $\ln G(t \to \infty) \to -t/\tau_{\dot{\gamma}}$. It is also found that when flow occurs, density fluctuations always decay, i.e. shear flow "interrupts aging". In the glassy region, i.e. $\varepsilon > 0$, a finite (nonzero) yield stress is predicted, and this prediction is (qualitatively) confirmed by the findings

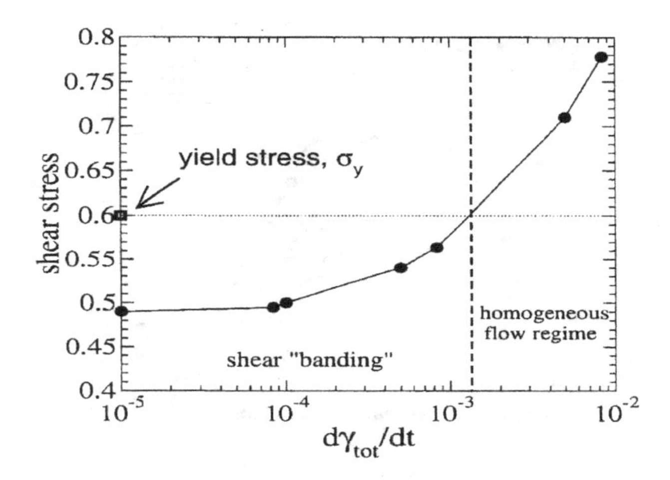

Fig. 6.22 Shear stress versus imposed shear rate, $d\gamma_{\rm tot}/dt$, under homogeneous flow conditions at $T = 0.2$ for the binary Lennard-Jones model in the same Couette cell as used in Figs. 6.20 and 6.21. For further explanations cf. text. From Varnik *et al.* (2004).

from MD simulations discussed above. The considerations based on the p-spin interaction model, however, do not predict a finite yield stress (Berthier *et al.* 2000). In contrast, Henrich *et al.* (2005) compared the temperature dependence of the (dynamic) yield strength predicted from MCT and from the simulations and found very good agreement. At T_c, this yield stress seems to exhibit a discontinuity (Varnik and Henrich 2006).

Of course, this is by no means the whole story: very interesting rheological behavior occurs in systems where gelation occurs, see e.g. Cates *et al.* (2004) and Laurati *et al.* (2009), but lack of space prevents us from giving a discussion of this rapidly evolving field. As has already been discussed in the previous subsection, gels and related complex fluids (foams, microemulsions, micellar aggregates, etc.) have a nontrivial structure, mesoscopic length scales, and this structure clearly is deformed and modified by shear flow, and hence the shear melting of frozen gels and foams can exhibit very rich features.

Another aspect that is out of consideration here is the response of glass-forming fluids and glasses to oscillatory shear (see, e.g., Crassous *et al.* 2008, Pham *et al.* 2006), and non-stationary relaxation processes after the sudden onset of a shear deformation (or when it is switched off) are also out of consideration. This topic has found much attention via experimental study of glassforming colloidal suspensions, MCT, and MD simulation (Zausch *et al.* 2008).

6.4.2 *Aging Dynamics of Glassy Systems*

If by lowering the temperature the glass transition is approached, the structural relaxation time τ increases up to macroscopocially large times, and cooling the system further, it falls out of equilibrium. In this subsection, we briefly discuss some of the many interesting features of the dynamics of the system if it is out of equilibrium.

Suppose we have suddenly cooled the system from some initial temperature T_0, at which the system was in equilibrium, to a temperature T below the glass transition, at which we wait for a waiting time t_w in order to equilibrate the system (or rather to anneal the system, since the equilibration time is too large). If we then study, e.g., the mean squared displacement of the particles, $g(t, t_w) = \langle[\vec{r}(t + t_w) - \vec{r}(t_w)]^2\rangle$, one finds that it depends on both times t and t_w, unlike equilibrium, where time-translation invariance (TTI) implies that $g(t, t_w)$ is independent of t_w. For example Varnik *et al.* (2004) showed that in the Kob-Andersen (1994) model at $T = 0.45$ their data converge indeed to a result that is independent of t_w if t_w is large, while for $T = 0.2$ no such convergence is observed. In this glassy state, one observes that $g(t, t_w)$ exhibits a plateau (which is interpreted as usual by the "caging effect"), but the lifetime of this "cage" increases systematically with t_w, irrespective how large values of t_w are tried, indicating that slow relaxation of this out-of-equilibrium glassy system is continuing. This behavior is termed "physical aging".

Such phenomena have most thoroughly been analyzed in the context of spin glasses, and we hence recall here the discussion on the extension of the fluctuation-dissipation-theorem (FDT) to out-of-equilibrium systems at the end of Sec. 4.5. The FDT related a response function $R(t, t')$, Eq. (4.248), to correlation functions $C(t, t')$, see Eq. (4.249). TTI would imply that both $R(t, t')$ and $C(t, t')$ depend on the difference in times, $t - t'$, only. Cugliandolo and Kurchan (1993) have suggested a generalization to out-of-equilibrium systems for large times t and t' (which neverthelesse must be much smaller than the times needed to reach full equilibrium). This so-called off-equilibrium fluctuation dissipation relation (OFDR) is, cf. Eq. (4.250),

$$k_B T R(t, t') = X\{C(t, t')\}\Theta(t - t')\partial C(t, t')/\partial t' \,. \tag{6.87}$$

where the function X, also calles "FDT-violation factor", is defined by this equation.

Defining then an integrated response

$$\chi(t, t_w) = \int_{t_w}^{t} R(t, t')dt' = (k_B T)^{-1} \int_{C(t,t_w)}^{1} du\, X(u), \tag{6.88}$$

where we have assumed a normalization such that $C(t, t' = t) = 1$, one finds that

$$k_B T\chi(t, t_w) = S[C(t, t_w)] \equiv \int_{C(t,t_w)}^{1} du\, X(u)\,. \tag{6.89}$$

If the system is in equilibrium, the FDT-violation factor $X = 1$, and one simply has

$$S(C) = 1 - C, k_B T\chi(t, t_w) = 1 - C(t, t_w) \tag{6.90}$$

In the out of equilibrium case, one can show for mean-field spin glass models that $C(t, t_w) = q$ in the limit $t \to \infty$, $t_w \to \infty$ (Cugliandolo and Kurchan 1993, Franz and Mézard 1994, Marinari *et al.* 1998), and hence

$$X[C(t, t_w)] \to x(q), \quad \text{where} \quad x(q) = \int_0^q dq' P(q')\,, \tag{6.91}$$

$P(q)$ being the distribution function of overlaps. Based on this result, Parisi *et al.* (1999) suggested the following classification (Crisanti and Ritort 2003): In models where no replica symmetry breaking occurs (e.g. in spin models such as randomly diluted ferromagnetic Ising models, or the random field Ising model [RFIM], etc.), $P(q)$ is a single delta function, $P(q) = \delta(q - q_{EA})$, with $0 < q_{EA} < 1$ in the low temperature phase. Then $S(C) = 1 - C$ for $C > q_{EA}$ and $S(C) = S(q_{EA})$ for $C < q_{EA}$. However, in the case of one-step replica symmetry breaking, where

$$P(q) = (1 - w)\delta(q) + w\delta(q - q_{EA}), \tag{6.92}$$

e.g. for the random energy model (Sec. 4.6.1), we have $S(C) = 1 - C$ for $q > q_{EA}$, while $S(C)$ for $q < q_{EA}$ is another straight line with a smaller slope. This type of behavior is also found for the binary Lennard-Jones mixture in the glassy state (Barrat and Kob 1999, Sciortino and Tartaglia 2001). For spin glass models with infinite number of replica symmetry breaking steps $S(C)$ for $q < q_{EA}$ is a nontrivial function of C. Thus we can conclude that *a priori* the out-of-equilibrium behavior can be used to characterize the type of glassy behavior. However, in reality this task is not that simple, since in practice one has to assure that the waiting times have indeed been chosen sufficiently large so that preasymptotic corrections can be excluded.

An important consequence of these results is that in the time region $t - t_w < t_w$ where $C > q_{EA}$ the FDT holds, $S(C) = 1 - C$: We have "quasi-equilibrium" behavior. The characteristic differences between types of glassforming systems arise thus only in the *aging regime*, defined by $t - t_w > t_w$ and $C < q_{EA}$. In the aging regime an effective temperature $\widehat{T}$ can be defined via (Cugliandolo *et al.* 1997, Crisanti and Ritort 2003)

$$\widehat{T} = T/X(C)\,. \tag{6.93}$$

For the models for which $S(C)$ is a constant for $C < q_{EA}$ this effective temperature becomes $\hat{T} = \infty$, however, while for the other types of models $\hat{T} > T$ takes nontrivial values. More details on this subject can be found in the review of Cugliandolo (2003).

For all these classes of models it is of interest to describe the processes that take place in this out-of-equilibrium behavior in more detail. Consider, e.g., a simple Ising model (with no disorder) which is quenched from $T_0 > T_c$ to $T < T_c$. One finds that after a short transient ordered domains have formed, of linear dimension $\ell(t)$, which grow with time according to a power-law, $\ell(t) \propto t^a$ with $a = 1/2$. The driving force for this domain growth is the minimization of interfacial free energy, and the growth law can be understood as the result of a random diffusion of interfaces (Bray 1994). When two interfaces locally meet, they annihilate each other.

While in such simple systems (without quenched disorder) the theory of aging behavior is rather well understood (Henkel *et al.* 2004), in disordered systems we still are far from a full understanding of all the rich features of aging behavior. While a large amount of experimental observations has been collected since a long time (Struik 1978), it has only recently been recognized that highly intermittent dynamics characterized by large fluctuations can in fact be found in real systems (Buisson *et al.* 2003) as well as in spin glass models (Sibani and Jensen 2005) which are related to the details of the OFDR discussed above. On the other hand, simple concepts such as "age-dependent relaxation times" are sometimes remarkably successful in accounting for experimental data (Lunkenheimer *et al.* 2005).

A concept vaguely related to the idea to describe the nonequilibrium character of a glass by the "effective temperature" $\widehat{T}$ is the idea of a "fictive temperature" T_{fict} already introduced by Tool (1946). This fictive temperature was defined as the temperature at which the glass would have been if the ordering behavior caused by varying the temperature would have continued below T_g in the same way as it proceeded above it. However, this concept needed to be abandoned since different observables for the same system yield somewhat different values for T_{fict} (see Leuzzi and Nieuwenhuizen 2008 for a more detailed discussion).

6.5 Concluding Remarks

In the present chapter, we have given a cursory review of some of the many attempts to develop a theory that describes the transition from the supercooled fluid to the glassy state, and we have tried to briefly summarize the concepts that "explain" how slow dynamics can emerge, without a significant change of static properties such as the structure factor. Since there is certainly no consensus in the literature which, if any, of these theories is the correct one, we have roughly followed the historical evolution, dealing with phenomenological descriptions such as the theory of Adam and Gibbs, the free volume theory, and the theory of Gibbs and DiMarzio in Sec. 6.1., and then we described the spin-facilitated kinetic Ising models (Sec. 6.1.3). Although these latter models lack a derivation from the microscopic Hamiltonian of an atomistically described fluid, they have the merit that they allow to show that slow relaxation with many features of real glassforming systems can result without the need for postulating any underlying static phase transition, unlike the theories mentioned above. In this context, we also mention evidence for the absence of static glass transitions in systems of polydisperse hard disks in $d = 2$ dimensions (Santen and Krauth 2000) or of binary hard disks (Donev *et al.* 2006). These models also exhibit many features of slow relaxation as other models for fluids near the glass transition do, and hence it is likely that at least some classes of glass-forming systems exist, for which the glass-like freezing in is a gradual process without underlying phase transition in the sense of thermodynamics. We also mention that the spin-facilitated Ising models have led to an interesting possibility for a theoretical description of dynamical heterogeneity (Sec. 6.2) and the associated (dynamical) correlation length that can be held responsible for the slow dynamics.

On the other hand, there are also good arguments in favor of the "random first-order phase transition" theory of Kirkpatrick *et al.* (1989), as described in more detail by Mézard and Parisi (2009) and Biroli and Bouchaud (2008). This theory suggests two successive phase transitions in the mean-field limit of infinite range forces: A dynamic ergodic-to-nonergodic transition at a higher temperature T_D, where an infinitely large number of metastable states with infinite lifetime appears (the complexity is then nonzero), and a static transition at a temperature $T_0 < T_D$, where the complexity vanishes and a static glass order parameter appears (which can be understood in terms of the response to the coupling of "real replicas", providing a formal description of what distinguishes a "frozen liquid" from a non-frozen liquid). Since the slowing down upon approaching T_D is described by mode coupling theory (MCT) (Chap. 5), which is very successful in describing the initial stages of slowing down in fragile glassforming liquids, there is indeed evidence also for the relevance of this scenario. Of course, real systems have finite range forces, so the sharp transition at T_D (or the T_c of MCT, respectively) is rounded off, metastable states for $T < T_c$ have finite lifetime, and the definition of "complexity" (if this concept is then useful at all) becomes fuzzy, and the description of the region for $T_0 < T < T_c$ in terms of the "mosaic theory" (Xia and Wolynes 2001, Biroli and Bouchaud 2004) becomes highly phenomenological, similar in many respects to the simplistic theory of Adam and Gibbs. As Biroli and Bouchaud (2008) put it, "Some of the outstanding questions that remain before we understand why glasses do not flow are the questions how relevant mean field ideas and models for real glasses are, and whether cooperativity is non-thermodynamical as in the spin-facilitated kinetic Ising models, or related to the metastable states of the mean-field models"; the crossover between the region described by MCT and the region where thermal activations prevail still is a challenge; a related issue is the character of the dynamical excitations, that effect these transitions, as well as the dynamical excitations that still occur in the glass state, etc.

At this point we have also to admit that we deliberately have ignored many other ideas and concepts that were also advocated in the literature: Tarjus and Kivelson (1995) and others have suggested that the glass transition results because an ordinary phase transition, that the interactions in the system could cause, is avoided because of "frustration" (like in spin glasses, where the onset of ferro-or antiferromagnetic order is suppressed if the exchange interactions are frustrated, see Chap. 4). This frustration in liquids is supposed to arise also from a competition between interactions of different range, and the resulting "frustration-limited domains" are then associated with the cooperative regions found near the glass transition.

However, this theory does not explain why one does not see any evidence (e.g., in static X-ray scattering, NMR, etc.) of short-range order associated with the character of the "avoided critical point". Also, if one tries to find simple explicit models that demonstrate this mechanism, e.g. an Ising model where short-range and long-range interactions compete (Viot and Tarjus 1998), one rather finds microphase separation and modulated order: Although simple ferro- or antiferromagnetic order is indeed avoided, the system does find more complex forms of standard long-range order. Only recent simulations on the hyperbolic plane (Sausset and Tarjus 2010) have given evidence that this frustration theory can indeed be realized for a particle system. However, to what extent these results are also relevant for real three dimensional systems remains open.

Other phenomenological models are referred to as "elastic models" (Dyre 2006) and start from the observation that on short times a supercooled liquid has an elastic response, like a solid, and they try to connect this high frequency response to the slow relaxation. Still other theories focus on the exploitation of Goldstein's (1969) energy landscape picture, see, e.g., Mauro *et al.* (2008) and refs. therein, or the modeling of activated barrier hopping, specifically in polymeric systems (Schweizer and Saltzman 2004, Saltzman and Schweizer 2007, Chen and Schweizer 2007). Since these topics are somewhat more specialized we did not attempt to discuss them here.

Last but not least, we have not considered the competition of glassification and crystal nucleation and time-temperature-transformation (TTT) curves that characterize cooling histories that avoid crystallization for particular glassforming fluids, see Gutzow and Schmeltzer (1995), Debenedetti (1997), Cavagna (2009), Kelton and Greer (2010) for corresponding discussions. Also the idea to describe driven glassforming fluids and granular matter on a unified level is out of our focus here.

Thus as summary we can conclude that the statistical mechanics of glass-forming systems has in the past been proven to be an extremely rich field. It can be expected that in the future this field will generate further interesting concepts that subsequently can be applied to other domains in science.

References

Adam, G., and Gibbs, J.H. (1965) *J. Chem. Phys.* **43**, 139.

Adler, J. (1991) *Physica* **A171**, 453.

Aichele, M., Gebremichael, Y., Starr, F.W., Baschnagel, J., and Glotzer, S.C. (2003) *J. Chem. Phys.* **119**, 5290.

Aizenman, M., and Lebowitz, J.L. (1988) *J. Phys.* **A21**, 3801.

Ajay, and Palmer, R.G. (1990) *J. Phys.* **A23**, 2139.

Alba, C., Busse, L.E., List, D.J., and Angell, C.A. (1990) *J. Chem. Phys.* **92**, 617.

Alcoutlabi, M., and McKenna, G.B. (2005) *J. Phys. Condens. Matter* **17**, R461.

Aldous, D., and Diaconis, P. (2002) *J. Stat. Phys.* **107**, 945.

Angell, C.A. (1997) *J. Res. NIST* **102**, 171.

Angell, C.A., and Borick, S. (2002) *J. Non-Cryst. Solids* **307-310**, 393.

Angell, C.A., Clarke, J.H.L., and Woodcock, L.V. (1981) *Adv. Chem. Phys.* **48**, 397.

Appignanesi, G.A., and Rodriguez Fris, J.A. (2009) *J. Phys.: Condens. Matter* **21**, 203103.

Appignanesi, G.A., Rodriguez Fris, J.A., Montani, R.A., and Kob, W. (2006) *Phys. Rev. Lett.* **96**, 057801.

Arndt, M., Stannarius, R., Grootheus, H., Hempel, E., and Kremer, F. (1997) *Phys. Rev. Lett.* **79**, 2077.

Asakura, S., and Oosawa, F. (1954) *J. Chem. Phys.* **22**, 1255.

Balogh, J., Bollobas, B., and Morris, R. (2009) *Ann. Prob.* **37**, 1329.

Barrat, J.-L., and Kob, W. (1999) *Europhys. Lett.*, **46**, 637.

Baschnagel, J., and Varnik, F. (2005) *J. Phys.: Condens. Matter* **17**, R851

Bässler, H. (1987) *Phys. Rev. Lett.* **58**, 767.

Bennemann, C., Donati, C., Baschnagel, J., and Glotzer, S.C. (1999) *Nature* **399**, 246.

Bernu, B., Hansen, J.P., Hiwatari, Y., Pastore, G. (1987) *Phys. Rev.* **A36**, 4891.

Berthier, L. (2003) *Phys. Rev. Lett.* **91**, 055701.

Berthier, L. (2004) *Phys. Rev.* **E69**, 020201.

Berthier, L., and Witten, T.A. (2009) *Phys. Rev.* **E80**, 021502.

Berthier, L., Barrat, J.-L. and Kurchan, J. (2000) *Phys. Rev.* **E61**, 5464.

Berthier, L., and Garrahan, J.P. (2003) *Phys. Rev.* **E68**, 041201.

Berthier, L., Biroli, G., Bouchaud, J.-P., Cipelleti, L., El Masri, L., L'Hôte, D., Ladieu, F., and Pierno, M. (2005a) *Science* **310**, 1797.

Berthier, L., Chandler, D., and Garrahan, J.P. (2005b) *Europhys. Lett.* **69**, 320.

Berthier, L., Biroli, G., Bouchard, J.-P., Kob, W. Myazaki, K., and Reichmann, D.R. (2007a) *J. Chem. Phys.* **126**, 184503.

Berthier, L., Biroli, G., Bouchard, J.-P., Kob, W. Myazaki, K., and Reichmann, D.R. (2007b) *J. Chem. Phys.* **126**, 184504.

Berthier, L., Biroli, G., Bouchaud, J.-P., Cipelletti, L., and van Saarloos W. (2010) (eds.) *Dynamical Heterogeneities in Supercooled Liquids, Colloids, and Granular Media* (Oxford University Press, Oxford).

Binder, K. (1981) *Z. Phys.* **B43**, 119.

Binder, K. (1983) in *Phase Transitions and Critical Phenomena, Vol 8* (C. Domb and L.J. Lebowitz, eds.) p. 1 (Academic, London).

Binder, K. (1984) *Phys. Rev.* **A29**, 341.

Binder, K. (1992) in *Computational Methods in Field Theory* (H. Gausterer and C.B. Lang, eds.) p. 59 (Springer, Berlin).

Binder, K. (2009) in *Kinetics of Phase Transitions* (S. Puri and V. Wadhawan, eds.) p. 63 (CRC Press, Boca Raton).

Binder, K., and Fratzl, P. (2001) in *Phase Transformations in Materials* (G. Kostorz, ed.) p. 409 (Wiley-VCH, Weinheim).

Binder, K., and Hohenberg, P.C. (1972) *Phys. Rev.* **B6**, 3461.

Binder, K., and Hohenberg, P.C. (1974) *Phys. Rev.* **B9**, 2194.

Binder, K., and Stauffer, D. (1976) *Adv. Phys.* **25**, 343.

Biroli, G., and Bouchaud, J.-P. (2004) *Europhys. Lett.* **67**, 21.

Biroli, G., Bouchaud, J.-P., Miyazaki, K., and Reichman, D.R. (2006) *Phys. Rev. Lett.* **97**, 195701.

Biroli, G., and Bouchaud, J.-P. (2007) *J. Phys.: Condens. Matter* **19**, 205101.

Biroli, G., and Bouchaud, J.-P. (2008) *Eur. Phys. J.* **B64**, 327.

Biroli, G., and Bouchaud, J.-P. (2009) arxiv: arXiv:0912.2542.

Biroli, G., Bouchaud, J.-P., Cavagna, A., Grigera, T.S., and Verrocchio, P. (2008) *Nature Physics* **4**, 771.

Boehmer, R., Hinze, G., Diezemann, G., Geil, B., Sillescu, H. (1996) *Europhys. Lett.* **36**, 55.

Böttcher, C.F.J., and Bordewijk, P. (1978) *Theory of Electric Polarization, Vol II* (Elsevier, Amsterdam).

Bonn, D., Tanaka, H., Wegdam, G., Kellay, H., and Meunier, J. (1999a) *Europhys. Lett.* **45**, 52.

Bonn, D., Kellay, H., Tanaka, H., Wegdam, G., and Meunier, J. (1999b) *Langmuir* **15**, 7534.

Bosse, J., and Wilke, S.D. (1998) *Phys. Rev. Lett.* **80**, 1260.

Bouchaud, J.-P., and Biroli, G. (2004) *J. Chem. Phys.* **121**, 7347.

Bouchaud, J.-P., Cugliandolo, L., Kurchan, J., and Mézard, M. (1996) *Physica* **A226**, 243.

Bouchaud, J.-P., Cugliandolo, L.F., Kurchan, J., and Mézard, M. (1998) in *Spin Glasses and Random Fields* (A.P. Young, ed.) p. 161, (World Scientific, Singapore).

Bray, A.J. (1994) *Adv. Phys.* **43**, 357.

Buhot, A., and Garrahan, J.P. (2001) *Phys. Rev.* **B64**, 021505.

Buission, I., Bellon, L., and Ciliberto, S. (2003) *J. Phys.: Condens. Matter* **15**, S1163.

Butler, S., and Harrowell, P. (1991) *J. Chem. Phys.* **95**, 4454.

Campbell, A.I., Anderson, V.I., van Duijneveldt, J.S., and Bartlett, P. (2005) *Phys. Rev. Lett.* **94**, 208301.

Cancrini, N., Martinelli, F., Roberto, C., and Toninelli, C. (2010) *Comm. Math. Phys.* **297**, 299.

Castellani, T., and Cavagna, A. (2005) *J. Stat.Mech.* P05012.

Cates, M.E., Fuchs, M., Kroy, K., Poon, W.C.K., and Puertas, A.M. (2004) *J. Phys.: Condens. Matter* **16**, S4861.

Cavagna, A. (2009) *Phys. Reps.* **476**, 51.

Cavallo, A, Müller, M., and Binder, K. (2006) *Macromolecules* **39**, 9539.

Chalupa, J., Leath, P.L., and Reich, G.R. (1979) *J. Phys. C: Solid State Phys.* **12**, L31.

Chandler, D., and Garrahan, J.P. (2010) *Ann. Rev. Phys. Chem.* **61**, 191.

Chang, I., and Sillescu, H. (1997) *J. Phys. Chem.* **B101**, 8794.

Chang, I., Fujara, F., Geil, B, Heuberger, G., Mangel, T., and Sillescu, H. (1994) *J. Non.-Cryst. Solids* **172-174**, 248.

Chaudhuri, P., Sastry, S., and Kob, W. (2008) *Phys. Rev. Lett.* **101**, 190601.

Chen, K., and Schweizer, K. (2007) *J. Chem. Phys.* **128**, 014904.

Chong, S.H., and Götze, W. (2000a) *Phys. Rev.* **E65**, 041503.

Chong, S.H., and Götze, W. (2000b) *Phys. Rev.* **E65**, 051201.

Chong, S.H., Moreno, A.J., Sciortino, F., and Kob, W. (2005) *Phys. Rev. Lett.* **94**, 215701.

Chong, S.H., and Kob, W. (2008) *Phys. Rev. Lett.* **102**, 025702.

Cicerone, M.T., and Ediger, M.D. (1995) *J. Chem. Phys.* **103**, 5684.

Cicerone, M.T., and Ediger, M.D. (1996) *J. Chem. Phys.* **104**, 7210.

Cicerone, M.T., Blackburn, F.R., and Ediger, M.D. (1995) *Macromolecules* **28**, 8224.

Cohen, M.H., and Grest, G.S. (1979) *Phys. Rev.* **B20**, 1077.

Cohen, M.H., and Grest, G.S. (1982) *Phys. Rev.* **B26**, 6313.

Cohen, M.H., and Turnbull, D. (1959) *J. Chem. Phys.* **31**, 1164.

Comez, L., Corezzi, S., Fioretto, D., Kriegs, H., Best, A., and Steffen, W. (2004) *Phys. Rev.* **B70**, 011504.

Coniglio, A. (1975) *J. Phys.* **A8**, 1773.

Coniglio, A., Peruggi, F., Nappi, C., and Russo, L. (1977) *J. Phys.* **A10**, 205.

Cooper, A.I. (2001) *Adv. Mater.* **13**, 1111.

Corezzi, S., Comez, L., Monaco, G., Verbeni, R., and Fioretto, D. (2006) *Phys. Rev. Lett.* **96**, 255702.

Coussot, P. (2005) *Rheometry of Pastes, Suspensions and Granular Materials* (Wiley-VCH, New York).

Crauste-Thibierge, C., Brun, C., Ladieu, F., L'Hôte, D., Biroli, G., and Bouchaud, J.-P. (2010) *Phys. Rev. Lett.* **104**, 165703.

Crassous, J.J., Siebenbürger, M., Ballauff, M.,Drechsler, M., Hajnal, D., Henrich, O., and Fuchs, M. (2008) *J. Chem. Phys.* **128**, 204902.

Crisanti, A., and Ritort, F. (2003) *J. Phys. A.: Math, Gen.* **36**, R181.

Cugliandolo, L.F. (2003) in *Lecture Notes for "Slow relaxations and nonequilibrium dynamics in condensed matter", Les Houches July,*

1–25, 2002; Les Houches Session LXXVII (J.-L. Barrat, M. Feigelman, J. Kurchan, and J. Dalibard, eds.) p. 367 (Springer, Berlin).

Cugliandolo, L.F., and Kurchan, J. (1993) *Phys. Rev. Lett.* **71**, 173.

Cugliandolo, L.F., Kurchan, J., and Peliti, L. (1997) *Phys. Rev.* **E55**, 3898.

Cummins, H.Z., Li, G., Hwang, Y.H., Shen, G.Q., Du, W.M., Herandez, J., and Tao, N.J. (1997) *Z. Phys. B: Condens. Matter* **103**, 501.

Dalle-Ferrier, C., Thibierge, C., Alba-Simionesco, C., Berthier, L., Biroli, G., Bouchaud, J.-P., Ladieu, F., L'Hôte, D., and Tarjus, G. (2007) *Phys. Rev.* **E76**, 041510.

Dasgupta, C., Indrani, A.V., Ramaswamy, S. and Phani, M.K. (1991) *Europhys. Lett.* **15**, 307.

Debenedetti, P.G. (1997) *Metastable Liquids* (Princeton University Press, Princeton).

de Candia, A., Del Gado, E., Fierro, A., Sator, N., and Coniglio, A. (2005) *Physica* **A358**, 239.

De Gregorio, P., Lawlor, A., Bradley, P., and Dawson, K.A. (2004) *Phys. Rev. Lett.* **93**, 025501.

Del Gado, E., and Kob, W. (2007) *Phys. Rev. Lett.* **98**, 028303.

Del Gado, E., Fierro, A., De Arcangelis, L., and Coniglio, A. (2003) *Europhys. Lett.* **63**, 1.

DiMarzio, E.A. (1997) *J. Res. Natl. Inst. Stand. Technol.* **102**, 135.

DiMarzio, E.A., and Gibbs, J.H. (1958) *J. Chem. Phys.* **28**, 807.

De Michele, C., and Leporini, D. (2001) *Phys. Rev.* **E63**, 036702.

Doliwa, B., and Heuer, A. (1998) *Phys. Rev. Lett.* **80**, 4915.

Donati, C., Douglas, J.F., Kob, W., Plimpton, S.W., Poole, P.H., and Glotzer, S.C. (1998) *Phys. Rev. Lett.* **80**, 2338.

Donati, C., Franz, S., Glotzer, S.C., and Parisi, G. (2002) *J. Non.-Cryst. Solids* **307**, 215.

Donati, C., Glotzer, S.C., and Poole, P.H. (1999) *Phys. Rev. Lett.* **82**, 5064.

Donev, A., Stillinger, F.H., and Torquato, S. (2006) *Phys. Rev. Lett.* **96**, 225502.

Donth, E., Huth, H., and Beiner, M. (2001) *J. Phys.: Condens. Matter* **13**, L451.

Doolittle, A.K. (1951) *J. Appl. Phys.* **22**, 1471.

Dyre, J. (2006) *Rev. Mod. Phys.* **78**, 953.

Eckmann, J.-P., and Ruelle, D. (1985) *Rev. Mod. Phys.* **57**, 617.

Ediger, M.D. (2000) *Ann. Rev. Phys. Chem.* **51**, 99.

Ehlich, D., and Sillescu, H. (1990) *Macromolecules* **23**, 1600.

Eisinger, S., and Jäckle, J. (1993) *J. Stat. Phys.* **73**, 643.

Ellison, C.J., and Torkelson, J.M. (2003) *Nat. Mater*, **2**, 695.

Erichsen, J., Kanzow, J., Schürmann, U., Dolgner, K., Schade, K.G., Strunskus, T., Zaporojtchenko, V., and Faupel, F. (2004) *Macromolecules* **37**, 1831.

Ernst, R.M., Nagel, S.R., and Grest, G.S. (1991) *Phys. Rev.* **B43**, 8070.

Ertel, W., Froböse, K., and Jäckle, J. (1988) *J. Chem. Phys.* **88**, 5027.

Evans, D.J., and Morris, G.P. (1990) *Statistical Mechanics of Nonequilibrium Liquids* (Academic Press, London, 1990)

Fielding, S.M., Sollich, P., and Cates, M.E. (2000) *J. Rheol.* **44**, 3233.

Forrest, J.A., and Dalnoki-Veress, K. (2001) *Adv. Colloid Interface Sci.* **94**, 167.

Fox, J.R., and Andersen, H.C. (1984) *J. Phys. Chem.* **88**, 4019.

Franz, S., Donati, C., Parisi, G., and Glotzer, S.C. (1999) *Philos. Mag.* **B79**, 1827.

Franz, S., and Mézard, M. (1994) *Europhys. Lett.* **26**, 209.

Franz, S., and Parisi, G. (2000) *J. Phys.: Condens. Matter*, 6335.

Franz, S., and Montanari, A. (2007) *J. Phys. A: Math. Theor.* **40**, F251.

Fredrickson, G.H., and Andersen, H.C. (1984) *Phys. Rev. Lett.* **53**, 1244.

Fredrickson, G.H., and Andersen, H.C. (1985) *J. Chem. Phys.* **83**, 5822.

Fredrickson, G.H., and Brawer, S.A. (1986) *J. Chem. Phys.* **84**, 3351.

Frick, B., Zorn, R., and Büttner, H. (eds.) (2000) *Proc. Workshop on Dynamics in Confinement J. Physique Coll. IV* **10**-3.

Frick, B., Koza, M., and Zorn, R. (eds.) (2003) *Proc. 2nd Int. Workshop on Dynamics in Confinement, Eur. Phys. J.* **E12** -5.

Fuchs, M., and Cates, M.E. (2002) *Phys. Rev. Lett.* **89**, 248304.

Fujara, F., Geil, B., Sillescu, H., and Fleischer, G. (1992) *Z. Phys.* **B88**, 195.

Garrahan, J.P., and Chandler, D. (2002) *Phys. Rev. Lett.* **89**, 035704.

Garrahan, J.P., and Chandler, D. (2003) *Proc. Natl. Acad. Sci. USA* **100**, 9710.

Garrahan, J.P., Jack, R.L., Lecomte, V., Pitard, E., van Duijvendijk, K., and van Wijland, F. (2007) *Phys. Rev. Lett.* **98**, 195702.

Garrahan, J.P., Jack, R.L., Lecomte, V. Pitard, E., van Duijvendijk, K., and van Wijland, F. (2009) *J. Phys. A: Math. Theor.* **42**, 075007.

Gaunt, D.S. (1967) *J. Chem. Phys.* **46**, 3237.

Gaunt, D.S., and Fisher, M.E. (1965) *J. Chem. Phys.* **43**, 2840.

Gebremichael, Y., Schrøder, T.V., Starr, F.W., and Glotzer, S.C. (2001) *Phys. Rev.* **E64**, 051503.

Gebremichael, Y., Vogel, M., and Glotzer, S.C. (2004) *J. Chem. Phys.* **120**, 061504.

Gibbs, J.H., and DiMarzio, E.A. (1958) *J. Chem. Phys.* **28**, 373.

Götze, W. (1991) in *Liquids, Freezing and the Glass Transition, Les Houches, Session LI, 1989* (J.-P. Hansen, D. Levesque and J. Zinn-Justin, eds.) p. 287 (North-Holland, Amsterdam).

Götze, W. (2009) *Complex Dynamics of Glass-Forming Liquids. A Mode Coupling Theory* (Oxford University Press, Oxford).

Götze, W., and Sjögren, J. (1992) *Rep. Progr. Phys.* **55**, 241.

Ghosh, S.S., and Dasgupta, C. (1996) *Phys. Rev. Lett.* **77**, 1310.

Glotzer, S.C. (2000) *J. Non-Cryst. Solids* **274**, 342.

Goldstein, M. (1969) *J. Chem. Phys.* **51**, 3728.

Graham, I.S., Piché, L., and Grant, M. (1997) *Phys. Rev.* **E55**, 2132.

Grigera, T.S., and Parisi, G. (2001) *Phys. Rev.* **E63**, 045102.

Gutzow, I., and Schmeltzer, J. (1995) *The Vitreous State: Thermodynamics, Structure, Rheology, and Crystallization* (Springer, Berlin).

Hayward, S., Heermann, D.W., and Binder, K. (1987) *J. Stat. Phys.* **49**, 1053.

Hedges, L.O., Jack, R.L., Garrahan, J.P. and Chandler, D. (2009) *Science* **323**, 5919.

Heermann, D.W., and Stauffer, D. (1981) *Z. Phys.* **44**, 339.

Henkel, M., Paessens, M., and Pleimling, M. (2004) *Phys. Rev.* **E69**, 056109.

Henrich, O., Varnik, F., and Fuchs, M. (2005) *J. Phys.: Condens. Matter* **17**, S3625.

Heuer, A. (2008) *J. Phys.: Condens. Matter* **20**, 373101.

Hikmet, R.M., Callister, S., and Keller, A. (1988) *Polymer* **29**, 1378.

Hohenberg, P.C., and Halperin, B.I. (1977) *Rev. Mod. Phys.* **49**, 435.

Horbach, J., and Kob, W. (1999) *Phys. Rev.* **B60**, 3169.

Horbach, J., and Kob, W. (2001) *Phys. Rev.* **E64**, 041503.

Hurley, M.M., and Harrowell, P. (1995) *Phys. Rev.* **E52**, 1694.

Jack, R.L., Garrahan, J.P., and Chandler, D. (2006) *J. Chem. Phys.* **125**, 184509.

Jäckle, J., and Eisinger, S. (1991) *Z. Phys.*, **B84**, 115.

Jäckle, J., Froböse, K., and Knödler, D. (1991) *J. Stat. Phys.* **63**, 249.

Jäckle, J. (2002) *J. Phys.: Condens. Matter* **14**, 1423.

Johari, G.P. (2000) *J. Chem. Phys.* **112**, 8958.

Jose, P.P., Chakrabarti, D., and Bagchi, B. (2006) *Phys. Rev.* **E73**, 031705.

Jones, R.A.L. (1999) *Curr. Opin. Colloid Interface Sci.* **4**, 153.

Jung, Y., Garrahan, J.P., and Chandler, D. (2004) *Phys. Rev.* **E69**, 061205.

Kämmerer, S., Kob, W., and Schilling, R. (1997) *Phys. Rev.* **56**, 5450.

Kämmerer, S., Kob, W., and Schilling, R. (1998a) *Phys. Rev.* **58**, 2131.

Kämmerer, S., Kob, W., and Schilling, R. (1998b) *Phys. Rev.* **58**, 2141.

Karmakar, S., Dasgupta, C., Sastry, S. (2009) *Proc. Natl. Acad. Sci. USA* **106**, 3675.

Kashchiev, D. (2000) *Nucleation: Basic Theory with Applications* (Butterworth-Heinemann, Oxford).

Kegel, W.K., and van Blaaderen, A. (2000) *Science* **287**, 290.

Kelton, K.F., and Greer, A.L. (2010) *Nucleation in Condensed Matter* (Elsevier, Amsterdam).

König, H., Hund, R., Zahn, K., and Maret, G. (2005) *Eur. Phys. J.* **E18**, 287.

Kirkpatrick, T.R., and Wolynes, P.G. (1987a) *Phys. Rev.* **A35**, 3072.

Kirkpatrick, T.R., and Wolynes, P.G. (1987b) *Phys. Rev.* **A36**, 8552.

Kirkpatrick, T.R., and Thirumalai, D. (1988a) *Phys. Rev.* **B37**, 5342.

Kirkpatrick, T.R., and Thirumalai, D. (1988b) *Phys. Rev.* **B37**, 4439.

Kirkpatrick, T.R., Thirumalai, D., and Wolynes, P.G. (1989) *Phys. Rev.* **A40**, 1045.

Kirkwood, G. (1950) *J. Chem. Phys.* **18**, 380.

Kob, W., and Andersen, H.C. (1993) *Phys. Rev.* **E48**, 4364.

Kob, W., and Andersen, H.C. (1994) *Phys. Rev. Lett.* **73**, 1376.

Kob, W., and Andersen, H.C. (1995a) *Phys. Rev.* **E51**, 4626.

Kob, W., and Andersen, H.C. (1995b) *Phys. Rev.* **E52**, 4134.

Kob, W., Donati, C., Plimpton, S.J., Glotzer, S.C., and Poole, P.H. (1997) *Phys. Rev. Lett.* **79**, 827.

Koza, M., Frick, B., and Zorn, R. (eds.) (2007) *3rd Internat. Workshop on Dynamics in Confinement, Eur. Phys. J. Special Topics* **141**.

Krause, B., Sijbesma, H.J.P., Münüklü, P., van der Vegt, N.F.A., and Wessling, M. (2001) *Macromolecules* **34**, 8792.

Kumar, S.K., and Douglas, J.F. (2001) *Phys. Rev. Lett.* **87**, 188301.

Kumar, S.K., Szamel, G., and Douglas, J.F. (2006) *J. Chem. Phys.* **124**, 214501.

Lacevic, N., Starr, F.W., Schrøder, T.B., and Glotzer, S.C. (2003) *J. Chem. Phys.* **119**, 7372.

Larson, R.G. (1999) *The Structure and Rheology of Complex Fluids* (Oxford University Press, New York).

Laurati, M., Petekidis, G., Koumakis, N., Cardinaux, F., Schofield, A.B., Brader, J.M., Fuchs, M., and Egelhaaf, S.U. (2009) *J. Chem. Phys.* **130**, 134907.

Lees, A.W. and Edwards, S.F. (1972) *J. Phys. C: Solid State Phys.* **5**, 1921.

Léonard, S., Mayer, P., Sollich, P., Berthier, L., and Garrahan J.P. (2007) *J. Stat. Mech. Theor. Exp.* P07017.

Leutheusser, E., and De Raedt, H. (1986) *Sol. State. Com.* **57**, 457.

Leuzzi, L., and Nieuwenhuizen, T.M. (2008) *Thermodynamics of the Glassy State* (Taylor & Francis, Boca Raton).

Likos, C.N. (2001) *Phys. Rep.* **348**, 267.

Lombardo, T.G., Debenedetti, P.G., and Stillinger, F.H. (2006) *J. Chem. Phys.* **125**, 174507.

Lo Verso, F., and Likos, C.N. (2008) *Polymer* **49**, 1425.

Lo Verso, F., Likos, C.N., Mayer, C., and Löwen, H. (2006a) *Phys. Rev. Lett.* **96**, 187808.

Lo Verso, F., Likos, C.N., Mayer, C., and Reatto, L. (2006b) *Mol. Phys.* **104**, 3523.

Lu, P.J., Conrad, J.C., Wyss, H.M., Schofield, A.B., and Weitz, D.A. (2006) *Phys. Rev. Lett.* **96**, 028106.

Lu, P.J., Zaccarelli, E., Ciulla, F., Schofield, A.B., Sciortino, F., and Weitz, D.A. (2008) *Nature*, **453**, 499.

Lubchenko V., and Wolynes, P.G. (2007) *Ann. Rev. Phys. Chem.* 58, 235.

Lunkenheimer, P., Wehn, R., Schneider, U., and Loidl, A. (2005) *Phys. Rev. Lett.* **95**, 055702.

Magill, J.H. (1967) *J. Chem. Phys.* **47**, 2802.

Marinari, E., Parisi, G., Ricci-Tersenghi, F., and Ruiz-Lorenzo, J.J. (1998) *J. Phys. A: Math. Gen.* **31**, 2611.

Mauch, F., and Jäckle, J. (1999) *Physica* **A262**, 98.

Mauro, J.C., Loucks, R.J., Varshneya, A.K., and Gupta, P.K. (2008) *Sci. Model Simul.* **15**, 241.

Merolle, M., Garrahan, J.P., and Chandler, D. (2005) *Proc. Natl. Acad. Sci. USA* **102**, 10837.

Mézard, M., and Parisi, G. (2009) arxiv: 0910.2838.

Michel, E., Filali, M., Aznar, R., Porte, G., and Appell, J. (2000) *Langmuir* **16**, 8702.

Milchev, A. (1983) *C.R. Acad. Bulg. Sci* **36**, 1415.

Milchev, A., Bhattacharya, A., and Binder, K. (2001) *Macromolecules* **34**, 1881.

Miyazaki, K., Reichman, D.R., and Yamamoto, R. (2004) *Phys. Rev.* **E70**, 011501.

Monthus, C., and Bouchaud, J.-P. (1996) *J. Phys. A: Math. Gen* **29**, 3847.

Moreno, A.J., Chong, S.H., Kob, W., and Sciortino, F. (2005) *J. Chem. Phys.* **123**, 204505.

Muranaka, T., and Hiwatari, Y. (1995) *Phys. Rev.* **E51**, R2735.

Nakanishi, H., and Takano, H. (1986) *Phys. Lett.* **A115**, 117.

Paluch, M., Casalini, R., and Roland, C.M. (2003) *Phys. Rev.* **E67**, 021508.

Parisi, G., Ricci-Tersenghi, F., and Ruiz-Lorenzo, J.J. (1999) *Eur. Phys. J.* **B11**, 317.

Pascault, J.P., Sauterau, H., Verdu, J., and Williams, R.J.J. (2002) *Thermosetting Polymers* (M. Dekker, New York).

Perera, D.N., and Harrowell, P. (1996) *Phys. Rev.* **E54**, 1652.

Perera, D.N., and Harrowell, P. (1999) *J. Chem. Phys.* **111**, 5441.

Peter, S., Meyer, H., and Baschnagel, J. (2006) *J. Polym. Sci., Part B, Polym. Phys.* **44**, 2951.

Pham, K.N., Petekidis, G., Vlassopoulos, D., Egelhaaf, S.U., Pusey, P., and Poon, W.C.K. (2006) *Europhys. Lett.* **75**, 624.

Pitts, S.J., and Andersen, H.C. (2001) *J. Chem. Phys.* **114**, 1101.

Pitts, S.J., Young, T., and Andersen, H.C. (2000) *J. Chem. Phys.* **113**, 8671.

Poon, W.C.K. (2002) *J. Phys.: Condens. Matter* **14**, R859.

Poon, W.C.K., and Pusey, P.N. (1995) in *Observation, Prediction and Simulation of Phase Transitions in Complex Fluids* (M. Baus, L.F. Rull, and J.P. Ryckaert, eds.) p. 3 (Kluwer Acad. Publ., Dordrecht).

Privmann, V.P. (1990) (ed.) *Finite Size Scaling and Numerical Simulation of Statistical Systems* (World Scientific, Singapore).

Puertas, A.M., Fuchs, M., and Cates, M.E. (2002) *Phys. Rev. Lett.* **88**, 098301.

Puertas, A.M., Fuchs, M., and Cates, M.E. (2003) *Phys. Rev.* **E67**, 031406.

Puri, S. (2009) in *Kinetics of Phase Transitions* (S. Puri and V. Wadhawan, eds.) p. 1 (CRC Press, Boca Raton).

Ray, P., and Binder, K. (1994) *Europhys. Lett.* **27**, 53.

Reiter, J. (1991) *J. Chem. Phys.* **95**, 544.

Reiter, J., Mauch, F., and Jäckle, J. (1992) *Physica* **A184**, 458.

Ricci, A., Nielaba, P., Sengupta, S., and Binder, K. (2007) *Phys. Rev.* **E75**, 011405.

Richert, R. (1994) *J. Non-Cryst. Solids* **172-174**, 209.

Richert, R. (2002) *J. Phys.: Condens. Matter* **14**, R 703.

Richert, R., and Angell, C.A. (1998) *J. Chem. Phys.* **108**, 9016.

Richert, R., and Bässler, H. (1990) *J. Phys.: Condens. Matter* **2**, 2273.

Ritort, F., and Sollich, P. (2003) *Adv. Phys.* **52**, 219.

Ritort, F., and Sollich, P. (2002) *Proceedings of conference "Glassy Dynamics in kinetically constrained models* Barcelona 2001, *J. Phys.: Condens. Matter* **14**, 1381.

Roland, C.M., Capaccioli, S., Lucchesi, M., and Casalini, R. (2004) *J. Chem. Phys.* **120**, 10640.

Roux, J.-N., Barrat, J.-L. and Hansen, J.-P. (1989) *J. Phys.: Condens. Matter* **17**, 7171.

Russo, J., Tartaglia, P., and Sciortino, F. (2009) *J. Chem. Phys.* **131**, 014504.

Russo, P.S. (1987) (ed.) *Reversible Polymeric Gels and Related Systems*, ACS Symposium Series 350 (American Chemical Society, Washington).

Ruzicka, B., Zulian, L., Angelini, R., Sztucki, M., Moussaid, A., and Ruocco, G. (2008) *Phys. Rev.* **E77**, 020402.

Saika-Voivod, I., Sciortino, F., and Poole, P.H. (2004) *Phys. Rev.* **E69**, 041503.

Saltzman, E.J., and Schweizer, K.S. (2007) *J. Phys.: Condens. Matter* **19**, 205129.

Santen, L., and Krauth, W. (2000) *Nature* **405**, 550.

Sausset, F., Tarjus, G., and Viot P. (2008) *Phys. Rev. Lett.* **101**, 155701.

Sausset, F., and Tarjus, G. (2010) *Phys. Rev. Lett.* **104**, 065701.

Saw, S., Ellegaard, N.L., Kob, W., and Sastry, S. (2009) *Phys. Rev. Lett.* **103**, 248305.

Scheidler, P., Kob, W., and Binder, K. (2000) *Europhys. Lett.* **52**, 277.

Scheidler, P., Kob, W., and Binder, K. (2002a) *Europhys. Lett.* **59**, 701.

Scheidler, P., Kob, W., Binder, K., and Parisi, G. (2002b) *Phil. Mag.* B**82**, 283.

Scheidler, P., Kob, W., and Binder, K. (2004) *J. Phys. Chem.* B **108**, 6673.

Schmidt-Rohr, K., and Spiess, H.W. (1991) *Phys. Rev. Lett.* **66**, 3020

Schneider, U., Lunkenheimer, P., Brand, R., and Loidl, A. (1999) *Phys. Rev.* **E 59**, 6924.

Schweizer, K.S., and Saltzmann, E.J. (2004) *J. Chem. Phys.* **121**, 1984.

Sciortino, F., and Tartaglia, P. (2001) *Phys. Rev. Lett.* **86**, 107.

Sciortino, F., Mossa, S., Zaccarelli, E., and Tartaglia, P. (2004) *Phys. Rev. Lett.* **93**, 055701.

Sciortino, F., Buldyrev, S.V., De Michele, C., Foffi, G., Ghofranhia, N., La Nave, E., Moreno, A., Mossa, S., Saika-Voivod, I., Tartaglia, P., and Zaccarelli, E. (2005) *Comp. Phys. Comm* **169**, 166.

Sellitto, M. (2002) *J. Phys.: Condens. Matter* **14**, 1455.

Semenov, A.N., and Rubinstein, M. (1998) *Macromolecules* **31**, 1373.

Sibani, P., and Jensen, H.J. (2005) *Europhys. Lett.* **69**, 563.

Sillescu, H. (1999) *J. Non-Cryst. Solids* **243**, 81.

Sollich, P. (1998) *Phys. Rev.* **E58**, 738.

Sollich, P., and Evans, M.R. (1999) *Phys. Rev. Lett.* **83**, 3238.

Sollich, P., and Evans, M.R. (2003) *Phys. Rev. Lett.* **68**, 031504.

Stafford, C.M., Russell, T.P., and McCarthy, T.J. (1999) *Macromolecules* **32**, 7610.

Stauffer, D., and Aharony, A. (1992) *Introduction to Percolation Theory* (Taylor and Francis, London).

Stein, R.S.L, and Andersen, H.C. (2008) *Phys. Rev. Lett.* **101**, 267802.

Stickel, F., Fischer, E.W., and Richert, R. (1995) *J. Chem. Phys.* **102**, 6251.

Stillinger, F.H. (1988) *J. Chem. Phys.* **88**, 7818.

Struik, L.C. (1978) *Physical Aging in Amorphous Polymers and Other Materials* (Elsevier, Amsterdam).

Tanaka, H. (2003) *Phys. Rev. Lett.* **90**, 055701.

Tarjus, G., and Kivelson, D. (1995) *J. Chem. Phys.* **103**, 3071.

Tarjus, G., Kivelson, D., and Viot, P. (2000) *J. Phys.: Condens. Matter* **12**, 6497.

Tölle, A. (2001) *Rep. Progr. Phys.* **64**, 1473.

Toninelli, C., Biroli, G., and Fisher, D.S. (2004) *Phys. Rev. Lett.* **92**, 185504.

Toninelli, C., Biroli, G., and Fisher, D.S. (2005) *J. Stat. Phys.* **120**, 167.

Toninelli, C., Biroli, G., and Fisher, D.S. (2006) *Phys. Rev. Lett.* **96**, 035702.

Tool, A.Q. (1946) *J. Amer. Ceram. Soc.* **29**, 240.

Torres, J.A., Nealey, P.F., and de Pablo, J.J. (2000) *Phys. Rev. Lett.* **85**, 3221.

Turnbull, D., and Cohen, M.H. (1961) *J. Chem. Phys.* **34**, 120.

Turnbull, D., and Cohen, M.H. (1970) *J. Chem. Phys.* **52**, 3038.

Tracht, U., Wilhelm, M., Heuer, A., and Spiess, H.-W. (1999) *J. Mag. Reson.* **140**, 460.

Tyrell, H.J.V., and Harris, K.R. (1984) *Diffusion in Liquids* (Butterworths, London).

Vallée, R.A.L., Van Der Auweraer, M., Paul, W., and Binder, K. (2006) *Phys. Rev. Lett.* **97**, 217801.

Vallée, R.A.L., Van Der Auweraer, M., Paul, W., and Binder, K. (2007a) *Europhys. Lett.* **79**, 46001.

Vallée, R.A.L., Paul, W., and Binder, K. (2007b) *J. Chem. Phys.* **127**, 154903.

van Blaaderen, A., and Wiltzius, P. (1995) *Science* **270**, 1177.

van Enter, A.C.D., Adler, J., and Duarte, J.A.M.A. (1990) *J. Stat. Phys.* **60**, 323.

Varnik, F. (2006) *J. Chem. Phys.* **125**, 164514 (2006).

Varnik, F., Baschnagel, J., and Binder, K. (2002a) *Eur. Phys. J.* **E8**, 175.

Varnik, F., Baschnagel, J., and Binder, K. (2002b) *Phys. Rev.* **E65**, 021507.

Varnik, F., Bocquet, L., Barrat, J.L., and Berthier, L. (2003) *Phys. Rev. Lett.* **90**, 095702.

Varnik, F., Bocquet, L., and Barrat, J.-L. (2004) *J. Chem. Phys.* **120**, 2788.

Varnik, F., and Henrich, O. (2006) *Phys. Rev.* **B73**, 174209.

Vink, R.L.C., Horbach, J., and Binder, K. (2005) *Phys. Rev.* **E71**, 011401.

Viot, P., and Tarjus, G. (1998) *Europhys. Lett.* **44**, 423.

Vlassopoulos, D., Pakula, T., Fytas, G., Pitsikalis, M., and Hadjichnistis, N. (1999) *J. Chem. Phys.* **111**, 1760.

Vlassopoulos, D., Fytas, G., Pakula, T., and Roovers, J. (2003) *J. Phys.: Condens. Matter* **13**, R855.

Vogel, M., Doliwa, B., Heuer, A., Glotzer, S.C. (2004) *J. Chem. Phys.* **120**, 4404.

Weeks, E.R., Crocker, J.C., Levit, A.C., Schofield, A., and Weitz, D.A. (2000) *Science*, **287**, 627.

Whitelam, S., and Garrahan, J.P. (2004) *J. Phys. Chem* B **108**, 6611.

Wolfgardt, M., Baschnagel, J., Paul, W., and Binder, K. (1996) *Phys. Rev.* **E54**, 1535.

Woodcock, L.V., and Angell, C.A. (1981) *Phys. Rev. Lett.* **47**, 1129.

Xia, X., and Wolynes, P.G. (2001) *Phys. Rev. Lett.* **80**, 5526.

Yamamoto, R., and Onuki, A. (1997) *J. Phys. Soc. Jpn.*, **66**, 2545.

Yamamoto, R., and Onuki, A. (1998) *Phys. Rev. Lett.* **81**, 4915.

Zaccarelli, E. (2007) *J. Phys.: Condens. Matter* **19**, 323101.

Zaccarelli, E., Buldyrev, S.V., La Nave, E., Moreno, A.J., Saika-Voivod, I., Sciortino, F., and Tartaglia, P. (2005) *Phys. Rev. Lett.* **94**, 218301.

Zaccarelli, E., Saika-Voivod, I., Moreno, A.J., Buldyrev, S.V., Tartaglia, P., and Sciortino, F. (2006) *J. Chem. Phys.* **124**, 124908.

Zausch, J., Horbach, J., Laurati, M., Egelhaaf, S.U., Brader, J.M., Voigtmann, T., and Fuchs, M. (2008) *J. Phys.: Condens. Matter* **20**, 404210.

Zettlemoyer, A.C. (1969) *Nucleation* (M. Dekker, New York).

Index